A Three-Phase FE-Model for Dewatering of Soils by Means of Compressed Air

Gerhard Öttl

innsbruck university press

Die Deutsche Bibliothek – CIP-Einheitsaufnahme
Ein Titeldatensatz für diese Publikation ist bei Der Deutschen Bibliothek erhältlich.

ISBN 3-901249-71-0

Herstellung: Books on Demand GmbH

© 2003 innsbruck university press
Alle Rechte vorbehalten

Universität Innsbruck, Innrain 52, A-6020 Innsbruck
http://www.university-press.at/

Preface

The work described in this thesis emerged from the participation in an Austrian Joint Research Initiative and from research work carried out as a University Assistant at the Institute for Structural Analysis and Strength of Materials at the University of Innsbruck.

I am indebted to many people for their interest and assistance during the course of the present work. In this regard, first of all I would like to express my very deep gratitude and sincere thanks to my supervisor o. Univ.-Prof. Dipl.-Ing. Dr. techn. Günter Hofstetter for his motivation to compile this thesis and for his enthusiastic supervision, patience, guidance and support in the course of this work.

I would also like to give sincere thanks to Univ.-Prof. Dipl.-Ing. Dr. techn. Günther Meschke, Head of the Institute for Structural Analysis at the Department of Civil Engineering of the Ruhr University Bochum, for his lively interest in my work and for delivering the second opinion to the thesis.

Additionally, I would like to thank all the staff at the Institute for Structural Analysis and Strength of Materials for their friendship and assistance. In particular, many thanks are due to ao. Univ.-Prof. Dipl.-Ing. Dr. techn. Rudolf Stark for his encouragement and many valuable and stimulating discussions and comments.

Part of the work was carried out within a research project supported by the Austrian Science Fund (FWF) under contract number S08005-TEC. This support is gratefully acknowledged.

Finally, I would like to thank my parents for their understanding and support in enabling my education and thus professional growth.

Innsbruck, April 2003.   G. ÖTTL

Abstract

Due to the fast development of very powerful computers in the last decades, the numerical solution of complex multi-phase problems has become possible. Thus, nowadays multi-phase formulations are applied to a broad variety of engineering tasks. Such analyses usually take into account a number of individual phases as well as thermal, chemical and biological processes.

The application of multi-phase formulations to geotechnical problems is shown in the present work. In particular, a numerical model for dewatering of soils by means of compressed air is developed, which primarily is of importance in the light of tunnelling in aquifers. To this end, a coupled three-phase model is employed, consisting of a deformable soil skeleton and two compressible barotropic fluid phases water and air.

Emanating from a very general formulation of the multi-phase balance laws, the balance equations of mass, linear momentum and angular momentum are derived on the basis of the averaging theory. In combination with a set of constitutive relations the governing equations of the coupled three-phase model are obtained. A number of special cases included in the three-phase formulation are discussed as well. The solution of the obtained coupled system of equations by means of the Finite Element Method is described hereafter.

For the verification of the model a number of example problems are investigated numerically and results are compared with analytical solutions, experimental data and numerical results documented in the literature. The examples deal with consolidation of soils, flow of compressed air through dry soil, seepage flow through earth dams and dewatering of soils due to gravitational effects or by means of compressed air. The latter is investigated in connection with a full-scale in-situ air permeability test and with tunnelling below the groundwater table using compressed air. The results of all numerical simulations indicate that the proposed coupled three-phase formulation and the included special cases allow for the analysis of a broad range of problems encountered in geotechnical engineering.

Kurzfassung

Die rasche Entwicklung sehr leistungsfähiger Computer in den letzten Jahrzehnten hat die numerische Lösung komplexer Mehrphasen-Probleme ermöglicht. So werden heute Mehrphasen-Formulierungen im Ingenieurwesen in einer breiten Palette von Aufgaben eingesetzt. Solche Berechnungen berücksichtigen üblicherweise eine Reihe unterschiedlicher Phasen sowie thermische, chemische und biologische Prozesse.

Die vorliegende Arbeit zeigt die Anwendung von Mehrphasen-Formulierungen auf geotechnische Problemstellungen. Insbesondere wird ein numerisches Modell zur Entwässerung von Böden mittels Druckluft entwickelt, das vor allem im Hinblick auf Tunnelvortriebe in wasserführenden Bodenschichten von Bedeutung ist. Dazu wird ein gekoppeltes Dreiphasen-Modell, bestehend aus einem deformierbaren Korngerüst und zwei kompressiblen, barotropen Fluiden Wasser und Luft, verwendet.

Ausgehend von einer sehr allgemeinen Formulierung der Mehrphasen-Bilanzgleichungen werden auf Basis der Mittelungstheorie die Bilanzgleichungen für Masse, Impuls und Drehimpuls abgeleitet. In Kombination mit konstitutiven Beziehungen ergeben sich daraus die grundlegenden Gleichungen für das gekoppelte Dreiphasen-Modell. Die in der Dreiphasen-Formulierung enthaltenen Spezialfälle werden ebenfalls hergeleitet. Schließlich wird die Lösung des erhaltenen gekoppelten Gleichungssystems im Rahmen der Methode der Finiten Elemente erläutert.

Die Verifikation des Modells erfolgt durch die numerische Simulation einer Reihe von Beispielen und einem Vergleich der Ergebnisse mit analytischen Lösungen, Versuchsdaten sowie aus der Literatur bekannten numerischen Ergebnissen. Die Beispiele umfassen Konsolidation, Strömung von Druckluft in trockenem Boden, Sickerströmung und Entwässerung von Böden infolge Gravitation oder durch Druckluft. Letztere wird in Zusammenhang mit einem in-situ Luftdurchlässigkeitsversuch sowie einem Tunnelvortrieb unterhalb des Grundwasserspiegels mittels Druckluft untersucht. Die Ergebnisse aller numerischen Simulationen zeigen, dass die vorliegende gekoppelte Dreiphasen-Formulierung mit den darin enthaltenen Sonderfällen für die Analyse einer Reihe von geotechnischen Aufgabenstellungen geeignet ist.

Contents

Preface iii

Abstract iv

Kurzfassung v

Contents vii

Notation xiii

1 Introduction 1
   1.1 General introduction 1
   1.2 Tunnelling below the groundwater level 4
   1.3 Outline of the thesis 7

2 State of the art 11
   2.1 Introduction 11
   2.2 Flow through porous media 12
      2.2.1 Introduction 12
      2.2.2 Two-phase problems 13
      2.2.3 Three-phase problems 16
      2.2.4 Multi-phase problems 18
   2.3 Numerical simulation of tunnelling 21
      2.3.1 Introduction 21
      2.3.2 Simulation of tunnelling in dry soil 22
      2.3.3 Simulation of tunnelling in aquifers 24

	2.4	Mathematical description of multi-phase media	33
		2.4.1 Introduction	33
		2.4.2 Classical theories of porous media	34
		2.4.3 Modern theory of porous media	40
		2.4.4 Averaging theory	47
3	**Mechanics of the coupled three-phase model**		**51**
	3.1	Introduction	51
	3.2	Kinematic equations	53
		3.2.1 Coordinates, velocities and accelerations	53
		3.2.2 Definition of material time derivatives	55
		3.2.3 Strain tensor for small displacements and small strains	58
		3.2.4 Some important integral theorems	59
	3.3	Averaging procedure	60
		3.3.1 Concept of volume fractions	60
		3.3.2 Definition of averaging operators	64
		3.3.3 Microscopic balance equations	67
		3.3.4 Macroscopic balance equations	72
	3.4	General formulation of the various balance equations	84
		3.4.1 Introduction	84
		3.4.2 Microscopic mass balance	85
		3.4.3 Macroscopic mass balance	86
		3.4.4 Microscopic linear momentum balance	88
		3.4.5 Macroscopic linear momentum balance	91
		3.4.6 Microscopic angular momentum balance	95
		3.4.7 Macroscopic angular momentum balance	99
	3.5	Constitutive equations for a partially saturated porous medium	100
		3.5.1 Introduction	100
		3.5.2 Averaged density of the three-phase mixture	101
		3.5.3 Stress tensor in the fluid phases	101
		3.5.4 Averaged stresses of the three-phase mixture	102
		3.5.5 Effective averaged stress tensor of the soil skeleton	103
		3.5.6 Soil suction and capillary stress	106
		3.5.7 Momentum exchange terms	112

		3.5.8	Elastic or elastic-plastic behaviour of the soil skeleton	113
		3.5.9	Drucker-Prager material model	117
		3.5.10	Equation of state for the water phase	121
		3.5.11	Equation of state for the air phase	125
		3.5.12	Capillary stress versus degree of water saturation	126
		3.5.13	Permeability coefficients for the fluid phases	133
	3.6	Field equations for the three-phase medium		142
		3.6.1	Introduction	142
		3.6.2	Mass balance for the soil skeleton	142
		3.6.3	Mass balance for the fluid phases	143
		3.6.4	Alternative derivation of the mass balance equations	145
		3.6.5	Linear momentum balance for the soil skeleton	146
		3.6.6	Linear momentum balance for the fluid phases	146
		3.6.7	Linear momentum balance for the mixture	148
		3.6.8	Alternative derivation of the equilibrium equations	149
	3.7	Special cases included in the three-phase formulation		152
		3.7.1	Introduction	152
		3.7.2	Dewatering under atmospheric conditions	153
		3.7.3	Consolidation	154
		3.7.4	Uncoupled approach	156
		3.7.5	Drained conditions	161

4 Numerical formulation for the three-phase model 163

	4.1	Introduction		163
	4.2	Weak formulation of the basic equations		164
		4.2.1	Introduction	164
		4.2.2	Formal statement of the problem	165
		4.2.3	Weak form of the linear momentum balance equation	168
		4.2.4	Weak form of the mass balance equation for the water phase	170
		4.2.5	Weak form of the mass balance equation for the air phase	171
		4.2.6	Summary of the weak formulation	172
	4.3	Spatial discretisation of the weak formulation		174
		4.3.1	Introduction	174
		4.3.2	Variables of the model	175

		4.3.3	Discretised linear momentum balance equation 177
		4.3.4	Discretised water mass balance equation 177
		4.3.5	Discretised air mass balance equation 178
		4.3.6	System of spatially discretised equations 178
	4.4	Numerical integration in the time domain 182
		4.4.1	Introduction 182
		4.4.2	Brief survey of the various time integration schemes 183
		4.4.3	Approximation of the variables in the current model 187
	4.5	Solution of the final coupled set of equations 190
		4.5.1	Introduction 190
		4.5.2	Finally obtained coupled system of equations 190
		4.5.3	Fixed-point iteration 193
		4.5.4	Newton method 194
		4.5.5	Suitable choice of finite elements 195
	4.6	Stability, consistency, accuracy 197
		4.6.1	Introduction 197
		4.6.2	Stability 197
		4.6.3	Consistency 200
		4.6.4	Accuracy 202

5 Verification of the three-phase formulation 205

	5.1	Introduction 205
	5.2	Terzaghi's problem – one-dimensional consolidation 207
		5.2.1 Description of the problem 207
		5.2.2 Analytical solution 209
		5.2.3 Numerical model 216
		5.2.4 Results of the numerical simulation 220
	5.3	Footing on a water saturated soil layer 225
		5.3.1 Description of the problem 225
		5.3.2 Analytical solution 226
		5.3.3 Numerical model 229
		5.3.4 Results of the numerical simulation 231
	5.4	Flow of compressed air through dry sand 243
		5.4.1 Description of the problem 243

	5.4.2	Numerical model	245
	5.4.3	Results of the numerical simulation	248

5.5 One-dimensional drainage of a soil column –
two-phase numerical simulation 254
    5.5.1 Description of the problem 254
    5.5.2 Numerical model 255
    5.5.3 Results of the numerical simulation 258

5.6 One-dimensional drainage of a soil column –
three-phase numerical simulation 267
    5.6.1 Numerical model 267
    5.6.2 Results of the numerical simulation 269

6 Application of the three-phase formulation   285

6.1 Introduction 285
6.2 Water flow through an earth dam 286
    6.2.1 Description of the problem 286
    6.2.2 Numerical model 288
    6.2.3 Results of the numerical simulation 292
6.3 In-situ air permeability test 307
    6.3.1 Description of the problem 307
    6.3.2 Numerical model 310
    6.3.3 Results of the numerical simulation 315
6.4 Tunnelling below the groundwater level 328
    6.4.1 Description of the problem 328
    6.4.2 Numerical model 330
    6.4.3 Results of the numerical simulation 336

7 Summary, conclusions and outlook   359

List of tables   365

List of figures   367

Bibliography   375

Notation

All notation and symbols are defined where they first appear in the text. For easy reference, the most frequently used symbols and their meanings are presented here.

Capital latin letters:

A	surface of the macroscopic domain of interest
A^+, A^-	surfaces of the two portions V^+ and V^- of V
A_a	part of A with prescribed air pressure boundary conditions
A_e	surface of the finite element e
A_n	amplification factor in the stability analysis
$A_\mathbf{u}$	part of A with prescribed soil displacement boundary conditions
A_w	part of A with prescribed water pressure boundary conditions
A^π	partial surface of the constituent π
A_a^q	part of A with prescribed surface flow of the air phase
$A_\mathbf{u}^q$	part of A with prescribed surface tractions
A_w^q	part of A with prescribed surface flow of the water phase
$(A_a^q)_e$	surface part of the finite element e with prescribed airflow
$(A_\mathbf{u}^q)_e$	surface part of the finite element e with prescribed tractions
$(A_w^q)_e$	surface part of the finite element e with prescribed water flow
$\mathbf{A}(\mathbf{Y}_{n+1})$	left hand side matrix of the final coupled system of equations
$\mathbf{A}_1, \mathbf{A}_2$	coefficient matrices of the coupled system of equations
$\mathbf{A}^\pi$	material acceleration of a particle with respect to the π-phase
B_1, B_2	parameters accounting for hardening
B^w, B^a	pore water, air pressure parameter for isotropic loading
$\mathbf{B}_p$	matrix of derivatives of the shape functions for the fluid stresses

$\mathbf{B_u}$	matrix of derivatives of the shape functions for the displacements
C	compressibility
C_{k+1}	truncation error coefficient
C^w, C^a	compressibility of the water, air phase
C^{wa}	compressibility of a water-air mixture
$\mathbf{C}^e$	elasticity tensor
$\mathbf{C}^{ep}$	elastic-plastic material tensor
$\mathbf{C}^{sa}$	global solid-air phase coupling matrix
$\mathbf{C}^{sw}$	global solid-water phase coupling matrix
$\mathbf{C}^{wa}$	global water-air phase coupling matrix
$\mathbf{C}_e^{sa}$	solid-air phase coupling matrix for a finite element e
$\mathbf{C}_{n+1}^{sa}$	solid-air phase coupling matrix at time t_{n+1}
$\mathbf{C}_e^{sw}$	solid-water phase coupling matrix for a finite element e
$\mathbf{C}_{n+1}^{sw}$	solid-water phase coupling matrix at time t_{n+1}
$\mathbf{C}_e^{wa}$	water-air phase coupling matrix for a finite element e
$\mathbf{C}_{n+1}^{wa}$	water-air phase coupling matrix at time t_{n+1}
D	diameter of tunnel cross section (in Chapter 2)
D	re-drying curve
D_0	initial drying curve
E	Young's modulus of the soil skeleton
E_{ij}	components of the Green strain tensor in Cartesian coordinates
E_s	constrained modulus of the soil
$\mathbf{E}$	Green strain tensor
$F(s), G(s)$	functions used for Heaviside expansion
$\mathbf{F}$	an arbitrary tensor
$G(\ldots)$	weak formulation of the linear momentum balance equation
G^s	shear modulus of the soil skeleton
H^1	Sobolev space of functions of the degree one
$H^a(\ldots)$	weak formulation of the mass balance equation for the air phase
$H^w(\ldots)$	weak formulation of the mass balance equation for the water phase
$\mathbf{H}^{aa}$	global air phase permeability matrix
$\mathbf{H}^{ww}$	global water phase permeability matrix
$\mathbf{H}_e^{aa}$	air phase permeability matrix for a finite element e
$\mathbf{H}_{n+1}^{aa}$	air phase permeability matrix at time t_{n+1}

$\mathbf{H}_e^{ww}$	water phase permeability matrix for a finite element e
$\mathbf{H}_{n+1}^{ww}$	water phase permeability matrix at time t_{n+1}
I_w	constant for the calculation of the settlement of a circular footing
$I_1^{\varepsilon,e}$	first invariant of the elastic strain tensor
$\tilde{I}_1'^{\sigma}$	first invariant of the effective averaged stress tensor
$\mathbf{I}$	second-order unit tensor
$\tilde{\mathbf{I}}'^{dev}$	deviatoric operator (fourth-order tensor)
J	Jacobian
K_0	coefficient of earth pressure at rest
K^s, K^w, K^a	bulk modulus of the soil skeleton, water, air phase
$\mathbf{K}$	linear elastic stiffness matrix of the soil skeleton
L	typical length of the domain under consideration
L_2	Hilbert space of square-integrable functions
$\mathbf{L}(\ldots)$	operator defined for accuracy analysis
$\mathbf{M}_n^A$	matrix similar to the amplification factor
NP_p	number of nodal points with hydrostatic stress degrees of freedom
NP_u	number of nodal points with soil displacement degrees of freedom
$N_p^{(i)}, N_p^{(j)}, \ldots$	individual shape functions for the fluid stresses
$N_u^{(i)}, N_u^{(j)}, \ldots$	individual shape functions for the soil displacements
$\mathbf{N}_p$	matrix of shape functions for the approximation of fluid stresses
$\mathbf{N_u}$	matrix of shape functions for the approximation of displacements
O	origin of the coordinate system
$\mathbf{O}$	order in a Taylor series with remainders
$\mathbf{P}^a$	global vector of nodal values of the hydrostatic air stresses
$\mathbf{P}^w$	global vector of nodal values of the hydrostatic water stresses
$\mathbf{P}_e^a$	element vector of nodal values of the hydrostatic air stresses
$\mathbf{P}_n^a, \mathbf{P}_{n+1}^a$	nodal values of the hydrostatic air stresses at time t_n, t_{n+1}
$\mathbf{P}_e^w$	element vector of nodal values of the hydrostatic water stresses
$\mathbf{P}_n^w, \mathbf{P}_{n+1}^w$	nodal values of the hydrostatic water stresses at time t_n, t_{n+1}
$\mathbf{Q}^{aa}$	global air phase compressibility matrix
$\mathbf{Q}^{ww}$	global water phase compressibility matrix
$\mathbf{Q}_e^{aa}$	air phase compressibility matrix for a finite element e
$\mathbf{Q}_{n+1}^{aa}$	air phase compressibility matrix at time t_{n+1}
$\mathbf{Q}_e^{ww}$	water phase compressibility matrix for a finite element e

$\mathbf{Q}_{n+1}^{ww}$	water phase compressibility matrix at time t_{n+1}
R	gas constant
REA	representative elementary area
REV	representative elementary volume
R_a	radius of tunnel excavation
R_i	radius of tunnel after placement of the lining
R_{min}, R, R_{max}	minimum, arbitrary, maximum pore radius in a soil
R_s	radius of the curvature of the meniscus in a capillary tube
$\mathbf{R}^\pi, \mathbf{R}^f$	resistivity tensor related to momentum exchange with respect to fluid phase π, f
S_e	effective degree of saturation
S^f	degree of saturation of the fluid phase f
S^w, S^a	degree of saturation of the water, air phase
S_k^w, S_{ek}	critical degree of water, effective saturation
S_r^w	residual degree of water saturation
S_s^w	maximum degree of water saturation
T	temperature (in Chapter 3), time domain
U	degree of consolidation
$\mathbf{U}$	global vector of nodal values of the soil displacements
$\mathbf{U}_e$	element vector of nodal values of the soil displacements
$\mathbf{U}_n, \mathbf{U}_{n+1}$	nodal values of the soil displacements at time t_n, t_{n+1}
V	volume of the macroscopic domain of interest
V^+, V^-	two portions of V forming a macroscopic surface of discontinuity
V_e	volume of the finite element e
V_0	volume of a body in its reference configuration
V^π	partial volume of the constituent π
$\mathbf{V}^\pi$	material velocity of a particle with respect to the π-phase
W	wetting curve
$X_1^\pi, X_2^\pi, X_3^\pi$	three coordinates of a point in the reference configuration with respect to the π-phase
$\mathbf{X}^\pi$	position vector of a point in the reference configuration with respect to the π-phase
$\mathbf{Y}(t_n)$	exact solution of a partial differential equation at time t_n
$\mathbf{Y}_n, \mathbf{Y}_{n+1}$	numerical approximation of the exact solution of a partial

	differential equation at time t_n, t_{n+1}
$\mathbf{Y}_{n+\alpha}$	numerical approximation of the exact solution of a partial differential equation at time $t_{n+\alpha}$
$\mathbf{Y}_{n+1}^i$	solution obtained after iteration step i

Small latin letters:

$\tilde{\mathbf{a}}$	acceleration vector
$\mathbf{a}^f$	acceleration of the fluid phase f
$\mathbf{a}^s, \mathbf{a}^w, \mathbf{a}^a$	acceleration of the solid, water, air phase
$\mathbf{a}^\pi$	spatial acceleration field with respect to the π-phase
$\overline{\mathbf{a}}^\pi$	mass averaged acceleration of the π-phase
b	half width of strip footing in Chapter 5
$\tilde{\mathbf{b}}$	body force vector
c	finite valued constant
c	cohesion of the soil (in Chapter 6)
c_v	consolidation coefficient
d	typical microscopic distance
da	averaging area REA (representative elementary area)
da^+, da^-	surfaces of the two portions dv^+ and dv^- of dv
$da^\pm$	summation of the two portions of da ($da^+ + da^-$)
da_{mic}	microscopic area element
$da^{(i)}$	surface elements normal to the coordinate axes x_i, $i = 1, 2, 3$
da^π	area of the constituent π within dv
$da^{\pi\alpha}$	interface between the π and the α-phase
dI_1^ε	infinitesimal increment of the first invariant of the strains
$dI_1^{\varepsilon,p}$	infinitesimal increment of the first invariant of the plastic strains
$d\tilde{I}_1^{\prime\sigma}$	infinitesimal increment of the first invariant of the effective averaged stress tensor
$d\tilde{m}$	mass of the averaged volume element
dt	infinitesimal time increment
dv	averaging volume REV (representative elementary volume)
dv^+, dv^-	two portions of dv forming a microscopic surface of discontinuity
$dv^\pm$	summation of the two portions of dv ($dv^+ + dv^-$)

dv_{mic}	microscopic volume element
dv^s, dv^w, dv^a	volume of the solid, water, air phase within dv
dv^α, dv^π	volume of the constituent α, π within dv
dx_1, dx_2, dx_3	lengths of the edges of the volume element dv
$d\varepsilon^{m,p}$	infinitesimal mean plastic strain increment
$d\varepsilon^{vol}$	infinitesimal increment of the volumetric strain
$d\lambda$	consistency parameter
$d\mathbf{e}$	infinitesimal increment of the deviator strains
$d\mathbf{e}^p$	infinitesimal increment of the plastic deviator strains
$d\tilde{\mathbf{R}}$	vector of resulting forces exerted on the averaged volume element
$d\tilde{\mathbf{s}}'$	infinitesimal increment of the deviatoric part of the effective averaged stress tensor
$d\boldsymbol{\varepsilon}$	infinitesimal increment of the strains
$d\boldsymbol{\varepsilon}^e$	infinitesimal increment of the elastic strains
$d\boldsymbol{\varepsilon}^p$	infinitesimal increment of the plastic strains
$d\tilde{\boldsymbol{\sigma}}'$	infinitesimal increment of the effective averaged stresses
e	void ratio of the soil
e_t	truncation error
e_ρ^π	density function for exchange of mass between constituent π and the other constituents α
$\mathbf{e}(t_n), \mathbf{e}(t_{n+1})$	error at time t_n, t_{n+1}
$\mathbf{e}_i, \mathbf{e}_j, \mathbf{e}_k$	orthogonal unit basis vectors in a rectangular Cartesian coordinate system
$\mathbf{e}^e$	deviatoric elastic strains
$\mathbf{e}_{\rho\dot{\mathbf{r}}}^\pi$	density function for exchange of linear momentum between constituent π and the other constituents α
$\mathbf{e}_{\rho\hat{\mathbf{r}}^\pi}^\pi$	density function for exchange of linear momentum due to deviation of the velocity $\dot{\mathbf{r}}$ from the mass averaged velocity $\overline{\mathbf{v}}^\pi$
$\mathbf{e}_{\rho\psi}^\pi$	density function for exchange of ψ between constituent π and the other constituents α
f	an arbitrary scalar
$f(\ldots)$	yield function
$f(r_1), f(r_2)$	pore water distribution functions
$f_{n+\alpha}$	right hand side of the modally decomposed system of equations

f^π	any differentiable function with respect to the π-phase
$\mathbf{f}$	an arbitrary vector
$\mathbf{f}(\mathbf{Y}_{n+1})$	right hand side vector of the final coupled system of equations
$\mathbf{f}_{n+\alpha}$	right hand side vector of the system of equations at time $t_{n+\alpha}$
$\mathbf{f}^{ex}$	global vector of external forces of the soil skeleton
$\mathbf{f}^{in}$	global vector of internal forces of the soil skeleton
$\mathbf{f}_e^{ex}$	vector of external forces of the soil skeleton for a finite element e
$\mathbf{f}_{n+1}^{ex}$	vector of external forces of the soil skeleton at time t_{n+1}
$\mathbf{f}_e^{in}$	vector of internal forces of the soil skeleton for a finite element e
$\mathbf{f}_n^{in}, \mathbf{f}_{n+1}^{in}$	vector of internal forces of the soil skeleton at time t_n, t_{n+1}
$\dot{\mathbf{f}}^a$	global air phase flow vector
$\dot{\mathbf{f}}^w$	global water phase flow vector
$\tilde{\mathbf{f}}^d$	drag forces
$\tilde{\mathbf{f}}^f$	averaged interaction force of the fluid phase f
$\tilde{\mathbf{f}}^s, \tilde{\mathbf{f}}^w, \tilde{\mathbf{f}}^a$	averaged interaction force of the solid, water, air phase
$\dot{\mathbf{f}}_e^a$	air phase flow vector for a finite element e
$\dot{\mathbf{f}}_{n+1}^a$	air phase flow vector at time t_{n+1}
$\dot{\mathbf{f}}_e^w$	water phase flow vector for a finite element e
$\dot{\mathbf{f}}_{n+1}^w$	water phase flow vector at time t_{n+1}
g	value of gravitational acceleration
$g(\ldots)$	plastic potential
g_i	components of the vector of gravitational acceleration
$\mathbf{g}$	vector of gravitational acceleration
h	coefficient of solubility (in Chapter 3)
h	thickness of consolidating soil layer (in Chapter 5)
h_c	height of the capillary rise in a cylindrical glass tube
h_{cds}	fully saturated capillary height when considering drainage
h_{cdx}	maximum capillary height when considering drainage
h_{cis}	minimum capillary height when considering imbibition
h_{cix}	maximum capillary rise when considering imbibition
h^w	depth below the groundwater table
i	imaginary unit
k	order of accuracy (rate of convergence)
k^{ow}, k^{oa}	isotropic permeability for a fully water saturated, dry soil

k^r	coefficient of relative permeability
k^{rw}, k^{ra}	relative permeability coefficient referring to the water, air phase
$k^{r\pi}, k^{rf}$	relative permeability coefficient referring to the fluid phase π, f
$k_x^{ow}, k_y^{ow}, k_z^{ow}$	hydraulic conductivity in the three coordinate directions x, y, z
k_{min}^{ra}	minimum relative air permeability coefficient
$\mathbf{k}^i$	intrinsic permeability of the soil
$\mathbf{k}^{ow}, \mathbf{k}^{oa}$	fluid permeability tensor for a fully water saturated, dry soil
$\mathbf{k}^{o\pi}, \mathbf{k}^{of}$	π-, f-phase permeability tensor for fully saturated conditions
$\mathbf{k}^\pi$	permeability tensor of the soil with respect to the fluid phase π
l	characteristic length of the average volume element
l_1, l_2	lengths of two cylindrical tubes
m, n	empirical parameters in $S^w(p^c)$-relationship of van Genuchten
m^a	mass of the air phase
m^π	mass of the constituent π within dv
n	porosity of the soil
n_{eq}	number of equations in a system of equations
n_{sd}	number of space dimensions
$\mathbf{n}$	outward unit vector normal to a surface
$\mathbf{n}^+, \mathbf{n}^-$	outward unit normal vectors to microscopic surface of discontinuity
$\mathbf{n}^{\pi\alpha}$	outward unit normal vector to $\pi\alpha$-interface
$\mathbf{n}\vert_-$	outward unit normal vector to macroscopic surface of discontinuity
p	an arbitrary power
p^a	hydrostatic pressure in the air phase
p^c	capillary stress
p^{ow}, p^{oa}	reference value for the hydrostatic stress in the water, air phase
p^s	total hydrostatic fluid stress exerted on the solid phase
p^w	hydrostatic pressure in the water phase
p^π, p^f	hydrostatic stress in the fluid phase π, f
p_e^a	hydrostatic air stresses at a point within the finite element e
p_{ex}^a	air pressure in excess of the water pressure
p_0^a	initial value of the hydrostatic air stress
p_b^c	air entry value (bubbling pressure)
p_e^w	hydrostatic water stresses at a point within the finite element e
p_0^w	initial value of the hydrostatic water stress

$\overline{p}^a$	specified hydrostatic air stresses on the boundary A_a
$\overline{p}^w$	specified hydrostatic water stresses on the boundary A_w
$q(x,y)$	surface load
q_0	uniform surface load
$\overline{q}^a$	prescribed airflow on the boundary A_a^q
$\overline{q}^w$	prescribed water flow on the boundary A_w^q
r_b	radius of a bulb
r_c	radius of a capillary tube
r_1, r_2	radii of two cylindrical tubes
$\mathbf{r}$	position of particle within dv referring to global coordinate system
$\dot{\mathbf{r}}$	microscopic velocity of the material present at a spatial point $\mathbf{r}$
$\ddot{\mathbf{r}}$	microscopic acceleration of the material present at a spatial point $\mathbf{r}$
$\overline{\dot{\mathbf{r}}}^\pi = \overline{\mathbf{v}}^\pi$	mass averaged velocity of the π-phase
$\hat{\dot{\mathbf{r}}}^\pi$	deviation of the velocity $\dot{\mathbf{r}}$ in the spatial point $\mathbf{r}$ from the mass averaged velocity $\overline{\mathbf{v}}^\pi$
s	matric suction
s	parameter in Laplace transform (in Chapter 5)
$s_1,\ldots,s_n$	zeros of function $G(s)$
$\tilde{\mathbf{s}}'$	deviatoric part of the effective averaged stress tensor
t	time
t_n, t_{n+1}	previous, current instant of time
t_s	surface tension in the contractile skin
$\overline{\mathbf{t}}$	vector of prescribed surface tractions on the boundary $A_\mathbf{u}^q$
$\mathbf{t}_{mic}$	microscopic stress vector
$\mathbf{t}^s, \mathbf{t}^f$	microscopic mechanical interactions referring to solid, fluid phase
$\mathbf{t}^\pi$	stress vector accounting for microscopic mechanical interactions
$\tilde{\mathbf{t}}^{(i)}(x_i)$	tractions acting on a surface of the averaged volume element
$\check{\mathbf{t}}^\pi$	partial (macroscopic) stress vector referring to the π-phase
u_i, u_j, u_k	displacement components of the soil skeleton with respect to the three coordinate directions
$\mathbf{u}$	vector of displacements of the soil skeleton
$\overline{\mathbf{u}}$	specified displacements of the soil skeleton on the boundary $A_\mathbf{u}$
$\mathbf{u}_e$	soil displacements at a point within the finite element e
$\mathbf{u}_0$	initial values of the displacements of the soil skeleton

$\mathbf{u}^\pi$	vector of displacements with respect to the π-phase
v^w, v^a	volume of water, air
v_i^f	components of the velocity of the fluid phase f
v_i^{fs}	components of the velocity of the fluid relative to the solid phase
v_i^s	components of the velocity of the solid phase s
$\tilde{v}_i^{fs}$	components of the artificial velocity for the fluid phase f
$\tilde{v}_z^{ws}$	artificial velocity for water in the vertical coordinate direction
$\mathbf{v}^f$	velocity of the fluid phase f
$\mathbf{v}^{fs}$	velocity of the fluid phase f relative to the solid phase s
$\mathbf{v}^s$	velocity of the solid phase s
$\mathbf{v}^\alpha$	spatial velocity field with respect to the α-phase
$\mathbf{v}^{\alpha\pi}$	relative velocity between the α and the π-phase
$\mathbf{v}^\pi$	spatial velocity field with respect to the π-phase
$\tilde{\mathbf{v}}^{fs}$	artificial velocity referring to the fluid phase f
$\tilde{\mathbf{v}}^{ws}, \tilde{\mathbf{v}}^{as}$	artificial velocity of the water, air phase
w	settlement
w_0, w_∞	initial, final settlement
$w(0, t)$	settlement of central point under load
$\mathbf{w}$	velocity of a microscopic surface of discontinuity
$\overline{\mathbf{w}}$	velocity of a macroscopic surface of discontinuity
$\mathbf{w}_b$	velocity of the boundary of the REV
$x_1^\pi, x_2^\pi, x_3^\pi$	three coordinates of a point in the current configuration with respect to the π-phase
$\mathbf{x}$	position vector denoting the centre of the REV referring to the global coordinate system
$\mathbf{x}^\pi$	position vector of a point in the current configuration with respect to the π-phase
y_n, y_{n+1}	solutions at time t_n, t_{n+1} of the modally decomposed system of equations
z	vertical coordinate

Capital greek letters:

Δh	typical measure of the element size

Δt	time increment
$\Delta \mathbf{U}_{n+1}$	unknown incremental nodal values of the soil displacements
$\Delta \mathbf{Y}_{n+1}$	difference between the solutions at time t_{n+1} and t_n
$\Delta \mathbf{P}^a_{n+1}$	unknown incremental nodal values of the hydrostatic air stresses
$\Delta \mathbf{P}^w_{n+1}$	unknown incremental nodal values of the hydrostatic water stresses
$\Delta \mathbf{Y}^i_{n+1}$	solution increment at iteration step i
$\Delta \mathbf{f}^{in}_{n+1}$	unknown increment of the internal forces of the soil skeleton
Σ	macroscopic surface of discontinuity
Σ_{mic}	microscopic surface of discontinuity
$\boldsymbol{\Phi}$	second-order tensor related to tractions on the microscale
$\boldsymbol{\Phi}_\pi$	second-order tensor related to flux density $\breve{\boldsymbol{\vartheta}}^\pi$
$\boldsymbol{\Psi}\left(\mathbf{Y}^i_{n+1}\right)$	residuum (deviation from exact solution) at iteration step i

Small greek letters:

α	parameter in the time integration scheme
α, π	an individual phase in the range from 1 to κ
α_c	contact angle between contractile skin and glass tube
β^w	coefficient of thermal expansion for the water phase
$\boldsymbol{\beta}$	microscopic vector of body forces within dv
γ^w	unit weight of water
γ^π	phase distribution function for the constituent π
δ_{ij}	Kronecker delta
δp^a	field of virtual hydrostatic stresses in the air phase
δp^w	field of virtual hydrostatic stresses in the water phase
δp^a_e	virtual hydrostatic air stresses at a point within the finite element e
δp^w_e	virtual hydrostatic water stresses at a point within the finite element e
$\delta \mathbf{U}_e$	element vector of the virtual nodal values of the soil displacements
$\delta \mathbf{P}^a_e$	element vector of the virtual nodal values of the hydrostatic air stresses
$\delta \mathbf{P}^w_e$	element vector of the virtual nodal values of the hydrostatic water stresses
$\delta \mathbf{u}$	field of virtual displacements of the soil skeleton

$\delta \mathbf{u}_e$	virtual soil displacements at a point within the finite element e
$\delta \boldsymbol{\varepsilon}$	field of virtual strains of the soil skeleton
$\delta \boldsymbol{\varepsilon}_e$	virtual soil strains at a point within the finite element e
ϵ_{ijk}	permutation symbol
ε_{ij}	components of the linearised strain tensor in Cartesian coordinates
$\varepsilon_{xx}, \varepsilon_{yy}, \varepsilon_{zz}$	strains in the coordinate directions x, y, z
$\varepsilon^{vol}, \dot{\varepsilon}^{vol}$	volumetric strain, volumetric strain rate
$\varepsilon^{vol,e}$	volumetric elastic strain
$\boldsymbol{\varepsilon}$	linearised strain tensor
$\boldsymbol{\varepsilon}_e$	soil strains at a point within the finite element e
$\boldsymbol{\varepsilon}^e$	linear elastic strains in the soil skeleton
$\boldsymbol{\varepsilon}^p$	plastic strains in the soil skeleton
ζ	typical microscopic quantity (field variable)
ζ_π	volume phase average of the quantity ζ
ζ^π	intrinsic volume phase average of the quantity ζ
$\overline{\overline{\zeta}}_\pi$	area average of the quantity ζ
$\overline{\zeta}^\pi$	mass average of the quantity ζ
$\hat{\zeta}^\pi$	deviation of a microscopic function ζ from its π-phase mass averaged quantity $\overline{\zeta}^\pi$
$\boldsymbol{\zeta}$	typical microscopic quantity with tensorial nature
η^s, η^w, η^a	volume fraction of the solid, water, air phase
η^π	volume fraction of the constituent π
$\overline{\eta^\pi}$	area fraction of the constituent π
$\theta(R), \theta_{sat}$	current, fully saturated water content
$\boldsymbol{\vartheta}$	microscopic vector of tractions acting on da
$\boldsymbol{\vartheta}^\pi$	density function for mechanical interactions between the π and all α-phases
$\check{\boldsymbol{\vartheta}}^\pi$	flux density for total macroscopically non-convective flux of ψ
κ	total number of constituents in a multi-phase medium
κ^s	parameter accounting for isotropic hardening
λ	pore size distribution index
$\lambda_{n_{eq}}$	maximum of the eigenvalues λ_{ni}
λ_{ni}	i-th-mode eigenvalue of the system
λ^s	Lame constant referring to the soil skeleton

μ	coefficient of friction (Drucker-Prager model)
μ^w, μ^a	dynamic viscosity of the water, air phase
μ^π, μ^f	dynamic viscosity of the fluid phase π, f
ν	Poisson's ratio of the soil skeleton
$\boldsymbol{\xi}$	position of a particle with respect to the centre of the REV
ρ	microscopic mass density
$\tilde{\rho}$	averaged density of a multi-phase medium
ρ_π	partial density referring to the π-phase
ρ^f	intrinsic density of the fluid phase f
ρ^{ow}	reference value for the density of the water phase
ρ^s, ρ^w, ρ^a	intrinsic density of the solid, water, air phase
ρ^π	intrinsic density referring to the π-phase
$\boldsymbol{\rho}^s$	parameter accounting for kinematic hardening
$(\sigma_{mic})_{jk}$	components of the microscopic stress tensor in Cartesian coordinates
$(\sigma_\pi)_{jk}$	components of the partial stress tensor in Cartesian coordinates
$\tilde{\sigma}_{ij}$	components of the total averaged stress tensor
$\tilde{\sigma}'_{ij}$	components of the effective averaged stress tensor
$\tilde{\sigma}'^m$	mean stress of the effective averaged stress tensor
$\tilde{\boldsymbol{\sigma}}$	total averaged stress tensor of a multi-phase medium
$\tilde{\boldsymbol{\sigma}}'$	effective averaged stress tensor of the soil skeleton
$\boldsymbol{\sigma}_{mic}$	microscopic stress tensor
$\boldsymbol{\sigma}_w, \boldsymbol{\sigma}_a$	partial stress tensor of the water, air phase
$\boldsymbol{\sigma}_\pi$	partial (macroscopic) stress tensor referring to the π-phase
$\boldsymbol{\sigma}^f$	intrinsic stress tensor of the fluid phase f
$\boldsymbol{\sigma}^s, \boldsymbol{\sigma}^w, \boldsymbol{\sigma}^a$	intrinsic stress tensor of the solid, water, air phase
$\boldsymbol{\sigma}^\pi$	intrinsic stress tensor referring to the π-phase
τ_F	initial yield stress in simple shear (Drucker-Prager model)
$\boldsymbol{\tau}(t_n)$	local truncation error at time t_n
φ	friction angle of the soil
χ	Bishop's parameter for the description of unsaturated soils
$\boldsymbol{\chi}^\pi$	mapping function referring to the phase π
ψ	typical thermodynamic property at the microscale
ω	typical microscopic quantity (field variable)

Subscripts and superscripts:

a	air phase of a mixture
e	linear elastic behaviour of the soil skeleton
e	a single finite element if used as a subscript
ep	elastic-plastic
ex	excess pressure or stress (used as a subscript in Chapter 2)
ex	external forces of the soil skeleton
f	fluid phase of a mixture
i	iteration step
i, j, k	three coordinate directions
in	internal forces of the soil skeleton
k	critical
m	mean value
$n, n+1$	previous, current point of time
p	plastic behaviour of the soil skeleton
p	referring to fluid pressure (if used as a subscript)
q	part of the surface with prescribed tractions or flows
(q)	q-th time derivative
r	relative
s	solid phase of a mixture
$\mathbf{u}$	referring to soil displacements (if used as a subscript)
vol	volumetric part of stress, strain
w	water phase of a mixture
α, π	an individual phase in the range from 1 to κ
0	initial
$1, 2, 3$	one, two or three dimensions of space, coordinate directions

Other symbols:

$\mathcal{B}$	reference configuration of the body under consideration
$\mathcal{C}_{pa}$	space of solutions for the hydrostatic air stresses
$\mathcal{C}_{pw}$	space of solutions for the hydrostatic water stresses
$\mathcal{C}_{\mathbf{u}}$	space of solutions for the soil skeleton displacements

$\mathcal{E}$	third-order permutation tensor
$\mathcal{L}(f), \mathcal{L}(p^w)$	Laplace transform of a function $f(t)$, pore water pressure p^w
$\mathcal{V}_{pa}$	space of weighting functions for the hydrostatic air stresses
$\mathcal{V}_{pw}$	space of weighting functions for the hydrostatic water stresses
$\mathcal{V}_{\mathbf{u}}$	space of weighting functions for the soil skeleton displacements
$\mathbb{R}$	one-dimensional Euclidean space
$\mathbb{R}_+$	positive one-dimensional Euclidean space
$\mathbb{R}^3$	three-dimensional Euclidean space
$\mathbb{R}^{n_{eq}}$	Euclidean space with eigenvectors as basis
$\mathbb{R}^{n_{sd}}$	n_{sd}-dimensional Euclidean space
$\mathbf{0}$	zero vector
$\mathbf{1}$	vector comparable to the Kronecker delta
$\sum_{\pi=s,w,a}$	summation over the solid, water and air phase
$\sum_{\pi=1}^{\kappa}$	summation over all κ constituents of the mixture
$\sum_{\alpha\neq\pi}^{\kappa}$	summation over all κ constituents except the πth
$\sum_{j=1}^{\infty}$	summation over an infinite number of terms
$\prod_{j=0}^{n}$	product of n terms
$\mapsto$	maps to
$\in$	belongs to
$\subset$	subset of
$\cup$	set union symbol
$\cap$	set intersection symbol
$\emptyset$	empty set
$\forall$	for all
∞	infinity
$\oplus$	addition of two portions of a domain
$[\ldots]_\Sigma$	jump of the respective function across a surface of discontinuity
$\vert_\alpha, \vert_\pi$	the preceding term has to be evaluated for the α, π-phase
$\Vert * \Vert$	norm of a matrix or vector
$*^T$	transpose of a matrix or vector
$\dot{*}$	time derivative of the respective quantity
$\tilde{*}$	averaged quantity
$\partial/\partial t$	partial time derivative
D/Dt	material time derivative

D^f/Dt	material time derivative referring to the fluid phase f
D^s/Dt	material time derivative referring to the solid phase s
D^α/Dt	material time derivative referring to the α-phase
D^π/Dt	material time derivative referring to the π-phase
$\mathbf{a} \cdot \mathbf{b}$	vector dot product (scalar product), $\mathbf{a} \cdot \mathbf{b} = a_i\, b_i$
$\mathbf{a} \times \mathbf{b}$	vector cross product (vector product), $\mathbf{a} \times \mathbf{b} = e_{ijk}\, a_j\, b_k$
$\mathbf{a} \otimes \mathbf{b}$	dyadic product (tensor product) of two vectors, $\mathbf{a} \otimes \mathbf{b} = A_{ij} = a_i\, b_j$
$\mathbf{A} : \mathbf{B}$	inner product (double contraction) of two tensors, $\mathbf{A} : \mathbf{B} = A_{ij}\, B_{ij}$
div	divergence operator
grad	gradient operator
∇	gradient operator

Chapter 1

Introduction

1.1 General introduction

Geotechnical engineering practice covers many diverse and important subjects. Typical problems range from utilising soil as a foundation for supporting structures and embankments, using soil as a construction material, designing structures for retaining soil from excavations and underground openings, constructing reservoirs for the storage of industrial fluids or waste, to investigating contaminant transport in or pollution of soils and proposing remediation processes. In a great many of these tasks the consideration of the soil as a multi-phase medium is advantageous, important or even inevitable for a realistic description of the particular problem as outlined subsequently.

Nearly every civil engineering structure – building, tower, bridge, highway, tunnel, channel or dam – has to be founded on or below the surface of the earth. Depending on the type of construction, the term 'shallow foundation' is used to describe an arrangement (footing, mat) where loads are carried by the soil directly under the structure, the notion 'deep foundation' expresses that piles, caissons or piers are employed to transfer the loads to firm soil at some depth. In connection with shallow foundations on water saturated soil layers, since the pioneering work of Terzaghi and Biot, the phenomenon of consolidation is well-known. For its mathematical description both phases, soil skeleton and pore water, have to be taken into account and thus, the soil is considered as a two-phase medium.

When soil is used as construction material in civil engineering, not only the proper type of soil must be selected but also the method of placement. Three typical examples where soil acts as construction material are: (i) embankments for highways or railways, (ii) earth dams and (iii) reclamation structures. In connection with the construction of earth dams, it is common to investigate the water flow through the dam (problem of seepage flow) which, mathematically speaking, requires a two-phase model for the soil, at least if the deformations of the soil are of interest as well.

When a soil surface is not horizontal, there is a component of gravity tending to move the soil downward. If along a potential slip surface in the soil the shear stress from gravity or any other source (such as the weight of an overlying structure or an earthquake) exceeds the strength of the soil along this surface, a shear rupture and movement can occur (landslide). Therefore, in natural slopes, compacted embankments and excavations for buildings, pipes or channels, frequently the stability of a slope has to be investigated by comparing the shear stress with the shear strength along a potential slip surface (stability analysis). In case of analysing the slope stability in combination with moving water or heavy rainfall, a two-phase soil model is advantageous.

Any structure built below the ground surface is subject to forces by the soil in contact with the structure. During construction of the building the boundaries of the excavation have to be supported by a system of earth retaining structures (such as sheet pile walls or anchored bulkheads) to take up these forces. Below the groundwater level the presence of water in the soil has to be taken into account. In this case the possible flow of water underneath the bottom end of the sheet pile wall (due to differing upstream and downstream hydraulic heads) has to be analysed in order to control the water inflow (making the construction process difficult) and to guarantee the stability of the excavation. A two-phase model serves as a very helpful tool for a coupled numerical investigation of this problem.

In connection with deep excavations for the construction of buildings or when dealing with underground openings for tunnels, powerhouses or large drainage structures, a temporary dewatering of the soil may become necessary in order to prevent massive water inflow during the construction period. A dewatering of soils can be accomplished either by means of pumping wells (drawdown of water table) or by the use of compressed air (displacement of groundwater). This topic is described in some more detail in Section 1.2, dealing with tunnelling below the groundwater level employing compressed air. Anyway, a two-phase model for dewatering with pumping wells or a three-phase model for compressed air dewatering is inevitable for a coupled numerical simulation of these kinds of processes.

While the loads under consideration are primarily static or at least quasi-static in the aforementioned examples, dynamic effects in combination with soils also play an important role in geotechnical engineering. Dynamic loads, for instance, stem from earthquakes, explosions or vibrations induced by large compressors or turbines. Coupled two-phase models are of importance for these tasks, e.g., when analysing the phenomenon of liquefaction of soils (loss of strength during cyclic loading) in combination with earthquakes.

Another problem of interest in geotechnical engineering refers to regional subsidence. Surface subsidence may occur due to changes in hydraulic equilibrium in systems comprising aquifers and aquitards and/or hydrocarbon reservoirs when there is extensive groundwater withdrawal and/or oil and gas pumpage. Typically, analyses of such problems are characterised by large domains and very long time spans. Coupled numerical simulations are performed on the basis of rather complex multi-phase models.

When frost susceptible soils are in contact with moisture and subject to freezing temperatures, they can imbibe water and undergo a very large expansion. These frost heaves exert forces onto adjacent structures, highways or airfield pavements, which may cause damage. On the other hand, soil freezing is employed as an alternative to the temporary dewatering of soils by means of pumping wells. For a coupled numerical simulation of these kinds of processes, of course, the temperature has to be taken into account and non-isothermal models have to be used.

Soil not only can be employed for the retention of water, but also for the construction of reservoirs to store industrial fluids. Structures of very large capacities can be built at very low costs compared to conventional steel tankage. Due to interfacial tension between water and certain industrial fluids (e.g., fuel oil), compacted, fine-grained, wet soil can be used to store such fluids with no leakage. Another example is the storage of refrigerated liquefied gas which can be stored in the soil because of the frozen pore water making the soil impermeable. In addition to that, very impervious soil or rock is also used for the storage of biological, chemical and nuclear waste. To simulate these processes numerically, very complex coupled non-isothermal multi-phase models have to be applied, taking into account several fluid phases present in the soil and their interactions under varying temperatures.

From an environmental point of view pollutant (contaminant) transport in the soil is an interesting and important topic nowadays. These contaminants are organic substances frequently used in industry as mineral fuels (e.g., fuel oil) and as solvents, cooling agents and detergents (e.g., chlorinated hydrocarbons). A common trait of these substances, termed Non-Aqueous Phase Liquids (NAPLs), is that they are slightly soluble in water. Additionally, they often contain easily volatile components which can escape via the soil air. Even in small amounts, these components can be toxic or change the taste and smell of groundwater (used as drinking water). Multi-phase models for numerical investigations of these problems have to take into account also chemical processes and are usually very sophisticated.

Similar models are also employed for the analysis of remediation processes dealing with the decontamination of the unsaturated and the saturated soil zone, the infiltration of dissolved substances or the supply of heat to increase mobility and vaporisation rate. The use of coupled multi-phase models in such a context enables a comparison of the efficiency of different remediation methods.

1.2 Tunnelling below the groundwater level

A major part of the present work emerged from the participation in the *Austrian Joint Research Initiative* on *Numerical Simulation in Tunnelling*, supported by the *Austrian Science Fund FWF*. A number of five Institutes from three Austrian Universities were involved in this Research Initiative, dealing with such different tasks as data acquisition on site, laboratory experiments, numerical simulation and visualisation [Beer(1999)]. In the sub-project entitled *Numerical simulation of the excavation of tunnels driven under compressed air*, the topic of tunnelling below the groundwater table was investigated. Since this thesis partly emerged from this research project, the main aspects of this tunnelling method are introduced subsequently.

The rapid increase of world population, and thus population density, as well as the increasing need of fast, comfortable and safe transport of people and goods require better and better infrastructure. Especially in highly populated areas the development of traffic routes meeting these requirements gets increasingly difficult and frequently only subsurface routes are possible. Therefore, not only in the course of the expansion of highway and railway networks or in connection with the relocation of existing routes into tunnels in the vicinity of populated areas, but also in urban subway construction shallow tunnelling becomes increasingly popular.

These shallow tunnels often have to be constructed in water-bearing soil layers, exhibiting low stability and high permeability with respect to water. However, water entering a tunnel during the construction period not only affects the construction process itself but also seriously influences the stability of the tunnel face and may even lead to collapse. For tunnels driven in aquifers, thus, the groundwater has to be displaced from the working area at the tunnel face, employing collateral measures dealing with the dewatering. Consequently, in this case deformations of the ground and surface settlements are not only caused by the advance of the tunnel face but also by dewatering of the soil.

Temporarily, water inflow into the tunnel may be prevented by (i) lowering of the groundwater table (sump drainage or lowering by pumping wells), (ii) groundwater cutoff (sheet pile walls, slurry walls, ground freezing) or (iii) displacement of the groundwater, e.g., by means of compressed air. However, a certain technique should be chosen carefully in accordance with the geological and hydrogeological information, with the likely method of tunnel driving and with environmental and economic facts. Frequently, lowering of the groundwater table or groundwater cutoff is not possible on legal grounds, for environmental reasons or due to overly large surface settlements. Hence, the temporary displacement of the groundwater by means of compressed air has become more important recently, because this technique least of all influences both the groundwater as well as the ground surface and surface settlements are reduced. Especially in urban areas this minimisation of surface settlements during the construction period of the tunnel plays a predominant role in order to avoid damage of existing buildings and infrastructure. Compared to lowering the groundwater table with pumping wells and driving the tunnel under atmospheric conditions, the application of compressed air for displacing the groundwater yields smaller settlements, as shown in a study conducted during the subway construction in Essen, Germany [Kramer(1987)]. This advantage is due to the air pressure and the drag forces of the airflow in the soil which counteract the deformations caused by dewatering and tunnel excavation.

Historically speaking, compressed air has been used for displacing the groundwater in combination with tunnel driving more or less since the beginning of tunnelling in the 19th century. Already in 1828 Colladon suggested the application of compressed air for the construction of the Thamse tunnel in London after a water inrush. However, Brunel, shield driving the tunnel, rejected his proposal. Eventually, for the first time, compressed air was employed in connection with tunnelling at construction sites in Antwerpen and New York in 1879 [Maidl(1995)].

Displacement of the groundwater by means of compressed air can be combined with a number of different tunnelling methods such as shield or pipe driving, the New Austrian

Tunnelling Method in combination with shotcrete or the cut-and-cover method. In connection with shield driving compressed air shields, membrane shields, hydraulic shields or earth-pressure shields can be employed. However, no matter what type of shield is used while driving the tunnel, for purposes of maintenance work or removal of boulders compressed air is employed temporarily.

An application of the displacement of groundwater by means of compressed air in combination with tunnel driving in aquifers is reasonable if (i) tunnels are located below waterbodies, (ii) lowering of the groundwater level is not possible for technical, environmental or economic reasons and (iii) lowering of the groundwater table results in large surface settlements and thus may cause damage of existing buildings. The major advantages of this method are its potential in combining it with a broad range of tunnel driving techniques and its easy adaptability to varying groundwater levels. However, the application of the method is limited by three facts, namely by (i) the air permeability of the soil (confer Section 2.3), (ii) the thickness of the soil layer overlying the tunnel (depending on the type of soil a thickness of one to two times the tunnel diameter is required) and (iii) the height of the excess air pressure necessary within the tunnel for the displacement of the groundwater (personnel has to work under this pressure).

Finally, it should be mentioned that for applying this method successfully a profound knowledge of the hydraulic and mechanical behaviour of the outcropping initially saturated soil, i.e. the soil subject to the airflow, is of primary importance. Since the stability of the tunnel face is considerably influenced by the thickness of the soil layer overlying the tunnel, high excess air pressures applied below relatively thin overlying soil layers may cause a loosening of the soil at the ground surface in the vicinity of the tunnel face, entailing a possible loss of stability at the tunnel face via so-called blowouts. However, despite such risks the number of applications of this method has increased in the last decades due to the increase of urban shallow tunnelling and the improvement of the involved technology.

Nevertheless, the design of tunnels driven below the groundwater table by means of compressed air even nowadays is primarily based on empirically founded calculation methods or uncoupled numerical investigations (flow and deformation problem are treated in two consecutive steps). Coupled numerical procedures are rarely used in practice. In this sense, the coupled model for the numerical simulation of tunnelling below the groundwater table by means of compressed air, developed in the present work, should be understood as a step in the direction of an improved treatment of such kinds of problems.

1.3 Outline of the thesis

In this thesis, a coupled three-phase model applicable to the numerical simulation of dewatering of soils, e.g., used in combination with tunnelling in aquifers where compressed air is employed for displacing the groundwater, as well as to a number of other tasks encountered in the realm of geotechnical engineering, such as dewatering under atmospheric conditions, seepage flow or consolidation, is developed. The model is implemented into an existing finite element programme [Carter(1995)] and validated by means of analytical solutions, numerical results available in the literature and experimental data.

A notion frequently used in the course of this work is the term 'model'. A model is defined as the result of an abstraction which, based on certain criteria, reduces the considered system to a few single but characteristic aspects. Different types of models can be distinguished and are addressed in the remainder of the work [Helmig(1997)]:

Conceptual model: A conceptual model is the description of a system which is able to represent those aspects of the system relevant for the model's application. Consequently, the degree of abstraction is of paramount importance. Sensible assumptions have to be made in order to maintain a balance between the necessary complexity and the system properties which are to be represented. Thus, the art of modelling is to choose the conceptual model in such a way that essential system properties remain clear.

Mathematical model: The mathematical model transfers the conceptual model to a mathematical formulation which includes the balance equations for mass, momentum and energy as well as the system-dependent equations of state. The derivations are usually based on continuum theory; in the present context a continuum corresponds to a representative elementary volume (confer Chapter 3). In addition to an adequate approximation of the processes chosen to be represented by the conceptual model, the integration of appropriate boundary and initial conditions, depending on the chosen primary variables, into the mathematical model is of particular importance. Concerning mathematical models a subdivision into *deterministic* and *stochastic models* is possible.

Numerical model: In general, for the majority of the cases mathematical models can only be treated analytically for certain simple boundary and initial conditions which strongly restricts their practical applicability. For a more flexible consideration, the balance equations have to be solved by means of numerical methods. Thus, the numerical model describes the transfer of the mathematical model to numerical algorithms. An appropriate numerical model must guarantee the solution of the constituting equations for different geometries as well as boundary and initial conditions with respect to the variables representing the state of the system.

The thesis is subdivided into seven chapters and a very brief overview of their contents is presented subsequently.

To provide a background to the individual chapters and to allow for a classification of this work in a scientific context, a review of several topics is given in Chapter 2. Since a three-phase formulation is used in the present approach, a range of widely different fields of application for multi-phase models is outlined. A discussion of numerical simulations in tunnelling is included in the chapter. Various possibilities for the mathematical description of multi-phase media are presented in combination with a short historical review.

In Chapter 3 the mathematical model for the three-phase formulation is presented. In the context of the averaging theory the individual balance equations and the involved constitutive equations are derived. Starting from the very general description of a multi-phase model, the equations employed for a three-phase formulation are specified and in a further step of simplification some special cases included in this formulation are obtained.

Chapter 4 starts with a summary of the governing equations together with the boundary and initial conditions of the three-phase formulation. Using this starting point, a weak formulation of the equations is derived which constitutes the basis for the spatially discretised set of equations. After a suitable choice of one of the many available time integration procedures, the final coupled set of equations is obtained. A description of iterative solution methods and considerations on stability, consistency and accuracy conclude the chapter.

The verification of the derived three-phase formulation is dealt with in Chapter 5. Starting with two-phase example problems describing the consolidation phenomenon (one-dimensional Terzaghi consolidation and two-dimensional consolidation) where analytical solutions are available for comparison, the complexity of the examples is increased step by step. The flow of compressed air through a dry soil is considered next, results are compared with experimental data. Unsaturated soil behaviour is addressed by means of the numerical simulation of the one-dimensional drainage of a soil column which concludes the chapter. This example is simulated using both a two-phase model (atmospheric air pressure assumed) and a three-phase model (hydrostatic stress in the air phase taken into account), respectively.

In Chapter 6 the two-phase model is applied to the coupled numerical simulation of water flow through an earth dam, an example of practical relevance in geotechnical engineering. In addition to that, two three-phase examples are presented. A large-scale in-situ air permeability test, conducted in connection with the subway construction in Essen, Germany, is dealt with first. Results of the calculation are compared with measurements obtained from the experiment. Finally, the applicability of the present three-phase formulation to the numerical simulation of tunnelling below the groundwater table using compressed air for dewatering of the soil is discussed in the last example of Chapter 6.

Chapter 7 completes the thesis by summarising the major findings of the work. The most important conclusions to be drawn are presented. A short outlook indicates tasks to be done for the future improvement of the model.

Chapter 2

State of the art

2.1 Introduction

In the present work a three-phase formulation is used for the numerical simulation of the dewatering of soils by means of compressed air and in particular, this topic is investigated in the light of the numerical simulation of tunnelling below the groundwater table by means of compressed air.

Since a coupled multi-phase approach is employed, some applications of multi-phase models are introduced first. However, no attempt is made to either give a comprehensive historical review or list all possible applications, rather a short overview is presented. The subsequent section starts with a discussion of two-phase formulations which have been used since the analytical treatment of the consolidation phenomenon by pioneers as Terzaghi and Biot. After the rise of the Finite Element Method these analytical solutions were frequently employed for verification purposes in connection with numerical models. Furthermore, two-phase formulations are also necessary for the coupled analysis of seepage flow. As computers became more and more powerful in the last decades, a broad range of problems were started being analysed on the basis of implementations of multi-phase models into a finite element framework. These formulations allow to take into account several fluid phases present in the soil (in addition to groundwater these fluid phases may be air, oil, gas or contaminants) as well as thermal or chemical processes. Several types of such topics are mentioned in the next section.

Since the three-phase model in the present work is also applied to tunnelling, a short section on numerical analyses of tunnelling is added. In this particular section, analyses of tunnels located in dry soil are distinguished from calculations taking into account the groundwater in the soil. While different approaches (e.g., the Finite and the Boundary Element Methods) and models (linear elastic-perfectly plastic, hypoplastic, multilaminate) exist for drained conditions, numerical simulations of tunnels located below the groundwater table are performed much more rarely. Although a couple of models allow to determine the shape of the groundwater table after the dewatering process, most of these models do either not take into account the excavation process or use an uncoupled approach, i.e. treat the flow and deformation problems sequentially. Coupled formulations for the particular problem of tunnelling below the groundwater level by means of compressed air, to the authors knowledge, barely exist.

In a third section an overview over different possibilities of the mathematical description of multi-phase flow through porous media is presented. After a short historical review, dealing with the development of today's modern theories from the beginnings as early as in the 18th century, the basic concepts and features of the modern approaches, i.e. of (i) the mixture theory extended by the concept of volume fractions and (ii) the averaging theory, are outlined in two individual sections. Thereby, the theory of mixtures extended by the volume fraction concept is treated in some more detail in this chapter, while the averaging theory, used in the present work, is discussed thoroughly in Chapter 3.

2.2 Flow through porous media

2.2.1 Introduction

Since a three-phase model is employed in the present work, a short overview over applications of multi-phase approaches is given in this section on flow through porous media. A broad range of entirely different topics in the realm of civil engineering are investigated with such kinds of models. Applications cover geotechnical problems, environmental engineering tasks or durability analyses, just to mention a few of the diverse subjects. Usually these models are characterised by the coupling of a flow problem with either a deformation problem, temperature effects, or chemical and biological processes, respectively, or even with a combination of those. From a mathematical point of view, the balance equations of mass, momentum and energy are used in combination with a number of more or less sophisticated constitutive relations, depending on the type of problem under consideration. For the solution of the obtained set of partial differential equations primarily the Finite Element Method is employed. Since the problems in general are both space- and time-dependent, fast and efficient numerical solution algorithms are required.

The subsequent sections are categorised according to the number of phases involved. The first section covers applications of two-phase formulations. In particular, topics encountered in geotechnical engineering are discussed. A short section on three-phase formulations follows because such an approach is used in the present work. Finally, diverse multi-phase models are introduced, dealing with problems including thermal, chemical and biological effects.

2.2.2 Two-phase problems

In the current section two-phase problems encountered in geotechnical engineering practice are introduced. Probably the two most important tasks where two-phase models are used, concern the phenomena of consolidation and seepage flow which are discussed subsequently in brief.

Consolidation

In civil engineering, two-phase formulations were first introduced when the theory of consolidation was developed by Terzaghi [Terzaghi(1923), Terzaghi(1925)]. The soil was assumed to be a two-phase material, consisting of an elastic skeleton and pores completely filled with water. Terzaghi's solution only applied to one-dimensional problems. Biot [Biot(1941)] extended Terzaghi's consolidation theory to the three-dimensional case. In this study, he presented a mathematical formulation of the fundamental equations for the consolidation of saturated elastic materials. In later papers, Biot generalised his approach by dealing with anisotropic soils [Biot(1954), Biot(1955), Biot(1956a)].

Since Biot's equations involve the complexities of an elastic problem coupled with a flow process, analytical solutions are quite difficult to establish. For linear elastic behaviour of the soil skeleton, the analytical solution of Booker [Booker(1974)] should be mentioned. When considering linear elastic-perfectly plastic constitutive behaviour of the soil skeleton, an analytical solution for the one-dimensional case was proposed by Pariseau [Pariseau(1999)], using the elastic-plastic models of Drucker-Prager and Mohr-Coulomb. However, nowadays primarily numerical methods are employed for solving consolidation problems and analytical solutions thus are basically of importance for the verification of computer codes.

A major step was achieved by Sandhu and Wilson [Sandhu(1969)] who first applied the Finite Element Method to the phenomenon of consolidation. A formulation based on a variational principle was developed for solving the problem of fluid flow in a saturated porous elastic material. The finite element equations were solved in the time domain using a time stepping technique in which the solution at a particular point of time was determined from the solution at the previous time step. Concerning elastic-plastic material behaviour the approach of Small et al. [Small(1976)] should be mentioned, combining consolidation and yielding of the soil. This formulation was derived using the principle of virtual work and implemented into a finite element framework. More recently, non-linear material behaviour of the soil skeleton was also investigated in the works of Siriwardane and Desai [Siriwardane(1981)], Prevost [Prevost(1982)], Buchmaier [Buchmaier(1985)] or Borja [Borja(1989)].

Of primary importance for a sufficiently accurate determination of the pressure in the fluid phase are both the method used for solving the problem in the time domain and the chosen time step size. Solutions obtained by means of time marching procedures, such as employed by Siriwardane and Wilson, are sometimes unstable. However, Booker and Small [Booker(1975)] examined various time integration techniques and determined conditions of stability for these schemes. In the sequel, a number of papers were published dealing with different time integration schemes as well as accuracy and stability of these schemes [Hughes(1983), Borja(1991a)]. More recently, an automatic time stepping procedure was proposed in [Sloan(1999a)] and [Sloan(1999b)] used in combination with elastic and elastic-plastic consolidation analyses. Thus, criteria for choosing an acceptable time increment size (which may vary by several orders of magnitude throughout the analysis) are available.

Other approaches deal with a number of special problems. For instance, the pore fluid compressibility was accounted for by Ghaboussi and Wilson [Ghaboussi(1973)], whereas in the original theory of Biot an incompressible fluid phase was assumed. Lewis et al. [Lewis(1976)] were the first to consider the aspect of permeability changes during consolidation. Furthermore, Zienkiewicz et al. [Zienkiewicz(1977)] introduced the compressibility of the solid grains. The development of a numerical model considering not only static but also dynamic behaviour of fully water saturated soils was described in [Zienkiewicz(1984)] and [Zienkiewicz(1990a)]. On the basis of a two-phase formulation numerical simulations were employed to investigate soils under earthquake loading. Dynamic models are of particular importance when dealing with the phenomenon of liquefaction (softening of soil occurring under earthquake loading). A model for fully saturated soil dynamics, based on the theory of mixtures, was put forward by Ehlers and Kubik [Ehlers(1994)]. Approaches presented by Carter et al. [Carter(1979)], Borja [Borja(1995), Borja(1998)] or Dluzewski [Dluzewski(2001)] deal with effects of finite deformations in combination with the consolidation problem. The phenomenon of secondary consolidation (creep in soils) is addressed in [Lewis(1989)] and [Lewis(1998)] which is of importance when considering very soft organic clays under compression.

Seepage flow

Two-phase formulations can also be employed for coupled numerical simulations of seepage flow. However, in 'classical' seepage analysis such coupled approaches are barely used. In general, seepage problems can be categorised into four groups: (i) steady confined, (ii) steady unconfined, (iii) transient confined and (iv) transient unconfined seepage flow. For steady seepage flow, the flow boundary conditions are independent of time, whereas for transient seepage flow, the flow boundary conditions change with time. Unconfined seepage flow occurs when the groundwater possesses a free surface along which the pore pressure is zero. When talking about confined seepage flow, the groundwater is confined in an aquifer where water along the boundaries has piezometric pressures which may differ from zero.

The various existing methods for the numerical analysis of seepage flow can be subdivided into three main categories: (i) analytical methods, (ii) analogue methods and (iii) numerical methods. Due to the limited usefulness of the former two for many practical cases and to the development of computers, in the last decades numerical methods, such as the Finite Difference Method [Herbert(1972)], the Finite Element Method [Bathe(1982), Cividini(1984)] or the Boundary Element Method [Brebbia(1978), Banerjee(1981)], played an increasingly important role for solving seepage flow tasks since these methods are also capable of handling complicated cases, exhibiting irregular geometry, anisotropy and non-homogeneity of the soil or a pronounced three-dimensional nature of the problem. When considering the Finite Element Method for the solution of unconfined seepage flow, 'variable mesh' and 'constant mesh' approaches exist.

Variable mesh approaches require iterative processes which modify the geometry of the mesh so that a part of its boundary coincides with the free surface until a convergent solution is reached. These methods usually give results of high accuracy since the free boundary position is well represented during the solution process. However, major drawbacks are the necessity of a new discretisation and the reassembly of the flow matrix at each iteration as well as the possible instability of the solution if a complex geometry is involved or the free surface intersects a boundary between different soil layers. Variable mesh approaches have been presented, e.g., in [Neuman(1971)], [Desai(1972)], [Gioda(1987)] or [Cividini(1989)].

When using constant mesh strategies, the solution process involves maintaining a constant geometry for the mesh and allowing the free surface to pass through the finite elements. Although these methods may yield slightly less accurate results than the variable mesh approaches, in general the computational effort is much less for the constant mesh methods and they can easily be applied to non-homogeneous or layered soils. These techniques are also more suitable if deformations of the soil are of concern. Constant mesh approaches are employed, for instance, in [Desai(1976)], [Bathe(1979)], [Oden(1980)], [Lacy(1987)] and [Borja(1991b)].

A more detailed description of numerical simulations of seepage flow problems can be found in [Gioda(1988)] or [Hsi(1992)].

2.2.3 Three-phase problems

Three-phase formulations are employed in the field of engineering, e.g., for the description of unsaturated soil behaviour or for the numerical simulation of the process of water injection into oil formations. These phenomena can be described by a soil model consisting of a deformable soil skeleton and pores filled with two immiscible fluid phases, a wetting and a non-wetting phase. The wetting phase is the groundwater present in the soil, the non-wetting phase is considered to be either air (when dealing with unsaturated soil problems) or oil (when investigating problems in connection with oil formations). Depending, to some extent, on the type of problem under consideration, different approaches exist which can be distinguished by the chosen primary variables. The majority of models is based on the choice of the displacements of the soil skeleton and the pressures in the two fluid phases as primary variables. However, formulations exist which use the soil displacements, the water pressure and either the capillary pressure or the degree of saturation as the primary unknowns.

Unsaturated soil behaviour

For the mechanical/mathematical description of unsaturated soil behaviour two possibilities exist: (i) The air pressure is kept at a constant reference pressure, e.g., at atmospheric air pressure. In this case a two-phase model, as discussed in Section 2.2.2, can be used where the degree of saturation varies with the negative water pressure. (ii) If the simultaneous flow of water and air in the deforming porous solid is of interest, a three-phase model has to be employed. In [Simoni(1991)] such an approach was investigated based on using the soil displacements along with water pressure and capillary pressure as primary variables. However, most of the models dealing with unsaturated soil behaviour rely on employing the two fluid pressures of the water and the air phase in combination with the displacements. Schrefler and Zhan [Schrefler(1993)] applied a three-phase approach with this particular set of variables for analysing footings on partially saturated soils and for modelling air storage in aquifers.

A model based on the same primary variables was also used by Alonso and Batlle [Alonso(1995)] for the numerical simulation of the construction and impoundment of an earth dam. The most pronounced difference between the latter two formulations lies in the assumed constitutive equation for the soil skeleton. While the former model uses the modified effective stress concept according to Bishop [Bishop(1959)], the latter approach is based on a state surface model where the void ratio depends on both net normal stress and capillary stress. Such state surface concepts, which employ two independent stress state variables (net normal stress and matric suction), have been widely used for modelling the behaviour of unsaturated soils in the last decades. The application of two stress state variables was first suggested by Coleman [Coleman(1962)] and Bishop and Blight [Bishop(1963)]. More recently, such models have been discussed by Fredlund and Morgenstern [Fredlund(1977)], Alonso et al. [Alonso(1990)], Fredlund and Rahardjo [Fredlund(1993)] or Wheeler and Karube [Wheeler(1995)]. However, these models seem to be of primary importance only for soils exhibiting pronounced swelling behaviour, for instance such as highly expansive clays.

Other topics using three-phase formulations were described by Zienkiewicz et al. [Zienkiewicz(1990b)], dealing with the fully dynamic behaviour of unsaturated soils in combination with earthquake loading, and by Klubertanz [Klubertanz(1999)], where a three-phase model was employed for the numerical investigation of landslides. Numerical simulations of the dewatering of soils by means of compressed air, as discussed in the present work, based on coupled three-phase models barely exist. The coupled approach of

Snee and Javadi [Snee(1996), Snee(1997), Javadi(2001)] should be mentioned in this context, dealing with airflow in partially saturated soils. However, this model is introduced in some more detail in the next section on the numerical simulation of tunnelling.

Modelling of oil reservoirs

As mentioned earlier, three-phase formulations are also employed in petroleum reservoir analysis. Three-phase models can be applied to the simulation of oil formations located in water saturated soil layers. Thus, such models are of particular importance for the prediction of surface subsidence and the influence of compaction on the reservoir performance. In [Li(1990)] and [Li(1992)] such an approach is proposed. The primary variables are chosen to be the displacements of the soil skeleton, the water pressure and the degree of water saturation. In these papers the mathematical model is discussed and numerical simulations are presented.

2.2.4 Multi-phase problems

Since computers became more and more powerful in the last decades, the solution of rather complex multi-phase problems is possible nowadays. This class of problems is characterised by taking into account the (simultaneous) flow of several fluid phases present in the pores, such as groundwater, air, oil, gas or contaminants, in combination with the deformations of the soil. Temperature effects as well as chemical or biological processes usually play a key role in the realm of these multi-phase tasks. The types of problems investigated concern oil or gas exploitation, the storage of industrial or nuclear waste, the durability of structures and, last but not least, contaminant transport in the field of environmental engineering, to mention a few.

Several fluid phases present in the soil

When dealing with deformation-fluid flow analyses in petroleum reservoir engineering, in addition to the deformations of the soil skeleton one usually has to take into account three fluid phases – water, oil and gas. From a mathematical point of view, the four fully coupled partial differential equations, including three immiscible and compressible fluid flow equations along with the equilibrium equations, are derived similar to the simpler two-phase formulations (which consist of one fluid flow equation and the equilibrium conditions). However, a number of additional constitutive equations are required. As primary variables commonly the soil displacements and the pressures in the individual fluid phases are chosen. Such approaches are described, e.g., in [Sukirman(1993)], [Lewis(1993)] or [Lewis(1998)]. Practical applications of these models mainly deal with the investigation of occurring surface subsidence due to the exploitation of oil and/or gas from reservoirs.

Non-isothermal processes

If the numerical model allows to take into account coupled heat and fluid flow in the deforming porous medium, a number of complex problems can be investigated. For instance, such an approach enables the analysis of land subsidence in connection with geothermal energy production for a given geothermal system. Models of this type can also be employed for designing thermal and/or hydraulic fracturing stimulation of oil reservoirs and for a more accurate interpretation of well tests when thermal effects are taken into account [Lewis(1998)]. Another important area of application of such non-isothermal models is the disposal of nuclear waste in geological formations. The decay of radioactive material produces heat, causing a temperature rise and an expansion of both the pore water and the soil skeleton. In general, the volume increase of the pore water is greater than the volume increase of the voids in the soil. Consequently, the pore pressure is increased and the effective stresses are reduced. If this stress reduction is too large, failure of the clay barrier may occur. In order to prevent such failure and to guarantee a safe storage of the nuclear material, predictive analyses are very important in this case.

From a mathematical/mechanical point of view, these so-called 'thermo-hydro-mechanical' models are based on the balance equations of mass (for the water and the air phase), momentum (equilibrium conditions) and energy. In addition to that, a number of constitutive equations are necessary, describing the behaviour of the individual constituents. Different choices for the primary variables exist. Models proposed in [Gawin(1995)] and [Gawin(1996)] use the soil displacements, the gas pressure, the capillary pressure and the temperature whereas approaches presented in [Thomas(1995), Thomas(1997), Thomas(1999)], in [Lewis(1998)] or in [Alonso(1998)], [Gens(1998)] and [Navarro(2000)] are formulated in terms of the soil displacements, the fluid pressures in the water and the air phase and the temperature. Furthermore, the most pronounced difference between the individual models concerns the constitutive behaviour of the soil skeleton. The approaches described in [Gawin(1995), Gawin(1996)] and [Lewis(1998)] are based on the concept of modified effective stresses (effective stresses of the soil skeleton are determined by subtracting the average pressure of the mixture of fluids surrounding the grains from the total stresses). On the contrary, the formulations presented in [Thomas(1995), Thomas(1997), Thomas(1999)] rely on the state surface concept and therefore use two independent stress state variables (net normal stress and matric suction). Both a state surface approach and an elastic-plastic constitutive model are investigated in the works of Alonso et al. [Alonso(1998)], Gens et al. [Gens(1998)] as well as Navarro and Alonso [Navarro(2000)].

Environmental engineering problems

Environmental problems are essentially multi-physics problems and often involve transport of substances. The transport of contaminants and other substances may occur in the atmosphere, in water bodies such as rivers, lakes or oceans, or in fluids filling the void spaces of porous media, i.e. water, water vapour and dry air. The subject which probably has received most attention is groundwater pollution. This is due to the fact that both the exploitation of underground water resources and the preservation of groundwater quality are of primary importance. Sources of groundwater pollution may be industrial and urban waste, agricultural pollutants or saltwater intrusion in costal aquifers. A major group of the existing flow and transport models describe groundwater flow and the convective-dispersive spreading of one or more components entirely dissolved in water. Some substances frequently used in industry, however, are hydrophobic, i.e. they are immiscible with water and only slightly soluble (e.g., halogenated hydrocarbons or petroleum products). In contrast to the transport of dissolved contaminants, these hydrophobic substances represent individual phases which have to be modelled. Thus, multi-phase approaches need to be applied. For the numerical simulation of thermal remediation processes, in addition to the movement of the individual phases, phase transitions and heat transfer processes have to be accounted for. An introductory treatment of these processes is given, amongst others, in books by Bear and Bachmat [Bear(1990)] or Helmig [Helmig(1997)]. Finite element applications may also be found in publications by, e.g., Schrefler et al. [Schrefler(1994), Schrefler(1995)] or Li et al. [Li(2000)].

Finally, a further problem with environmental implications involving pollutant transport in porous media is the durability of concrete and other porous building materials, such as bricks. The durability of concrete mainly depends on its ability to (i) resist penetration by aggressive substances from the environment and (ii) protect the embedded steel reinforcement. The diffusion of carbon dioxide or oxygen in dry pores is of concern as well as the diffusion of chloride or sulphate ions in pore water. Complex thermo-hydro-mechanical multi-phase models, described, e.g., in [Schrefler(1995)] or in [Grasberger(2000), Grasberger(2002)] and [Meschke(2001)], enable numerical investigations on the durability of concrete structures considering damage, moisture transport, chemical dissolution processes and thermal effects.

2.3 Numerical simulation of tunnelling

2.3.1 Introduction

Due to the increasing demand of modern traffic/transport routes and to the rising population density, tunnelling has become a more and more important task in civil engineering in the last decades. Nowadays, tunnels have to be driven in a broad variety of soils, reaching from solid rock to soft soil, deep below the ground surface or with a small overburden layer. Thus, the construction of tunnels is a challenging task, not only for the engineer on site but also for the simulation specialist in charge in the civil engineering office. Since tunnels can be located in widely different types of soil, such as solid or jointed rock or soft soil (dry or bearing groundwater), soil models have to be able to deal with a variety of phenomena.

A huge number of more or less sophisticated numerical models and approaches describing such problems have been proposed in the literature. Even so, more advanced formulations are still not routinely employed for solving practical problems. The reason for this is that either these models are sometimes very difficult to understand and therefore not accepted in practice or the computational effort is too high, or both. However, as commercial software packages increasingly include more advanced models together with fast solution algorithms and in addition to that very powerful computers are available, the advantages of sophisticated constitutive models over, e.g., the simpler linear elastic-perfectly plastic formulations may be more and more appreciated in practice in the future. It should be emphasised that only by employing these models for the solution of real life problems their limitations become obvious and the formulations can be improved.

A brief overview over different types of models is presented in the subsequent two sections, dealing with the numerical simulation of tunnelling under drained conditions or in water bearing soil layers (aquifers).

2.3.2 Simulation of tunnelling in dry soil

The numerical simulation of tunnelling offers a particular challenge to both the model applied and the civil engineer in charge. On the one hand, tunnels are driven in a soil/rock mass which for all practical purposes can be considered to be infinite or semi-infinite and may be highly inhomogeneous or jointed. On the other hand, one usually has to consider a complex sequence of excavation and installation of support measures (such as shotcrete or rock bolts) as well as non-linear material behaviour of soil/rock and shotcrete. For the numerical treatment of tunnel excavation problems nowadays two methods are very popular, namely the Finite Element Method (FEM) and, more recently, also the Boundary Element Method (BEM).

The Finite Element Method is ideally suited for such tasks and is therefore widely used in practice. However, in the majority of all cases tunnels driven in either drained soil or rock (may be jointed) are analysed. In addition to that, only two-dimensional analyses are performed frequently although the problem under consideration clearly possesses a pronounced three-dimensional nature. The reason for relying on two-dimensional calculations probably is the vast effort needed for both generating the mesh and computing results when dealing with three-dimensional numerical simulations of tunnelling. Even when using modern and efficient hardware and software, 3D computations may take from several hours up to days [Golser(1999)].

This expensive computation cost is mainly due to three facts: (i) Because of the inability of the FEM to model infinite or semi-infinite domains and consequently the need to truncate the mesh, the number of finite elements required tends to be large (in the order of 10000 to 100000 elements), especially if sequential excavation and construction processes have to be modelled in detail (50 to 100 construction stages are not unusual at all). (ii) Additionally, an iterative solution procedure may be necessary due to the consideration of non-linear material behaviour of soil and shotcrete. (iii) Storage requirement can be in the order of Gbytes, where most of the storage is used for keeping element results. To mention two examples of three-dimensional finite element analyses of tunnelling, the works of Golser [Golser(1999)], investigating the performance of various existing models when doing numerical simulations on site, and Kropik [Kropik(1994)], dealing with elasto-viscoplastic deformation and stress analyses, can be cited. In practice, only rather complex tasks such as tunnel junctions, cross drifts or tunnels located in areas sensitive to surface settlements are analysed by means of three-dimensional models.

In every day civil engineering practice, analyses of tunnels are mainly performed employing two-dimensional finite element models. The constitutive behaviour of the soil is commonly described using the linear elastic-perfectly plastic models of Mohr-Coulomb and Drucker-Prager, respectively, or one of the well-known cap models. A study using these kinds of linear elastic-perfectly plastic models for the soil in combination with two-dimensional tunnel analyses is presented in [Oettl(1998)]. An entirely different approach is described, e.g., in [Schuller(1999)] and [Schweiger(2000)], where a multilaminate model is proposed for the analysis of shallow tunnels located in soft soil.

Because of the pronounced three-dimensional nature of the problem, at least in the vicinity of the tunnel face, two-dimensional models in general cannot yield reliable predictions for this region. Usually a number of (more or less arbitrary) parameters (stress relief factor) have to be introduced in order to correct results obtained from two-dimensional analyses. However, based on the experience of the personnel on site such an adaptation of results is often successful.

As mentioned above, as soon as one has to tackle the problem of three-dimensional tunnel analysis (especially when taking into account material non-linearities and the construction process in detail), use of the Finite Element Method becomes somewhat cumbersome and both modelling effort and storage requirement increase substantially. Therefore, it is thought that the Boundary Element Method can be used successfully when adding some necessary problem specific adaptations [Beer(2000)]. The BEM offers the following advantages: (i) Infinite or semi-infinite domains can easily be dealt with since the shape functions used to describe the variation of the displacements automatically satisfy the infinity condition. (ii) The effort in mesh generation is reduced significantly because only the surface of the excavation needs to be described by elements (thus only a few hundred elements are required). (iii) The results are more accurate than the ones obtained with the FEM since no truncation error is introduced (the infinite domain is considered in the BEM implicitly) and the equilibrium conditions inside the domain are satisfied exactly. (iv) Storage requirements (results can be specified where desired) and computation times are smaller [Beer(2000)].

The modifications and adaptations of the basic BEM necessary in order to allow for a consideration of plastic material behaviour and sequential excavation and construction, two features of primary importance for the numerical simulation of tunnelling, are described subsequently in brief. Plasticity is commonly introduced into the BEM by providing a volume mesh in zones where plastic behaviour is expected. Although no additional unknowns are associated with this mesh, the effort for computing internal results at the nodes of the volume mesh may become significant. Furthermore, making an educated guess at the extent of the plastic zones constitutes a difficulty. Thus, in [Beer(2000)] an adaptive volume meshing technique is proposed where the volume mesh grows together with the zone of plasticity. This scheme is based on the assumption that plasticity sets out from a boundary of the domain which seems to be justified for the numerical simulation of tunnelling problems. A capability for dealing with sequential excavation and construction offers the so-called multi-region BEM. This method also allows to consider piecewise non-homogeneous soil masses or jointed rock. In a multi-region analysis the considered domain is divided into a number of subregions which are connected to each other [Beer(2000)]. The modelling of the shotcrete lining is usually realised by employing finite shell elements for the shotcrete and combining them with the boundary elements [Beer(1992)]. However, for a successful and efficient use of the Boundary Element Method in combination with the numerical simulation of tunnelling still a considerable amount of research work has to be performed [Beer(2000)].

2.3.3 Simulation of tunnelling in aquifers

If a tunnel has to be driven in an aquifer, i.e. the tunnel is either partly or completely located below the groundwater level, during the construction process the groundwater has to be displaced from the working area at the tunnel face. To achieve this groundwater displacement in the working area, two possibilities exist. (i) The groundwater table can be lowered with pumping wells and the tunnel is then driven under atmospheric conditions. Since the soil in a rather large region is influenced if lowering of the groundwater over a significant height is necessary, for environmental reasons this method clearly should only be applied if the tunnel is located in the aquifer just partly. (ii) The application of compressed air for displacing the groundwater from the working area at the tunnel face influences the surrounding soil much less and, as shown in a study [Kramer(1987)] conducted during the subway construction in Essen, Germany, or, more recently, by measurements of surface settlements performed in connection with driving of the tunnel Siegaue [Schmettow(2002)], yields smaller settlements than a lowering of the groundwater. This advantage, of particular importance in urban areas, is due to the air pressure and the drag forces from the airflow which counteract the deformations caused by dewatering and tunnel excavation. In the present work the application of compressed air for dewatering of soils is investigated numerically.

Principle of dewatering of soils by means of compressed air

The stability of the tunnel face during tunnelling in aquifers crucially depends on groundwater present in the working area. The method of dewatering by means of compressed air is based on the displacement of the groundwater in the soil in the vicinity of the tunnel face. Hence, during the construction period a permanent airflow through the tunnel face (and through cracks in the lining occurring in the already secured part of the tunnel subject to compressed air) is encountered, as shown in Figure 2.1. To achieve a dewatering of the soil, this airflow has to be maintained and the air loss has to be replaced continuously. The amount of compressed air necessary for the dewatering depends on (i) the permeability of the soil, (ii) the geometry (diameter and length) of the tunnel and (iii) the type of tunnel lining.

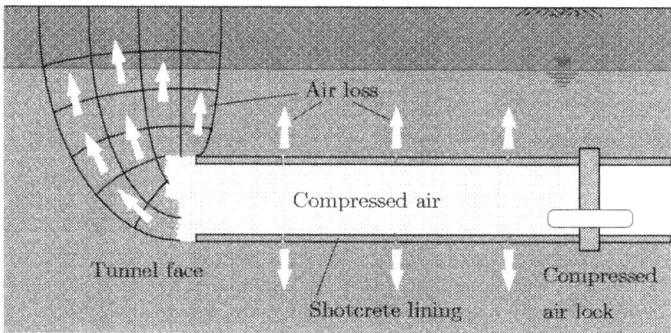

Figure 2.1: Tunnelling below the groundwater table by means of compressed air.

The pressure level of the compressed air in the tunnel usually is determined according to the hydrostatic water pressure present in the soil at the depth of the tunnel invert (confer Figure 2.2). Since the air pressure is constant along the height of the tunnel whereas on the contrary the water pressure in the soil varies linearly with depth (linear increase of water pressure below the groundwater table), the air pressure exceeds the water pressure. This excess of the air pressure over the water pressure increases from $p_{ex}^a = 0$ at the invert of the tunnel to a value of $p_{ex}^a = D \cdot \gamma^w$ at the crown of the tunnel, D being the diameter of the tunnel and γ^w denoting the unit weight of water, and causes a flow of compressed air through the tunnel face (confer Figure 2.2). In the soil in the vicinity of the tunnel face in the course of time an air pressure field develops

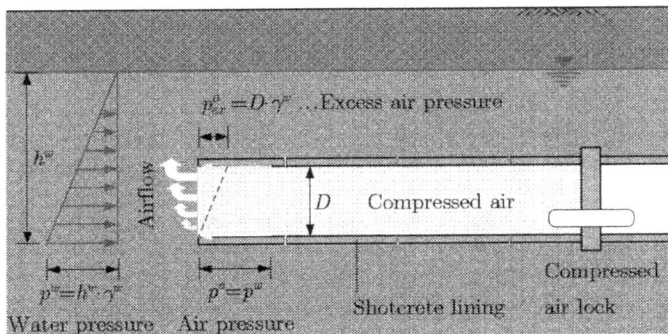

Figure 2.2: Assumed distribution of hydrostatic fluid stresses.

which exhibits a gradient in the direction of the groundwater table or the ground surface (due to the atmospheric air pressure prevailing at these levels). Thus, the compressed air yields a displacement of the groundwater in the pores of the soil and the initially water saturated soil changes to a partially saturated state. In this region of the soil a flow of both fluid phases water and compressed air is encountered.

If the air pressure in the tunnel is chosen too large compared to the overburden of the tunnel (stresses due to the overlying soil layer), the large air pressure gradient may cause a loosening of the soil at the ground surface above the tunnel face. As a result the increasing air loss in the tunnel (blowouts) can yield to instability of the tunnel face and to a possible water inrush. Therefore, a sufficiently large overlying soil layer and blowout safety have to be proved in any case of practical interest.

Tunnel driving methods used in combination with compressed air

As briefly mentioned in Chapter 1, a major advantage of the displacement of the groundwater by means of compressed air is that it can be combined with a number of different tunnelling methods such as the New Austrian Tunnelling Method (NATM) in combination with shotcrete, shield driving or the cut-and-cover method. Many completed building projects document the successful application of these methods in combination with compressed air dewatering of the soil during construction.

Dewatering by means of compressed air in combination with the New Austrian Tunnelling Method was employed for the first time at subway construction sites in Munich, Germany, in 1981 [Kramer(1989)]. Figure 2.1 shows a schematic diagram of the method, indicating the air loss at the tunnel face (necessary for dewatering of the soil) and through cracks in the shotcrete lining. It is obviously these cracks which result in an increase of the air loss with increasing tunnel length (length of the contract section). More recently, the NATM in combination with compressed air has been proposed for contract section U3-1 of the Munich subway, as reported by Prof. Semprich, Head of the Institute for Soil Mechanics and Foundation Engineering at Graz University of Technology (2002).

A second method quite frequently used in combination with dewatering by means of compressed air is shield driving. When shield driving a tunnel, the securing is usually achieved by a sequential lining. The compressed air lock in this case may be located

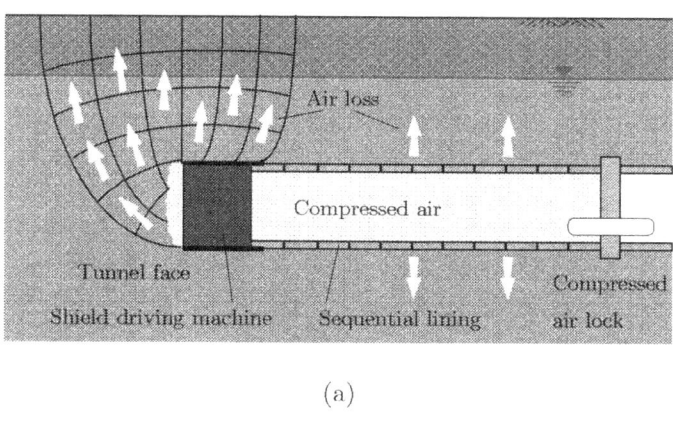

(a)

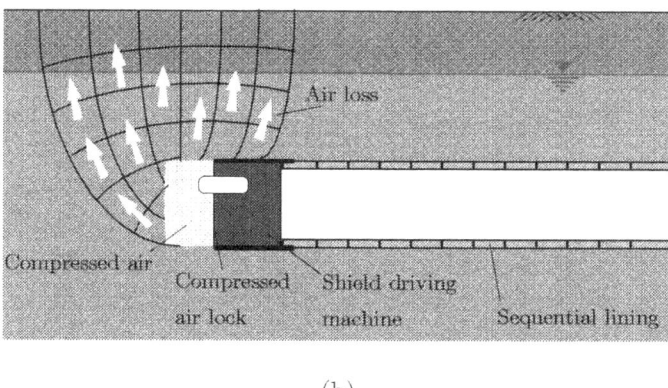

(b)

Figure 2.3: Shield driving method in combination with compressed air; air lock (a) behind and (b) in front of the shield driving machine.

behind the shield driving machine, i.e. in the part of the tunnel secured by the lining (confer Figure 2.3(a)), or in front of the shield driving machine (Figure 2.3(b)). In the latter case only a short distance of the tunnel is subject to compressed air which reduces the air loss (no air loss through the lining). The method shown in Figure 2.3(a) has been used recently, e.g., for the construction of a large waste water sewer built in Braila, Romania [Bally(2002)]. The method depicted in Figure 2.3(b) was employed for the construction of the fourth tube of the Elbe tunnel in Hamburg, Germany, as reported by Prof. Semprich (2002).

A third method used in combination with dewatering by means of compressed air is the cut-and-cover method. An application of this technique requires the groundwater table to be located somewhere at the height of the tunnel (see Figure 2.4). First, two

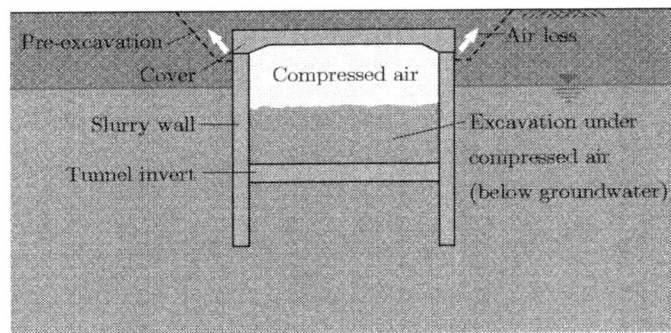

Figure 2.4: Cut-and-cover method in combination with dewatering by means of compressed air.

slurry walls and the cover are installed and the excavation of the part of the tunnel located above the groundwater level is performed. The soil in the remainder of the tunnel cross section is finally removed under compressed air, where the compressed air yields a downward displacement of the groundwater. Recently, this method has been applied, for instance, for the construction of the 'Audi-Tunnel' near Ingolstadt, Germany (report Prof. Semprich (2002)), or the railway tunnel Emmequerung in Switzerland [Harsch(2002)].

A large number of other building projects using one of the above described methods may be found at the web-site http://www.structurae.de/.

Limits for the application of compressed air

The displacement of the groundwater by means of compressed air is limited in its application by two facts, (i) by the maximum air pressure which can be employed and (ii) by the type of soil. The German guideline 'Verordnung über Arbeiten in Druckluft (Druckluftverordnung)' governs the level of air pressure to which personnel working in tunnels can be exposed and prescribes a number of additional measures to reduce health risks. According to these instructions, a maximum excess air pressure of 3.6 bar is permitted for a maximum length of stay of two hours.

From a technical point of view, the type of soil also limits the application of compressed air. In particular the permeabilities of the outcropping soil layers play a key role. According to Figure 2.5, dewatering by means of compressed air can be employed for soils such as sands and silts, exhibiting coefficients of permeability with respect to the water phase in the range of 10^{-4} m/s to 10^{-8} m/s. In clays dewatering is not necessary due to the very low permeability of these soils. In sands and gravels with permeabilities $k^{ow} \geq 5 \cdot 10^{-4}$ m/s additional measures such as sealing injections may be required, depending on the overburden of the tunnel and on the level of air pressure in the tunnel.

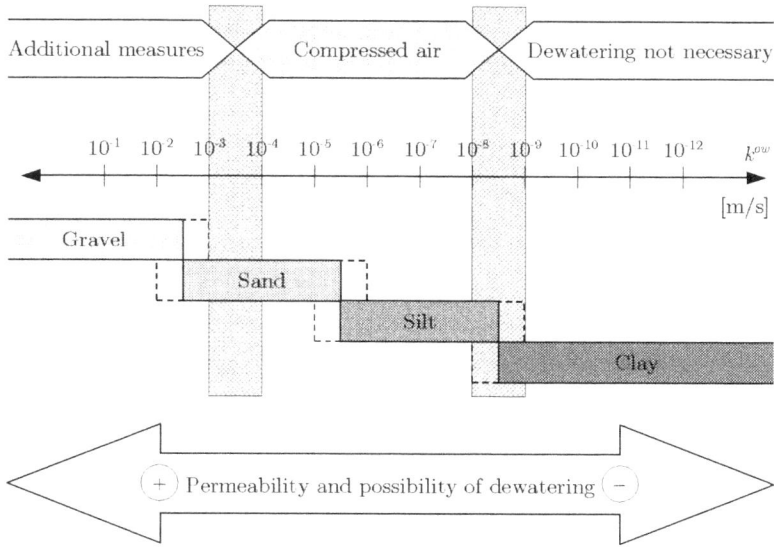

Figure 2.5: Applicability of dewatering by means of compressed air depending on the type of soil [Arz(1994)].

Numerical models for tunnelling in aquifers

If compressed air is used for the dewatering of soils in connection with tunnelling in aquifers, the necessary excess air pressure and the amount of compressed air which has to be supplied by compressors, even nowadays are primarily calculated by means of empirical formulae (based on laboratory or field tests) and experience of the engineer in charge. A number of empirical approaches for estimating the required excess air pressure is given in [Snee(1994)]. Usually the air pressure is chosen to balance the water pressure at the tunnel invert. However, as long as there does not exist a significant water flow in the direction of the tunnel face, the excess air pressure can be somewhat smaller since the soil at the tunnel face not necessarily has to be drained completely [Snee(1994)]. The amount of compressed air, which has to be supplied to the tunnel by compressors, can be estimated according to a range of different formulae given, e.g., in [Kramer(1989), Jodl(1995), Hochguertel(1998), Kammerer(2000)], depending on the type of soil and on the used tunnelling method (shield driving or NATM). These approaches basically contain the hydraulic conductivity, the hydraulic gradient and the size of the tunnel face.

In the late eighties, first attempts were made to improve calculation methods for tunnelling below the groundwater table taking into account compressed air as a means for dewatering the soil. Basically, there exist two different solution strategies for the numerical simulation of tunnelling under compressed air, an uncoupled and a coupled one.

If an uncoupled approach is chosen, the flow of water and compressed air in the soil and the deformations of the soil, caused by dewatering as well as by the advance of the tunnel face, are treated in two consecutive steps. First, the fluid flow in the soil is investigated, assuming a rigid soil skeleton, and second, the deformations of the soil are computed by making use of the results of the flow analysis. Consequently, any interactions between the fluid flow through the pores and the deformations of the soil skeleton are neglected.

When using a coupled solution procedure, the fluid flow in the soil and the deformations of the soil are determined simultaneously. In this case, the soil is considered as a three-phase medium, consisting of a deformable soil skeleton and the two fluid phases water and compressed air. From a mechanical point of view, only this approach allows to properly take into account the intrinsic coupling of the process of dewatering with the deformations of the soil in a physically consistent manner. From a practical point of view, such a coupled model offers the advantage that all quantities are computed simultaneously on the basis of one coupled numerical model and no switching between flow and deformation analyses has to be performed.

In the early numerical models only the flow problem was treated. Becker and Baumann [Becker(1986)] performed a parametric study to investigate the effect of the excess air pressure in the tunnel on the fluid flow in the soil in connection with the stability of the tunnel face in highly inhomogeneous soils. A three-dimensional finite difference model was used, however, the capillary pressure was not taken into account. Thus, the transition between air and water phase is simulated as a strict borderline and the air permeability in the flow field is overestimated.

Strobl [Strobl(1986), Strobl(1991)] employed a three-dimensional finite element model to calculate the distribution of the potential in the airflow field arising from the application of compressed air. He only investigated the airflow in the soil and iteratively adjusted the finite element mesh to the borderline between fully saturated and partially saturated parts of the domain (lowered groundwater table). For performing a three-dimensional elastic analysis Strobl mentioned the possibility of calculating drag forces from the distribution of the potential. Fuchsberger et al. [Fuchsberger(1988)] used this model to calculate the air pressure distribution in the soil in connection with a contract section of the Vienna subway.

Similar investigations based on the three-dimensional steady state potential flow were performed by Chen et al. [Chen(1990), Chen(1991)] to determine the lowered groundwater level and the airflow in the soil due to the application of compressed air in combination with tunnelling employing the Boundary Element Method. These calculations use the assumption of a pure airflow field above the lowered groundwater level. The mathematical model consists in employing the mass balance equation as well as Darcy's law and Boyle-Mariotte's law and thus possesses some similarity to the present approach.

The above mentioned numerical approaches are characterised by considering only one-phase flow in the soil. Gülzow [Guelzow(1994)] proposed a model for tunnelling below the groundwater table by means of compressed air in which two-phase flow, i.e. the simultaneous flow of an incompressible water phase and a compressible air phase, is taken into account. His model is based on the two mass balance equations for the fluid phases, Darcy's law and Boyle-Mariotte's law for an ideal gas. This model was implemented into a finite element programme and verified by means of analytical solutions and experimental data. Hence, this approach is rather similar to the fluid flow part of the present work, however, the model of Gülzow is not coupled and deformations of the soil skeleton are not taken into account.

Snee and Javadi [Snee(1996), Snee(1997), Javadi(2001)] developed a three-dimensional numerical approach based on the Finite Element Method to simulate the airflow in partially saturated porous media. The model for the determination of the airflow field uses the mass balance equation, Darcy's law and Boyle-Mariotte's law, similar to the approaches proposed by Gülzow or Chen. However, not the flow of two fluids in the porous medium is considered but pure airflow. In analogy to Strobl and Chen, an iterative procedure to locate the free surface of the groundwater (groundwater table) is applied. The iteration targets in attaining a steady state for the potential flow of the air phase. Deformations of the soil are determined by means of a constitutive model for partially saturated soils. The results of the numerical simulation of the airflow in the soil are used for calculating the changes in the shear strength and the displacements of the soil due to the airflow. The volume change of the partially saturated soil due to airflow is taken into account employing a state surface for the void ratio as a function of capillary stress and net normal stress [Fredlund(1979)]. However, the model is simplified by neglecting the influence of the capillary stress on the void ratio. In addition to that, in their approach only deformations due to the airflow are considered, deformations resulting from the excavation are neglected. Despite that, this is obviously a first attempt to model the airflow in the soil and the deformations due to this airflow in a coupled manner.

Hochgürtel [Hochguertel(1998)] extended the approach proposed by Gülzow [Guelzow(1994)] (which is characterised by a rigid soil skeleton) to allow for a calculation of the deformations of the soil. To this end, constitutive models for the soil skeleton are implemented into a three-dimensional finite element framework and results of the two-phase flow calculation of Gülzow are taken into account. Thus, both finite element programmes are used in combination to investigate the deformations of the soil due to the flow of the fluids in the soil and the tunnel excavation process. Two entirely different types of constitutive behaviour are investigated to compute the deformations: (i) a model based on the effective stress concept as suggested by Bishop [Bishop(1959)] in combination with the linear elastic-perfectly plastic model of Mohr-Coulomb and (ii) a model on the basis of two independent stress state variables (net normal stress and capillary stress) proposed by Fredlund et al. [Fredlund(1993), Fredlund(1996)]. According to a comparison of results obtained by Hochgürtel, for soils of interest in connection with tunnelling below the groundwater table by means of compressed air (confer Figure 2.5) the difference between the two types of models is negligibly small in all investigated cases [Hochguertel(1998)]. The approach employed by Hochgürtel is very similar to the present work, especially concerning the fluid flow part and the use of the effective stress concept according to Bishop. However, the model of Hochgürtel is not coupled in the sense of the approach chosen for the present work, since the fluid flow in the soil and the deformations of the soil are computed in two consecutive steps.

2.4 Mathematical description of multi-phase media

2.4.1 Introduction

When dealing with the description of a body, a determination of its composition is of primary importance, i.e. it has to be investigated how many physically and chemically different materials constitute the body.

In classical continuum mechanics a body is formed by one single constituent, either solid or fluid. An extension of this approach towards the description of multi-phase media is provided by the theory of mixtures where an arbitrary number of solids and/or fluids with independent degrees of freedom can be taken into account. Both classical continuum mechanics and mixture theory are macroscopic theories, i.e. the description of a body is based on its macroscale and no further attention is paid to the actual microstructure. However, the internal pore structure of a porous skeleton material is easily recognised. Thus, a continuum theory for a *porous* medium obviously should result from a macroscopic model substituting the exact physical situation by an idealised statistical model. In order to account for the immiscible constituents, volume fractions are introduced for the individual phases.

A second approach to describe porous media is given by the averaging theory which proceeds from the microscopic level. Referring to the microscopic volume element, the so-called averaging process links the microstructure to the macrostructure by means of averaging theorems. This procedure results in an averaged continuum. The use of the averaging theory offers possibilities for a better understanding of the microscopic situation and its relation to the macrorange. However, the obtained governing equations of the macroscale can easily be related to those resulting from mixture theory extended by the concept of volume fractions.

From a classical point of view Biot's well-known consolidation theory should be mentioned, dealing with the description of a soil skeleton fully saturated with one single fluid. A survey of the different approaches for the mathematical description of porous media is given in [Boer(1991b)]. In the subsequent sections an overview over the development of classical theories is presented, followed by a survey over the modern porous media theory and the averaging theory, respectively.

2.4.2 Classical theories of porous media

Considering the historical development of the porous media theory three major periods of time can be distinguished [Boer(1991a), Boer(1996)]. In the early era, covering the 18th and 19th century, the concept of volume fractions was stated and the classical mixture theory was founded. In the period between 1910 and 1960 first attempts to clarify the mechanical interaction between fluids and a porous solid were made. Rigid as well as deformable media were treated. The development of modern theories for immiscible mixtures started in the 1970s. Some important steps of the first two eras are highlighted in the following paragraphs. The last era, concerning modern porous media theory, is dealt with in a separate section.

According to [Boer(1991a), Boer(1996)], Euler (1762) was one of the first scientists working on the description of a porous body. In his definition of the term 'pore' he already clearly differentiated between free and trapped pores. Another important step was set by Woltman (1794). For his investigation on the earth pressure theory he distinguished four types of soil: sand, lime, clay and compost-earth. He constituted friction as a common characteristic property of all kinds of soil. For him, friction also marked the difference between soils and fluids. In connection with his discussion of the mechanical behaviour of mud, Woltman not only spoke of a mixture but also introduced the concept of volume fractions (the volume fraction relates the volume of a single constituent to the volume of the bulk body) being one of the main ingredients of modern porous media theory [Boer(1991a)].

Furthermore, in this early era, some fundamental laws for the development of modern porous media theory were discovered. Delesse (1848) succeeded in proving that under certain circumstances the surface fractions are equal to the volume fractions. Fick (1855), studying the problem of diffusion, made first attempts to establish a phenomenological theory for the description of gas mixtures. Finally, he derived a differential equation for the diffusion stream being similar to Fourier's equation of heat propagation. Today this relation is known as Fick's second law of diffusion. Another rather important law also dates from that time – Darcy's law (1856), stating the proportionality between the total volume of water flowing through the soil and the loss of pressure. Even today Darcy's findings are essential for a mechanical treatment of the motion of a liquid in a porous solid [Boer(1996)].

In the 19th century the development of the second main ingredient of the porous media theory, namely the mixture theory, started. An approach to the diffusion problem completely different to the one proposed by Fick was founded by Maxwell. He was the first scientist to develop the hydrodynamic equations for gas mixtures from the basic principles of mechanics. These investigations finally led to the creation of the kinetic gas theory. Based on Maxwell's approach, Stefan (1871) accomplished a decisive step towards a continuum mechanical theory of mixtures. He stated that "equations must be set up which contain the conditions of equilibrium and the laws of motion for every individual constituent in the mixture". He also introduced interaction forces between the constituents: "In a mixture each particle of a gas, if it is in motion, suffers from each of the other gases a resistance which is proportional to the density of this gas and the relative velocity of both" [Boer(1996)]. After the derivation of the equations of equilibrium and of mass balance, Stefan investigated solutions for special initial conditions and compared the results with experimental data. He also employed his theory for describing the diffusion of a gas through a porous diaphragm. In this case, he substituted one gas by a porous substance in such a manner that he assumed its particles to be fixed. The ratio between the free gas pressure and the partial gas pressure inside the pores was stated by using the porosity of the porous solid. Thus, Stefan, for the first time, applied mixture theory extended by the concept of volume fractions to a two-constituent problem within the framework of continuum mechanics [Boer(1991a), Boer(1996)].

In the second era, i.e. the period between 1910 and 1960, decisive progress was made towards creating a consistent porous media theory. Two notable steps should be pointed out in this period: (i) Important mechanical effects in a liquid-saturated *rigid* solid were discovered, e.g., phenomena such as uplift, friction and capillarity and the concept of effective stresses; and (ii) attempts were carried out to investigate saturated *deformable* soils. The most important scientists with respect to these two steps were Terzaghi and Fillunger as well as their successors Biot, Heinrich and Frenkel.

It was Fillunger [Fillunger(1914)] who performed pioneering investigations concerning the uplift problem in rigid saturated porous materials. In connection with the analysis of gravity dams, he discussed various possibilities for calculating the uplift force. However, it was Terzaghi who published the correct formula for the uplift force in water saturated sand in his book entitled "Erdbaumechanik auf bodenphysikalischer Grundlage" [Terzaghi(1925)]. Compared to the findings of Fillunger the formula of Terzaghi was based on a different definition of the porosity which Terzaghi determined by means of theoretical investigations in strength of materials and by experiments on concrete specimens.

The first scientist working on the friction phenomenon which occurs during the flow of a liquid through a saturated porous medium, again was Fillunger [Fillunger(1914)]. In this paper he proposed a formula for the frictional force referring to the volume element under consideration.

In connection with the uplift problem also a theoretical treatment of the capillary forces in saturated porous media was performed. Terzaghi first dealt with the capillary problem in his famous book [Terzaghi(1925)], later he developed a proportional relationship between the hydrostatic stress in the pore-fluid and the product of the specific weight of the fluid and the capillary rise. Both Terzaghi and Fillunger agreed in the main statement that the capillary suction in the liquid causes an additional pressure for the solid phase.

First reflections on the concept of the effective stresses were already made by scientists like Lyell in 1871, Boussinesq in 1876 and Reynolds in 1886 as well as Föppl and Rudeloff in 1900 and 1912, respectively, as mentioned in [Boer(1991a), Boer(1996)]. However, once again, Fillunger (1913) seems to be the first author to state that the pore water pressure does not have any influence on the strength of the porous solid. Further investigations on the partitioning of the stress state were made by Fillunger, and also by Terzaghi, in connection with the analysis of the consolidation phenomenon.

Terzaghi started the development of the idea of effective stresses [Terzaghi(1923)]. He divided the total stresses into two portions: The first part, called the neutral stress or pore water pressure, was assumed to act in the water and in the solid in every direction with equal intensity while the second part, representing an excess over the neutral stress, was located exclusively in the solid phase of the soil. This principle is of paramount importance in soil mechanics and its realisation is entirely due to Terzaghi. His use of the equation $\sigma' = \sigma - u$, i.e. the effective stress is equal to the difference between the total stress and the pore water pressure, marked a primary step in the development of modern soil mechanics in general and porous media theory in particular.

Despite of the progress made in the description of a saturated *rigid* porous body, the theory of porous media remained incomplete since deformations and the determination of the stress state in a saturated *deformable* porous body were not taken into account. The investigations on the mathematical treatment of a typical problem in engineering practice – the consolidation phenomenon – may be considered as the second important topic in the period of time from 1910 to 1960 [Boer(1991a), Boer(1996)]. A fluid saturated solid, subject to a given load, does not deflect under that load immediately but settles gradually at a variable rate, i.e. the settlement is caused by a gradual adaptation of the soil to the load variation. The phenomenon is very apparent in clays and sands saturated with water.

First attempts for a mathematical description of the consolidation problem were made by Terzaghi [Boer(1991a)]. He performed experiments on clay prisms to investigate the permeability coefficient of the soil. The analysis of settlements was not his primary interest. After loading a homogeneous soil layer, the deformation gradually increased for a certain time after the load was applied. The physical reason for this phenomenon was that the load induced a hydrostatic fluid stress within the water saturated soil which caused the pore water to flow towards the surface of the column until stress equilibrium was reached and the hydrostatic fluid stress decreased to zero [Terzaghi(1923)].

Inspired by such kind of experiments, Terzaghi [Terzaghi(1923)] intended to develop a relationship between surface pressure, water content and time. He assumed that the particles constituting the soil were more or less bound together by certain molecular forces. The soil was defined as a porous solid with elastic properties and the voids of the skeleton were filled with water. For deriving his well-known differential equation for the bulk body, neither did he separate the constituents solid and water nor did he explicitly use the concept of volume fractions for defining mechanical quantities. Moreover, considering only the one-dimensional case and just small pressure changes resulted in two further restricting assumptions. Although Terzaghi found his differential equation for the problem of consolidation more intuitively than by applying established mechanical principles, he was the first author to successfully describe the water saturated deformable porous solid and his equation is still used in civil engineering practice to determine the time-dependent settlements of water saturated soils [Boer(1986a), Boer(1991a)].

Considering the viewpoint of mathematical physics a more substantiated consolidation theory was developed by Fillunger [Boer(1991a)]. He proceeded from a two-phase system which he described using ensured mechanical axioms and principles. He stated that consolidation was a matter of two coupled flows, i.e. the pore water flow in the upward direction and the porous soil flow with the settlement rate downwards. He then, for the first time, introduced the balance equations of momentum and mass for each constituent to mathematically formulate the one-dimensional consolidation process. He took into account an interaction force in the balance of momentum and admitted finite deformations, only mass exchange was excluded from his model. Moreover, Fillunger also employed the concept of volume fractions using the porosity of the soil. The two constituents were assumed to be incompressible. The interaction force was determined by means of Darcy's law [Fillunger(1936)]. The drawbacks of this derivation of the interaction force were obviously entirely known to Fillunger because he indicated that inherent in his approach for deriving the interaction force were the assumptions of steady-state and homogeneous flow, a rigid soil skeleton with its velocity equal to zero and a constant porosity.

Fillunger suggested several generalisations of his model: (i) the treatment of three-dimensional consolidation with a viscous fluid (he restricted his investigations to the one-dimensional case as Terzaghi did), (ii) the introduction of further permeability coefficients, e.g., different values for horizontal and vertical coordinate directions, (iii) the consideration of the self-weight of the water and the solid phase and (iv) the use of a constitutive equation for the skeleton material [Fillunger(1936)]. Summarising Fillunger's ideas, he created the basic concept for the modern description of porous media, namely combining elements of mixture theory with the concept of volume fractions.

Biot generalised Terzaghi's theory of consolidation by extending it to the three-dimensional case and by establishing equations valid for arbitrary loads variable with time [Boer(1991a)]. He assumed the following basic properties for the soil: "(1) isotropy of the material, (2) reversibility of stress-strain relations under final equilibrium conditions, (3) linearity of stress-strain relations, (4) small strains, (5) the water contained in the pores is incompressible, (6) the water may contain air bubbles, (7) the water flows through the porous skeleton according to Darcy's law" [Biot(1941)].

For the derivation of the soil stresses, Biot considered "a small cubic element taken to be large enough compared to the size of the pores so that it may be treated as homogeneous, and at the same time small enough, compared to the scale of the macroscopic phenomena in which we are interested, so that it may be considered as infinitesimal in the mathematical treatment" [Biot(1941)]. The average stress condition in the soil was represented by forces distributed uniformly on the faces of the volume element and the stresses had to satisfy the equilibrium conditions for a stress field. Physically, Biot thought of these stresses as being composed of two parts: one caused by the hydrostatic pressure of the water filling the pores and the other one caused by the averaged stress in the skeleton. Thus, he considered the stresses in the soil as being carried partly by the water and partly by the solid phase. The water pressure in the pores was assumed constant throughout the soil element. In order to completely describe the macroscopic situation of the soil the variation of the water content in the pores was chosen as an additional variable [Biot(1941)]. Biot then used the condition of a zero variation in water content to calculate the water pressure immediately after loading of the soil sample. However, he mentioned that this variable would be more significant for soils not completely saturated with water but also containing air bubbles.

Using Hooke's law for an isotropic elastic body Biot derived a relationship between the strains and the stresses in the soil, also taking into account the influence of the water pressure in the pores. An additional equation was developed for the dependence of the increment of water content on the normal stresses and on the water pressure. These equations contained a total number of five distinct physical constants. Due to the assumption of reversibility of the stress-strain relations, Biot stated the existence of a potential energy of the soil which resulted in a reduction of the number of physical constants to four, being the shear modulus, Poisson's ratio and two coefficients measuring the compressibility of the soil for a change in water pressure and the change in water content for a given change in water pressure, respectively. The final equations of Biot's consolidation theory [Biot(1941)] described the settlement of the soil and the change of the water pressure, both depending on time. It should be mentioned that the equations are coupled.

In his further investigations, Biot generalised his theory for the cases of a porous anisotropic solid [Biot(1955)] and for a porous viscoelastic anisotropic solid [Biot(1956a)]. In the latter paper, the term 'viscoelasticity' included phenomena like the thermoelastic effect (the heat exchange due to additional local heating and cooling) or the 'sponginess effect' (the walls of the main pores contain cracks or micropores into which the fluid will seep in or out).

Furthermore, Biot also developed a theory for the propagation of stress waves in a porous elastic solid fully saturated with a compressible viscous fluid [Biot(1956b)]. The fluid could flow relative to the solid, causing friction to arise. Changes of the mass density of both constituents and of the porosity were considered to be small and thus negligible, i.e. the partial mass densities were constant. In order to allow for a description of the two-phase system with different states of motion of the phases, Biot assumed different displacement vectors for the soil skeleton and the fluid phase. For deriving the equations of motion for the system, a Lagrangian viewpoint was used together with the concept of generalised coordinates which were chosen as the six displacement components of the solid and the fluid phases. The flow of the fluid through the pores relative to the solid was described as Poiseuille type flow. Dissipation, depending on the relative motion between the fluid and the solid, was introduced by means of a dissipation function. In the case of a zero relative motion, dissipation vanished [Biot(1956b)].

Although it could be annotated critically that Biot's theory was not derived from fundamental axioms and principles of mechanics and thermodynamics in all respects (see [Boer(1991a), Boer(1996)]), his extensive investigations on the treatment of the fluid saturated deformable porous body marked a very important step in the development of the modern porous media theory.

Following the theory of Fillunger, Heinrich and Desoyer published two papers on steady and unsteady groundwater flow through anisotropic soils. First, they investigated a porous body consisting of grains being in contact with each other only at single points, and then, these considerations were extended to general solid matrices. In the early 1960s they also developed a consolidation theory [Heinrich(1961)]. The obtained equations were solved by means of Laplace transformations. Other contributions were due to Frenkel who investigated "seismic and seismoelectric phenomena in a moist soil". With Frenkel's derivations the second historical era can be regarded as completed [Boer(1991a), Boer(1996)].

2.4.3 Modern theory of porous media

This section is devoted to the development of the modern theory of porous media in the last few decades, beginning in the 1970s. In the present context, porous media theory is understood as the mixture theory extended by the volume fraction concept [Bedford(1983), Boer(1996)].

Considering the development of the mixture theory in the 1960s and the 1970s, several papers on porous solids filled with fluids were published using mixture theory without the volume fraction concept [Green(1966), Green(1970), Bowen(1976)]. According to [Boer(1996)], the fundamental contributions of Fillunger [Fillunger(1936)] and his followers were obviously not taken into account by these scientists.

It was Mills [Mills(1966)] who used the concept of volume fractions again when investigating incompressible mixtures of two separated Newtonian fluids. In the same paper, he also formulated the incompressibility condition in such a way that he assumed the intrinsic densities of the constituents to be constant and that he considered this fact in the volume fraction condition, i.e. all volume fractions sum up to one. However, this form of the incompressibility condition reflected only the volume fraction condition as an internal constraint and not the incompressibility of the individual constituents. Assuming a constant intrinsic density is a sufficient condition for the incompressibility of a constituent only in the case of thermodynamic processes without any mass exchange as proved some thirty years later [Bluhm(1997b)]. Mills' formulation of the incompressibility condition was adopted by several other authors of his time.

Goodman and Cowin [Goodman(1972)] developed a theory for granular materials with interstitial voids using formal arguments of continuum mechanics. For instance, the deformation gradient, Jacobian or material time derivatives were introduced in their familiar forms. In order to overcome the drawback of one missing field equation, they stated a so-called balance of equilibrated forces and equilibrated inertia. However, they presented the correct formulation of the incompressibility condition of the solid material in kinematic terms. This constraint was then considered in the Clausius-Duhem inequality (provides a lower bound for the increase of entropy accompanying the actual accumulation of heat at various temperatures in an actual process) to gain restrictions for the constitutive equations of an incompressible porous material.

Morland's investigations focussed on "a simple constitutive theory for a fluid saturated porous solid" [Morland(1972)]. He described the framework of mixture theory as being one of the most convenient for the treatment of fluid saturated porous media. Before introducing the volume fraction concept, Morland considered the kinematics of the single constituents. The state of stress and the balance of momentum were formulated in the usual way. In the section on the constitutive theory he expressed the partial density (referring to a unit volume of the total body) of each constituent by means of its volume fraction and the intrinsic density (referring to the unit volume of the constituent). Similarly, the partial stress tensor of each constituent was formulated in terms of the surface fractions and the intrinsic stress tensor. He also used a decomposition of the partial deformation gradient into a spherical part and a partial density preserving part. The intrinsic stress tensor was determined by a constitutive equation which was assumed to be a functional depending on the deformation gradient. Morland's further treatment of the fluid saturated porous solid was directed to the geometrically linear theory.

Further contributions to the development of the porous media theory are due to Kenyon [Kenyon(1976a), Kenyon(1976b)] and Passman [Passman(1977)]. Kenyon employed mixture theory together with a constitutive equation for the volume fractions to investigate thermostatics of solid-fluid mixtures. Passman used the approach of Goodman and Cowin to describe mixtures of granular materials. He wrote balance laws for each constituent and for the mixture and derived a constitutive theory for a mixture of two dry granular materials.

In 1980, Nunziato and Walsh [Nunziato(1980)] published a review of the different approaches to describe multi-phase mixtures. Additionally, they extended the theory to include chemically reacting materials. They adhered to the balance equation of equilibrated forces as proposed by Goodman and Cowin and tried to motivate this equation as being necessary in order to account for the dynamic effects associated with changes in the volume fractions of the constituents. According to [Boer(1996)], the correct number of necessary constraints in fluid saturated granular materials with incompressible constituents was first stated in [Nunziato(1981)]. A number of κ Lagrangian multipliers was inserted into the entropy inequality in order to gain restrictions for the constitutive relations. The volume fraction condition was considered in the entropy inequality as well.

Incompressible and compressible porous media models were investigated in [Bowen(1980)] and [Bowen(1982)], respectively. Bowen used the mixture theory extended by the volume fraction concept. Compared to other papers on multi-phase media at that time, he chose a different strategy to describe the thermodynamic behaviour of porous media. He also omitted additional balance equations like, e.g., the balance of equilibrated forces.

In the first paper [Bowen(1980)], thermodynamics of mixtures with incompressible constituents were treated. Bowen pointed out that the formalism of the modern theory of mixtures which originally arose from the classical theories of gas mixtures, also applies to porous media modelling. He stated that the assumption of incompressibility implies a certain constraint among the variables and causes an indeterminacy to exist among the constitutive variables. The volume fraction concept containing the saturation condition, i.e. the volume fractions of all constituents sum up to one, was introduced as a constraint. In the following, Bowen investigated the constitutive equations of an incompressible elastic porous solid filled with $\kappa - 1$ incompressible fluids on the basis of second-order materials (the gradient of the deformation gradient is also contained in the set of variables). From the Clausius-Duhem inequality Bowen derived many restrictions for the constitutive relations [Bowen(1980)]. However, two weak points in Bowen's theory should be mentioned: (i) He did not consider all constraints; besides the saturation condition there are κ additional incompressibility conditions. (ii) Concerning the thermodynamic description of incompressible porous media, there is no need to proceed from a basis of second-order materials. According to [Boer(1996)], Bowen had to use second-order materials due to his introduction of a free energy function per unit volume into the Clausius-Duhem inequality.

In the second paper [Bowen(1982)], an extension to compressible porous media was given, i.e. a model for a mixture consisting of a compressible elastic solid and an arbitrary number of compressible immiscible fluids was treated. The volume fraction of each fluid appeared as an independent variable in the constitutive equations and was assumed to obey a certain rate type constitutive equation. A specialisation of the model to one where the volume fractions adjust instantaneously to a value determined by the local state of the mixture was possible. After summarising all basic relations of the mixture theory extended by the concept of volume fractions, Bowen mentioned that the mixture is incompressible whenever the true (microscopic) densities of the individual constituents are constant. In this case, in his opinion, the volume fraction concept represented "a constraining relationship between the bulk densities". This, however, is not true since the volume fraction concept is always a constraint independent of the assumption of compressibility or incompressibility of the constituents [Boer(1996)]. Finally, Bowen [Bowen(1982)] specialised his general theory for a solid-fluid mixture with linear dissipation and for several classical models.

In the second half of the 1980s and in the 1990s, the research on porous media theory mainly focussed on three different topics: (i) the implementation of the developed porous media models into the finite element framework, (ii) the incorporation of more sophisticated constitutive relations into the mathematical models and (iii) the investigation of special phenomena appearing in saturated and empty porous solids. The numerical models are treated in the current work in a separate section. Concerning elastic-plastic material behaviour, e.g., [Boer(1986b)] and [Boer(1991a)] should be mentioned. [Boer(1990a)] and [Boer(1990b)] provide an extensive discussion of various phenomena in saturated porous media. For instance, uplift, friction and capillarity are treated in [Boer(1990a)] and the concept of effective stresses in [Boer(1990b)].

Finally, the current state of the theory of porous media, i.e. mixture theory extended by the concept of volume fractions, remains to be discussed. Subsequently, the characteristic features are highlighted in brief rather than giving a large body of mathematical derivations which can be found in the literature; e.g., [Ehlers(1989)], [Boer(1991a)] or [Bluhm(1997a)], just to mention three references representative for a broad range of publications.

When using an approach based on the porous media theory as described in this section, one has to proceed from the macroscopic situation. The so-called concept of volume fractions is used to distribute the masses of the individual constituents upon the total control space. Consequently, 'smeared' substitute continua with reduced densities for the different phases arise which simultaneously fill the domain under consideration and which can be treated using the tools of mixture theory. Inherent in this concept are the assumptions that (i) the porous solid always determines the control space and (ii) only the fluid phases contained in the pores can leave this space. Furthermore, the pores are conceived to be distributed statistically and in the reference state also homogeneously. An arbitrary volume element in the current configuration is composed of the microscopic volume elements of the constituents. Finally, the porous medium is described by a model of a macroscopic body, where neither a geometrical interpretation of the pore structure nor the exact location of the individual components of the body are considered. In order to formulate the concept of volume fractions in the current configuration, the existence of volume fractions as being functions of the position vector of the spatial point and of time has to be stated. Each point is simultaneously occupied by material points of all constituents. The volume fractions satisfy the so-called volume fraction condition, i.e. the sum of all volume fractions is equal to one. This condition plays an important role as a constraint in the porous media theory of saturated (all volume fractions sum up to one) porous solids [Ehlers(1989), Boer(1991a)].

Because of the volume fraction concept all geometrical and physical quantities such as motion, deformation or stress, are defined in the total control space, and thus, they can be interpreted as average values of the real quantities. A saturated (all volume fractions sum up to one) porous medium is treated as an immiscible mixture of all constituents. Since each spatial point of the current configuration is simultaneously occupied by particles of all constituents, each phase is assigned its own independent motion function. This function is postulated to be unique and uniquely invertible at any point of time. Different material derivatives have to be formulated as well because the individual constituents in general follow different motions. For the description of the compressibility or the incompressibility of the phases, a consideration of the microscale is needed, i.e. the incompressibility condition has to be formulated by means of physical quantities on the microscale. However, it is advisable to transfer the microscopic deformation behaviour to the macroscale which can be achieved by a multiplicative decomposition of the deformation gradient on the macroscale. A parallel exists in the treatment of metal plasticity, where a multiplicative decomposition of the deformation gradient into an elastic and a plastic part is widely used. The plastic part of the deformation on the macroscale results from dislocations on the microscale. In general, these microscopic dislocations are also represented by incompatible strains on the macroscale. Therefore, the reason for this multiplicative decomposition of the deformation gradient on the macroscale lies in transferring physical phenomena from the microscopic to the macroscopic level [Ehlers(1989), Boer(1991a), Boer(1996)].

The balance equations in the porous media theory, i.e. the balance of mass, linear momentum, angular momentum and energy, must be established for each constituent, considering all interactions and external supply terms. The sum of the partial balances of all constituents has to result in the balance laws for the mixture which, formally, must be equivalent to the balance laws of a one-component material. This requirement involves a constraint for the introduced interaction terms. Truesdell [Truesdell(1984)] summarised these fundamental ideas in his three well-known 'metaphysical principles':

"1. All properties of the mixture must be mathematical consequences of properties of the constituents.
2. So as to describe the motion of a constituent, we may in imagination isolate it from the rest of the mixture, provided we allow properly for the actions of the other constituents upon it.
3. The motion of the mixture is governed by the same equations as is a single body."

Truesdell [Truesdell(1984)] provided an interpretation of these principles as follows: "The first principle asserts, roughly, that the whole is no more than the sum of its parts, and the third, that in its motion as a whole a body does not know whether it is a mixture or not. The second is an extension of the familiar principle of solidification, by which the parts of a body occupying different regions of space are imagined cut asunder in geometry but united in physics by suitable forces or energies. Here we distinguish in imagination different constituents occupying the same region of space." A detailed derivation of the balance equations of mass, linear momentum, angular momentum and energy within the framework of the porous media theory can be found, e.g., in [Boer(1986a), Ehlers(1989), Boer(1991a)].

The second law of thermodynamics (also termed entropy inequality) represents a condition for general irreversible processes. In modern porous media theory, no longer are separate inequalities for all the constituents employed because this is regarded as a sufficient but overly rigid requirement. In fact, the postulate of one common entropy inequality is simultaneously a necessary and a sufficient condition for the existence of dissipation mechanisms within the mixture. This inequality is used to gain thermodynamic restrictions for the constitutive equations. A detailed derivation is given in [Ehlers(1989), Boer(1991a)].

When considering the mixture theory, it is apparent that it is closed, i.e. the number of variables is equal to the sum of balance equations and constitutive relations, as can easily be proved. Because of the introduction of the volume fraction concept in order to obtain 'smeared' continua which can be treated by continuum mechanical methods, the problem arises that a number of $\kappa - 1$ field equations are missing in case the volume fraction condition is taken into account. Thus, in order to achieve closure, additional field equations have to be formulated which is done by setting up constitutive relations as is usual in continuum mechanics of one-component media [Boer(1996)]. These constitutive relations have to be formulated for partial quantities and for several coupling mechanisms between the different phases. The development of the equations for saturated (all volume fractions sum up to one) porous media is based on the same principles that govern the derivation of constitutive relations in classical continuum mechanics of single-phase materials: (i) determinism, (ii) equipresence, (iii) local action, (iv) material frame-indifference and (v) dissipation. Additionally, the postulated constitutive relations have to fulfill the thermodynamic restrictions which result from the exploitation of the entropy inequality. These restrictions are derived on the basis of a certain number of chosen independent thermodynamic variables. Details on the proper choice of these variables can be found in [Ehlers(1989)]. A variety of different constitutive models are treated, e.g., in [Ehlers(1989), Boer(1991a), Boer(1996), Bluhm(1996), Bluhm(1997a)].

In conclusion of this section, it should be mentioned that if one is not particularly interested in the microscopic situation, the modern porous media theory – mixture theory extended by the volume fraction concept – provides a valuable and ensured basis for the mathematical description of media containing a certain number of constituents which all have their own degrees of freedom. However, it should be noted that the assumption of the classical theory of mixtures for gases and liquids of complete miscibility is not valid for a porous medium filled with fluids. Thus, this approach does not solve the exact problem but only the problem for the substitute continuum model [Boer(1996)].

2.4.4 Averaging theory

In addition to the theory of porous media discussed in the previous section, a further approach for the mathematical description of multi-phase media exists, taking into account the microscopic situation to a greater extent. Employing the so-called averaging theory for the description of a multi-phase system, one has to start at the microscopic level. Considering the historical development of the averaging theory, three fundamental shortcomings of these early attempts should be discussed.

Firstly, since these works focussed on particular types of multi-phase systems, assumptions and microscopic constitutive relations were usually introduced into the derivations at a very early stage, and thus, restricted the equations to special cases (such as incompressibility, steady-state, etcetera) or to certain materials. Additionally, the steps for a generalisation of the models were sometimes rather obscure. Secondly, another problem was the proper definition of macroscopic (averaged) quantities: (i) They were either connected to the special case, (ii) not physically motivated enough or (iii) not measurable in experiments. And thirdly, only one averaging operator was employed to obtain averaged values of completely different physical quantities; e.g., volume averaging operators were not only used to determine the averaged mass density but also for the definition of the averaged stress vector. The different nature of these quantities, i.e. the mass density being based on a unit volume element, the stress vector on a unit area element, was not taken into account. A more rational procedure would seem to be one which uses volume averaging for quantities defined per unit volume and area averaging for those defined per unit area [Hassanizadeh(1979a)].

A considerable improvement to overcome these shortcomings and to provide a systematic framework for the use of the averaging theory represents the general formulation proposed by Hassanizadeh and Gray [Hassanizadeh(1979a), Hassanizadeh(1979b), Hassanizadeh(1980)]. The basic steps of this approach are summarised subsequently.

To derive the governing equations for multi-phase media, the technique of local volume averaging is employed. The body is conceived to be composed of interpenetrating continua, each occupying only part of the volume and being separated by highly irregular interfaces. The usual quantities associated with each phase are continuous for that particular phase but discontinuous over the entire domain under consideration. The state of the system is governed by the general balance laws of classical continuum mechanics (conservation of mass, momentum and energy, second law of thermodynamics) along with appropriate interfacial and boundary conditions. Since the solution of these equations at the microscopic level would be an overwhelming task, the scale at which the system is described is altered by averaging the equations over some local representative volume element. The resulting averaged (also called macroscopic) equations are written in terms of macroscopic quantities, i.e. quantities which are actually measurable in experiments. The constitutive assumptions for the phases are not included in the formulation on the microscopic level but rather at the macroscale, either by direct postulation of desired relations or by following the Coleman and Noll method [Coleman(1963)], i.e. the systematic exploitation of the second law of thermodynamics. Therefore, two important conditions to provide a systematic framework within which the averaging procedure may be applied are (i) the existence of a set of criteria, based on physical principles, to deduce macroscopic variables, corresponding to actually measurable quantities, from particular microscopic variables and (ii) the employment of general forms of the balance equations.

The set of equations obtained by this averaging process can be employed for a broad range of problems such as flow through porous media, solid-fluid mixtures, flow of immiscible fluids and drying. Furthermore, the local volume averaging technique also provides some insight into the theory of mixtures since (i) assumptions inherent in the latter approach are explicitly revealed and (ii) postulates (such as boundedness and objectivity of several macroscopic properties) are deduced using the local averaging procedure [Trapp(1976), Hassanizadeh(1979a)].

In conclusion, this averaging technique seems to be a powerful framework for the understanding of multi-phase flow through porous media also taking into account the microscopic situation. Recent advances in the description of multi-phase media by means of the averaging theory focussed on several improvements discussed subsequently.

A general set of macroscopic conservation equations has been derived to take into account the effects of the interfaces. Thus, not only for the individual phases of the system but also for the interfaces all the classical balance equations are formulated [Marle(1982), Hassanizadeh(1990)]. In earlier approaches [Hassanizadeh(1979a), Hassanizadeh(1979b), Hassanizadeh(1980)] the interfaces have been simply modelled as surfaces of discontinuity in phase properties, undergoing jump conditions, unable to affect thermodynamic processes on their own. These newer theories have been employed to derive the basic equations describing two-phase flow in porous media [Gray(1991a), Gray(1991b), Hassanizadeh(1993)].

A more general expression of Darcy's law has been developed which allows for an intrinsic coupling between the motion of various fluids in the porous medium [Hassanizadeh(1993)]. The influence of the consideration of this coupling on a correct description of multi-phase flow has also been investigated in a series of papers by Rose, e.g., [Rose(1989)].

In a rather detailed model proposed by Gray and Hassanizadeh [Gray(1998)] and by Gray [Gray(1999)], not only the phases and interfaces have their own sets of balance equations but additionally, so-called common lines (regions of transition among three neighbouring phases) and common points (existing, e.g., in a four-phase system where the different common lines come together) are introduced as independent constituents of the formulation having their own balance equations. The common lines are modelled as one-dimensional regions with assigned thermodynamic properties such as mass, velocity, energy and entropy. In a three-phase system, e.g., a porous medium containing two fluids, only one common line type may exist whereas in a four-phase system four different common line types are possible and they come together at a common point. In cases of practical interest the amount of mass associated with an interface or a common line may be negligibly small compared to the mass storage and transport within the phases and thus these constituents can be treated as massless. Nevertheless, they sustain surface or line forces and also have their own energy. Therefore, they still influence the dynamics of the complete system considerably [Gray(1998)].

The incorporation of the averaging theory into a thermodynamic context is discussed thoroughly for the special case of a three-phase formulation in a paper by Gray [Gray(1999)]. Within the framework of thermodynamics, consistent and systematic postulates concerning the dependence of the internal energy on independent variables for phases, interfaces and common lines have to be made. These postulates are embedded into the entropy inequality. Then, from these postulated forms, relations among variables and insights into the system behaviour can be obtained. For the development of macroscale thermodynamics, the approach of Callen [Callen(1985)] is employed which gives thermodynamic relations appropriate for the macroscopic description of a porous medium. Furthermore, the determination of mechanical equilibrium constraints and their incorporation into the entropy inequality as well as the exploitation of this inequality to obtain relations for a description of the behaviour of the equilibrium system are investigated in [Gray(1999)].

Since the current work relies on the theoretical basis of the averaging theory as originally proposed in [Hassanizadeh(1979a), Hassanizadeh(1979b), Hassanizadeh(1980)], a detailed mathematical derivation is given in Chapter 3.

Chapter 3

Mechanics of the coupled three-phase model

3.1 Introduction

This chapter deals with the derivation of the governing equations for the behaviour of a *partially saturated porous medium*. In particular, a porous solid with voids containing the fluids water and air is considered. It should be mentioned that the term 'partially saturated' is not understood here as used in mixture theory but rather in a soil mechanics context. The voids are not fully saturated with water but also air is present in the pores and thus the term partially saturated is employed in geotechnics. However, the volume fractions of the three constituents – solid, water and air – sum up to one. Therefore, in the literature on multi-phase mixtures the medium would be called saturated. On the other hand, in an unsaturated mixture the volume fractions add up to less than one. According to this definition, the latter medium always possesses one volume fraction (which, however, is not counted) whose space is massless [Hutter(1999)].

As indicated in the previous chapter, the description of multi-phase systems consisting of interpenetrating continuous bodies is today based either on the mixture theory extended by the concept of volume fractions or on the averaging theory. From a more classical point of view Biot's theory can be employed. In the work presented here the governing equations are derived using the *averaging theory* as originally proposed in [Hassanizadeh(1979a), Hassanizadeh(1979b), Hassanizadeh(1980)].

To start this chapter, the most important kinematic equations are introduced briefly, together with a summary on definitions of different material time derivatives. The main part of the averaging theory, i.e. the averaging procedure, is dealt with next: After discussing the concept of volume fractions the so-called averaging operators are defined. When employing the averaging theory, one has to proceed from the balance equations on the microscopic level. Since all balance equations (mass, linear and angular momentum, energy and entropy) are of the same mathematical structure, it seems to be convenient to introduce one general microscopic balance law (might be called a 'microscopic master balance law') which is then used as a starting point to explain the procedure of deriving a general macroscopic balance equation ('macroscopic master balance law'). A systematic averaging procedure is used to obtain this general macroscopic balance equation as well as the corresponding general macroscopic jump condition.

From these general equations one proceeds to the specific macroscopic balance laws for mass, linear momentum, angular momentum, energy and entropy. This set of equations then provides the basis for the analysis of multi-phase systems. Its application to a particular type of problem, however, requires the general forms to be supplemented by appropriate constitutive equations. Of primary importance is the choice of the macroscopic variables in such a way that they correspond to real measurable quantities of laboratory tests.

Before describing the averaging process three different levels frequently used within the formulation should be defined:

Microscopic level: On this level the real non-homogeneous structure of the porous medium is encountered. The scale of the inhomogeneity is of the same order of magnitude as the dimensions of a pore or a grain, say d (Figure 3.1(a)). Attention is focussed on what happens at a mathematical point within a single phase. The state of that phase is described by field variables which are only defined at the points occupied by the particular phase.

Macroscopic level: The real multi-phase system being present in the porous medium domain is substituted by a model in which each phase fills the entire domain. Thus, each constituent, having a reduced density according to its percentile share, is spread over the total control space and all the phases are assumed to occupy every spatial point simultaneously (concept of overlapping continua). On this level with a typical dimension of l, $l \gg d$, the tools of classical continuum mechanics can be used to describe the continuously distributed constituents (Figure 3.1(b)). Although usually homogeneous

media are considered at this level, non-homogeneities may still be present, e.g., strata.
Megascopic level: At this level, having a characteristic dimension of an order of L, $L \gg l$, the conditions are similar to those at the macroscopic level. When considering, for instance, a surface subsidence problem where usually a very large domain has to be modelled, the mathematical description may be stated in a domain with fewer dimensions than the real domain. Hence, the three-dimensional nature of the original problem is reduced to two dimensions with field variables averaged over the thickness. Numerical solutions of this type of mathematical model can be found in the literature [Simoni(1989), Lewis(1998)].

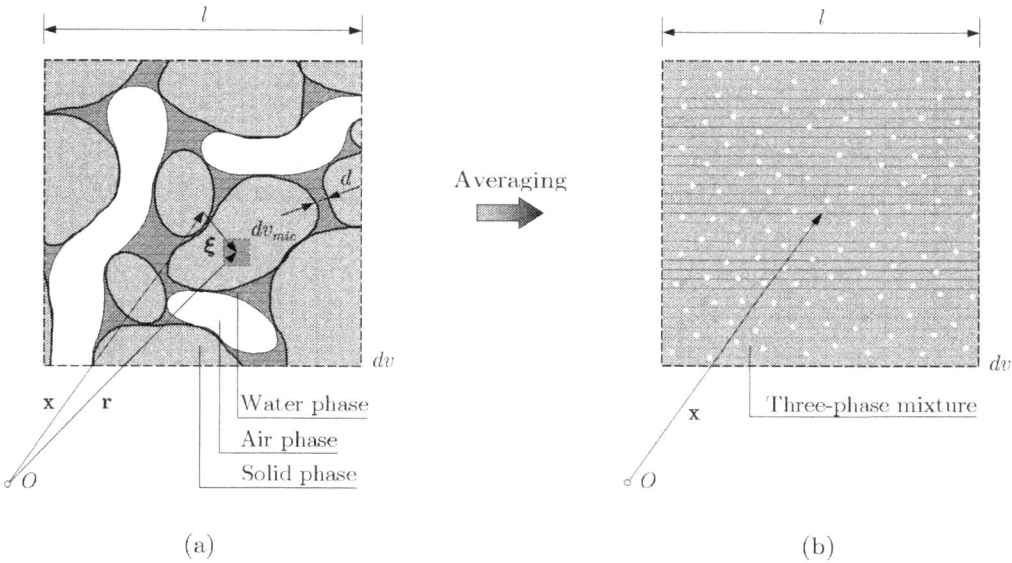

Figure 3.1: Averaging procedure for a three-phase medium: volume element (a) before and (b) after the averaging process.

3.2 Kinematic equations

3.2.1 Coordinates, velocities and accelerations

As indicated in the introduction (Section 3.1), a multi-phase medium can be described as the superposition of a number of κ constituents (continua), i.e. each point of the domain is simultaneously occupied by all of the phases (Figure 3.1(b)), having of course reduced densities according to their percentile share (concept of overlapping continua). However, the state of motion of each constituent is formulated independently.

Let a *reference configuration* $\mathcal{B}$ be chosen for the body under consideration. By this, a set in $\mathbb{R}^3$ with piecewise smooth boundary is understood in the present context. A *configuration* of $\mathcal{B}$ is a mapping $\chi^\pi : \mathcal{B} \mapsto \mathbb{R}^3$ that is sufficiently smooth, orientation preserving and invertible. Points in $\mathcal{B}$ are denoted $\mathbf{X}^\pi = (X_1^\pi, X_2^\pi, X_3^\pi) \in \mathcal{B}$ and are called *material points*, while points in $\mathbb{R}^3$ are denoted $\mathbf{x}^\pi = (x_1^\pi, x_2^\pi, x_3^\pi) \in \mathbb{R}^3$ and are called *spatial points*. The particular constituent is indicated by the superscript π, $\pi = 1, 2, \ldots, \kappa$. The relationship between material points and spatial points can be expressed as

$$\mathbf{x}^\pi = \chi^\pi (\mathbf{X}^\pi) . \tag{3.1}$$

A *motion* of $\mathcal{B}$ is a time-dependent family of configurations. In a Lagrangian or material description of motion the position $\mathbf{x}^\pi$ of each material point at a time t is a function χ^π of its placement in a chosen reference configuration $\mathbf{X}^\pi$:

$$\mathbf{x}^\pi = \chi^\pi (\mathbf{X}^\pi, t) . \tag{3.2}$$

It should be mentioned that it is also quite common to write $\mathbf{x}^\pi$ instead of introducing a separate function χ^π on the right hand side of the equations (3.1) and (3.2) [Marsden(1983), Holzapfel(2000)]. The relationship between the reference configuration and the current configuration can be described by means of the displacement vector $\mathbf{u}^\pi(\mathbf{X}^\pi, t)$ as

$$\mathbf{x}^\pi(\mathbf{X}^\pi, t) = \mathbf{X}^\pi + \mathbf{u}^\pi(\mathbf{X}^\pi, t) . \tag{3.3}$$

The material velocity $\mathbf{V}^\pi(\mathbf{X}^\pi, t)$ of a particle is given as

$$\mathbf{V}^\pi(\mathbf{X}^\pi, t) = \dot{\mathbf{x}}^\pi(\mathbf{X}^\pi, t) = \frac{\partial \mathbf{x}^\pi(\mathbf{X}^\pi, t)}{\partial t} \tag{3.4}$$

and regarded as a vector emanating from the point $\mathbf{x}^\pi$. The material acceleration $\mathbf{A}^\pi(\mathbf{X}^\pi, t)$ reads as

$$\mathbf{A}^\pi(\mathbf{X}^\pi, t) = \ddot{\mathbf{x}}^\pi(\mathbf{X}^\pi, t) = \frac{\partial^2 \mathbf{x}^\pi(\mathbf{X}^\pi, t)}{\partial t^2} . \tag{3.5}$$

Since χ^π maps the reference configuration to the current configuration uniquely at any time t, its inverse function $(\chi^\pi)^{-1}$ can be introduced to formulate the Eulerian or spatial description of motion:

$$\mathbf{X}^\pi = (\chi^\pi)^{-1} (\mathbf{x}^\pi, t) . \tag{3.6}$$

By inserting (3.6) into the material description of the velocity field (3.4) the spatial formulation of the velocity field is obtained as

$$\mathbf{v}^\pi(\mathbf{x}^\pi, t) = \dot{\mathbf{x}}^\pi(\mathbf{X}^\pi(\mathbf{x}^\pi, t), t) = \frac{\partial \mathbf{x}^\pi(\mathbf{X}^\pi(\mathbf{x}^\pi, t), t)}{\partial t} . \tag{3.7}$$

To derive the spatial description of the acceleration field from equation (3.7), the chain rule of differentiation has to be taken into account because $\mathbf{x}^\pi$ depends on the time t as well. This yields

$$\mathbf{a}^\pi(\mathbf{x}^\pi, t) = \frac{\partial \mathbf{v}^\pi(\mathbf{x}^\pi, t)}{\partial t} + \frac{\partial \mathbf{v}^\pi(\mathbf{x}^\pi, t)}{\partial \mathbf{x}^\pi(\mathbf{X}^\pi(\mathbf{x}^\pi, t), t)} \cdot \frac{\partial \mathbf{x}^\pi(\mathbf{X}^\pi(\mathbf{x}^\pi, t), t)}{\partial t} . \tag{3.8}$$

Since $\partial \mathbf{v}^\pi/\partial \mathbf{x}^\pi = \mathrm{grad}\,\mathbf{v}^\pi$ denotes the spatial gradient of the velocity field, equation (3.8) can be rewritten as

$$\mathbf{a}^\pi(\mathbf{x}^\pi, t) = \frac{\partial \mathbf{v}^\pi(\mathbf{x}^\pi, t)}{\partial t} + \mathrm{grad}\,\mathbf{v}^\pi(\mathbf{x}^\pi, t) \cdot \mathbf{v}^\pi(\mathbf{x}^\pi, t) = \frac{D^\pi[\mathbf{v}^\pi(\mathbf{x}^\pi, t)]}{Dt} . \tag{3.9}$$

Hence, the acceleration is obtained from the spatial description of the velocity by means of the material time derivative.

In porous media theory it is both customary and advantageous to describe the motion of the fluid phases in terms of velocities/accelerations relative to the moving soil skeleton. The velocities of the fluid particles can be written with reference to those of corresponding solid points, once relative velocities are introduced, as

$$\mathbf{v}^{\alpha\pi}(\mathbf{x}^\pi, t) = \mathbf{v}^\alpha(\mathbf{x}^\pi, t) - \mathbf{v}^\pi(\mathbf{x}^\pi, t) , \tag{3.10}$$

where $\mathbf{v}^\alpha$ might be specified as the velocity $\mathbf{v}^f$ of the fluid phase and $\mathbf{v}^\pi$ as the velocity $\mathbf{v}^s$ of the solid phase, respectively, thus yielding the velocity $\mathbf{v}^{fs}$ of the fluid relative to the moving solid.

3.2.2 Definition of material time derivatives

In this section some frequently used material time derivatives of any differentiable function $f^\pi(\mathbf{x}^\pi, t)$, given in its spatial coordinates $\mathbf{x}^\pi$ and depending on time t, are defined. Additionally, the material time derivative of a volume integral of the function $f^\pi(\mathbf{x}^\pi, t)$ is introduced.

According to (3.9), the material time derivative D^π/Dt of the function $f^\pi(\mathbf{x}^\pi, t)$ referring to a moving particle of the π-phase is given as [Malvern(1969)]

$$\frac{D^\pi[f^\pi(\mathbf{x}^\pi, t)]}{Dt} = \frac{\partial f^\pi(\mathbf{x}^\pi, t)}{\partial t} + \mathrm{grad}\,f^\pi(\mathbf{x}^\pi, t) \cdot \mathbf{v}^\pi(\mathbf{x}^\pi, t) , \tag{3.11}$$

where $\partial/\partial t$ denotes the partial time derivative and grad is the gradient operator. If the derivative refers to a particle of the α-phase rather than the π-phase, then

$$\frac{D^\alpha[f^\pi(\mathbf{x}^\pi,t)]}{Dt} = \frac{\partial f^\pi(\mathbf{x}^\pi,t)}{\partial t} + \operatorname{grad} f^\pi(\mathbf{x}^\pi,t) \cdot \mathbf{v}^\alpha(\mathbf{x}^\pi,t) \qquad (3.12)$$

is valid [Malvern(1969)]. Using equation (3.10) for the relative velocity, the difference between the two equations (3.12) and (3.11) is obtained as

$$\frac{D^\alpha[f^\pi(\mathbf{x}^\pi,t)]}{Dt} = \frac{D^\pi[f^\pi(\mathbf{x}^\pi,t)]}{Dt} + \operatorname{grad} f^\pi(\mathbf{x}^\pi,t) \cdot \mathbf{v}^{\alpha\pi}(\mathbf{x}^\pi,t) \ . \qquad (3.13)$$

Subsequently, the material time derivative of a volume integral of the function $f^\pi(\mathbf{x}^\pi,t)$ is defined [Marsden(1983)]:

$$\frac{D^\pi}{Dt}\int_V f^\pi(\mathbf{x}^\pi,t)\,dV = \int_V \left[\frac{D^\pi[f^\pi(\mathbf{x}^\pi,t)]}{Dt} + f^\pi(\mathbf{x}^\pi,t)\operatorname{div}\mathbf{v}^\pi(\mathbf{x}^\pi,t)\right]dV =$$

$$= \int_V \left[\frac{\partial f^\pi(\mathbf{x}^\pi,t)}{\partial t} + \operatorname{div}[f^\pi(\mathbf{x}^\pi,t)\mathbf{v}^\pi(\mathbf{x}^\pi,t)]\right]dV \ , \qquad (3.14)$$

where V denotes the volume of the body in its actual configuration and div is the divergence operator. To prove the relationship (3.14), material coordinates are introduced for the spatial coordinates on the left hand side of equation (3.14) by using $\mathbf{x}^\pi = \mathbf{x}^\pi(\mathbf{X}^\pi,t)$ and the volume element dV in the current configuration is replaced by

$$dV = J(\mathbf{X}^\pi,t)\,dV_0 \ , \qquad (3.15)$$

with dV_0 denoting the volume element in the reference configuration and $J(\mathbf{X}^\pi,t)$ being the Jacobian. This yields

$$\frac{D^\pi}{Dt}\int_V f^\pi(\mathbf{x}^\pi,t)\,dV = \frac{D^\pi}{Dt}\int_{V_0} f^\pi(\mathbf{x}^\pi(\mathbf{X}^\pi,t),t)J(\mathbf{X}^\pi,t)\,dV_0 =$$

$$= \int_{V_0}\left[\frac{\partial f^\pi(\mathbf{x}^\pi(\mathbf{X}^\pi,t),t)}{\partial t}J(\mathbf{X}^\pi,t) + f^\pi(\mathbf{x}^\pi(\mathbf{X}^\pi,t),t)\frac{\partial J(\mathbf{X}^\pi,t)}{\partial t}\right]dV_0 \ , \qquad (3.16)$$

and using the relationship [Marsden(1983)]

$$\frac{\partial J(\mathbf{X}^\pi,t)}{\partial t} = J(\mathbf{X}^\pi,t)\operatorname{div}\mathbf{v}^\pi(\mathbf{x}^\pi,t) \ , \qquad (3.17)$$

the second part of equation (3.16) can be rewritten as

$$\int_{V_0} \left[\frac{\partial f^\pi(\mathbf{x}^\pi(\mathbf{X}^\pi,t),t)}{\partial t} J(\mathbf{X}^\pi,t) + f^\pi(\mathbf{x}^\pi(\mathbf{X}^\pi,t),t) \frac{\partial J(\mathbf{X}^\pi,t)}{\partial t} \right] dV_0 =$$

$$= \int_{V_0} \left[\frac{\partial f^\pi(\mathbf{x}^\pi(\mathbf{X}^\pi,t),t)}{\partial t} + f^\pi(\mathbf{x}^\pi(\mathbf{X}^\pi,t),t) \operatorname{div} \mathbf{v}^\pi(\mathbf{x}^\pi,t) \right] J(\mathbf{X}^\pi,t) \, dV_0 \, . \qquad (3.18)$$

Changing variables back to the spatial coordinates $\mathbf{x}^\pi$ and taking into account equation (3.15) results in [Marsden(1983)]

$$\frac{D^\pi}{Dt} \int_V f^\pi(\mathbf{x}^\pi,t) \, dV =$$

$$= \int_{V_0} \left[\frac{\partial f^\pi(\mathbf{x}^\pi(\mathbf{X}^\pi,t),t)}{\partial t} + f^\pi(\mathbf{x}^\pi(\mathbf{X}^\pi,t),t) \operatorname{div} \mathbf{v}^\pi(\mathbf{x}^\pi,t) \right] J(\mathbf{X}^\pi,t) \, dV_0 =$$

$$= \int_V \left[\frac{D^\pi[f^\pi(\mathbf{x}^\pi,t)]}{Dt} + f^\pi(\mathbf{x}^\pi,t) \operatorname{div} \mathbf{v}^\pi(\mathbf{x}^\pi,t) \right] dV =$$

$$= \int_V \left[\frac{\partial f^\pi(\mathbf{x}^\pi,t)}{\partial t} + \operatorname{div} \left[f^\pi(\mathbf{x}^\pi,t) \mathbf{v}^\pi(\mathbf{x}^\pi,t) \right] \right] dV \, , \qquad (3.19)$$

where the last line has been obtained by making use of the following relationship,

$$\operatorname{div} \left[f^\pi(\mathbf{x}^\pi,t) \mathbf{v}^\pi(\mathbf{x}^\pi,t) \right] = f^\pi(\mathbf{x}^\pi,t) \operatorname{div} \mathbf{v}^\pi(\mathbf{x}^\pi,t) + \operatorname{grad} f^\pi(\mathbf{x}^\pi,t) \cdot \mathbf{v}^\pi(\mathbf{x}^\pi,t) \, . \qquad (3.20)$$

Finally, equation (3.17) is still left to be proved. This, however, can be done most conveniently by just calculating both sides of the equation. For the sake of simplicity, the subsequent proof is performed for a two-dimensional problem and the superscript π, indicating the particular constituent, is omitted. The time derivative of $J(\mathbf{X}^\pi,t)$ on the left hand side of equation (3.17) is obtained from

$$J = \frac{\partial x_1}{\partial X_1} \frac{\partial x_2}{\partial X_2} - \frac{\partial x_1}{\partial X_2} \frac{\partial x_2}{\partial X_1} \qquad (3.21)$$

by taking into account the relationship $\dot{x}_i = \dot{u}_i$ which follows from equation (3.3):

$$\frac{\partial J}{\partial t} = \frac{\partial \dot{x}_1}{\partial X_1} \frac{\partial x_2}{\partial X_2} + \frac{\partial x_1}{\partial X_1} \frac{\partial \dot{x}_2}{\partial X_2} - \ldots = \frac{\partial \dot{u}_1}{\partial X_1} \frac{\partial x_2}{\partial X_2} + \frac{\partial x_1}{\partial X_1} \frac{\partial \dot{u}_2}{\partial X_2} - \ldots \qquad (3.22)$$

Calculation of the right hand side of equation (3.17) yields

$$J \operatorname{div} \mathbf{v} = J \operatorname{div} \dot{\mathbf{u}} = \left(\frac{\partial x_1}{\partial X_1} \frac{\partial x_2}{\partial X_2} - \frac{\partial x_1}{\partial X_2} \frac{\partial x_2}{\partial X_1} \right) \left(\frac{\partial \dot{u}_1}{\partial x_1} + \frac{\partial \dot{u}_2}{\partial x_2} \right) =$$

$$= \frac{\partial x_1}{\partial X_1} \frac{\partial x_2}{\partial X_2} \frac{\partial \dot{u}_1}{\partial x_1} + \ldots$$

$$= \frac{\partial \dot{u}_1}{\partial X_1} \frac{\partial x_2}{\partial X_2} + \ldots \qquad (3.23)$$

which is the same result as obtained in (3.22) and completes the proof.

3.2.3 Strain tensor for small displacements and small strains

The relationship between the displacements and the strains in a continuous medium is in general described by means of a second-order tensor $\mathbf{E}$ called the Green strain tensor. In rectangular Cartesian coordinates its components E_{ij} may be written as [Mang(2000)]

$$E_{ij} = \frac{1}{2} \left(\frac{\partial u_i}{\partial X_j} + \frac{\partial u_j}{\partial X_i} + \frac{\partial u_k}{\partial X_i} \frac{\partial u_k}{\partial X_j} \right). \qquad (3.24)$$

If the derivatives of the displacements, $\partial u_i / \partial X_j$, are very small compared to one, the quadratic terms in equation (3.24) can be neglected and thus only the linear terms are taken into account. Additionally, in this case the difference between the derivatives with respect to the Lagrangian coordinates X_j and those with respect to the Eulerian coordinates x_j can be neglected according to (3.3). These two conditions are formulated mathematically in the following equations as

$$\frac{\partial u_i}{\partial X_j} \ll 1 \quad \text{and} \quad \frac{\partial u_i}{\partial X_j} \approx \frac{\partial u_i}{\partial x_j}. \qquad (3.25)$$

Taking into account these assumptions, equation (3.24) yields the components of the linearised strain tensor ε_{ij} which provides a relationship between the displacements and the strains in a continuum for the special case of small displacements and small strains. In index notation it reads as

$$\varepsilon_{ij} = \frac{1}{2} \left(\frac{\partial u_i}{\partial x_j} + \frac{\partial u_j}{\partial x_i} \right), \qquad (3.26)$$

and in a more general vector notation, employing the gradient operator, it may be written as

$$\boldsymbol{\varepsilon} = \frac{1}{2} \left[\operatorname{grad} \mathbf{u} + (\operatorname{grad} \mathbf{u})^T \right] = \frac{1}{2} \left[\boldsymbol{\nabla} \mathbf{u} + (\boldsymbol{\nabla} \mathbf{u})^T \right]. \qquad (3.27)$$

It should be mentioned that in the current work only the case of small displacements and small strains is considered which seems to be appropriate for the type of problems investigated here.

3.2.4 Some important integral theorems

This section is devoted to the introduction of some important integral theorems for an arbitrary scalar f, a vector $\mathbf{f}$ and a tensor $\mathbf{F}$, respectively. Subsequently, these theorems are necessary for the derivation of the various balance equations.

One of the most frequently employed integral theorems is the well-known divergence theorem which is used to convert an integral over the surface A of a domain into an integral over its volume V. For a vector $\mathbf{f}$ and a tensor $\mathbf{F}$, respectively, the divergence theorem can be formulated as [Malvern(1969)]

$$\int_A \mathbf{f} \cdot \mathbf{n} \, dA = \int_V \operatorname{div} \mathbf{f} \, dV \quad \text{and} \quad \int_A \mathbf{F} \cdot \mathbf{n} \, dA = \int_V \operatorname{div} \mathbf{F} \, dV \,, \tag{3.28}$$

where $\mathbf{n}$ denotes the unit vector normal to the surface A of the domain under consideration.

Furthermore, two extended versions of the divergence theorem are also necessary for the subsequent derivations. In case of the inclusion of a vector cross product in the surface integral, the following relationship can be applied to obtain an integral over the volume [Holzapfel(2000)],

$$\int_A \mathbf{f} \times (\mathbf{F} \cdot \mathbf{n}) \, dA = \int_V \left(\mathbf{f} \times \operatorname{div} \mathbf{F} + \mathcal{E} : \mathbf{F}^T \right) dV \,, \tag{3.29}$$

where $\mathcal{E}$ is the third-order permutation tensor which in index notation may be written as

$$\mathcal{E} = \epsilon_{ijk} \, \mathbf{e}_i \otimes \mathbf{e}_j \otimes \mathbf{e}_k \,. \tag{3.30}$$

In (3.30) $\mathbf{e}_i$, $\mathbf{e}_j$ and $\mathbf{e}_k$ are the three orthogonal unit basis vectors in a rectangular Cartesian coordinate system and ϵ_{ijk} is the permutation symbol defined as

$$\epsilon_{ijk} = \begin{cases} 1 & \text{for even permutations of } (i,j,k), \text{ (i.e. 123, 231, 312)} \,, \\ -1 & \text{for odd permutations of } (i,j,k), \text{ (i.e. 132, 213, 321)} \,, \\ 0 & \text{if there is a repeated index} \,. \end{cases} \tag{3.31}$$

If an integration over two different surfaces/volumes (differing in size) has to be performed, the divergence theorem as proposed by Eringen and Suhubi [Eringen(1964)] can be employed:

$$\int_A \left[\frac{1}{a} \int_a \mathbf{f} \cdot \mathbf{n} \, da \right] dA = \int_V \left[\frac{1}{v} \int_v \operatorname{div} \mathbf{f} \, dv \right] dV = \int_V \operatorname{div} \left[\frac{1}{v} \int_v \mathbf{f} \, dv \right] dV , \qquad (3.32)$$

with v and a denoting, e.g., the volume and surface of a representative volume element and V and A the global volume and area of the domain, respectively.

Finally, the so-called Reynolds' transport theorem [Gray(1998), Holzapfel(2000)] can be obtained when using the divergence theorem (3.28_1) in combination with the material time derivative of a volume integral (3.14) as

$$\frac{D}{Dt} \int_V f \, dV = \int_V \frac{\partial f}{\partial t} \, dV + \int_A f \mathbf{v} \cdot \mathbf{n} \, dA . \qquad (3.33)$$

Equation (3.33) expresses the fact that the time rate of change of the integral over the scalar field f on the left hand side is equal to the local time rate of change of f within the region V plus the rate of transport (outward normal flux) of f across the surface A of the domain where $\mathbf{v}$ denotes the velocity.

3.3 Averaging procedure

3.3.1 Concept of volume fractions

For the equations derived in this chapter a multi-phase system is considered which occupies a volume V, $V \subset \mathcal{B}$, and is bounded by a surface A. Each point in V is the centroid of an averaging volume dv. These volumes are defined such that their shape and size are identical at all locations and for every point of time. In a similar manner, averaging areas da, associated with each point on A, are defined. An averaged quantity is then obtained by integrating a microscopic quantity over an averaging volume (*representative elementary volume*, REV) or an averaging area (*representative elementary area*, REA). This process is used to develop a field of macroscopic variables for each phase. In the macroscopic field an average volume represents a physical point. Thus, it may as well be termed an infinitesimal element of volume. Analogously, the average area is referred to as an infinitesimal element of area [Hassanizadeh(1979a)].

A proper choice of the size of the REV is of primary importance. Averaged quantities have to be independent of the size of the representative elementary volume and continuous in space and time. Thus, the averaging region is required to possess certain characteristics. Figure 3.2 shows the graph of a computed averaged quantity using averaging regions that vary from a very small to a very large size. When the domain is very small, the quantity may be finite or zero. Then, as the region size increases, fluctuations will occur in the computed averaged value because relatively large portions of one phase may become included in the averaging region. However, these fluctuations tend to diminish as the size of the region further increases and eventually within some interval of region sizes the averaged quantities are independent of the dimensions of the domain. Gross inhomogeneities of the medium may affect the stability of the averaged quantities if the region becomes even larger.

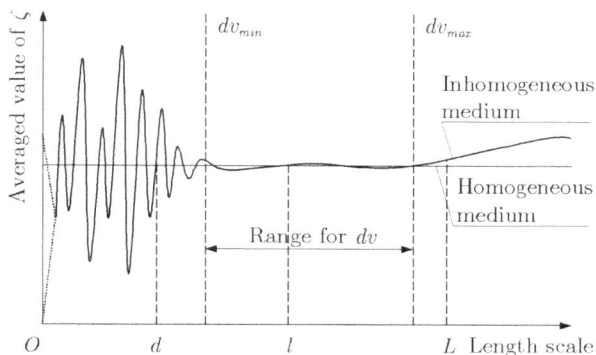

Figure 3.2: Averaged quantity ζ versus characteristic length of the averaging volume dv [Hassanizadeh(1979a)].

Therefore, to be considered as mathematically infinitesimal (meaning that the partial derivatives make sense), the REV has to be selectively small enough and at the same time large enough with respect to the heterogeneities of the material. The characteristic length l of the averaging region has to be such that the resulting averaged quantity is insensitive to small changes in this length. In order to obtain meaningful averaged properties, the characteristic length of the infinitesimal volume or area element has to satisfy the inequality

$$d \ll l \ll L, \quad (3.34)$$

with d, l and L being a typical microscopic distance, a characteristic dimension of the average volume element and a typical length of the entire domain (Figure 3.2), respectively.

In a *global frame of reference* the centre of the REV is denoted by the position vector **x** (Figure 3.1). The vector **r**, indicating the position of a particle within dv with respect to the global coordinate system (Figure 3.1(a)), i.e. the position of a microscopic volume element dv_{mic}, is defined as

$$\mathbf{r} = \mathbf{x} + \boldsymbol{\xi} , \tag{3.35}$$

where $\boldsymbol{\xi}$ describes the position of the particle with respect to the centre of the representative elementary volume (Figure 3.1(a)).

The multi-phase medium under consideration is assumed to consist of a number of κ constituents with partial volumes V^π, bounded by surfaces A^π, $\pi = 1, 2, \ldots, \kappa$. The volume dv of the REV is composed of the individual volume elements dv^π such that

$$dv = \sum_{\pi=1}^{\kappa} dv^\pi . \tag{3.36}$$

A phase distribution function $\gamma^\pi(\mathbf{r}, t)$ may be introduced for each constituent π as

$$\gamma^\pi(\mathbf{r}, t) = \begin{cases} 1 & \text{for} \quad \mathbf{r} \in dv^\pi, \quad \pi = 1, 2, \ldots, \kappa, \\ 0 & \text{for} \quad \mathbf{r} \in dv^\alpha, \quad \alpha, \pi = 1, 2, \ldots, \kappa, \quad \pi \neq \alpha . \end{cases} \tag{3.37}$$

$\gamma^\pi(\mathbf{r}, t)$ is a function of the position vector **r** of the microscopic volume element dv_{mic} and of the time t. This function is used to obtain the part of dv occupied by the π-phase. Hence, the average volume element dv^π can be calculated as

$$dv^\pi(\mathbf{x}, t) = \int_{dv} \gamma^\pi(\mathbf{r}, t) \, dv_{mic} , \tag{3.38}$$

depending on the position vector **x** (describing the centre of the REV) and on the time t. Equation (3.38) can be interpreted as follows: The integration is performed over the volume dv of the REV. dv_{mic} represents the microscopic volume element meaning that the integration refers to the microscopic local coordinate system $\boldsymbol{\xi}$ with its origin in **x** (Figure 3.1(a)). The distribution function $\gamma^\pi(\mathbf{r}, t)$ catches all the microscopic volume elements of the π-phase within the REV.

According to Delesse's law (1848) [Boer(1996), Lewis(1998)], the distribution of the area elements is stated to be similar to the distribution of the volume elements. This assumption may be justified when thinking of the individual constituents as being statistically distributed in the representative elementary volume. Thus, a typical cross sectional area da of the REV consists of parts da^π, $\pi = 1, 2, \ldots, \kappa$, in a manner similar to (3.36). Analogous to (3.38) an average area element da^π can then be defined as

$$da^\pi(\mathbf{x}, t) = \int_{da} \gamma^\pi(\mathbf{r}, t)\, da_{mic}\,, \tag{3.39}$$

where da_{mic} denotes the microscopic area element.

Although the values of dv^π and da^π depend on the sizes of the averaging volume and area, respectively, the ratio of each of these quantities to dv and da is a macroscopic variable which exhibits the same type of behaviour as depicted in Figure 3.2. Accordingly, the following definitions can be introduced for the volume and area fractions: The *volume fraction* of the π-phase, η^π, denoting the part of dv occupied by the respective phase is given as

$$\eta^\pi(\mathbf{x}, t) = \frac{dv^\pi}{dv} = \frac{1}{dv} \int_{dv} \gamma^\pi(\mathbf{r}, t)\, dv_{mic}\,. \tag{3.40}$$

The volume fractions η^π of all constituents π, $\pi = 1, 2, \ldots, \kappa$, sum up to one, i.e. they are subject to the constraint

$$\sum_{\pi=1}^{\kappa} \eta^\pi = 1 \quad \text{with} \quad 0 \leq \eta^\pi \leq 1\,. \tag{3.41}$$

The *area fraction* of the π-phase, $\overline{\eta^\pi}$, is the part of da which intersects the π-phase,

$$\overline{\eta^\pi}(\mathbf{x}, t) = \frac{da^\pi}{da} = \frac{1}{da} \int_{da} \gamma^\pi(\mathbf{r}, t)\, da_{mic}\,, \tag{3.42}$$

and all area fractions are constrained by

$$\sum_{\pi=1}^{\kappa} \overline{\eta^\pi} = 1 \quad \text{with} \quad 0 \leq \overline{\eta^\pi} \leq 1\,. \tag{3.43}$$

It should be mentioned that in general η^π and $\overline{\eta^\pi}$ need not be equal. However, some arguments have been presented in the literature [Morland(1972)] which support the intuitively appealing identity between the volume and area fractions. Hence, this assumption is also employed in the current work.

To conclude this section, the application of the volume fraction concept to a three-phase medium, e.g., a porous solid filled with two fluid phases, is discussed. The percentage of voids in the unit volume dv is denoted by the porosity n, being defined as

$n = (dv^w + dv^a)/dv$, where the sum $(dv^w + dv^a)$ describes the void space. Hence, the part $(1-n)$ of the volume element is occupied by the solid phase. From relating the void space to the volume of the solid phase follows the definition of the void ratio e, $e = (dv^w + dv^a)/dv^s$. Therefore, void ratio and porosity are connected with each other by $e = n/(1-n)$. The pore spaces are assumed to be filled with two immiscible fluids, e.g., water w and air a. The degree of saturation S^f, $f = w, a$, describes the portion of each fluid present in the voids, i.e. $S^w = dv^w/(dv^w + dv^a)$ and $S^a = dv^a/(dv^w + dv^a)$. Clearly, the condition $S^w + S^a = 1$ must hold, i.e. the pores are completely filled with the two fluids. For the three-phase medium the product nS^f defines the portion of the fluid phase f referring to the total volume element under consideration. Summing up the volume fractions of the three constituents finally yields:

$$\sum_{\pi=s,w,a} \eta^\pi = (1-n) + nS^w + nS^a = (1-n) + n(S^w + S^a) = 1 . \quad (3.44)$$

For easy reference some of the most important relations concerning the volume fractions in a three-phase medium are summarised in Table 3.1. In particular, volume fractions, porosity and the degrees of saturation are given:

Constituent	Constituent's volume	Volume fraction	Useful definitions
Solid	dv^s	$\dfrac{dv^s}{dv} = \eta^s = (1-n)$	$n = \dfrac{dv^w + dv^a}{dv}$
Water	dv^w	$\dfrac{dv^w}{dv} = \eta^w = nS^w$	$S^w = \dfrac{dv^w}{dv^w + dv^a}$
Air	dv^a	$\dfrac{dv^a}{dv} = \eta^a = nS^a$	$S^a = \dfrac{dv^a}{dv^w + dv^a}$

Table 3.1: Definition of the volume fractions in a three-phase medium.

3.3.2 Definition of averaging operators

The microscopic quantities are transformed to macroscopic values using so-called *averaging operators*. For a given size of the averaging region different averaging operators can be defined. The intrinsic nature of the particular property should be taken into account when choosing a certain averaging operator. For example, it seems to be reasonable that an averaged stress, which acts per unit area, should be obtained from a different averaging operator than the averaged density being a mass related quantity. This need for various types of averaging operators is expressed in the following four criteria [Hassanizadeh(1979a)]:

Criterion I: When an averaging procedure involves integration, the integrand multiplied by the infinitesimal element of integration has to be an additive quantity, where additive means that a quantity of the total system can be obtained as the sum of the individual quantities related to any partition of the system into a number of subsystems. The definition of an average in terms of a quantity which is not additive would not be physically meaningful. For instance, when considering an integral over a volume, the internal energy density function, defined as internal energy per unit mass [energy/mass], is not additive, as well as the internal energy density function multiplied by the infinitesimal element of volume. Since the latter product possesses units of [(energy/mass) × volume], an integration over the volume would not result in the correct physical unit of energy. On the other hand, the total internal energy, defined as density times internal energy density function with units [(mass/volume) × (energy/mass)], is an additive quantity and an integration over the volume yields the correct physical dimension of energy.

Criterion II: The macroscopic quantities should exactly account for the total amount of the corresponding microscopic quantities. For example, the macroscopic flux through a given boundary must be equal to the total microscopic flux through that boundary.

Criterion III: The concept of a physical quantity as first introduced in 'classical' continuum mechanics must be preserved by a proper definition of the respective macroscopic quantity. For instance, heat is a mode of transfer of energy through a boundary; therefore, the definition of a macroscopic heat flux must also be a mode of energy transfer.

Criterion IV: The way of defining averaged values of microscopic quantities (the type of averaging operator applied to a certain microscopic quantity) must be closely related to the kind of quantities most widely observed and measured in field situations or laboratory practice. For example, velocities obtained from experiments are usually mass averaged quantities; hence, macroscopic velocities should also be mass averaged quantities.

In order to adhere to the above introduced guidelines, three different types of averaging operators are required, namely for volume, mass and area related quantities, respectively. Subsequently, these averaging operators are defined and applied to a function $\zeta(\mathbf{r}, t)$ which stands for a typical microscopic field variable.

First, two different kinds of volume average operators are introduced: The *volume phase average* $\zeta_\pi(\mathbf{x}, t)$ is given as

$$\zeta_\pi(\mathbf{x}, t) = \frac{1}{dv} \int_{dv} \zeta(\mathbf{r}, t)\, \gamma^\pi(\mathbf{r}, t)\, dv_{mic} \,. \tag{3.45}$$

Here, the microscopic field variable $\zeta(\mathbf{r}, t)$ is related to the volume dv of the REV. The *intrinsic volume phase average* $\zeta^\pi(\mathbf{x}, t)$ is defined as

$$\zeta^\pi(\mathbf{x}, t) = \frac{1}{dv^\pi} \int_{dv^\pi} \zeta(\mathbf{r}, t)\, dv_{mic} \,, \tag{3.46}$$

where the field variable $\zeta(\mathbf{r}, t)$ is referred to the average volume element dv^π of the constituent π.

Using the phase distribution function $\gamma^\pi(\mathbf{r}, t)$ as defined in equation (3.37), the integral in (3.46) can be rewritten as

$$\int_{dv^\pi} \zeta(\mathbf{r}, t)\, dv_{mic} = \int_{dv} \zeta(\mathbf{r}, t)\, \gamma^\pi(\mathbf{r}, t)\, dv_{mic} \,. \tag{3.47}$$

After inserting (3.47) into equation (3.46) the intrinsic volume phase average is obtained as

$$\zeta^\pi(\mathbf{x}, t) = \frac{1}{dv^\pi} \int_{dv} \zeta(\mathbf{r}, t)\, \gamma^\pi(\mathbf{r}, t)\, dv_{mic} \,. \tag{3.48}$$

Comparing equations (3.45) and (3.48) and employing the definition of the volume fraction as stated in (3.40) yields the following relationship between the volume phase average and the intrinsic volume phase average of a field variable:

$$\zeta_\pi(\mathbf{x}, t) = \eta^\pi(\mathbf{x}, t)\, \zeta^\pi(\mathbf{x}, t) \,. \tag{3.49}$$

Second, a *mass average operator* is defined. If the microscopic mass density $\rho(\mathbf{r}, t)$ is used as a weighting function, a mass average operator $\bar{\zeta}^\pi(\mathbf{x}, t)$ can be formulated as

$$\bar{\zeta}^\pi(\mathbf{x}, t) = \frac{\int_{dv} \rho(\mathbf{r}, t)\, \zeta(\mathbf{r}, t)\, \gamma^\pi(\mathbf{r}, t)\, dv_{mic}}{\int_{dv} \rho(\mathbf{r}, t)\, \gamma^\pi(\mathbf{r}, t)\, dv_{mic}} \,. \tag{3.50}$$

By means of the above defined volume phase average (equation (3.45)), applied to the microscopic mass density (that is $\zeta(\mathbf{r}, t) = \rho(\mathbf{r}, t)$), relation (3.50) may be rewritten as

$$\bar{\zeta}^\pi(\mathbf{x}, t) = \frac{1}{\rho_\pi(\mathbf{x}, t)\, dv} \int_{dv} \rho(\mathbf{r}, t)\, \zeta(\mathbf{r}, t)\, \gamma^\pi(\mathbf{r}, t)\, dv_{mic} \,. \tag{3.51}$$

If $\rho(\mathbf{r},t)$ is assumed to be constant, equation (3.50) can also be formulated as

$$\overline{\zeta}^\pi(\mathbf{x},t) = \frac{\int_{dv} \zeta(\mathbf{r},t)\,\gamma^\pi(\mathbf{r},t)\,dv_{mic}}{\int_{dv} \gamma^\pi(\mathbf{r},t)\,dv_{mic}}. \qquad (3.52)$$

Making use of the definition of the average volume element dv^π, as given in (3.38), yields

$$\overline{\zeta}^\pi(\mathbf{x},t) = \frac{1}{dv^\pi} \int_{dv} \zeta(\mathbf{r},t)\,\gamma^\pi(\mathbf{r},t)\,dv_{mic}, \qquad (3.53)$$

and with the volume phase average (equation (3.45)) the mass averaging operator is obtained as

$$\overline{\zeta}^\pi(\mathbf{x},t) = \frac{1}{dv^\pi}\,\zeta_\pi(\mathbf{x},t)\,dv. \qquad (3.54)$$

Employing the volume fraction definition (equation (3.40)) and the relationship (3.49) between the volume phase average and the intrinsic volume phase average one finally gets for a constant microscopic mass density, i.e. for $\rho(\mathbf{r},t) = const.$,

$$\overline{\zeta}^\pi(\mathbf{x},t) = \frac{1}{\eta^\pi(\mathbf{x},t)}\,\zeta_\pi(\mathbf{x},t) = \zeta^\pi(\mathbf{x},t). \qquad (3.55)$$

Hence, for a constant microscopic mass density the mass average of a quantity may be related to the volume phase average of the respective quantity by means of the volume fraction of the particular phase. Additionally, equation (3.55) indicates that for a constant microscopic mass density the mass average is equal to the intrinsic volume phase average.

Third, an *area average operator* $\overline{\overline{\zeta}}_\pi(\mathbf{x},t)$ is defined as

$$\overline{\overline{\zeta}}_\pi(\mathbf{x},t) = \frac{1}{da}\int_{da} \zeta(\mathbf{r},t)\cdot\mathbf{n}(\mathbf{r},t)\,\gamma^\pi(\mathbf{r},t)\,da_{mic}. \qquad (3.56)$$

It should be mentioned that the variable $\zeta(\mathbf{r},t)$ in equation (3.56) is of tensorial nature. $\mathbf{n}(\mathbf{r},t)$ denotes the outward normal unit vector to the area element da_{mic}. Note that the area averaging operator reduces the tensorial nature of the quantity averaged by one order.

3.3.3 Microscopic balance equations

On the microscopic level the body under consideration is a heterogeneous medium composed of a number of κ different constituents as depicted in Figure 3.1(a) for the special case of a three-phase system, e.g., a porous medium consisting of a solid phase and two fluids. The individual phases are supposed to be immiscible and to have distinct thermodynamic properties. Each constituent π, $\pi = 1, 2, \ldots, \kappa$, occupies only a portion dv^π of the control space dv. As is evident from Figure 3.1(a), dv^π is the union of a large number of isolated subregions within which π forms a continuum. These subregions are separated from each other by interfaces to which, in the current model, no thermomechanical quantities are attributed. However, the material properties and thermodynamic variables of the individual phases may exhibit discontinuities at these boundaries. The existence of such interfaces is conceivable for solid-solid, fluid-solid and immiscible fluid-fluid systems.

For the mathematical description of the microscopic situation the governing equations in each subregion or phase π are the classical balance laws of continuum mechanics for a single-phase medium all of which may be represented by one general form as introduced subsequently. For the particular interfaces, which merely act as surfaces of discontinuity due to the neglect of their thermodynamic quantities, the standard jump conditions for the conservation equations are formulated.

Consider an open system in a single-phase continuum, i.e. a system consisting of a fixed amount of volume of a properly selected region, occupying a control volume dv which is enclosed by a control surface da, independent of time t. The boundary of dv, which both mass and energy can cross, may have a velocity $\mathbf{w}_b$. For a typical thermodynamic property ψ at the microscale the global general conservation equation for a single-phase medium can then be formulated as [Eringen(1989), Jouanna(1995), Holzapfel(2000)]

$$\frac{D}{Dt}\int_{dv} \rho\,\psi\,dv_{mic} = -\int_{da} \rho\,\psi \otimes (\dot{\mathbf{r}} - \mathbf{w}_b) \cdot \mathbf{n}\,da_{mic} + \int_{da} \boldsymbol{\vartheta}\,da_{mic} + \int_{dv} \rho\,\boldsymbol{\beta}\,dv_{mic}\,, \qquad (3.57)$$

where ρ and $\dot{\mathbf{r}}$ denote the density and the velocity of the phase, respectively. $\mathbf{n}$ is the unit normal vector to da, pointing out of dv. $\boldsymbol{\vartheta}$ may be interpreted as a vector of tractions acting on the surface da and $\boldsymbol{\beta}$ accounts for any body forces within dv. It should be noted that the property ψ is conceived to be either a scalar or a vector, depending on the particular type of balance law under consideration. For the case of a scalar ψ, however, the last two integrals in (3.57) vanish as will be shown subsequently.

Equation (3.57) can be viewed as a mathematical statement of the physical principle that the rate of change of some property $(\rho\,\psi)$ in a volume dv is equal to the flux of that quantity across the boundary da of the volume under consideration plus the tractions acting on da plus any body forces being present within dv. In terms of energy the latter would more likely be called an external supply. It should be mentioned that when considering the second law of thermodynamics (entropy principle) an additional term has to be added to the balance law (3.57) in order to account for the net production of the property ψ within dv. The negative sign of the flux term results from a positive defined inward flux.

The global (integral) form of the general balance equation (3.57) is now modified in order to obtain a local (differential) form of that equation. The term on the left hand side of (3.57) can be rewritten by means of the Reynolds' transport theorem (3.33) as

$$\frac{D}{Dt}\int_{dv} \rho\,\psi\,dv_{mic} = \int_{dv} \frac{\partial(\rho\,\psi)}{\partial t}\,dv_{mic} + \int_{da} \rho\,\psi\,(\mathbf{w}_b \cdot \mathbf{n})\,da_{mic}\,. \tag{3.58}$$

For the traction vector $\boldsymbol{\vartheta}$ a relationship of the type

$$\boldsymbol{\vartheta} = \boldsymbol{\Phi} \cdot \mathbf{n} \tag{3.59}$$

may be defined. Equation (3.59) states that the traction vector $\boldsymbol{\vartheta}$ is related to a second-order tensor $\boldsymbol{\Phi}$ and that $\boldsymbol{\vartheta}$ linearly depends on the outward unit normal $\mathbf{n}$ of the surface da. Relation (3.59) together with the divergence theorem (3.28$_2$) can be employed to formulate the second integral on the right hand side of (3.57) in terms of a volume integral as

$$\int_{da} \boldsymbol{\vartheta}\,da_{mic} = \int_{da} \boldsymbol{\Phi} \cdot \mathbf{n}\,da_{mic} = \int_{dv} \operatorname{div} \boldsymbol{\Phi}\,dv_{mic}\,. \tag{3.60}$$

Inserting (3.60) and (3.58) into equation (3.57) and dividing the first integral on the right hand side of (3.57) into two parts yields

$$\int_{dv} \frac{\partial(\rho\,\psi)}{\partial t}\,dv_{mic} + \int_{da} \rho\,\psi\,(\mathbf{w}_b \cdot \mathbf{n})\,da_{mic} = -\int_{da} \rho\,\psi\,(\dot{\mathbf{r}} \cdot \mathbf{n})\,da_{mic} +$$

$$+ \int_{da} \rho\,\psi\,(\mathbf{w}_b \cdot \mathbf{n})\,da_{mic} + \int_{dv} \operatorname{div} \boldsymbol{\Phi}\,dv_{mic} + \int_{dv} \rho\,\boldsymbol{\beta}\,dv_{mic}\,. \tag{3.61}$$

where the two terms containing the boundary velocity $\mathbf{w}_b$ of the domain cancel out. Applying the divergence theorem to the first term on the right hand side of (3.61) results in the following final form of the global general balance law,

$$\int_{dv} \frac{\partial(\rho\,\psi)}{\partial t}\,dv_{mic} = -\int_{dv} \operatorname{div}(\rho\,\psi \otimes \dot{\mathbf{r}})\,dv_{mic} + \int_{dv} \operatorname{div} \boldsymbol{\Phi}\,dv_{mic} + \int_{dv} \rho\,\boldsymbol{\beta}\,dv_{mic}\,. \tag{3.62}$$

Since equation (3.62) is assumed to hold for an arbitrary microscopic volume element, the local form of the general microscopic balance law is obtained as

$$\frac{\partial(\rho\,\psi)}{\partial t} + \mathrm{div}(\rho\,\psi \otimes \dot{\mathbf{r}}) - \mathrm{div}\,\mathbf{\Phi} - \rho\,\boldsymbol{\beta} = \mathbf{0}\;. \tag{3.63}$$

At a surface of discontinuity and in particular on the interfaces the general balance law (3.63) does not hold. However, a so-called jump discontinuity can be established from the integral form (3.57). To this end, two terms have to be added to the balance law (3.57) in order to account for the surface flux and the tractions on the surface of discontinuity. Since the microscopic surface of discontinuity Σ_{mic} divides the volume under consideration into two portions dv^+ and dv^-, the balance equation has to be formulated for both parts, e.g., for dv^+ it reads

$$\frac{D}{Dt}\int_{dv^+}\rho\,\psi\,dv_{mic} = -\int_{da^+}\rho\,\psi \otimes (\dot{\mathbf{r}} - \mathbf{w}_b)\cdot\mathbf{n}\,da_{mic} - \int_{\Sigma_{mic}}\rho\,\psi \otimes (\dot{\mathbf{r}} - \mathbf{w})\cdot\mathbf{n}^+\,da_{mic} +$$

$$+\int_{da^+}\mathbf{\Phi}\cdot\mathbf{n}\,da_{mic} + \int_{\Sigma_{mic}}\mathbf{\Phi}\cdot\mathbf{n}^+\,da_{mic} + \int_{dv^+}\rho\,\boldsymbol{\beta}\,dv_{mic}\;, \tag{3.64}$$

where $\mathbf{w}$ denotes the velocity of the surface of discontinuity Σ_{mic} and $\mathbf{n}^+$ is its unit normal vector, pointing out of dv^+. A similar equation can also be written for the part dv^- of the domain. After adding up the equations for the two portions, a balance law is obtained which is valid for the whole domain, including the surface of discontinuity:

$$\frac{D}{Dt}\int_{dv^\pm}\rho\,\psi\,dv_{mic} + \int_{da^\pm}\rho\,\psi \otimes (\dot{\mathbf{r}} - \mathbf{w}_b)\cdot\mathbf{n}\,da_{mic} - \int_{da^\pm}\mathbf{\Phi}\cdot\mathbf{n}\,da_{mic} - \int_{dv^\pm}\rho\,\boldsymbol{\beta}\,dv_{mic} =$$

$$= -\int_{\Sigma_{mic}}\rho\,\psi \otimes (\dot{\mathbf{r}} - \mathbf{w})\cdot\mathbf{n}^+\,da_{mic} - \int_{\Sigma_{mic}}\rho\,\psi \otimes (\dot{\mathbf{r}} - \mathbf{w})\cdot\mathbf{n}^-\,da_{mic} +$$

$$+\int_{\Sigma_{mic}}\mathbf{\Phi}\cdot\mathbf{n}^+\,da_{mic} + \int_{\Sigma_{mic}}\mathbf{\Phi}\cdot\mathbf{n}^-\,da_{mic}\;, \tag{3.65}$$

where $dv^\pm$ and $da^\pm$ denote the summation over the two individual portions of the domain, $dv^+ + dv^-$ and $da^+ + da^-$, respectively. It should be noted that for the unit normal vectors of the surface of discontinuity the relation $\mathbf{n}^+ = -\mathbf{n}^-$ holds. Whereas the left hand side of equation (3.65) expresses the global form of the continuum balance law, similar to equation (3.57) for a domain without discontinuity, the right hand side of (3.65) describes the balance law for the surface of discontinuity Σ_{mic}. Both parts of the equation have to vanish [Hassanizadeh(1979a)] which yields the global general balance law (jump condition) for Σ_{mic},

$$\int_{\Sigma_{mic}} [\mathbf{\Phi} - \rho\boldsymbol{\psi} \otimes (\dot{\mathbf{r}} - \mathbf{w})] \cdot \mathbf{n}^+ \, da_{mic} + \int_{\Sigma_{mic}} [\mathbf{\Phi} - \rho\boldsymbol{\psi} \otimes (\dot{\mathbf{r}} - \mathbf{w})] \cdot \mathbf{n}^- \, da_{mic} = \mathbf{0} \,. \quad (3.66)$$

For an arbitrary surface of discontinuity Σ_{mic} or an interface the local balance equation is then obtained from the global form (3.66) as

$$[\mathbf{\Phi} - \rho\boldsymbol{\psi} \otimes (\dot{\mathbf{r}} - \mathbf{w})] \cdot \mathbf{n}^+ + [\mathbf{\Phi} - \rho\boldsymbol{\psi} \otimes (\dot{\mathbf{r}} - \mathbf{w})] \cdot \mathbf{n}^- = \mathbf{0} \,. \quad (3.67)$$

Since the right hand side of (3.67) is zero, no thermodynamic properties of the particular surface are taken into account. However, an exchange of mass, momentum and energy between the different constituents is possible through the interface.

To gain the particular balance equations for mass, linear momentum and angular momentum, the variables $\boldsymbol{\psi}$, $\boldsymbol{\vartheta}$, $\boldsymbol{\beta}$ and $\boldsymbol{\Phi}$ in the general forms (3.57), (3.63), (3.66) or (3.67) have to be chosen in a certain way [Hassanizadeh(1979b), Holzapfel(2000)]. Table 3.2 summarises the necessary specifications of these variables which are used in subsequent sections of this chapter. The individual values of $\boldsymbol{\Phi}$ are introduced in these later sections. It should be mentioned that also the energy and entropy principles can be derived from the general conservation laws. However, these two principles are not considered in the present work. In Table 3.2 $\mathbf{t}_{mic}$ denotes the microscopic stress vector and $\mathbf{g}$ is the vector of gravitational acceleration.

Theoretically, after setting up a certain number of constitutive laws, one should be able to obtain a solution of equation (3.63) subject to the interfacial relation (3.67) and

Balance equation	$\boldsymbol{\psi}$	$\boldsymbol{\vartheta}$	$\boldsymbol{\beta}$
mass	1	0	0
linear momentum	$\dot{\mathbf{r}}$	$\mathbf{t}_{mic}$	$\mathbf{g}$
angular momentum	$\mathbf{r} \times \dot{\mathbf{r}}$	$\mathbf{r} \times \mathbf{t}_{mic}$	$\mathbf{r} \times \mathbf{g}$

Table 3.2: Specification of the general balance equation variables to obtain particular balance laws.

to appropriate boundary and initial conditions. However, an analytical application of the interface condition is virtually impossible because of the in general highly complicated geometry of these surfaces. Furthermore, microscopic quantities are not measurable in most cases and thus any constitutive postulates would not be subject to experimental verification. In fact, only average (macroscopic) values of the microscopic quantities are measured and of interest in practice.

3.3.4 Macroscopic balance equations

This section deals with the derivation of general forms of the macroscopic balance equation for the generic quantity $\boldsymbol{\psi}(\mathbf{r},t)$ and of the corresponding jump condition. These equations constitute the basis for the specification of the general macroscopic balance laws for the special case of a three-phase medium, a specification which is performed subsequently in two steps: First, the general equations are reformulated for the individual balances of mass, linear momentum and angular momentum which, second, are then specified for the constituents of the three-phase medium. It should be noted that the balance equations are derived in this section in a material-free manner. Any constitutive assumptions are introduced later when deriving the individual balances.

The general macroscopic balance equation is obtained from the microscopic balance law (3.63) by performing three steps: First, equation (3.63) has to be multiplied by the phase distribution function $\gamma^\pi(\mathbf{r},t)$ defined in (3.37). Second, the resulting product must be integrated over the volume dv of the REV and afterwards divided by dv. Finally, an integration has to be performed over all averaging volumes covering the space V. Applying this procedure to the microscopic balance law (3.63) yields

$$\int_V \left[\frac{1}{dv} \int_{dv} \frac{\partial}{\partial t}[\rho(\mathbf{r},t)\boldsymbol{\psi}(\mathbf{r},t)]\gamma^\pi(\mathbf{r},t)\,dv_{mic}\right]dV +$$

$$+ \int_V \left[\frac{1}{dv} \int_{dv} \mathrm{div}[\rho(\mathbf{r},t)\boldsymbol{\psi}(\mathbf{r},t)\otimes\dot{\mathbf{r}}(\mathbf{r},t)]\gamma^\pi(\mathbf{r},t)\,dv_{mic}\right]dV -$$

$$- \int_V \left[\frac{1}{dv} \int_{dv} \mathrm{div}[\boldsymbol{\Phi}(\mathbf{r},t)]\gamma^\pi(\mathbf{r},t)\,dv_{mic}\right]dV -$$

$$- \int_V \left[\frac{1}{dv} \int_{dv} \rho(\mathbf{r},t)\boldsymbol{\beta}(\mathbf{r},t)\gamma^\pi(\mathbf{r},t)\,dv_{mic}\right]dV = \mathbf{0}\,, \tag{3.68}$$

which is the most general form of a macroscopic balance equation for a continuous domain. It should be mentioned that an additional term, similar to the last integral on the left hand side of the equation and accounting for the production of $\boldsymbol{\psi}(\mathbf{r},t)$, has to be included in (3.68) for deducing the entropy inequality. To develop a more suitable formulation of (3.68), several modifications are performed subsequently. For this purpose a set of necessary integral expressions and identities is first introduced.

In this elaboration of the balance equations macroscopic quantities are obtained by means of the previously defined averaging operators. Hence, considering (3.45) and (3.56), expressions of the following type appear in the derivation when performing an integration over the volume,

$$\int_V \left[\frac{1}{dv} \int_{dv} \zeta(\mathbf{r},t)\, \gamma^\pi(\mathbf{r},t)\, dv_{mic} \right] dV = \int_V \zeta_\pi(\mathbf{x},t)\, dV \; , \qquad (3.69)$$

or an integration over the area of the control space,

$$\int_A \left[\frac{1}{da} \int_{da} \zeta(\mathbf{r},t) \cdot \mathbf{n}(\mathbf{r},t)\, \gamma^\pi(\mathbf{r},t)\, da_{mic} \right] dA = \int_A \overline{\overline{\zeta}}_\pi(\mathbf{x},t)\, dA \; . \qquad (3.70)$$

To convert an integral over dv including a time derivative of the integrand $\zeta(\mathbf{r},t)$ into the derivative of the integral, the following equation may be useful [Gray(1977), Hassanizadeh(1979a)],

$$\int_V \left[\frac{1}{dv} \int_{dv} \frac{\partial \zeta(\mathbf{r},t)}{\partial t}\, \gamma^\pi(\mathbf{r},t)\, dv_{mic} \right] dV =$$

$$= \int_V \frac{\partial}{\partial t} \left[\frac{1}{dv} \int_{dv} \zeta(\mathbf{r},t)\, \gamma^\pi(\mathbf{r},t)\, dv_{mic} \right] dV -$$

$$- \int_V \left[\frac{1}{dv} \sum_{\alpha \neq \pi}^{K} \int_{da^{\pi\alpha}} \zeta(\mathbf{r},t)\, \mathbf{w}(\mathbf{r},t) \cdot \mathbf{n}^{\pi\alpha}(\mathbf{r},t)\, da_{mic} \right] dV \; , \qquad (3.71)$$

where $da^{\pi\alpha}$ denotes the interface with the unit normal vector $\mathbf{n}^{\pi\alpha}(\mathbf{r},t)$ between the π and the α-phase and $\mathbf{w}(\mathbf{r},t)$ is the velocity of the interface. The last integral in (3.71) can be interpreted as the flux of $\zeta(\mathbf{r},t)$ across all the $\pi\alpha$-interfaces inside the averaging volume (integration over the $\pi\alpha$-interface $da^{\pi\alpha}$ and summation over all phases within the REV). It should be noted that this equation is obtained from applying Reynolds' transport theorem in the form (3.33) to the first term on the right hand side of (3.71). The constant factor $(1/dv)$ can be put in front of the time derivative and thus it also appears as a constant in the other two integral terms. When moving the last integral on the right hand side of (3.71) to the left, an interpretation of this relationship according to the one given in connection with equation (3.33) is easily possible.

For the divergence of a vector or tensor $\boldsymbol{\zeta}(\mathbf{r},t)$ and its subsequent integration over dv the equation

$$\int_V \left[\frac{1}{dv} \int_{dv} \operatorname{div}[\boldsymbol{\zeta}(\mathbf{r},t)] \, \gamma^\pi(\mathbf{r},t) \, dv_{mic} \right] dV =$$

$$= \int_V \left[\frac{1}{dv} \int_{dv} \operatorname{div}[\boldsymbol{\zeta}(\mathbf{r},t) \, \gamma^\pi(\mathbf{r},t)] \, dv_{mic} \right] dV +$$

$$+ \int_V \left[\frac{1}{dv} \sum_{\alpha \neq \pi}^{K} \int_{da^{\pi\alpha}} \boldsymbol{\zeta}(\mathbf{r},t) \cdot \mathbf{n}^{\pi\alpha}(\mathbf{r},t) \, da_{mic} \right] dV \quad (3.72)$$

is needed [Gray(1977), Hassanizadeh(1979a)]. Proofs for the above two relationships (3.71) and (3.72) can be found, e.g., in [Gray(1977)]. Equation (3.72) can be rewritten by means of the divergence theorem of Eringen and Suhubi (relationship (3.32)). Introducing the volume dv and the surface da of the REV for v and a, respectively, and inserting $\mathbf{f} = \boldsymbol{\zeta}(\mathbf{r},t) \, \gamma^\pi(\mathbf{r},t)$ into (3.32) yields for the first part of equation (3.32)

$$\int_V \left[\frac{1}{dv} \int_{dv} \operatorname{div}[\boldsymbol{\zeta}(\mathbf{r},t) \, \gamma^\pi(\mathbf{r},t)] \, dv_{mic} \right] dV = \int_A \left[\frac{1}{da} \int_{da} \boldsymbol{\zeta}(\mathbf{r},t) \cdot \mathbf{n}(\mathbf{r},t) \, \gamma^\pi(\mathbf{r},t) \, da_{mic} \right] dA \; . \quad (3.73)$$

Substituting equation (3.73) into (3.72) results in

$$\int_V \left[\frac{1}{dv} \int_{dv} \operatorname{div}[\boldsymbol{\zeta}(\mathbf{r},t)] \, \gamma^\pi(\mathbf{r},t) \, dv_{mic} \right] dV =$$

$$= \int_A \left[\frac{1}{da} \int_{da} \boldsymbol{\zeta}(\mathbf{r},t) \cdot \mathbf{n}(\mathbf{r},t) \, \gamma^\pi(\mathbf{r},t) \, da_{mic} \right] dA +$$

$$+ \int_V \left[\frac{1}{dv} \sum_{\alpha \neq \pi}^{K} \int_{da^{\pi\alpha}} \boldsymbol{\zeta}(\mathbf{r},t) \cdot \mathbf{n}^{\pi\alpha}(\mathbf{r},t) \, da_{mic} \right] dV \; . \quad (3.74)$$

Chapter 3. Mechanics of the coupled three-phase model

In addition to the above listed integral expressions it should be emphasised that for an integration over the volume element dv averaged quantities are constant in dv. Furthermore, the following identities will prove useful during the manipulation of the equations. The deviation $\hat{\zeta}^\pi(\mathbf{x}, \boldsymbol{\xi}, t)$ of a microscopic function $\zeta(\mathbf{r}, t)$ at a point $\mathbf{r}$ ($\mathbf{r} \in dv$, centred at $\mathbf{x}$, where $\boldsymbol{\xi} = \mathbf{r} - \mathbf{x}$ according to (3.35)) from its π-phase mass averaged quantity $\overline{\zeta}^\pi(\mathbf{x}, t)$ at the point $\mathbf{x}$ is defined as

$$\hat{\zeta}^\pi(\mathbf{x}, \boldsymbol{\xi}, t) = \zeta(\mathbf{r}, t) - \overline{\zeta}^\pi(\mathbf{x}, t) \ . \tag{3.75}$$

Velocities usually obtained from experiments are regarded as mass averaged quantities. Therefore, a mass averaged velocity $\overline{\mathbf{v}}^\pi(\mathbf{x}, t)$ of the π-phase is introduced according to (3.51) as

$$\overline{\dot{\mathbf{r}}}^\pi(\mathbf{x}, t) = \overline{\mathbf{v}}^\pi(\mathbf{x}, t) = \frac{1}{\rho_\pi(\mathbf{x}, t)\, dv} \int_{dv} \rho(\mathbf{r}, t)\, \dot{\mathbf{r}}(\mathbf{r}, t)\, \gamma^\pi(\mathbf{r}, t)\, dv_{mic} \ . \tag{3.76}$$

Since a mass averaged quantity $\overline{\zeta}^\pi(\mathbf{x}, t)$ is constant in an integral over dv or over da, two further identities must hold. Applying the definition of the mass averaging operator as given in (3.51) to the deviation $\hat{\zeta}^\pi(\mathbf{x}, \boldsymbol{\xi}, t)$ of a microscopic function from its mass averaged quantity (equation (3.75)) yields

$$\overline{\hat{\zeta}}^\pi(\mathbf{x}, \boldsymbol{\xi}, t) = \frac{1}{\rho_\pi(\mathbf{x}, t)\, dv} \int_{dv} \rho(\mathbf{r}, t) \left[\zeta(\mathbf{r}, t) - \overline{\zeta}^\pi(\mathbf{x}, t) \right] \gamma^\pi(\mathbf{r}, t)\, dv_{mic} =$$

$$= \frac{1}{\rho_\pi(\mathbf{x}, t)\, dv} \int_{dv} \rho(\mathbf{r}, t)\, \zeta(\mathbf{r}, t)\, \gamma^\pi(\mathbf{r}, t)\, dv_{mic} -$$

$$- \frac{1}{\rho_\pi(\mathbf{x}, t)\, dv} \overline{\zeta}^\pi(\mathbf{x}, t) \int_{dv} \rho(\mathbf{r}, t)\, \gamma^\pi(\mathbf{r}, t)\, dv_{mic} =$$

$$= \overline{\zeta}^\pi(\mathbf{x}, t) - \overline{\zeta}^\pi(\mathbf{x}, t) = 0 \ . \tag{3.77}$$

To arrive at the last line of equation (3.77), the definitions of the mass averaging operator (3.51) and of the volume phase averaging operator (3.45) have been employed.

Another identity for the case of two microscopic functions $\zeta(\mathbf{r}, t)$ and $\omega(\mathbf{r}, t)$ being present is given as

$$\int_A \left[\frac{1}{da} \int_{da} \left[\rho(\mathbf{r},t) \overline{\omega}^\pi(\mathbf{r},t) \hat{\boldsymbol{\zeta}}^\pi(\mathbf{x},\boldsymbol{\xi},t) \right] \cdot \mathbf{n}(\mathbf{r},t) \gamma^\pi(\mathbf{r},t) \, da_{mic} \right] dA =$$

$$= \int_V \left[\frac{1}{dv} \int_{dv} \mathrm{div} \left[\rho(\mathbf{r},t) \overline{\omega}^\pi(\mathbf{r},t) \gamma^\pi(\mathbf{r},t) \hat{\boldsymbol{\zeta}}^\pi(\mathbf{x},\boldsymbol{\xi},t) \right] dv_{mic} \right] dV =$$

$$= \int_V \mathrm{div} \left[\frac{1}{dv} \int_{dv} \rho(\mathbf{r},t) \overline{\omega}^\pi(\mathbf{r},t) \gamma^\pi(\mathbf{r},t) \hat{\boldsymbol{\zeta}}^\pi(\mathbf{x},\boldsymbol{\xi},t) \, dv_{mic} \right] dV =$$

$$= \int_V \mathrm{div} \left[\overline{\omega}^\pi(\mathbf{r},t) \rho_\pi(\mathbf{x},t) \overline{\hat{\boldsymbol{\zeta}}}^\pi(\mathbf{x},\boldsymbol{\xi},t) \right] dV = 0 \ . \tag{3.78}$$

In (3.78) application of the divergence theorem proposed by Eringen and Suhubi (relationship (3.32)) results in the third line and using the mass averaging operator (3.51) yields the last line of the equation. The last integral has to be zero due to identity (3.77).

At this point, having a couple of integral expressions and identities handy as necessary and useful tools, we can start manipulating the individual terms in the general macroscopic balance equation (3.68).

The first term of the general equation (3.68), containing the time derivative of the generic variable $\boldsymbol{\psi}(\mathbf{r},t)$, can be split into two parts using expression (3.71):

$$\int_V \left[\frac{1}{dv} \int_{dv} \frac{\partial}{\partial t} [\rho(\mathbf{r},t) \boldsymbol{\psi}(\mathbf{r},t)] \gamma^\pi(\mathbf{r},t) \, dv_{mic} \right] dV =$$

$$= \int_V \frac{\partial}{\partial t} \left[\frac{1}{dv} \int_{dv} \rho(\mathbf{r},t) \boldsymbol{\psi}(\mathbf{r},t) \gamma^\pi(\mathbf{r},t) \, dv_{mic} \right] dV -$$

$$- \int_V \left[\frac{1}{dv} \sum_{\alpha \neq \pi}^K \int_{da^{\pi\alpha}} \rho(\mathbf{r},t) \boldsymbol{\psi}(\mathbf{r},t) \otimes \mathbf{w}(\mathbf{r},t) \cdot \mathbf{n}^{\pi\alpha}(\mathbf{r},t) \, da_{mic} \right] dV \ . \tag{3.79}$$

Employing the definition of the mass averaging operator (3.51) for rewriting the first integral on the right hand side of (3.79),

$$\overline{\boldsymbol{\psi}}^\pi(\mathbf{x},t) \rho_\pi(\mathbf{x},t) \, dv = \int_{dv} \rho(\mathbf{r},t) \boldsymbol{\psi}(\mathbf{r},t) \gamma^\pi(\mathbf{r},t) \, dv_{mic} \ , \tag{3.80}$$

finally yields the following form for the first term in the general macroscopic balance equation (3.68):

$$\int_V \left[\frac{1}{dv} \int_{dv} \frac{\partial}{\partial t} [\rho(\mathbf{r},t)\,\boldsymbol{\psi}(\mathbf{r},t)]\,\gamma^\pi(\mathbf{r},t)\,dv_{mic} \right] dV =$$

$$= \int_V \frac{\partial}{\partial t} \left[\rho_\pi(\mathbf{x},t)\,\overline{\boldsymbol{\psi}}^\pi(\mathbf{x},t) \right] dV -$$

$$- \int_V \left[\frac{1}{dv} \sum_{\alpha \neq \pi}^{K} \int_{da^{\pi\alpha}} \rho(\mathbf{r},t)\,\boldsymbol{\psi}(\mathbf{r},t) \otimes \mathbf{w}(\mathbf{r},t) \cdot \mathbf{n}^{\pi\alpha}(\mathbf{r},t)\,da_{mic} \right] dV . \quad (3.81)$$

The second term of the macroscopic balance equation (3.68), containing the divergence of the quantity $\boldsymbol{\psi}(\mathbf{r},t)$, is decomposed into two parts by means of the previously derived relation (3.74):

$$\int_V \left[\frac{1}{dv} \int_{dv} \mathrm{div}[\rho(\mathbf{r},t)\,\boldsymbol{\psi}(\mathbf{r},t) \otimes \dot{\mathbf{r}}(\mathbf{r},t)]\,\gamma^\pi(\mathbf{r},t)\,dv_{mic} \right] dV =$$

$$= \int_A \left[\frac{1}{da} \int_{da} \rho(\mathbf{r},t)\,\boldsymbol{\psi}(\mathbf{r},t) \otimes \dot{\mathbf{r}}(\mathbf{r},t) \cdot \mathbf{n}(\mathbf{r},t)\,\gamma^\pi(\mathbf{r},t)\,da_{mic} \right] dA +$$

$$+ \int_V \left[\frac{1}{dv} \sum_{\alpha \neq \pi}^{K} \int_{da^{\pi\alpha}} \rho(\mathbf{r},t)\,\boldsymbol{\psi}(\mathbf{r},t) \otimes \dot{\mathbf{r}}(\mathbf{r},t) \cdot \mathbf{n}^{\pi\alpha}(\mathbf{r},t)\,da_{mic} \right] dV . \quad (3.82)$$

By means of the definition (3.75) for the deviation of a microscopic quantity from its mass averaged value the product $\boldsymbol{\psi}(\mathbf{r},t)\,\dot{\mathbf{r}}(\mathbf{r},t)$ of the second term in equation (3.82) can be subdivided into four parts,

$$\boldsymbol{\psi}(\mathbf{r},t) \otimes \dot{\mathbf{r}}(\mathbf{r},t) = \left[\overline{\boldsymbol{\psi}}^\pi(\mathbf{x},t) + \hat{\boldsymbol{\psi}}^\pi(\mathbf{x},\boldsymbol{\xi},t) \right] \otimes \left[\overline{\dot{\mathbf{r}}}^\pi(\mathbf{x},t) + \hat{\dot{\mathbf{r}}}^\pi(\mathbf{x},\boldsymbol{\xi},t) \right] =$$

$$= \overline{\boldsymbol{\psi}}^\pi(\mathbf{x},t) \otimes \overline{\dot{\mathbf{r}}}^\pi(\mathbf{x},t) + \overline{\boldsymbol{\psi}}^\pi(\mathbf{x},t) \otimes \hat{\dot{\mathbf{r}}}^\pi(\mathbf{x},\boldsymbol{\xi},t) +$$

$$+ \hat{\boldsymbol{\psi}}^\pi(\mathbf{x},\boldsymbol{\xi},t) \otimes \overline{\dot{\mathbf{r}}}^\pi(\mathbf{x},t) + \hat{\boldsymbol{\psi}}^\pi(\mathbf{x},\boldsymbol{\xi},t) \otimes \hat{\dot{\mathbf{r}}}^\pi(\mathbf{x},\boldsymbol{\xi},t) . \quad (3.83)$$

Inserting these four terms of equation (3.83) into the integral over the representative elementary area da of (3.82) yields five integrals for the second term of the general macroscopic balance law (3.68), namely:

$$\int_V \left[\frac{1}{dv}\int_{dv} \mathrm{div}[\rho(\mathbf{r},t)\,\boldsymbol{\psi}(\mathbf{r},t)\otimes\dot{\mathbf{r}}(\mathbf{r},t)]\,\gamma^\pi(\mathbf{r},t)\,dv_{mic}\right]dV =$$

$$=\int_A \left[\frac{1}{da}\int_{da} \rho(\mathbf{r},t)\,\overline{\boldsymbol{\psi}}^\pi(\mathbf{x},t)\otimes\overline{\dot{\mathbf{r}}}^\pi(\mathbf{x},t)\cdot\mathbf{n}(\mathbf{r},t)\,\gamma^\pi(\mathbf{r},t)\,da_{mic}\right]dA +$$

$$+\int_A \left[\frac{1}{da}\int_{da} \rho(\mathbf{r},t)\,\overline{\boldsymbol{\psi}}^\pi(\mathbf{x},t)\otimes\hat{\dot{\mathbf{r}}}^\pi(\mathbf{x},\boldsymbol{\xi},t)\cdot\mathbf{n}(\mathbf{r},t)\,\gamma^\pi(\mathbf{r},t)\,da_{mic}\right]dA +$$

$$+\int_A \left[\frac{1}{da}\int_{da} \rho(\mathbf{r},t)\,\hat{\boldsymbol{\psi}}^\pi(\mathbf{x},\boldsymbol{\xi},t)\otimes\overline{\dot{\mathbf{r}}}^\pi(\mathbf{x},t)\cdot\mathbf{n}(\mathbf{r},t)\,\gamma^\pi(\mathbf{r},t)\,da_{mic}\right]dA +$$

$$+\int_A \left[\frac{1}{da}\int_{da} \rho(\mathbf{r},t)\,\hat{\boldsymbol{\psi}}^\pi(\mathbf{x},\boldsymbol{\xi},t)\otimes\hat{\dot{\mathbf{r}}}^\pi(\mathbf{x},\boldsymbol{\xi},t)\cdot\mathbf{n}(\mathbf{r},t)\,\gamma^\pi(\mathbf{r},t)\,da_{mic}\right]dA +$$

$$+\int_V \left[\frac{1}{dv}\sum_{\alpha\neq\pi}^{\kappa}\int_{da^{\pi\alpha}} \rho(\mathbf{r},t)\,\boldsymbol{\psi}(\mathbf{r},t)\otimes\dot{\mathbf{r}}(\mathbf{r},t)\cdot\mathbf{n}^{\pi\alpha}(\mathbf{r},t)\,da_{mic}\right]dV . \qquad (3.84)$$

However, the second and the third term on the right hand side of equation (3.84) are zero because identity (3.78) must hold. Rewriting the first term on the right hand side of (3.84) by means of the divergence theorem proposed by Eringen and Suhubi (relationship (3.32)) and employing the definition of the volume phase average operator (3.45) for the density $\rho(\mathbf{r},t)$ hereafter (note that the mass averaged quantities $\overline{\boldsymbol{\psi}}^\pi(\mathbf{x},t)$ and $\overline{\dot{\mathbf{r}}}^\pi(\mathbf{x},t)$ are constant in the integral over dv and thus can be put in front of it) results in

$$\int_A \left[\frac{1}{da}\int_{da} \rho(\mathbf{r},t)\,\overline{\boldsymbol{\psi}}^\pi(\mathbf{x},t)\otimes\overline{\dot{\mathbf{r}}}^\pi(\mathbf{x},t)\cdot\mathbf{n}(\mathbf{r},t)\,\gamma^\pi(\mathbf{r},t)\,da_{mic}\right]dA =$$

$$=\int_V \mathrm{div}\left[\frac{1}{dv}\int_{dv} \rho(\mathbf{r},t)\,\overline{\boldsymbol{\psi}}^\pi(\mathbf{x},t)\otimes\overline{\dot{\mathbf{r}}}^\pi(\mathbf{x},t)\,\gamma^\pi(\mathbf{r},t)\,dv_{mic}\right]dV =$$

$$=\int_V \mathrm{div}\left[\rho_\pi(\mathbf{x},t)\,\overline{\boldsymbol{\psi}}^\pi(\mathbf{x},t)\otimes\overline{\dot{\mathbf{r}}}^\pi(\mathbf{x},t)\right]dV . \qquad (3.85)$$

Finally, the second term of the general balance equation (3.68) can be written as

$$\int_V \left[\frac{1}{dv} \int_{dv} \mathrm{div}[\rho(\mathbf{r},t)\,\boldsymbol{\psi}(\mathbf{r},t) \otimes \dot{\mathbf{r}}(\mathbf{r},t)]\,\gamma^\pi(\mathbf{r},t)\,dv_{mic} \right] dV =$$

$$= \int_V \mathrm{div}\left[\rho_\pi(\mathbf{x},t)\,\overline{\boldsymbol{\psi}}^\pi(\mathbf{x},t) \otimes \overline{\dot{\mathbf{r}}}^\pi(\mathbf{x},t)\right] dV +$$

$$+ \int_A \left[\frac{1}{da} \int_{da} \rho(\mathbf{r},t)\,\hat{\boldsymbol{\psi}}^\pi(\mathbf{x},\boldsymbol{\xi},t) \otimes \hat{\dot{\mathbf{r}}}^\pi(\mathbf{x},\boldsymbol{\xi},t) \cdot \mathbf{n}(\mathbf{r},t)\,\gamma^\pi(\mathbf{r},t)\,da_{mic} \right] dA +$$

$$+ \int_V \left[\frac{1}{dv} \sum_{\alpha\neq\pi}^{\kappa} \int_{da^{\pi\alpha}} \rho(\mathbf{r},t)\,\boldsymbol{\psi}(\mathbf{r},t) \otimes \dot{\mathbf{r}}(\mathbf{r},t) \cdot \mathbf{n}^{\pi\alpha}(\mathbf{r},t)\,da_{mic} \right] dV . \qquad (3.86)$$

The third term of the balance law (3.68), containing the divergence of $\boldsymbol{\Phi}(\mathbf{r},t)$ where $\boldsymbol{\Phi}(\mathbf{r},t)$ is associated with the traction vector $\boldsymbol{\vartheta}$, is divided into two parts in a manner similar to the second term of (3.68) using the integral expression (3.74):

$$\int_V \left[\frac{1}{dv} \int_{dv} \mathrm{div}\,\boldsymbol{\Phi}(\mathbf{r},t)\,\gamma^\pi(\mathbf{r},t)\,dv_{mic} \right] dV =$$

$$= \int_A \left[\frac{1}{da} \int_{da} \boldsymbol{\Phi}(\mathbf{r},t) \cdot \mathbf{n}(\mathbf{r},t)\,\gamma^\pi(\mathbf{r},t)\,da_{mic} \right] dA +$$

$$+ \int_V \left[\frac{1}{dv} \sum_{\alpha\neq\pi}^{\kappa} \int_{da^{\pi\alpha}} \boldsymbol{\Phi}(\mathbf{r},t) \cdot \mathbf{n}^{\pi\alpha}(\mathbf{r},t)\,da_{mic} \right] dV . \qquad (3.87)$$

Finally, the definition of the mass averaging operator (3.51) is applied to the last term of the general macroscopic balance law (3.68), containing the body forces $\boldsymbol{\beta}(\mathbf{r},t)$, which yields

$$\int_V \left[\frac{1}{dv} \int_{dv} \rho(\mathbf{r},t)\,\boldsymbol{\beta}(\mathbf{r},t)\,\gamma^\pi(\mathbf{r},t)\,dv_{mic} \right] dV = \int_V \rho_\pi(\mathbf{x},t)\,\overline{\boldsymbol{\beta}}^\pi(\mathbf{x},t)\,dV . \qquad (3.88)$$

The above performed modifications of the four terms of the general macroscopic balance equation (3.68), i.e. the formulae (3.81), (3.86), (3.87) and (3.88), are now substituted into (3.68). After rearranging some of the terms the following form of the macroscopic balance equation (3.68) for the general thermodynamic property $\boldsymbol{\psi}(\mathbf{r},t)$ is obtained:

$$\int_V \frac{\partial}{\partial t}\left[\rho_\pi(\mathbf{x},t)\overline{\boldsymbol{\psi}}^\pi(\mathbf{x},t)\right]dV + \int_V \mathrm{div}\left[\rho_\pi(\mathbf{x},t)\overline{\boldsymbol{\psi}}^\pi(\mathbf{x},t)\otimes\overline{\dot{\mathbf{r}}}^\pi(\mathbf{x},t)\right]dV -$$

$$-\int_V \left[\frac{1}{dv}\sum_{\alpha\neq\pi}^\kappa \int_{da^{\pi\alpha}} \rho(\mathbf{r},t)\boldsymbol{\psi}(\mathbf{r},t)\otimes[\mathbf{w}(\mathbf{r},t)-\dot{\mathbf{r}}(\mathbf{r},t)]\cdot\mathbf{n}^{\pi\alpha}(\mathbf{r},t)\,da_{mic}\right]dV -$$

$$-\int_V \left[\frac{1}{dv}\sum_{\alpha\neq\pi}^\kappa \int_{da^{\pi\alpha}} \boldsymbol{\Phi}(\mathbf{r},t)\cdot\mathbf{n}^{\pi\alpha}(\mathbf{r},t)\,da_{mic}\right]dV -$$

$$-\int_A \left[\frac{1}{da}\int_{da}\left[\boldsymbol{\Phi}(\mathbf{r},t)-\rho(\mathbf{r},t)\hat{\boldsymbol{\psi}}^\pi(\mathbf{x},\boldsymbol{\xi},t)\otimes\hat{\dot{\mathbf{r}}}^\pi(\mathbf{x},\boldsymbol{\xi},t)\right]\cdot\mathbf{n}(\mathbf{r},t)\gamma^\pi(\mathbf{r},t)\,da_{mic}\right]dA -$$

$$-\int_V \rho_\pi(\mathbf{x},t)\overline{\boldsymbol{\beta}}^\pi(\mathbf{x},t)\,dV = \mathbf{0}\,. \tag{3.89}$$

Further modifications of (3.89) can be achieved by introducing the three subsequent definitions concerning the third, fourth and fifth term. The third integral of (3.89) accounts for the flux of $\rho(\mathbf{r},t)\boldsymbol{\psi}(\mathbf{r},t)$ between the π-phase and all the other phases α within the REV which is expressed by first integrating over the $\pi\alpha$-interface $da^{\pi\alpha}$ and then summing up over all the constituents κ. If the $\pi\alpha$-interface is a surface such that there is no flux through it, the product $[\mathbf{w}(\mathbf{r},t)-\dot{\mathbf{r}}(\mathbf{r},t)]\cdot\mathbf{n}^{\pi\alpha}(\mathbf{r},t)\,da_{mic}$, and thus also the integral, will be identical to zero. According to [Hassanizadeh(1979a)], this term includes "interphase transport as well as change of state, e.g., evaporation". The density function $\mathbf{e}_{\rho\psi}^\pi(\mathbf{x},t)$ for the exchange of $\boldsymbol{\psi}(\mathbf{r},t)$ between the constituent π and the other constituents α is stated as

$$\mathbf{e}_{\rho\psi}^\pi(\mathbf{x},t) = \frac{1}{\rho_\pi(\mathbf{x},t)\,dv}\sum_{\alpha\neq\pi}^\kappa \int_{da^{\pi\alpha}} \rho(\mathbf{r},t)\boldsymbol{\psi}(\mathbf{r},t)\otimes[\mathbf{w}(\mathbf{r},t)-\dot{\mathbf{r}}(\mathbf{r},t)]\cdot\mathbf{n}^{\pi\alpha}(\mathbf{r},t)\,da_{mic}\,. \tag{3.90}$$

The notation is herein adopted that $\mathbf{e}_{\rho\psi}^\pi(\mathbf{x},t)$ has the same tensorial character as the subscript product $\rho\boldsymbol{\psi}$, e.g., $\mathbf{e}_{\rho\dot{\mathbf{r}}}^\pi$ is a vector as $\dot{\mathbf{r}}$ is.

The second-order tensor $\boldsymbol{\Phi}(\mathbf{r},t)$ in the fourth integral of (3.89) is associated with the microscopic traction vector $\boldsymbol{\vartheta}$ (equation (3.59)). Hence, this term accounts for the mechanical interaction between the π-phase and all the other phases α (again there are the integration over the $\pi\alpha$-interface $da^{\pi\alpha}$ and the subsequent summation to be performed). A density function $\vartheta^\pi(\mathbf{x},t)$ for the mechanical interaction between the π and all α-phases may be defined as

$$\vartheta^\pi(\mathbf{x},t) = \frac{1}{\rho_\pi(\mathbf{x},t)\,dv} \sum_{\alpha\neq\pi}^{\kappa} \int_{da^{\pi\alpha}} \mathbf{\Phi}(\mathbf{r},t) \cdot \mathbf{n}^{\pi\alpha}(\mathbf{r},t)\,da_{mic}\ . \qquad (3.91)$$

The fifth integral contains two portions, both related to the boundary da of the representative elementary volume. One part is associated with the tractions acting on the boundary, the other part denotes the flux of $\rho(\mathbf{r},t)\,\hat{\psi}^\pi(\mathbf{x},\boldsymbol{\xi},t)$ across the boundary da of the REV. The latter arises from the deviation of a quantity defined in $\mathbf{r}$ (arbitrary point within the REV) from the mass average of the respective quantity referred to $\mathbf{x}$ (centre of the REV, see Figure 3.1(a)). Due to the presence of the phase distribution function $\gamma^\pi(\mathbf{r},t)$ in the integrand the integration over da actually catches the area da^π covered by the constituent π. According to [Hassanizadeh(1979a)], the integral over the area da of the REV in the fifth term of (3.89) represents the "total, macroscopically non-convective flux of the property $\psi(\mathbf{r},t)$". Hence, it is assumed that there exists an average flux density $\check{\vartheta}^\pi(\mathbf{x},t)$, defined as

$$\check{\vartheta}^\pi(\mathbf{x},t) = \frac{1}{da}\int_{da} \left[\mathbf{\Phi}(\mathbf{r},t) - \rho(\mathbf{r},t)\,\hat{\psi}^\pi(\mathbf{x},\boldsymbol{\xi},t) \otimes \hat{\mathbf{r}}^\pi(\mathbf{x},\boldsymbol{\xi},t)\right]\cdot \mathbf{n}(\mathbf{r},t)\,\gamma^\pi(\mathbf{r},t)\,da_{mic}\ , \qquad (3.92)$$

which is conceived to be linearly related to the unit normal vector $\mathbf{n}(\mathbf{x},t)$ of a macroscopic boundary by means of a second-order tensor $\mathbf{\Phi}_\pi(\mathbf{x},t)$:

$$\check{\vartheta}^\pi(\mathbf{x},t) = \mathbf{\Phi}_\pi(\mathbf{x},t)\cdot\mathbf{n}(\mathbf{x},t)\ . \qquad (3.93)$$

A proof of the existence of the relationship (3.93) is to be found in the literature [Hassanizadeh(1979a)].

After the above considerations about some of the terms in (3.89) a brief summarising interpretation of the general macroscopic balance equation (3.89) can be given. The time rate of change of a property $\rho_\pi(\mathbf{x},t)\,\overline{\psi}^\pi(\mathbf{x},t)$ within the volume V, expressed by the first integral, has to be balanced by several portions: a first part accounting for the flux of the quantity $\rho_\pi(\mathbf{x},t)\,\overline{\psi}^\pi(\mathbf{x},t)$ across the boundary A of the volume V (second integral – note that the divergence theorem has already been applied), a second part (third integral) which regards for transport phenomena between the individual phases on the microscale, a third part (fourth integral) taking into account mechanical interactions between the constituents on the microscale, a fourth part (fifth integral) addressing existing tractions and surface fluxes with respect to the boundary of the REV and finally a fifth part (last integral) which accounts for any body forces present in the domain. It should be noted that an additional term reflecting the production of the quantity $\psi(\mathbf{r}, t)$ within V has to be included in equation (3.89) when deriving the entropy inequality from this general balance law.

Employing the definitions (3.90), (3.91) and (3.92) to the general macroscopic balance equation (3.89) results in its following more concise form,

$$\int_V \frac{\partial}{\partial t} \left[\rho_\pi(\mathbf{x}, t) \overline{\psi}^\pi(\mathbf{x}, t) \right] dV + \int_V \operatorname{div} \left[\rho_\pi(\mathbf{x}, t) \overline{\psi}^\pi(\mathbf{x}, t) \otimes \bar{\mathbf{r}}^\pi(\mathbf{x}, t) \right] dV -$$

$$- \int_V \rho_\pi(\mathbf{x}, t) \mathbf{e}^\pi_{\rho\psi}(\mathbf{x}, t) \, dV - \int_V \rho_\pi(\mathbf{x}, t) \vartheta^\pi(\mathbf{x}, t) \, dV -$$

$$- \int_A \check{\vartheta}^\pi(\mathbf{x}, t) \, dA - \int_V \rho_\pi(\mathbf{x}, t) \overline{\boldsymbol{\beta}}^\pi(\mathbf{x}, t) \, dV = \mathbf{0} \, . \qquad (3.94)$$

Making use of the relationship (3.93) and the divergence theorem (3.28$_2$) to reformulate the integral over the surface A,

$$\int_A \check{\vartheta}^\pi(\mathbf{x}, t) \, dA = \int_A \boldsymbol{\Phi}_\pi(\mathbf{x}, t) \cdot \mathbf{n}(\mathbf{x}, t) \, dA = \int_V \operatorname{div} \boldsymbol{\Phi}_\pi(\mathbf{x}, t) \, dV \, , \qquad (3.95)$$

finally yields the general global form of the macroscopic balance equation which serves as a basis for deducing all the individual balance principles for a continuous domain consisting of several phases π:

$$\int_V \left[\frac{\partial}{\partial t} \left[\rho_\pi(\mathbf{x}, t) \overline{\psi}^\pi(\mathbf{x}, t) \right] + \operatorname{div} \left[\rho_\pi(\mathbf{x}, t) \overline{\psi}^\pi(\mathbf{x}, t) \otimes \bar{\mathbf{r}}^\pi(\mathbf{x}, t) \right] \right] dV -$$

$$- \int_V \left[\operatorname{div} \boldsymbol{\Phi}_\pi(\mathbf{x}, t) + \rho_\pi(\mathbf{x}, t) \left[\mathbf{e}^\pi_{\rho\psi}(\mathbf{x}, t) + \vartheta^\pi(\mathbf{x}, t) + \overline{\boldsymbol{\beta}}^\pi(\mathbf{x}, t) \right] \right] dV = \mathbf{0} \, . \qquad (3.96)$$

Next, the microscopic interfacial condition (equation (3.67)) is utilised to obtain more information about the exchange terms appearing in (3.96). An integration of (3.67) first over $da^{\pi\alpha}$ and $da^{\alpha\pi}$, respectively, and then over V yields terms similar to the third and fourth integral in equation (3.89) accounting for interfacial phenomena:

$$\int_V \left[\frac{1}{dv} \sum_{\alpha \neq \pi}^{\kappa} \int_{da^{\pi\alpha}} \rho(\mathbf{r},t)\, \boldsymbol{\psi}(\mathbf{r},t) \otimes [\mathbf{w}(\mathbf{r},t) - \dot{\mathbf{r}}(\mathbf{r},t)] \cdot \mathbf{n}^{\pi\alpha}(\mathbf{r},t)\, da_{mic} \right] dV +$$

$$+ \int_V \left[\frac{1}{dv} \sum_{\alpha \neq \pi}^{\kappa} \int_{da^{\pi\alpha}} \boldsymbol{\Phi}(\mathbf{r},t) \cdot \mathbf{n}^{\pi\alpha}(\mathbf{r},t)\, da_{mic} \right] dV +$$

$$+ \int_V \left[\frac{1}{dv} \sum_{\alpha \neq \pi}^{\kappa} \int_{da^{\alpha\pi}} \rho(\mathbf{r},t)\, \boldsymbol{\psi}(\mathbf{r},t) \otimes [\mathbf{w}(\mathbf{r},t) - \dot{\mathbf{r}}(\mathbf{r},t)] \cdot \mathbf{n}^{\alpha\pi}(\mathbf{r},t)\, da_{mic} \right] dV +$$

$$+ \int_V \left[\frac{1}{dv} \sum_{\alpha \neq \pi}^{\kappa} \int_{da^{\alpha\pi}} \boldsymbol{\Phi}(\mathbf{r},t) \cdot \mathbf{n}^{\alpha\pi}(\mathbf{r},t)\, da_{mic} \right] dV = \mathbf{0} \,. \qquad (3.97)$$

It should be noted that the integration over $da^{\pi\alpha}$ and $da^{\alpha\pi}$, respectively, results from considering both phases π and α constituting the $\pi\alpha$-interface. Summation over all κ constituents together with an invocation of (3.90) and (3.91) yields

$$\int_V \sum_{\pi=1}^{\kappa} \rho_\pi(\mathbf{x},t) \left[\mathbf{e}_{\rho\psi}^\pi(\mathbf{x},t) + \boldsymbol{\vartheta}^\pi(\mathbf{x},t) \right] dV = \mathbf{0} \,. \qquad (3.98)$$

This equation, derived from the microscopic interfacial condition, asserts the conservation of the thermodynamic property $\boldsymbol{\psi}(\mathbf{r},t)$ for the whole body.

The general global balance equations (3.96) and (3.98) are valid only if there is no macroscopic surface of discontinuity present in the domain under consideration. In this case a differential form of the balance law (3.96) can be derived under certain smoothness conditions. However, if a macroscopic surface of discontinuity Σ exists in the body, the volume V is divided into regions V^+ and V^- and the area A into parts A^+ and A^-. Equation (3.96) then has to be employed for both regions V^+ and V^-. Similar to the microscopic case these equations are added up and after the application of a generalised transport theorem a general balance equation valid for the parts V^+ and V^- and a jump condition for Σ are obtained. Hence, in the presence of macroscopic discontinuities the balance equations can be expressed according to the following theorem [Hassanizadeh(1979a)]:

Theorem I: A κ-phase system, occupying the space V and being bounded by a surface A, can be considered as the superposition of κ continua which exchange thermomechanical properties through phase changes and mechanical interactions. At every point $\mathbf{x}$ in the domain the thermomechanical quantity $\psi(\mathbf{r},t)$ is conserved according to the following equation:

$$\frac{\partial}{\partial t}\left[\rho_\pi(\mathbf{x},t)\overline{\psi}^\pi(\mathbf{x},t)\right] + \mathrm{div}\left[\rho_\pi(\mathbf{x},t)\overline{\psi}^\pi(\mathbf{x},t) \otimes \overline{\dot{\mathbf{r}}}^\pi(\mathbf{x},t)\right] - \mathrm{div}\,\boldsymbol{\Phi}_\pi(\mathbf{x},t) -$$
$$- \rho_\pi(\mathbf{x},t)\left[\mathbf{e}^\pi_{\rho\psi}(\mathbf{x},t) + \boldsymbol{\vartheta}^\pi(\mathbf{x},t) + \overline{\boldsymbol{\beta}}^\pi(\mathbf{x},t)\right] = \mathbf{0} \qquad \forall\ \mathbf{x} \in V^+ \oplus V^-, \qquad (3.99)$$

subject to

$$\sum_{\pi=1}^{\kappa} \rho_\pi(\mathbf{x},t)\left[\mathbf{e}^\pi_{\rho\psi}(\mathbf{x},t) + \boldsymbol{\vartheta}^\pi(\mathbf{x},t)\right] = \mathbf{0} \qquad \forall\ \mathbf{x} \in V^+ \oplus V^- \qquad (3.100)$$

and

$$\left[\rho_\pi(\mathbf{x},t)\overline{\psi}^\pi(\mathbf{x},t) \otimes \left[\overline{\dot{\mathbf{r}}}^\pi(\mathbf{x},t) - \overline{\mathbf{w}}(\mathbf{x},t)\right] - \boldsymbol{\Phi}_\pi(\mathbf{x},t)\right]_\Sigma \cdot \mathbf{n}|_- = 0 \qquad \forall\ \mathbf{x} \in \Sigma. \qquad (3.101)$$

In (3.99) and (3.100) the $\oplus$ sign denotes that the particular equation is valid in both continuous parts V^+ and V^- of the domain. In (3.101) the quantity $\overline{\mathbf{w}}(\mathbf{x},t)$ is the velocity of the macroscopic surface of discontinuity Σ. $[\ldots]_\Sigma$ describes the jump of the respective function across the surface of discontinuity Σ and $\mathbf{n}|_-$ is the unit normal vector of Σ pointing out of V^+ and into V^-. The general balance equation (3.99) does not include terms accounting for the production of the property $\psi(\mathbf{r},t)$ which would be necessary in order to derive the entropy inequality from (3.99). Additionally, it should be noted that equation (3.100) does not contain any term due to surface exchange of the property $\psi(\mathbf{r},t)$ between the phases. This is in accordance with the assumption that the interfaces do not possess thermomechanical properties. Otherwise, the right hand side of the relationship (3.100) would be non-zero. As a concluding remark it should be mentioned that the derived general balance laws correspond to those found in the continuum theory of mixtures [Boer(1991b)] where the balance equations have one of the forms (3.96) or (3.99).

3.4 General formulation of the various balance equations

3.4.1 Introduction

In this section the general global and local forms of the microscopic as well as the macroscopic balance equations and jump conditions are specialised for mass, linear momentum and angular momentum. The appropriate definitions for the thermodynamic quantity ψ and its corresponding properties are stated for each case and inserted into the equations (3.57), (3.63), (3.67) and (3.94), (3.99), (3.100), (3.101) to obtain the microscopic and the macroscopic balance equations and jump conditions, respectively. Detailed derivations of the energy balance and the entropy inequality can be found in the literature [Hassanizadeh(1979b)] and are omitted here. Only the balance laws necessary for the finite element treatment of the particular three-phase problem are given. It should be mentioned that the formulation in this section is not yet associated with a particular material, i.e. no specific assumptions for the constitutive behaviour are introduced. However, in this and in the subsequent parts of the work, it is no longer explicitly indicated that all the various quantities are functions of the coordinates $\mathbf{r}$ or $\mathbf{x}$ (remember that $\mathbf{r}$ indicates an arbitrary point within the representative elementary volume and $\mathbf{x}$ refers to the REVs' centre, see Figure 3.1(a)) and of the time t.

3.4.2 Microscopic mass balance

When taking a look at the general microscopic balance equation (3.57) and having in mind that the mass balance law is an equation relating scalar quantities, it becomes obvious that ψ has to be a scalar for the deduction of the mass balance law from (3.57). Additionally, the mass balance equation describes the time rate of change of the density of a constituent within a certain fixed amount of volume. Consequently, considering the left hand side integral of (3.57) or (3.62), the property ψ has to be chosen as

$$\psi = 1 \ . \tag{3.102}$$

Since any tractions and body forces present in the domain do not enter into the mass balance, the conditions

$$\boldsymbol{\vartheta} = \mathbf{0} \tag{3.103}$$

for the tractions and

$$\boldsymbol{\beta} = \mathbf{0} \tag{3.104}$$

for the body forces have to hold. For the choices (3.102) to (3.104) confer also Table 3.2. Taking into account the relationship (3.59), the second-order tensor $\boldsymbol{\Phi}$ which is related to the traction vector $\boldsymbol{\vartheta}$, has to vanish according to (3.103),

$$\boldsymbol{\Phi} = \mathbf{0} \ . \tag{3.105}$$

Using the equations (3.102) to (3.105) in combination with (3.62) results in the global form of the microscopic mass balance for the π-phase (which is the same for all constituents),

$$\int_{dv} \frac{\partial \rho}{\partial t} dv_{mic} = -\int_{dv} \text{div}(\rho \dot{\mathbf{r}}) dv_{mic} . \qquad (3.106)$$

The local form of the microscopic mass balance equation for the π-phase is obtained from (3.63) as

$$\frac{\partial \rho}{\partial t} + \text{div}(\rho \dot{\mathbf{r}}) = 0 \qquad (3.107)$$

and the corresponding jump condition on the interface between the π and α-phases from (3.67) as

$$[\rho(\mathbf{w} - \dot{\mathbf{r}})]\big|_{\pi} \cdot \mathbf{n}^{\pi\alpha} + [\rho(\mathbf{w} - \dot{\mathbf{r}})]\big|_{\alpha} \cdot \mathbf{n}^{\alpha\pi} = 0 , \qquad (3.108)$$

where $|_\pi$ and $|_\alpha$ indicate that the preceding terms have to be evaluated for the π and the α-phase, respectively.

3.4.3 Macroscopic mass balance

Two basic statements valid for the microscopic mass balance also apply to the derivation of the macroscopic mass balance law from (3.94): (i) The macroscopic mass balance equation contains scalar quantities and (ii) it describes the time rate of change of the density within a fixed volume, however, in this case it is the volume phase average of the density of the constituent π. Consequently, $\overline{\psi}^\pi$ also has to be chosen as a scalar, similar to the microscopic case,

$$\overline{\psi}^\pi = 1 . \qquad (3.109)$$

For convenience the substitution

$$\overline{\dot{\mathbf{r}}}^\pi = \overline{\mathbf{v}}^\pi \qquad (3.110)$$

is made for the mass averaged velocity in the macroscopic balance equations. The third integral in (3.94), accounting for transport phenomena between the individual constituents on the microscale, has to be considered in the mass balance equation. Inserting (3.102) into the definition (3.90) yields

$$e_\rho^\pi = \frac{1}{\rho_\pi dv} \sum_{\alpha \neq \pi}^{\kappa} \int_{da^{\pi\alpha}} \rho(\mathbf{w} - \dot{\mathbf{r}}) \cdot \mathbf{n}^{\pi\alpha} da_{mic} , \qquad (3.111)$$

where ρ and $\dot{\mathbf{r}}$ denote the density and the velocity of the material present at a spatial point $\mathbf{r}$ on the microscale, respectively. Equation (3.111) describes the flux of ρ (or the transport of mass) between the phase π and all the other constituents α across the $\pi\alpha$-interfaces on the microscopic level. However, the ratio $1/(\rho_\pi \, dv)$ relates this flux to the total mass of the π-phase within the REV which is easily recognised when taking into account the equations (3.45) and (3.47): $\rho_\pi \, dv = \int_{dv} \rho\gamma^\pi \, dv_{mic} = \int_{dv^\pi} \rho \, dv_{mic} = m^\pi$, the mass of the constituent π within dv. The fourth integral in (3.94) contains the traction vector $\boldsymbol{\vartheta}^\pi$ defined in (3.91) which is related to the second-order tensor $\boldsymbol{\Phi}$ of the microscale. According to (3.105), $\boldsymbol{\Phi}$ is zero for the mass balance equation which, from (3.91), yields

$$\boldsymbol{\vartheta}^\pi = \mathbf{0} \tag{3.112}$$

and thus the fourth integral in (3.94) vanishes. Consideration of the definition of $\check{\boldsymbol{\vartheta}}^\pi$ in (3.92) allows for an interpretation of the surface integral in the general balance law (3.94). $\boldsymbol{\Phi}$ in (3.92) again is zero and, since $\psi = 1 = const.$ is valid for the balance of mass, the deviation $\hat{\psi}^\pi$ of the quantity ψ between the spatial points $\mathbf{r}$ and $\mathbf{x}$ vanishes. Therefore,

$$\check{\boldsymbol{\vartheta}}^\pi = \mathbf{0} \tag{3.113}$$

and hence the integral over the surface A does not enter into the mass balance equation. The last integral in (3.94) referring to body forces present in the domain V does not have to be taken into account for deducing the mass balance equation either which means

$$\overline{\boldsymbol{\beta}}^\pi = \mathbf{0} \ . \tag{3.114}$$

When making use of the above expressions (3.109) to (3.114) in connection with the general balance law (3.94), the global formulation of the macroscopic mass balance equation for the constituent π in a κ-phase system occupying a volume V can be written as

$$\int_V \frac{\partial \rho_\pi}{\partial t} \, dV + \int_V \mathrm{div}(\rho_\pi \overline{\mathbf{v}}^\pi) \, dV - \int_V \rho_\pi \, e_\rho^\pi \, dV = 0 \ , \tag{3.115}$$

where e_ρ^π is defined according to (3.111). The local macroscopic mass balance equation, again employing (3.109) to (3.114), is deduced from (3.99) as

$$\frac{\partial \rho_\pi}{\partial t} + \mathrm{div}(\rho_\pi \overline{\mathbf{v}}^\pi) = \rho_\pi \, e_\rho^\pi \qquad \forall \ \mathbf{x} \in V^+ \oplus V^- \ , \tag{3.116}$$

where the second-order tensor $\boldsymbol{\Phi}_\pi$ in (3.99) is assumed to be zero,

$$\boldsymbol{\Phi}_\pi = \mathbf{0} \ , \tag{3.117}$$

due to definition (3.93) and relation (3.113). Equation (3.116), however, is subject to the constraint

$$\sum_{\pi=1}^{\kappa} \rho_\pi \, e_\rho^\pi = 0 \qquad \forall \ \mathbf{x} \in V^+ \oplus V^- \ , \tag{3.118}$$

which is obtained from the local form (3.100). Relationship (3.118) reflects the fact that the microscale transport of mass between all κ constituents within the domain has to sum up to zero. For a macroscopic surface of discontinuity Σ, moving with the velocity $\overline{\mathbf{w}}$, it follows from (3.101) that

$$\left[\rho_\pi(\overline{\mathbf{v}}^\pi - \overline{\mathbf{w}})\right]_\Sigma \cdot \mathbf{n}|_- = 0 \qquad \forall \ \mathbf{x} \in \Sigma \ . \tag{3.119}$$

When applying the definition of the material time derivative to the averaged density, i.e. formulating equation (3.11) with $f^\pi = \rho_\pi$, and employing the identity

$$\mathrm{div}(\rho_\pi \overline{\mathbf{v}}^\pi) = \rho_\pi \, \mathrm{div}\, \overline{\mathbf{v}}^\pi + \mathrm{grad}\, \rho_\pi \cdot \overline{\mathbf{v}}^\pi \tag{3.120}$$

for the second term in (3.116), the local macroscopic mass balance equation (3.116) can be rewritten as

$$\frac{D^\pi \rho_\pi}{Dt} + \rho_\pi \, \mathrm{div}\, \overline{\mathbf{v}}^\pi = \rho_\pi \, e_\rho^\pi \qquad \forall \ \mathbf{x} \in V^+ \oplus V^- \ . \tag{3.121}$$

Inserting the relationship between the volume phase average ρ_π and the intrinsic volume phase average ρ^π for the density, as stated in equation (3.49), finally yields

$$\frac{D^\pi (\eta^\pi \rho^\pi)}{Dt} + \eta^\pi \rho^\pi \, \mathrm{div}\, \overline{\mathbf{v}}^\pi = \eta^\pi \rho^\pi \, e_\rho^\pi \qquad \forall \ \mathbf{x} \in V^+ \oplus V^- \ . \tag{3.122}$$

3.4.4 Microscopic linear momentum balance

Contrary to the mass balance equations in the previous sections the linear momentum balance laws are equations containing vector quantities. Thus, for instance (3.57) straightforward indicates that $\boldsymbol{\psi}$ has to be a vector when deriving the microscopic linear momentum balance equation from (3.57). In addition to that, in basic physics the linear momentum of a body is defined as the product of mass and velocity. Consequently, considering the

Chapter 3. Mechanics of the coupled three-phase model

integral on the left hand side of the equations (3.57) or (3.62), ψ has to be equal to the velocity of a point $\mathbf{r}$ on the microscale,

$$\psi = \dot{\mathbf{r}} . \tag{3.123}$$

The second and the third integral on the right hand side of (3.57) are referred to as terms arising from surface tractions and body forces present in the domain, respectively. Of course such contributions enter into the linear momentum balance. Hence, the integral over the surface da of the representative elementary volume implies ϑ to be interpreted as a vector $\mathbf{t}_{mic}$ of surface tractions on the microscopic level,

$$\vartheta = \mathbf{t}_{mic} , \tag{3.124}$$

whereas in the last integral over the volume dv of the REV the vector $\boldsymbol{\beta}$ symbolises the gravitational acceleration $\mathbf{g}$,

$$\boldsymbol{\beta} = \mathbf{g} , \tag{3.125}$$

and the term accounts for the body forces present in the domain. For the substitutions (3.123) to (3.125) confer also Table 3.2. According to equation (3.59), the microscopic stress vector $\mathbf{t}_{mic}$ is related to a second-order tensor $\boldsymbol{\Phi}$ which might be viewed as the microscopic stress tensor $\boldsymbol{\sigma}_{mic}$, that is $\boldsymbol{\Phi} = \boldsymbol{\sigma}_{mic}$, and therefore,

$$\mathbf{t}_{mic} = \boldsymbol{\sigma}_{mic} \cdot \mathbf{n} . \tag{3.126}$$

Keeping in mind the above considerations and employing the expressions (3.123) to (3.125), a global form of the microscopic linear momentum balance equation for the constituent π can be deduced from (3.57) as

$$\frac{D}{Dt} \int_{dv} \rho \dot{\mathbf{r}} \, dv_{mic} = -\int_{da} \rho \dot{\mathbf{r}} \otimes (\dot{\mathbf{r}} - \mathbf{w}_b) \cdot \mathbf{n} \, da_{mic} + \int_{da} \mathbf{t}_{mic} \, da_{mic} + \int_{dv} \rho \mathbf{g} \, dv_{mic} . \tag{3.127}$$

A global formulation containing the microscopic stress tensor $\boldsymbol{\sigma}_{mic}$ instead of the traction vector $\mathbf{t}_{mic}$ is obtained from (3.62) as

$$\int_{dv} \frac{\partial (\rho \dot{\mathbf{r}})}{\partial t} \, dv_{mic} = -\int_{dv} \operatorname{div}(\rho \dot{\mathbf{r}} \otimes \dot{\mathbf{r}}) \, dv_{mic} + \int_{dv} \operatorname{div} \boldsymbol{\sigma}_{mic} \, dv_{mic} + \int_{dv} \rho \mathbf{g} \, dv_{mic} . \tag{3.128}$$

Using (3.123) to (3.126) in combination with equation (3.63) yields a local form of the microscopic linear momentum balance law for the π-phase,

$$\frac{\partial (\rho \dot{\mathbf{r}})}{\partial t} + \operatorname{div}(\rho \dot{\mathbf{r}} \otimes \dot{\mathbf{r}}) - \operatorname{div} \boldsymbol{\sigma}_{mic} - \rho \mathbf{g} = \mathbf{0} . \tag{3.129}$$

The corresponding jump condition for the interface between the two constituents π and α is obtained from the general equation (3.67) as

$$[\boldsymbol{\sigma}_{mic} - \rho\dot{\mathbf{r}} \otimes (\dot{\mathbf{r}} - \mathbf{w})]\big|_{\pi} \cdot \mathbf{n}^{\pi\alpha} + [\boldsymbol{\sigma}_{mic} - \rho\dot{\mathbf{r}} \otimes (\dot{\mathbf{r}} - \mathbf{w})]\big|_{\alpha} \cdot \mathbf{n}^{\alpha\pi} = \mathbf{0} \,, \tag{3.130}$$

with $|_\pi$ and $|_\alpha$ indicating that the preceding terms correspond to the π and the α-phase, respectively.

The local microscopic balance law (3.129) may be rearranged as follows: The time derivative of the first term is calculated as

$$\frac{\partial(\rho\dot{\mathbf{r}})}{\partial t} = \frac{\partial \rho}{\partial t}\dot{\mathbf{r}} + \rho\frac{\partial \dot{\mathbf{r}}}{\partial t} \,. \tag{3.131}$$

According to equation (3.9), writing the microscopic velocity $\dot{\mathbf{r}}$ instead of $\mathbf{v}^\pi$ and the microscopic acceleration $\ddot{\mathbf{r}}$ instead of $\mathbf{a}^\pi$, the partial time derivative of $\dot{\mathbf{r}}$ may be formulated as

$$\frac{\partial \dot{\mathbf{r}}}{\partial t} = \ddot{\mathbf{r}} - \operatorname{grad}\dot{\mathbf{r}} \cdot \dot{\mathbf{r}} \,. \tag{3.132}$$

Inserting (3.132) into (3.131) yields for the first term of equation (3.129)

$$\frac{\partial(\rho\dot{\mathbf{r}})}{\partial t} = \frac{\partial \rho}{\partial t}\dot{\mathbf{r}} + \rho\ddot{\mathbf{r}} - \rho\operatorname{grad}\dot{\mathbf{r}} \cdot \dot{\mathbf{r}} \,. \tag{3.133}$$

The second term of the local microscopic linear momentum balance law (3.129) is written as

$$\operatorname{div}(\rho\dot{\mathbf{r}} \otimes \dot{\mathbf{r}}) = \dot{\mathbf{r}} \otimes \dot{\mathbf{r}} \cdot \operatorname{grad}\rho + \rho\operatorname{grad}\dot{\mathbf{r}} \cdot \dot{\mathbf{r}} + \rho\dot{\mathbf{r}}\operatorname{div}\dot{\mathbf{r}} \,. \tag{3.134}$$

Substituting the relationship

$$\operatorname{div}(\rho\dot{\mathbf{r}}) = \rho\operatorname{div}\dot{\mathbf{r}} + \operatorname{grad}\rho \cdot \dot{\mathbf{r}} \tag{3.135}$$

into the microscopic mass balance law (3.107) and multiplying the resulting equation by $\dot{\mathbf{r}}$ yields

$$\frac{\partial \rho}{\partial t}\dot{\mathbf{r}} + \rho\dot{\mathbf{r}}\operatorname{div}\dot{\mathbf{r}} + \dot{\mathbf{r}} \otimes \dot{\mathbf{r}} \cdot \operatorname{grad}\rho = \mathbf{0} \,. \tag{3.136}$$

When using (3.133), (3.134) and (3.136) to rearrange equation (3.129), finally the local form of the microscopic linear momentum balance law is obtained as

$$\operatorname{div}\boldsymbol{\sigma}_{mic} + \rho\,(\mathbf{g} - \ddot{\mathbf{r}}) = \mathbf{0} \,, \tag{3.137}$$

which is of course identical to the familiar form of the linear momentum balance for a single-phase material.

3.4.5 Macroscopic linear momentum balance

Similar to the microscopic case, the macroscopic linear momentum balance law is an equation containing vector quantities. Additionally, the physical definition of the linear momentum as a product of mass times velocity has to be satisfied. Therefore, the first integral in equation (3.94) immediately yields the expression for $\overline{\psi}^\pi$,

$$\overline{\psi}^\pi = \overline{\mathbf{v}}^\pi \;, \tag{3.138}$$

which means that $\overline{\psi}^\pi$ is equal to the mass averaged velocity of the π-phase. Like in the macroscopic mass balance the choice

$$\overline{\dot{\mathbf{r}}}^\pi = \overline{\mathbf{v}}^\pi \tag{3.139}$$

is convenient here as well. Considering the third integral in (3.94) an examination of the definition (3.90) for $\mathbf{e}^\pi_{\rho\psi}$ provides more insight into this term. For the linear momentum balance (3.90) becomes

$$\mathbf{e}^\pi_{\rho\psi} = \mathbf{e}^\pi_{\rho\dot{\mathbf{r}}} = \frac{1}{\rho_\pi \, dv} \sum_{\alpha \neq \pi}^{\kappa} \int_{da^{\pi\alpha}} \rho \, \dot{\mathbf{r}} \otimes (\mathbf{w} - \dot{\mathbf{r}}) \cdot \mathbf{n}^{\pi\alpha} \, da_{mic} \;. \tag{3.140}$$

The velocity $\dot{\mathbf{r}}$ in an arbitrary point $\mathbf{r}$ on the microscopic level can be split into two parts when taking into account definition (3.75),

$$\dot{\mathbf{r}} = \overline{\dot{\mathbf{r}}}^\pi + \hat{\dot{\mathbf{r}}}^\pi = \overline{\mathbf{v}}^\pi + \hat{\dot{\mathbf{r}}}^\pi \;. \tag{3.141}$$

Substituting (3.141) into (3.140) yields the two portions of $\mathbf{e}^\pi_{\rho\dot{\mathbf{r}}}$,

$$\mathbf{e}^\pi_{\rho\dot{\mathbf{r}}} = \frac{\overline{\mathbf{v}}^\pi}{\rho_\pi \, dv} \sum_{\alpha \neq \pi}^{\kappa} \int_{da^{\pi\alpha}} \rho \, (\mathbf{w} - \dot{\mathbf{r}}) \cdot \mathbf{n}^{\pi\alpha} \, da_{mic} + \frac{1}{\rho_\pi \, dv} \sum_{\alpha \neq \pi}^{\kappa} \int_{da^{\pi\alpha}} \rho \, \hat{\dot{\mathbf{r}}}^\pi \otimes (\mathbf{w} - \dot{\mathbf{r}}) \cdot \mathbf{n}^{\pi\alpha} \, da_{mic} \;, \tag{3.142}$$

where for obtaining the first term in (3.142) the fact that a mass averaged quantity is constant within an integral over a volume or an area has been considered. The two parts in equation (3.142) can be interpreted as follows: The first term describes the flux of ρ (or the transport of mass) across all the $\pi\alpha$-interfaces within the representative elementary volume; this amount of mass is multiplied by the mass averaged velocity $\overline{\mathbf{v}}^\pi$ which yields a linear momentum term. The second part of (3.142), called an 'intrinsic momentum' term in [Hassanizadeh(1979b)], accounts for the additional linear momentum due to the deviation of the velocity in the spatial point $\mathbf{r}$ from its mass averaged counterpart defined in $\mathbf{x}$. Both terms are related to the mass of the constituent π within the REV. According to (3.142) and (3.111), the subsequent short notation may be introduced,

$$\mathbf{e}^\pi_{\rho\dot{\mathbf{r}}} = e^\pi_\rho \overline{\mathbf{V}}^\pi + \mathbf{e}^\pi_{\rho\hat{\dot{\mathbf{r}}}^\pi} \ . \tag{3.143}$$

To investigate the fourth integral in (3.94), the definition (3.91) should be considered. Using the specification (3.126) for the linear momentum balance, it reads as

$$\boldsymbol{\vartheta}^\pi = \frac{1}{\rho_\pi \, dv} \sum_{\alpha \neq \pi}^\kappa \int_{da^{\pi\alpha}} \boldsymbol{\sigma}_{mic} \cdot \mathbf{n}^{\pi\alpha} \, da_{mic} \ , \tag{3.144}$$

which easily implies the interpretation that this term accounts for mechanical interactions between the π and α-phases along the $\pi\alpha$-interfaces within the REV, i.e. the mechanical interactions on the microscopic level. For the sake of a concise notation in the final balance equation a stress vector $\mathbf{t}^\pi$ may be defined according to (3.144),

$$\boldsymbol{\vartheta}^\pi = \mathbf{t}^\pi \ . \tag{3.145}$$

The fifth integral in (3.94) over the surface A can be interpreted as follows. Employing the expression (3.92) the integrand $\check{\boldsymbol{\vartheta}}^\pi$ can be written as

$$\check{\boldsymbol{\vartheta}}^\pi = \frac{1}{da} \int_{da} \left(\boldsymbol{\sigma}_{mic} - \rho\, \hat{\dot{\mathbf{r}}}^\pi \otimes \hat{\dot{\mathbf{r}}}^\pi \right) \cdot \mathbf{n}\, \gamma^\pi \, da_{mic} = \frac{1}{da} \int_{da^\pi} \left(\boldsymbol{\sigma}_{mic} - \rho\, \hat{\dot{\mathbf{r}}}^\pi \otimes \hat{\dot{\mathbf{r}}}^\pi \right) \cdot \mathbf{n}\, da_{mic} \ , \tag{3.146}$$

where the definition (3.37) for the phase distribution function has been applied. Due to the presence of the phase distribution function γ^π in (3.146$_1$) the integration only covers the area da^π of the constituent π. Considering (3.146) it becomes obvious that $\check{\boldsymbol{\vartheta}}^\pi$ accounts for the mechanical interaction between the portion of the π-phase lying inside the representative elementary volume and its counterpart outside the REV. This interaction term $\check{\boldsymbol{\vartheta}}^\pi$ is constituted by two parts: one due to the presence of microscopic stresses (microscopic stress tensor $\boldsymbol{\sigma}_{mic}$), the other one due to the flux of $\rho\, \hat{\dot{\mathbf{r}}}^\pi$ across the boundary da^π of the π-phase (linear momentum arising from the deviation $\hat{\dot{\mathbf{r}}}^\pi$ of the microscopic velocity $\dot{\mathbf{r}}$ in an arbitrary point $\mathbf{r}$ from the mass averaged velocity $\overline{\dot{\mathbf{r}}}^\pi$ referring to the point $\mathbf{x}$ defining the centre of the REV). When considering the macroscopic linear momentum balance, $\check{\boldsymbol{\vartheta}}^\pi$ can be viewed as a partial (macroscopic) stress vector $\check{\mathbf{t}}^\pi$,

$$\check{\boldsymbol{\vartheta}}^\pi = \check{\mathbf{t}}^\pi \ . \tag{3.147}$$

According to relationship (3.93), a partial stress tensor $\boldsymbol{\sigma}_\pi$ is connected to the stress vector $\check{\mathbf{t}}^\pi$ such that $\check{\mathbf{t}}^\pi$ is a linear function of the normal vector $\mathbf{n}$ of the surface [Hassanizadeh(1979b)],

$$\check{\mathbf{t}}^\pi = \boldsymbol{\sigma}_\pi \cdot \mathbf{n} \ . \tag{3.148}$$

Finally, equation (3.88) can be employed to gain more information about the last term (sixth integral) in (3.94). Thus, $\overline{\boldsymbol{\beta}}^\pi$ is calculated as

$$\overline{\boldsymbol{\beta}}^\pi = \frac{1}{\rho_\pi\, dv} \int\limits_{dv} \rho\, \mathbf{g}\, \gamma^\pi \, dv_{mic} = \frac{1}{\rho_\pi} \mathbf{g}\, \rho_\pi = \mathbf{g} \ , \tag{3.149}$$

where $\boldsymbol{\beta} = \mathbf{g}$ has been substituted according to (3.125) and the definition of the volume phase average (3.45) has been applied with $\zeta = \rho$.

By means of the above expressions (3.138), (3.139), (3.143), (3.145), (3.147) and (3.149) the global form of the macroscopic linear momentum balance equation for the constituent π is obtained from (3.94) as

$$\int\limits_V \frac{\partial(\rho_\pi \overline{\mathbf{v}}^\pi)}{\partial t}\, dV + \int\limits_V \operatorname{div}(\rho_\pi \overline{\mathbf{v}}^\pi \otimes \overline{\mathbf{v}}^\pi)\, dV - \int\limits_V \rho_\pi \left(e_\rho^\pi \overline{\mathbf{v}}^\pi + \mathbf{e}_{\rho\hat{\mathbf{r}}^\pi}^\pi \right) dV -$$

$$- \int\limits_V \rho_\pi \mathbf{t}^\pi\, dV - \int\limits_A \check{\mathbf{t}}^\pi\, dA - \int\limits_V \rho_\pi \mathbf{g}\, dV = \mathbf{0} \ . \tag{3.150}$$

Using relationship (3.148), the integral over the surface A in (3.150) may be transformed into an integral over the volume V according to the divergence theorem as given in (3.95),

$$\int\limits_A \check{\mathbf{t}}^\pi\, dA = \int\limits_A \boldsymbol{\sigma}_\pi \cdot \mathbf{n}\, dA = \int\limits_V \operatorname{div} \boldsymbol{\sigma}_\pi\, dV \ . \tag{3.151}$$

Substituting (3.151) into (3.150) and reordering terms finally yields the following global macroscopic linear momentum balance equation,

$$\int\limits_V \left[\frac{\partial(\rho_\pi \overline{\mathbf{v}}^\pi)}{\partial t} + \operatorname{div}(\rho_\pi \overline{\mathbf{v}}^\pi \otimes \overline{\mathbf{v}}^\pi) \right] dV - \int\limits_V \operatorname{div} \boldsymbol{\sigma}_\pi\, dV -$$

$$- \int\limits_V \rho_\pi \left(e_\rho^\pi \overline{\mathbf{v}}^\pi + \mathbf{e}_{\rho\hat{\mathbf{r}}^\pi}^\pi + \mathbf{t}^\pi + \mathbf{g} \right) dV = \mathbf{0} \ . \tag{3.152}$$

Since (3.152) has to hold for an arbitrary volume V, the local (differential) macroscopic linear momentum balance equation can be deduced from (3.152) as

$$\frac{\partial(\rho_\pi \overline{\mathbf{v}}^\pi)}{\partial t} + \mathrm{div}(\rho_\pi \overline{\mathbf{v}}^\pi \otimes \overline{\mathbf{v}}^\pi) - \mathrm{div}\,\boldsymbol{\sigma}_\pi -$$

$$- \rho_\pi \left(e_\rho^\pi \overline{\mathbf{v}}^\pi + \mathbf{e}_{\rho\hat{\mathbf{r}}^\pi}^\pi + \mathbf{t}^\pi + \mathbf{g} \right) = 0 \qquad \forall\, \mathbf{x} \in V^+ \oplus V^-, \qquad (3.153)$$

which of course can be compared to the general local balance law (3.99). Equation (3.153) is subject to the following constraint resulting from (3.100),

$$\sum_{\pi=1}^{\kappa} \rho_\pi \left(e_\rho^\pi \overline{\mathbf{v}}^\pi + \mathbf{e}_{\rho\hat{\mathbf{r}}^\pi}^\pi + \mathbf{t}^\pi \right) = 0 \qquad \forall\, \mathbf{x} \in V^+ \oplus V^-. \qquad (3.154)$$

According to (3.101), for a macroscopic surface of discontinuity Σ the relation

$$\left[\rho_\pi \overline{\mathbf{v}}^\pi \otimes (\overline{\mathbf{v}}^\pi - \overline{\mathbf{w}}) - \boldsymbol{\sigma}_\pi \right]_\Sigma \cdot \mathbf{n}\big|_- = 0 \qquad \forall\, \mathbf{x} \in \Sigma \qquad (3.155)$$

has to hold.

A rearrangement procedure similar to the one performed for the microscopic linear momentum balance equation is employed to gain a more convenient form of the macroscopic linear momentum balance law (3.153). The first term is differentiated as

$$\frac{\partial(\rho_\pi \overline{\mathbf{v}}^\pi)}{\partial t} = \frac{\partial \rho_\pi}{\partial t} \overline{\mathbf{v}}^\pi + \rho_\pi \frac{\partial \overline{\mathbf{v}}^\pi}{\partial t}. \qquad (3.156)$$

Equation (3.9) may be written in terms of mass averaged quantities $\overline{\mathbf{v}}^\pi$ and $\overline{\mathbf{a}}^\pi$ as

$$\frac{\partial \overline{\mathbf{v}}^\pi}{\partial t} = \overline{\mathbf{a}}^\pi - \mathrm{grad}\,\overline{\mathbf{v}}^\pi \cdot \overline{\mathbf{v}}^\pi. \qquad (3.157)$$

Substitution of (3.157) into (3.156) yields

$$\frac{\partial(\rho_\pi \overline{\mathbf{v}}^\pi)}{\partial t} = \frac{\partial \rho_\pi}{\partial t} \overline{\mathbf{v}}^\pi + \rho_\pi \overline{\mathbf{a}}^\pi - \rho_\pi \mathrm{grad}\,\overline{\mathbf{v}}^\pi \cdot \overline{\mathbf{v}}^\pi. \qquad (3.158)$$

The second term of (3.153) is split into three parts as

$$\mathrm{div}(\rho_\pi \overline{\mathbf{v}}^\pi \otimes \overline{\mathbf{v}}^\pi) = \overline{\mathbf{v}}^\pi \otimes \overline{\mathbf{v}}^\pi \cdot \mathrm{grad}\,\rho_\pi + \rho_\pi \mathrm{grad}\,\overline{\mathbf{v}}^\pi \cdot \overline{\mathbf{v}}^\pi + \rho_\pi \overline{\mathbf{v}}^\pi \mathrm{div}\,\overline{\mathbf{v}}^\pi. \qquad (3.159)$$

The mass balance law (3.116) is employed similar to the microscopic case. Insertion of the relationship (3.120) into the macroscopic mass balance equation (3.116) and a subsequent multiplication by $\overline{\mathbf{v}}^\pi$ results in the identity

$$\frac{\partial \rho_\pi}{\partial t} \overline{\mathbf{v}}^\pi + \rho_\pi \overline{\mathbf{v}}^\pi \mathrm{div}\,\overline{\mathbf{v}}^\pi + \overline{\mathbf{v}}^\pi \otimes \overline{\mathbf{v}}^\pi \cdot \mathrm{grad}\,\rho_\pi = \rho_\pi e_\rho^\pi \overline{\mathbf{v}}^\pi. \qquad (3.160)$$

Chapter 3. Mechanics of the coupled three-phase model

Finally, the macroscopic linear momentum balance equation is obtained from (3.153) by means of the relations (3.158) to (3.160) as

$$\text{div}\,\boldsymbol{\sigma}_\pi + \rho_\pi(\mathbf{g} - \overline{\mathbf{a}}^\pi) + \rho_\pi\left(\mathbf{e}^\pi_{\hat{\rho}\hat{\mathbf{r}}^\pi} + \mathbf{t}^\pi\right) = \mathbf{0} \qquad \forall\,\mathbf{x} \in V^+ \oplus V^-. \tag{3.161}$$

In equation (3.161) $\text{div}\,\boldsymbol{\sigma}_\pi$ denotes the divergence of the partial stress tensor for the π-phase, $\rho_\pi \mathbf{g}$ and $-\rho_\pi \overline{\mathbf{a}}^\pi$ account for gravitational and inertia effects, respectively. These three terms usually constitute the linear momentum balance for a single-phase medium. The last term in brackets on the left hand side of (3.161) refers to certain interaction effects due to the microstructure of the multi-phase medium under consideration which have already been explained in some more detail in connection with the relations (3.142) and (3.144). Using the volume and area fractions (3.40) and (3.42) together with equation (3.49) yields

$$\text{div}\left(\overline{\eta^\pi}\boldsymbol{\sigma}^\pi\right) + \eta^\pi\rho^\pi(\mathbf{g} - \overline{\mathbf{a}}^\pi) + \eta^\pi\rho^\pi\left(\mathbf{e}^\pi_{\hat{\rho}\hat{\mathbf{r}}^\pi} + \mathbf{t}^\pi\right) = \mathbf{0} \qquad \forall\,\mathbf{x} \in V^+ \oplus V^-. \tag{3.162}$$

Note that the area fraction $\overline{\eta^\pi}$ was employed in (3.49) instead of the volume fraction η^π for rewriting the area-related partial stress tensor $\boldsymbol{\sigma}_\pi$ present in (3.161).

3.4.6 Microscopic angular momentum balance

For the partially saturated porous medium under consideration all the constituents are assumed to be microscopically non-polar materials, i.e. no internal spin momenta and no body or surface couples act on the domain. For this case the angular momentum (sometimes also called moment of momentum or rotational momentum) balance, in combination with the linear momentum and the mass balance equations, may be employed to prove the symmetry of the stress tensor.

The angular momentum is defined as a momentum related to a fixed point in space (could, e.g., be the origin O in Figure 3.1). The distance between this reference position and any arbitrary location in the domain may be denoted by $\mathbf{r}$ (see Figure 3.1(a)). From a mathematical point of view the angular momentum balance equation contains vector quantities similar to the linear momentum. However, all the individual momentum terms in this case have to be combined with the distance $\mathbf{r}$ between the microscopic volume element and the reference point which, to conserve the vector nature of the equation, is to be done by a vector cross product.

To deduce the microscopic angular momentum balance equation from the general microscopic balance law (3.57), the specifications of the quantities ψ, ϑ and β employed for the linear momentum can be used in a cross product with the vector **r** describing the distance of the particle from the origin O (Figure 3.1(a)). Thus, according to the equations (3.123) to (3.125) of the linear momentum balance, one obtains for the thermodynamic property ψ,

$$\psi = \mathbf{r} \times \dot{\mathbf{r}}, \qquad (3.163)$$

for the traction vector ϑ,

$$\vartheta = \mathbf{r} \times \mathbf{t}_{mic}, \qquad (3.164)$$

and for β,

$$\beta = \mathbf{r} \times \mathbf{g}. \qquad (3.165)$$

The expressions (3.163) to (3.165) are also summarised in Table 3.2. Insertion of (3.163) to (3.165) into the general global balance law (3.57) yields

$$\frac{D}{Dt} \int_{dv} \rho (\mathbf{r} \times \dot{\mathbf{r}}) \, dv_{mic} = -\int_{da} \rho (\mathbf{r} \times \dot{\mathbf{r}}) \otimes (\dot{\mathbf{r}} - \mathbf{w}_b) \cdot \mathbf{n} \, da_{mic} +$$

$$+ \int_{da} (\mathbf{r} \times \mathbf{t}_{mic}) \, da_{mic} + \int_{dv} \rho (\mathbf{r} \times \mathbf{g}) \, dv_{mic}. \qquad (3.166)$$

The material time derivative of the integral on the left hand side of (3.166) can be rewritten by means of the Reynolds' transport theorem (3.33):

$$\frac{D}{Dt} \int_{dv} \rho (\mathbf{r} \times \dot{\mathbf{r}}) \, dv_{mic} = \int_{dv} \frac{\partial [\rho (\mathbf{r} \times \dot{\mathbf{r}})]}{\partial t} \, dv_{mic} + \int_{da} \rho (\mathbf{r} \times \dot{\mathbf{r}}) \otimes \mathbf{w}_b \cdot \mathbf{n} \, da_{mic}. \qquad (3.167)$$

The first integral on the right hand side of (3.166) is split into two parts: The one containing the velocity $\mathbf{w}_b$ of the boundary cancels with the second term in (3.167), the other one is reformulated employing the divergence theorem,

$$-\int_{da} \rho (\mathbf{r} \times \dot{\mathbf{r}}) \otimes \dot{\mathbf{r}} \cdot \mathbf{n} \, da_{mic} = -\int_{dv} \operatorname{div} [\rho \dot{\mathbf{r}} \otimes (\mathbf{r} \times \dot{\mathbf{r}})] \, dv_{mic}. \qquad (3.168)$$

According to (3.126), the microscopic stress vector $\mathbf{t}_{mic}$ is related to the microscopic stress tensor $\boldsymbol{\sigma}_{mic}$. In combination with the extended version of the divergence theorem (3.29) this relationship may be used to transform the second integral on the right hand side of (3.166) over the surface da into an integral over the volume dv of the representative elementary volume,

$$\int_{da} (\mathbf{r} \times \mathbf{t}_{mic}) \, da_{mic} = \int_{da} \mathbf{r} \times (\boldsymbol{\sigma}_{mic} \cdot \mathbf{n}) \, da_{mic} = \int_{dv} (\mathbf{r} \times \operatorname{div} \boldsymbol{\sigma}_{mic} + \mathcal{E} : \boldsymbol{\sigma}_{mic}^T) \, dv_{mic}, \quad (3.169)$$

where $\mathcal{E}$ is the third-order permutation tensor defined in (3.30). Substituting (3.167) to (3.169) into (3.166) yields the following global form of the microscopic angular momentum balance equation,

$$\int_{dv} \frac{\partial [\rho (\mathbf{r} \times \dot{\mathbf{r}})]}{\partial t} \, dv_{mic} = - \int_{dv} \operatorname{div} [\rho \dot{\mathbf{r}} \otimes (\mathbf{r} \times \dot{\mathbf{r}})] \, dv_{mic} +$$

$$+ \int_{dv} \mathbf{r} \times \operatorname{div} \boldsymbol{\sigma}_{mic} \, dv_{mic} + \int_{dv} \mathcal{E} : \boldsymbol{\sigma}_{mic}^T \, dv_{mic} + \int_{dv} \rho (\mathbf{r} \times \mathbf{g}) \, dv_{mic}. \quad (3.170)$$

Since (3.170) has to hold for any arbitrary volume dv, the local form of the microscopic angular momentum balance is obtained from (3.170) as

$$\frac{\partial [\rho (\mathbf{r} \times \dot{\mathbf{r}})]}{\partial t} + \operatorname{div}[\rho \dot{\mathbf{r}} \otimes (\mathbf{r} \times \dot{\mathbf{r}})] - \mathbf{r} \times \operatorname{div} \boldsymbol{\sigma}_{mic} - \mathcal{E} : \boldsymbol{\sigma}_{mic}^T - \mathbf{r} \times \rho \mathbf{g} = 0. \quad (3.171)$$

It should be mentioned that the additional term in (3.170) compared to the general microscopic balance law (3.62) arises from the cross product present in the second integral on the right hand side of (3.166) over the surface da which yields two terms instead of one when applying the divergence theorem as a comparison of (3.169) with (3.60) shows. The jump condition at a microscopic surface of discontinuity Σ_{mic} is derived by establishing the angular momentum balance for both individual parts of the domain adjacent to the surface Σ_{mic}. A summation of these equations finally results in the following local form of the jump condition at the $\pi\alpha$-interface,

$$[\mathbf{r} \times (\boldsymbol{\sigma}_{mic} \cdot \mathbf{n}^{\pi\alpha}) - \rho (\mathbf{r} \times \dot{\mathbf{r}}) \otimes (\dot{\mathbf{r}} - \mathbf{w}) \cdot \mathbf{n}^{\pi\alpha}]\big|_{\pi} +$$

$$+ [\mathbf{r} \times (\boldsymbol{\sigma}_{mic} \cdot \mathbf{n}^{\alpha\pi}) - \rho (\mathbf{r} \times \dot{\mathbf{r}}) \otimes (\dot{\mathbf{r}} - \mathbf{w}) \cdot \mathbf{n}^{\alpha\pi}]\big|_{\alpha} = 0. \quad (3.172)$$

To arrive at the desired result of this section, i.e. the proof of the symmetry of the microscopic stress tensor $\boldsymbol{\sigma}_{mic}$, some modifications of the local microscopic angular momentum balance equation (3.171) are performed subsequently. The time derivative of the first term in (3.171) is calculated as

$$\frac{\partial [\rho (\mathbf{r} \times \dot{\mathbf{r}})]}{\partial t} = \frac{\partial \rho}{\partial t} (\mathbf{r} \times \dot{\mathbf{r}}) + \rho \left(\frac{\partial \mathbf{r}}{\partial t} \times \dot{\mathbf{r}} \right) + \rho \left(\mathbf{r} \times \frac{\partial \dot{\mathbf{r}}}{\partial t} \right), \quad (3.173)$$

and the first divergence term reads as

$$\text{div}[\rho \dot{\mathbf{r}} \otimes (\mathbf{r} \times \dot{\mathbf{r}})] = \rho \,\text{div}\,\dot{\mathbf{r}}\,(\mathbf{r} \times \dot{\mathbf{r}}) + (\text{grad}\,\rho \cdot \dot{\mathbf{r}})(\mathbf{r} \times \dot{\mathbf{r}}) +$$
$$+ \rho \,(\text{grad}\,\mathbf{r} \cdot \dot{\mathbf{r}}) \times \dot{\mathbf{r}} + \rho \,(\text{grad}\,\dot{\mathbf{r}} \cdot \dot{\mathbf{r}}) \times \mathbf{r} \,. \tag{3.174}$$

Substituting the relations (3.173) and (3.174) into the balance law (3.171) yields

$$\frac{\partial \rho}{\partial t}(\mathbf{r} \times \dot{\mathbf{r}}) + \rho \left(\frac{\partial \mathbf{r}}{\partial t} \times \dot{\mathbf{r}} \right) + \rho \left(\mathbf{r} \times \frac{\partial \dot{\mathbf{r}}}{\partial t} \right) + \rho \,\text{div}\,\dot{\mathbf{r}}\,(\mathbf{r} \times \dot{\mathbf{r}}) +$$
$$+ (\text{grad}\,\rho \cdot \dot{\mathbf{r}})\,(\mathbf{r} \times \dot{\mathbf{r}}) + \rho\,(\text{grad}\,\mathbf{r} \cdot \dot{\mathbf{r}}) \times \dot{\mathbf{r}} + \rho\,(\text{grad}\,\dot{\mathbf{r}} \cdot \dot{\mathbf{r}}) \times \mathbf{r} -$$
$$- \mathbf{r} \times \text{div}\,\boldsymbol{\sigma}_{mic} - \mathcal{E} : \boldsymbol{\sigma}_{mic}^T - \mathbf{r} \times \rho \mathbf{g} = \mathbf{0} \,. \tag{3.175}$$

The second and the sixth term in (3.175) are zero because of $(\dot{\mathbf{r}} \times \dot{\mathbf{r}}) = \mathbf{0}$. Then, terms can be collected to obtain the more convenient form

$$\left(\frac{\partial \rho}{\partial t} + \rho \,\text{div}\,\dot{\mathbf{r}} + \text{grad}\,\rho \cdot \dot{\mathbf{r}} \right)(\mathbf{r} \times \dot{\mathbf{r}}) + \rho \mathbf{r} \times \left(\frac{\partial \dot{\mathbf{r}}}{\partial t} + \text{grad}\,\dot{\mathbf{r}} \cdot \dot{\mathbf{r}} \right) -$$
$$- \mathbf{r} \times \text{div}\,\boldsymbol{\sigma}_{mic} - \mathcal{E} : \boldsymbol{\sigma}_{mic}^T - \mathbf{r} \times \rho \mathbf{g} = \mathbf{0} \,. \tag{3.176}$$

Using the equations (3.132) and (3.135), the balance law (3.176) can be written as

$$\left[\frac{\partial \rho}{\partial t} + \text{div}(\rho \dot{\mathbf{r}}) \right](\mathbf{r} \times \dot{\mathbf{r}}) - \mathbf{r} \times [\text{div}\,\boldsymbol{\sigma}_{mic} + \rho\,(\mathbf{g} - \ddot{\mathbf{r}})] - \mathcal{E} : \boldsymbol{\sigma}_{mic}^T = \mathbf{0} \,. \tag{3.177}$$

In (3.177) the first term in square brackets denotes the microscopic mass balance equation (3.107), the second term in square brackets is identical to the microscopic linear momentum balance equation (3.137). Thus, these two terms are equal to zero and the local microscopic angular momentum balance law reduces to

$$\mathcal{E} : \boldsymbol{\sigma}_{mic}^T = \mathbf{0} \,, \tag{3.178}$$

which holds at each point of the domain and for all times t. In index notation the double contraction $\mathcal{E} : \boldsymbol{\sigma}_{mic}^T$ yields a vector with the components

$$\epsilon_{ijk}\,(\sigma_{mic})_{kj} = 0 \,, \tag{3.179}$$

ϵ_{ijk} being the permutation symbol defined in (3.31). For the individual values of i

equation (3.179) can be written as follows:

$$i = 1: \quad \sigma_{32} - \sigma_{23} = 0 ,$$
$$i = 2: \quad \sigma_{13} - \sigma_{31} = 0 , \qquad (3.180)$$
$$i = 3: \quad \sigma_{21} - \sigma_{12} = 0 .$$

The relations (3.178) or (3.179) are satisfied if and only if the microscopic stress tensor $\boldsymbol{\sigma}_{mic}$ is symmetric:

$$\boldsymbol{\sigma}_{mic} = \boldsymbol{\sigma}_{mic}^T \quad \text{or} \quad (\sigma_{mic})_{jk} = (\sigma_{mic})_{kj} . \qquad (3.181)$$

Equation (3.181) is also known as Cauchy's second law of motion [Malvern(1969), Holzapfel(2000)]. Note that for a polar material the symmetry property (3.181) does not hold any longer.

3.4.7 Macroscopic angular momentum balance

To prove the symmetry of the stress tensor on the macroscopic level, various approaches are possible. A procedure similar to the one for the microscopic stress tensor (Section 3.4.6) may be employed, i.e. establishing the angular momentum balance equation for the domain and then using the mass and linear momentum balance laws to simplify this equation. A detailed derivation of the symmetry properties of the macroscale stress tensor according to this approach can be found in [Hassanizadeh(1979b)]. Alternatively, the fact that the microscopic stress tensor is symmetric for microscopically non-polar media (as proved in the previous Section 3.4.6) may be exploited directly. This second approach is discussed subsequently.

As used in the linear momentum balance equation a partial stress tensor $\boldsymbol{\sigma}_\pi$ on the macroscale is defined according to the relations (3.146) to (3.148). When considering (3.146), the twofold composition of $\boldsymbol{\sigma}_\pi$ becomes obvious, i.e. the microscopic stress part and the surface flux part which both constitute the macroscopic stress tensor. Hence, if $\boldsymbol{\sigma}_{mic}$ in (3.146) is symmetric for a microscopically non-polar medium then, because of the symmetry of the flux term $\rho\,\hat{\mathbf{r}}^\pi \otimes \hat{\mathbf{r}}^\pi$, the macroscopic stress tensor has to be symmetric as well such that

$$\boldsymbol{\sigma}_\pi = \boldsymbol{\sigma}_\pi^T \quad \text{or} \quad (\sigma_\pi)_{jk} = (\sigma_\pi)_{kj} \qquad (3.182)$$

in vector or index notation are valid for all locations $\mathbf{x}$ of the domain and for all points of time t [Gray(1998)].

3.5 Constitutive equations for a partially saturated porous medium

3.5.1 Introduction

After the derivation of the individual balance equations for mass, linear momentum and angular momentum the specification of a set of constitutive relations is now needed in order to complete the description of the mechanical behaviour of the multi-phase medium in general and the three-phase model in particular. The general formulation of the balance equations in the previous Section 3.4 allows for an introduction of quite elaborate constitutive theories to take into account different phenomena such as temperature or chemical effects or second-grade materials for the solid phase, depending on the type of problem under consideration. An even more sophisticated model would be one with balance equations not only for the bulk material but also for the interfaces or common lines [Gray(1998)].

Of primary importance is a proper formulation of the constitutive models, i.e. they have to be based on quantities measurable in laboratory tests or field situations. Since the current work is more application oriented, the constitutive theories are introduced by direct postulation of appropriate relationships rather than by a systematic but lengthy exploitation of the entropy inequality. The latter approach is treated in detail in the respective literature, e.g., [Gray(1999)]. However, most of the relations introduced in the current section correspond to reductions of more complex arguments obtained from the entropy inequality.

For the description of a partially saturated porous medium under isothermal conditions the following assumptions are inherent in the constitutive relations introduced subsequently: (i) The constituents are assumed to be immiscible and chemically non-reacting. (ii) The voids of the porous solid are completely filled with the fluids. (iii) While the solid phase is conceived to be incompressible, the compressibilities of both fluid phases are taken into account. (iv) Contrary to the solid phase the soil skeleton is considered to be deformable which results in a coupling between the solid and the fluid phases. (v) The fluid flow may result from pressure gradients or capillary effects. (vi) For the stresses in the soil skeleton as well as for the hydrostatic stresses in the fluid phases tensile components are considered to be positive quantities.

3.5.2 Averaged density of the three-phase mixture

To define an averaged value for the density of the three-phase medium, the intrinsic densities of the individual constituents ρ^s, ρ^w and ρ^a (equation (3.46)) and the corresponding volume fractions η^s, η^w and η^a (equation (3.40)) are necessary. The values of the densities are usually determined from laboratory tests or taken from the literature. The volume fractions have been derived in Section 3.3.1. They depend on the porosity n of the soil and on the degree of saturation S^f, $f = w, a$, of the particular fluid phase (see Table 3.1).

For the general case of a number of κ constituents the averaged value of the density $\tilde{\rho}$ for the κ-phase medium is calculated using the equation

$$\tilde{\rho} = \sum_{\pi=1}^{\kappa} \rho_\pi = \sum_{\pi=1}^{\kappa} \eta^\pi \rho^\pi \ . \tag{3.183}$$

Thus, the intrinsic density of each phase is weighted with the respective volume fraction and the resulting products are summed up for all the constituents. The relationship (3.49) between the volume phase average and the intrinsic volume phase average has been used in the above equation to correlate the partial density ρ_π to the intrinsic density ρ^π of the constituent π. For the special case of a three-phase medium considered in the present work (3.183) becomes

$$\tilde{\rho} = (1-n)\rho^s + nS^w \rho^w + nS^a \rho^a \ , \tag{3.184}$$

where the volume fractions of the three constituents solid, water and air have been used as summarised in Table 3.1.

3.5.3 Stress tensor in the fluid phases

The stress tensor in the fluid phase π is assumed to be a hydrostatic one. Hence, the intrinsic stress tensor $\boldsymbol{\sigma}^\pi$ for the π-phase, related to the area element da^π of the respective constituent, is defined as

$$\boldsymbol{\sigma}^\pi = p^\pi \mathbf{I} \ , \tag{3.185}$$

where p^π is the hydrostatic stress and $\mathbf{I}$ denotes the second-order unit tensor. The latter guarantees that only the normal stress components are non-trivial in $\boldsymbol{\sigma}^\pi$ while all shear stresses are equal to zero.

By means of the area fraction $\overline{\eta^\pi}$ (equation (3.42)) the stress tensor of the individual fluid phase π may be referred to the representative elementary area da. Thus, the partial stress tensor $\boldsymbol{\sigma}_\pi$ can be written as

$$\boldsymbol{\sigma}_\pi = \overline{\eta^\pi}\,\boldsymbol{\sigma}^\pi\,. \tag{3.186}$$

Inserting equation (3.185) into (3.186) yields the following formulation of the partial stress tensor $\boldsymbol{\sigma}_\pi$ for the constituent π:

$$\boldsymbol{\sigma}_\pi = \overline{\eta^\pi}\,p^\pi\,\mathbf{I}\,. \tag{3.187}$$

The specification of equation (3.187) for the case of a partially saturated porous medium, i.e. for a solid containing voids filled with the two fluid phases water w and air a, results in

$$\boldsymbol{\sigma}_w = nS^w p^w\,\mathbf{I} \quad \text{and} \quad \boldsymbol{\sigma}_a = nS^a p^a\,\mathbf{I}\,, \tag{3.188}$$

where the area fractions given in Table 3.1 have been used (remember that according to Delesse's law the area fractions are assumed to be equal to the volume fractions, see Section 3.3.1).

3.5.4 Averaged stresses of the three-phase mixture

The averaged stress tensor $\tilde{\boldsymbol{\sigma}}$ for a multi-phase medium can be introduced similar to the definition of the averaged density for a κ-phase mixture (equation (3.183)) as

$$\tilde{\boldsymbol{\sigma}} = \sum_{\pi=1}^{\kappa} \boldsymbol{\sigma}_\pi = \sum_{\pi=1}^{\kappa} \overline{\eta^\pi}\,\boldsymbol{\sigma}^\pi\,, \tag{3.189}$$

where $\boldsymbol{\sigma}_\pi$ and $\boldsymbol{\sigma}^\pi$ are the partial and the intrinsic stress tensors of the constituent π, respectively. These two quantities are related by means of the area fraction $\overline{\eta^\pi}$ as defined in equation (3.42). For the special case of a three-phase medium formula (3.189) reads as

$$\tilde{\boldsymbol{\sigma}} = (1-n)\,\boldsymbol{\sigma}^s + nS^w\boldsymbol{\sigma}^w + nS^a\boldsymbol{\sigma}^a\,. \tag{3.190}$$

The stresses in the two fluid phases, $\boldsymbol{\sigma}^w$ and $\boldsymbol{\sigma}^a$, are defined in equation (3.185) of Section 3.5.3 and can be inserted into (3.190) which yields

$$\tilde{\boldsymbol{\sigma}} = (1-n)\,\boldsymbol{\sigma}^s + n(S^w p^w + S^a p^a)\,\mathbf{I}\,. \tag{3.191}$$

Chapter 3. Mechanics of the coupled three-phase model   103

The second term in brackets in (3.191), representing the total hydrostatic stress of the two fluid phases exerted on the solid phase, may be denoted as p^s,

$$p^s = S^w p^w + S^a p^a \ . \tag{3.192}$$

Insertion of this definition into (3.191) results in

$$\tilde{\boldsymbol{\sigma}} = (1-n)\,\boldsymbol{\sigma}^s + n p^s \,\mathbf{I} \ , \tag{3.193}$$

and after some rearrangement

$$\tilde{\boldsymbol{\sigma}} = (1-n)(\boldsymbol{\sigma}^s - p^s \,\mathbf{I}) + p^s \,\mathbf{I} \tag{3.194}$$

is obtained. Equation (3.193) expresses the fact that the total stresses in the three-phase medium are composed of one part bound to the solid phase of the domain and another part related to its void space.

3.5.5 Effective averaged stress tensor of the soil skeleton

The effective stress concept is well established for the description of completely water saturated soils. According to this concept, the effective stresses, resulting in deformations of the soil skeleton, can be calculated as the difference between the total stresses and the pore water pressure. Evidence has shown that only one stress state variable, namely the effective stress, is required to describe the mechanical behaviour of a saturated soil. The effective stress is applicable to all types of soil, for instance, sands, silts or clays, because it is independent of the soil properties. The volume change and the shear strength characteristics of a saturated soil are both controlled by the effective stress.

For unsaturated soils the establishment of suitable stress state variables is a much more difficult task. Although steady progress has been made in recent years in the effort to identify the stress state variables governing partially saturated soil behaviour, still some unanswered questions exist due to a lack of fundamental experimental data [Fredlund(1993)]. However, one of the best-known expressions for the description of partially saturated soil behaviour is the relationship proposed by Bishop [Bishop(1959)]. He was probably the first author to suggest an equation taking into account the pressures in both fluid phases water and air explicitly. Employing the terminology of the current work it reads

$$\tilde{\boldsymbol{\sigma}}' = (\tilde{\boldsymbol{\sigma}} - \boldsymbol{\sigma}^a) + \chi\,(\boldsymbol{\sigma}^a - \boldsymbol{\sigma}^w) \ , \tag{3.195}$$

where $\tilde{\boldsymbol{\sigma}}'$ denotes the effective averaged stresses and $\tilde{\boldsymbol{\sigma}}$ are the total averaged stresses in the soil as introduced in Section 3.5.4. The quantity χ in (3.195) is the so-called Bishop's parameter which is related to the degree of water saturation of the soil. The magnitude of χ is unity for a saturated soil and zero for a dry one. The relationship between the Bishop's parameter χ and the degree of water saturation S^w of the soil has to be determined experimentally. Laboratory tests have been performed on different types of soil and under various conditions. Figures 3.3 show some results of these tests [Fredlund(1993)] and demonstrate the influence of the type of soil on the value of χ. The dashed line in each of the two figures represents the case $\chi = S^w$, i.e. the Bishop's parameter is equal to the degree of water saturation.

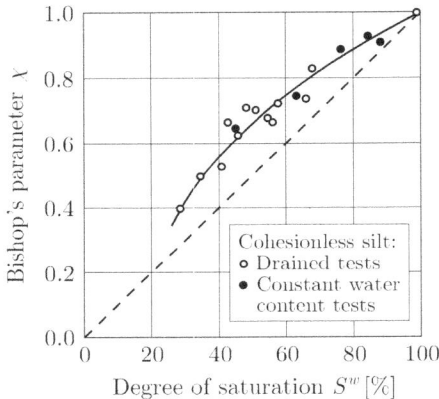
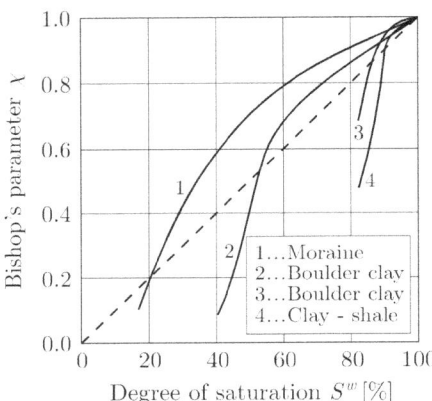

Figure 3.3: Experimental data of the relationship between the Bishop's parameter χ and the degree of water saturation S^w for a cohesionless silt and for different types of compacted soils.

Equation (3.195) represents an attempt to describe the mechanical behaviour of unsaturated soils by combining total stress $\tilde{\boldsymbol{\sigma}}$, hydrostatic water stress $\boldsymbol{\sigma}^w$ and hydrostatic air stress $\boldsymbol{\sigma}^a$ within a single effective stress $\tilde{\boldsymbol{\sigma}}'$. Various drawbacks of this approach are mentioned in the literature: (i) While the effective stress is a stress variable and hence related to the equilibrium conditions alone, equation (3.195) contains a parameter, χ, depending on the constitutive behaviour, i.e. on the type of soil as is easily recognised from Figure 3.3 [Fredlund(1993)]. (ii) Bishop's law (3.195) may not provide an adequate relationship between the volume change and the effective stresses for soils with particularly low degrees of saturation, i.e. for values below a critical degree of saturation which is estimated to be approximately 20 % for silts or sands and as high as 90 % for clays [Fredlund(1993)]. (iii) As reported in [Wheeler(1995)], the above definition of an effective stress (3.195) is not capable for describing some of the most fundamental features of unsaturated soil behaviour: While it is relatively easy to relate the shear strength of a partially saturated soil to a single stress parameter involving $\tilde{\boldsymbol{\sigma}}$, $\boldsymbol{\sigma}^w$ and $\boldsymbol{\sigma}^a$, the volumetric behaviour is not controlled by the same stress parameter or by any other single stress variable. In particular, it has proved impossible to represent the complex pattern of wetting-induced swelling and collapse in terms of a single effective stress.

Despite these drawbacks and although more sophisticated models are currently available for the description of the behaviour of unsaturated soils (which also involve more effort in determining the material parameters), equation (3.195) is employed in the present approach. Concerning the above statement (iii), it should be mentioned that soils exhibiting a pronounced swelling behaviour such as clays are not considered in the current work. Furthermore, the parameter χ in (3.195) is assumed to be equal to the degree of water saturation S^w which seems to be a justified approximation according to the experimental data shown in Figure 3.3. Thus, the relationship (3.195) can be written as

$$\tilde{\boldsymbol{\sigma}}' = (\tilde{\boldsymbol{\sigma}} - \boldsymbol{\sigma}^a) + S^w(\boldsymbol{\sigma}^a - \boldsymbol{\sigma}^w) \,, \qquad (3.196)$$

or in a rearranged form as

$$\tilde{\boldsymbol{\sigma}}' = \tilde{\boldsymbol{\sigma}} - S^w \boldsymbol{\sigma}^w - (1 - S^w) \boldsymbol{\sigma}^a \,. \qquad (3.197)$$

After the introduction of the saturation condition, i.e. of $S^w + S^a = 1$, one obtains

$$\tilde{\boldsymbol{\sigma}}' = \tilde{\boldsymbol{\sigma}} - S^w \boldsymbol{\sigma}^w - S^a \boldsymbol{\sigma}^a \,. \qquad (3.198)$$

For the intrinsic stresses in the two fluid phases, $\boldsymbol{\sigma}^w$ and $\boldsymbol{\sigma}^a$, definition (3.185) can be employed which yields

$$\tilde{\boldsymbol{\sigma}}' = \tilde{\boldsymbol{\sigma}} - (S^w p^w + S^a p^a)\mathbf{I} = \tilde{\boldsymbol{\sigma}} - p^s \mathbf{I} \,, \qquad (3.199)$$

where the relationship (3.192) describing the hydrostatic stress p^s of the fluid phases exerted on the solid phase has been used. From (3.199) it follows that

$$\tilde{\boldsymbol{\sigma}} = \tilde{\boldsymbol{\sigma}}' + p^s \mathbf{I} \,, \qquad (3.200)$$

and a comparison of this equation with (3.194) yields

$$\tilde{\boldsymbol{\sigma}}' = (1 - n)(\boldsymbol{\sigma}^s - p^s \mathbf{I}) \,. \qquad (3.201)$$

Hence, the total stress tensor $\tilde{\boldsymbol{\sigma}}$ in the three-phase mixture may be decomposed into two parts, one being the effective averaged stress tensor $\tilde{\boldsymbol{\sigma}}'$ resulting in deformations of the soil skeleton and the other one being the hydrostatic stress p^s due to the two fluid phases present in the voids. The former stresses are introduced into the constitutive law describing the behaviour of the soil skeleton. It should be noted that due to the assumed incompressibility of the solid phase only the effective stresses yield deformations of the soil skeleton.

3.5.6 Soil suction and capillary stress

One of the most characteristic phenomena of unsaturated soil behaviour is the so-called soil suction. The total suction is composed of two components, namely the osmotic suction and the matric suction. The former is more closely related to the diffuse double layer around the clay particles, whereas the latter is mainly associated with the water-air interface. Any change in suction affects the overall equilibrium of the soil mass and may be caused by a change in either one or both components of the total suction. Figure 3.4 shows total, osmotic and matric suction measurements for a typical soil (compacted clay) [Fredlund(1993)].

The osmotic suction is related to the salt content in the pore water and thus may be encountered in both saturated and unsaturated soils. Osmotic suction changes, i.e.

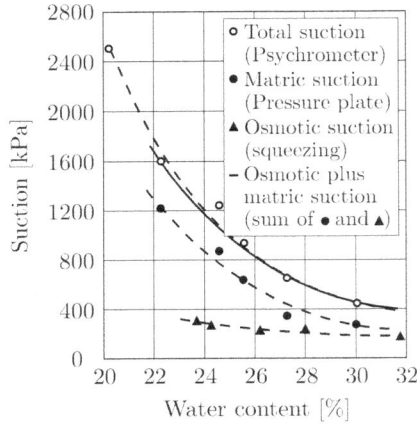

Figure 3.4: Total, osmotic and matric suction measurements on a compacted clay [Fredlund(1993)].

changes in the salt content, have an effect on the mechanical behaviour of the soil and especially on its shear strength. However, osmotic suction is usually not taken into account in most engineering problems because they rather deal with situations involving a varying water content due to environmental aspects. The changes in osmotic suction for a varying water content are generally less important than those in matric suction as Figure 3.4 indicates. Only in the case where the salt content of the soil is altered by chemical contamination, the effect of osmotic suction changes may influence the soil behaviour significantly.

The matric suction component is commonly associated with the capillary phenomenon arising from the surface tension of water. Surface tension is considered to be a typical property of the water-air interface (contractile skin) and results from intermolecular forces acting on the molecules in this skin. While a molecule in the interior of the water experiences equal forces in all directions, i.e. there are no unbalanced forces, a molecule within the contractile skin is subject to an unbalanced force towards the interior of the water. Thus, in order to maintain equilibrium a tensile pull is built up along the contractile skin which is possible due to the existence of the property called surface tension. The surface tension can be thought of as a tensile force per unit length of the contractile skin acting in tangential direction and depending on the temperature.

The capillary phenomenon is usually illustrated by the rise of a water surface in a so-called capillary tube. After insertion of a small cylindrical glass tube into water under atmospheric conditions the tube is considered to be filled with the two immiscible fluids water and air (Figure 3.5). Due to the different electromolecular forces between the water phase and the tube and the air phase and the tube, respectively, a curved interface (meniscus) develops between the fluid phases. Since the water phase has a higher affinity to the tube than the air phase, the interface always takes a shape concave to the air phase and convex to the water phase. This phenomenon is commonly referred to as wetting. The fluid water, forming an acute angle α_c with the capillary tube, is called a wetting fluid while the air phase with its obtuse angle $(\pi - \alpha_c)$ is called non-wetting. This angle α_c, termed contact angle, is a measure of the wettability of a material. Its magnitude depends on the adhesion between the molecules in the contractile skin and the material comprising the tube. The better the wettability, the smaller is the contact angle α_c.

Due to the concave shape of the water-air interface and to the presence of surface tension, along the interface the hydrostatic pressure in the water phase is less than the hydrostatic pressure in the air phase, i.e. with respect to the atmospheric air pressure

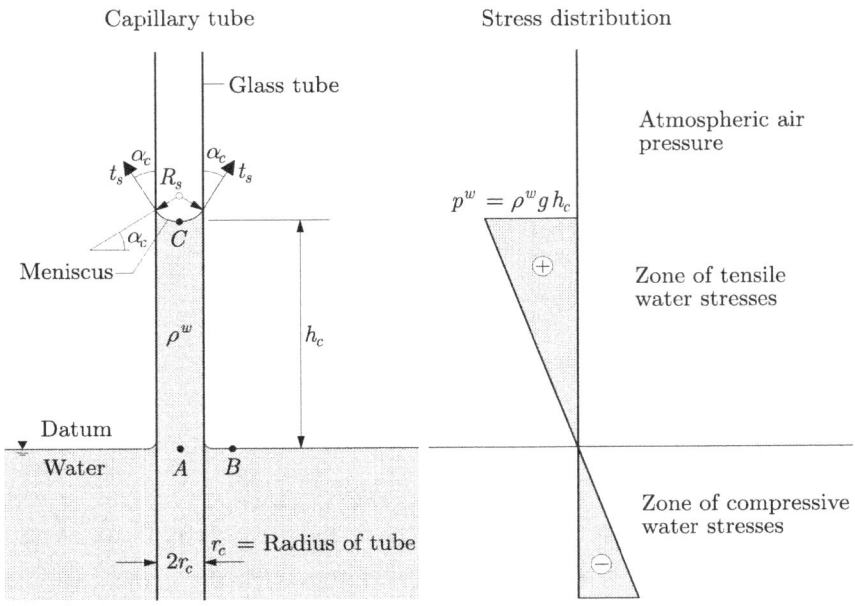

Figure 3.5: Physical model of the capillary phenomenon and corresponding stress distribution in the fluids.

tensile stresses are prevailing in the water phase. These tensile stresses yield a rise of the water surface in the capillary tube. The height of the capillary rise, h_c, can be calculated by investigating the equilibrium of the vertical forces in the capillary tube whereby two portions have to be distinguished (Figure 3.5): (i) the vertical resultant of the surface tension in the meniscus, $2\pi r_c t_s \cos \alpha_c$, which is responsible for holding the weight of the water column and (ii) the weight of the water with the height h_c, amounting to $\pi r_c^2 h_c \rho^w g$. The equilibrium can then be expressed in the following equation:

$$2\pi r_c t_s \cos \alpha_c = \pi r_c^2 h_c \rho^w g \ , \qquad (3.202)$$

where r_c is the radius of the capillary tube, t_s denotes the surface tension, ρ^w is the intrinsic density of the water and g is the gravitational acceleration. Equation (3.202) can be rearranged to obtain the maximum height of water in the capillary tube, h_c, as

$$h_c = \frac{2 t_s \cos \alpha_c}{\rho^w g \, r_c} = \frac{2 t_s}{\rho^w g R_s} \ , \qquad (3.203)$$

where R_s describes the radius of the curvature of the meniscus, i.e. $R_s = r_c / \cos \alpha_c$. The contact angle α_c between the contractile skin for pure water and clean glass is zero and thus the radius of curvature R_s is equal to the radius of the capillary tube r_c. Therefore, the capillary height of pure water in a clean glass amounts to

$$h_c = \frac{2\,t_s}{\rho^w g\,r_c} \,. \tag{3.204}$$

From equation (3.204) it follows that the smaller the radius of the tube, the higher will be the capillary rise.

When taking a look at Figure 3.5, the points A, B and C have to be in hydrostatic equilibrium. The water stresses at the locations A and B are atmospheric, that is equal to zero. The elevation of these points is assumed to be the datum of the system. As a result the hydraulic heads at points A and B are equal to zero. Point C is located at a height h_c above the datum and thus its elevation head is equal to h_c. The hydrostatic equilibrium among the points A, B and C requires that the hydraulic heads at all three points are equal, that is zero. Consequently, the pressure head at C equals the negative value of the elevation head and thus the hydrostatic stress in the water phase can be calculated as

$$p^w = \rho^w g\,h_c \,. \tag{3.205}$$

The water stresses above point A in the capillary tube are tensile stresses as shown in Figure 3.5 which are assumed to be positive quantities in the present work. The stress in the air phase at C is zero. Hence, the matric suction s at point C, defined as $s = p^a - p^w$, can be expressed as

$$s = p^a - p^w = -\rho^w g\,h_c \,. \tag{3.206}$$

Substituting equation (3.204) into (3.206) yields the matric suction being written in terms of the surface tension for a zero contact angle:

$$s = p^a - p^w = -\frac{2\,t_s}{r_c} \,. \tag{3.207}$$

The radius r_c of the capillary tube may be compared with the pore radius in soils. Thus, the development of the capillary height in a soil is affected by the pore size distribution. However, compared to a capillary tube the geometrical situation of the pore spaces in a soil is much more complex. Above the groundwater table usually a so-called capillary fringe is encountered where, with increasing height above the water level, the amount of water in the pores decreases and the pore fluid is increasingly trapped in menisci in the vicinity of the grain to grain contacts. The vertical extent of the capillary fringe depends on the type of process under consideration, that is either drainage or imbibition, and on the size of the soil voids. Since soils possess a broad range of void sizes, the capillary height also varies within certain limits. Therefore, there exists no unique capillary height of a soil but more likely limiting values of the particular quantity may be defined [Lambe(1979)].

Figure 3.6(a) shows a plot of the degree of water saturation versus the distance above the water table. If an initially water saturated soil is allowed to drain until steady state conditions are reached, the distribution of moisture above the water table could be represented by curve A. If, on the other hand, a dry soil is exposed to water, curve B could describe the distribution of moisture above the water surface at equilibrium. Curves A and B thus constitute two limiting conditions of the pore liquid decrease above the water table. When considering the drainage curve A, two critical points are of interest, namely point a indicating the highest elevation to which any continuous channel of water exists above the free water surface (called maximum capillary height h_{cdx}) and point b describing the highest elevation at which complete saturation occurs (saturation capillary height h_{cds}). On the distribution curve B for the capillary rise (imbibition) also two points are of particular interest, a point c marking the distance from the water surface to the highest elevation to which capillary water would rise in a dry soil (termed

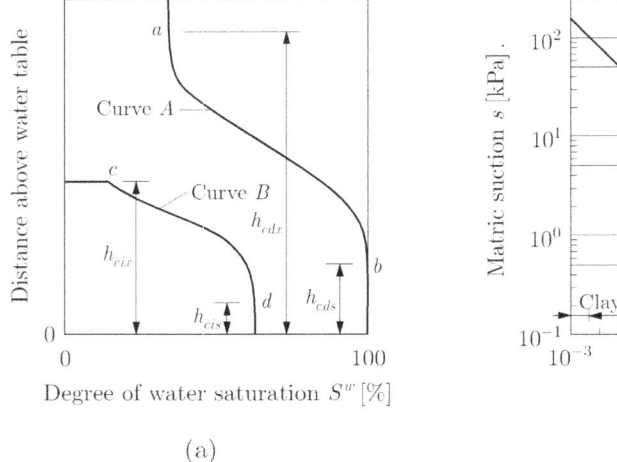
(a)

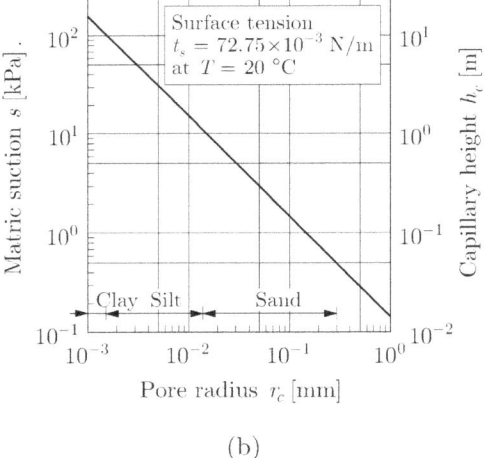
(b)

Figure 3.6: (a) Various capillary heights of a soil and (b) relationship among pore radius, capillary height and matric suction.

capillary rise h_{cix}) and a point d indicating the distance from the water level to the highest elevation at which the maximum degree of saturation exists (minimum capillary height h_{cis}). The four above mentioned capillary heights are limits of the possible range of capillary heights which may be encountered for a particular soil. Hence, any capillary height associated with drainage would lie between h_{cdx} and h_{cds} and any connected to capillary rise between h_{cix} and h_{cis}, respectively [Lambe(1979)].

However, for engineering purposes a description of the capillary phenomenon in soils for convenience should be based rather on matric suction than on the capillary height. From equation (3.207) it becomes obvious that the smaller the pore radii in the soil, the higher the matric suction can be. Figure 3.6(b) illustrates the relationship among pore size, capillary height and matric suction.

With respect to the air pressure, which is generally atmospheric in a field situation, the capillary water in the menisci exhibits a tensile hydrostatic stress state (see equation (3.205)). At low degrees of saturation for very small pore sizes the tensile pore water stresses can assume values as high as 7000.0 kPa. In this case the adsorptive forces between soil particles are believed to play an important role in sustaining the highly tensile pore water stresses in the soil [Fredlund(1993)]. In general, in an unsaturated soil matric suction acts like an additional hydrostatic pressure exerted on the soil skeleton and consequently increases the shear strength of the soil. A consideration of equation (3.195) for the effective stresses in the soil skeleton in combination with the definition of the hydrostatic stress state in the fluid phases (3.185) indicates this fact quite clearly:

$$\begin{aligned}\tilde{\boldsymbol{\sigma}}' &= (\tilde{\boldsymbol{\sigma}} - \boldsymbol{\sigma}^a) + \chi \left(\boldsymbol{\sigma}^a - \boldsymbol{\sigma}^w\right) \\ &= (\tilde{\boldsymbol{\sigma}} - \boldsymbol{\sigma}^a) + \chi \left(p^a - p^w\right) \mathbf{I} \\ &= (\tilde{\boldsymbol{\sigma}} - \boldsymbol{\sigma}^a) + \chi \, s \, \mathbf{I} \, .\end{aligned} \qquad (3.208)$$

In (3.208) the definition (3.206$_1$) for the matric suction s has been introduced to obtain the last line of the equation. If the air pressure $\boldsymbol{\sigma}^a$ is atmospheric and the total stresses $\tilde{\boldsymbol{\sigma}}$ are assumed to be compressive (i.e. negative), a negative value of the suction s would cause an increase in the negative value of $\tilde{\boldsymbol{\sigma}}'$ and thus the suction s can be regarded as an additional hydrostatic pressure acting on the soil skeleton.

If the pressure in the air phase differs from the atmospheric one, it may be appropriate to replace the term 'matric suction' by the term 'capillary stress'. The capillary stress p^c is defined as the difference between the hydrostatic stresses in the gas phase and the water phase as

$$p^c = p^a - p^w \ . \tag{3.209}$$

As shown in the literature [Gray(1991a)], equation (3.209) is not just a definition but a derived relationship between two independent quantities at equilibrium, namely p^c and $p^a - p^w$.

3.5.7 Momentum exchange terms

Considering slow phenomena for the transport of both fluids water and air, Darcy's law is assumed to be valid, however, generalised by introducing the relative permeability to account for the flow of two phases. The macroscopic linear momentum balance equation (3.162) can be used to derive Darcy's law for the respective fluid phase water or air. In this case the momentum exchange term due to mechanical interaction in (3.162) has to be specified appropriately as [Prevost(1980)]

$$\eta^\pi \rho^\pi \mathbf{t}^\pi = -p^\pi \operatorname{grad} \eta^\pi - \eta^\pi \mathbf{R}^\pi \cdot \overline{\mathbf{v}}^{\pi\alpha} \ . \tag{3.210}$$

It represents the part of the fluid-solid exchange of momentum containing the spatial variation of the volume fraction and thus also of the porosity (first term in (3.210)) and the viscosity of the fluid phase (second term in (3.210)). Sometimes the latter term is also called 'Stokes drag'. The symmetric, positive definite second-order resistivity tensor $\mathbf{R}^\pi$ is assumed to be invertible and its inverse may be defined as

$$(\mathbf{R}^\pi)^{-1} = \frac{\mathbf{k}^\pi}{\mu^\pi \eta^\pi} \ . \tag{3.211}$$

In equation (3.211) $\mathbf{k}^\pi$ denotes the permeability tensor of the soil with respect to the fluid phase π and μ^π is the dynamic viscosity of the fluid.

If the voids of the soil are filled with more than one fluid, momentum exchange terms arise not only at fluid-solid but also at fluid-fluid interfaces. Therefore, the permeability of the fully saturated case has to be modified. This modification depends on the volume fractions of the other phases, on capillary effects and on the fluid-fluid momentum exchange. At the macroscopic level the permeability $\mathbf{k}^\pi$ may be expressed as

$$\mathbf{k}^\pi = \mathbf{k}^i k^{r\pi} = \frac{\mathbf{k}^{o\pi} \mu^\pi}{\rho^\pi g} k^{r\pi} \ , \tag{3.212}$$

where $\mathbf{k}^i$ is the intrinsic permeability of the soil, $k^{r\pi}$ is the relative permeability coefficient, a dimensionless parameter varying between zero and one, and $\mathbf{k}^{o\pi}$ is the permeability tensor with respect to the π-phase for fully saturated conditions. Insertion of equation (3.212) into (3.211) yields the inverse of the resistivity tensor $\mathbf{R}^\pi$:

$$(\mathbf{R}^\pi)^{-1} = \frac{\mathbf{k}^i k^{r\pi}}{\mu^\pi \eta^\pi} = \frac{\mathbf{k}^{o\pi} k^{r\pi}}{\rho^\pi g \, \eta^\pi} \ . \tag{3.213}$$

The permeability of an unsaturated soil varies with both the void ratio and the degree of saturation. However, for many situations changes due to an altering void ratio are of secondary importance and the permeability coefficient may be satisfactorily defined as a function of the degree of saturation [Fredlund(1993), Lewis(1998)]. For a particular porous medium the relation $k^{r\pi}(S^w)$ is either predicted by empirical relations (see Section 3.5.13), based on certain models for the matric suction versus degree of saturation curve, or determined experimentally.

3.5.8 Elastic or elastic-plastic behaviour of the soil skeleton

An appropriate stress-strain relation for the description of the mechanical behaviour of the soil skeleton is of primary importance in order to provide a realistic solution to problems in geotechnical engineering. The choice of a certain type of constitutive relationship may have a significant influence on the numerical results obtained. In this section equations are introduced considering the soil skeleton as a linear elastic or an elastic-plastic material. Since the solid phase, constituted by the particles of the soil, is assumed to be incompressible, only the effective stresses cause deformations of the soil skeleton (these deformations result from movements of the rigid particles relative to each other; in the case of a compressible solid phase also the deformations of the particles itself would have to be taken into account). Consequently, the constitutive equations may be formulated here in terms of the effective stresses (see Section 3.5.5).

Linear elastic constitutive behaviour of the soil skeleton can be expressed by means of a linear relationship between the effective averaged stresses $\tilde{\boldsymbol{\sigma}}'$ and the strains $\boldsymbol{\varepsilon} = \boldsymbol{\varepsilon}^e$ in the soil skeleton as

$$\tilde{\boldsymbol{\sigma}}' = \mathbf{C}^e : \boldsymbol{\varepsilon}^e \ , \tag{3.214}$$

where $\mathbf{C}^e$ denotes the elasticity tensor. The superscript of $\boldsymbol{\varepsilon}^e$ indicates that the strains are elastic strains. For a two-dimensional plane strain case, e.g., the elasticity tensor takes the familiar form

$$\mathbf{C}^e = \frac{E(1-\nu)}{(1+\nu)(1-2\nu)} \begin{bmatrix} 1 & \frac{\nu}{1-\nu} & 0 \\ & 1 & 0 \\ \text{symm.} & & \frac{1-2\nu}{2(1-\nu)} \end{bmatrix} . \tag{3.215}$$

For linear elastic isotropic behaviour of the material the number of independent constants of the tensor $\mathbf{C}^e$ is two. However, different pairs of such constants may be chosen to describe the material. Probably the most widely used formulation is based on Young's modulus E and Poisson's ratio ν as given in (3.215). Alternatively, the so-called Lamé constants λ^s and G^s may be employed (for reasons of clearness the superscript s indicates that these constants refer to the soil skeleton) which can be written in terms of E and ν as

$$\lambda^s = \frac{E\nu}{(1+\nu)(1-2\nu)} \tag{3.216}$$

and

$$G^s = \frac{E}{2(1+\nu)} , \tag{3.217}$$

respectively. G^s is called shear modulus. In soil mechanics usually the shear modulus G^s together with the bulk modulus K^s are preferred. The latter relates the volumetric strain to the hydrostatic portion of the stress state. It can be expressed by means of E and ν as

$$K^s = \frac{E}{3(1-2\nu)} . \tag{3.218}$$

Considering elastic-plastic behaviour of the soil skeleton three main ingredients are necessary: (i) the yield function (yield surface), (ii) the flow rule and (iii) the hardening law. The strains in the soil skeleton are decomposed into two parts, namely the elastic (reversible) one and the plastic (irreversible) one. In rate form this additive decomposition may be expressed as

$$\dot{\boldsymbol{\varepsilon}} = \dot{\boldsymbol{\varepsilon}}^e + \dot{\boldsymbol{\varepsilon}}^p , \tag{3.219}$$

where $\boldsymbol{\varepsilon}^e$ and $\boldsymbol{\varepsilon}^p$ are the elastic and plastic parts of the total strains $\boldsymbol{\varepsilon}$, respectively. Multiplication of (3.219) by an infinitesimal time increment dt and substitution of the resulting equation into the infinitesimal incremental form of (3.214) yields the following relationship between the effective averaged stresses and the strains:

$$d\tilde{\boldsymbol{\sigma}}' = \mathbf{C}^e : (d\boldsymbol{\varepsilon} - d\boldsymbol{\varepsilon}^p) . \tag{3.220}$$

To determine whether a certain stress state causes just elastic or elastic and plastic strains in the soil, a yield function may be defined as

$$f(\tilde{\boldsymbol{\sigma}}', \boldsymbol{\rho}^s, \kappa^s) = 0 \ . \tag{3.221}$$

This function depends on the effective averaged stresses $\tilde{\boldsymbol{\sigma}}'$ of the soil skeleton and on the hardening parameters $\boldsymbol{\rho}^s$ and κ^s where the former accounts for kinematic and the latter for isotropic hardening of the soil skeleton, respectively. The values of the yield function are either less than zero, indicating that the stress point lies inside the yield surface and thus the material behaviour is linear elastic, or equal to zero, denoting a stress state on the yield surface which causes plastic material behaviour. Values of the yield function greater than zero are not admissible when dealing with rate independent plasticity.

For the description of the evolution of the plastic strains a flow rule is defined which is a mathematical expression for the rate of the plastic strains $\dot{\boldsymbol{\varepsilon}}^p$ or the infinitesimal increment of the plastic strains $d\boldsymbol{\varepsilon}^p$:

$$d\boldsymbol{\varepsilon}^p = d\lambda \frac{\partial f}{\partial \tilde{\boldsymbol{\sigma}}'} \quad \text{or} \quad d\boldsymbol{\varepsilon}^p = d\lambda \frac{\partial g}{\partial \tilde{\boldsymbol{\sigma}}'} \ . \tag{3.222}$$

The first of the equations (3.222) is known as the normality rule because the plastic strain increment vector $d\boldsymbol{\varepsilon}^p$ is proportional to the gradient of the yield surface in an n-dimensional stress space. $d\lambda$ denotes a constant called consistency parameter, yet undetermined. This flow rule is also called associated since it depends on the yield function $f(\tilde{\boldsymbol{\sigma}}', \boldsymbol{\rho}^s, \kappa^s)$. However, in order to obtain better correspondence between numerical results and experimental data, the yield function f in (3.222_1) can be replaced by a so-called plastic potential $g(\tilde{\boldsymbol{\sigma}}', \boldsymbol{\rho}^s, \kappa^s)$ yielding the flow rule (3.222_2). Especially in soil mechanics it is quite common to specify the flow rule with respect to a plastic potential function. In this case the flow rule is said to be non-associated because it does not depend on the yield function but on the plastic potential. In the above definition the gradient of the yield surface, $\partial f/\partial \tilde{\boldsymbol{\sigma}}'$, or the gradient of the plastic potential, $\partial g/\partial \tilde{\boldsymbol{\sigma}}'$, indicate the direction of the vector $d\boldsymbol{\varepsilon}^p$ while the scalar factor $d\lambda$ gives its length.

The plastic strain increment $d\boldsymbol{\varepsilon}^p$ will occur only if the elastic stress increment, that is $d\tilde{\boldsymbol{\sigma}}' = \mathbf{C}^e : d\boldsymbol{\varepsilon}^e$, tends to put the stress outside the yield surface, i.e. in the case of plastic loading. If, on the other hand, the stress change is such that it causes unloading, then of course no plastic straining will be present. For plastic loading both conditions $f = 0$ and $df = 0$ hold which means that the stress point $\tilde{\boldsymbol{\sigma}}'$ lies on the yield surface and remains on the yield surface. Hence, in order to determine the value of $d\lambda$ the derivative of equation (3.221) is calculated and set equal to zero:

$$df(\tilde{\boldsymbol{\sigma}}', \boldsymbol{\rho}^s, \kappa^s) = \frac{\partial f}{\partial \tilde{\boldsymbol{\sigma}}'} : d\tilde{\boldsymbol{\sigma}}' + \frac{\partial f}{\partial \boldsymbol{\rho}^s} : d\boldsymbol{\rho}^s + \frac{\partial f}{\partial \kappa^s} d\kappa^s = 0 \ . \qquad (3.223)$$

For the terms on the right hand side of (3.223) several substitutions are useful. Insertion of the definition (3.222$_2$) for the plastic strain increment into equation (3.220) for the effective averaged stress increment yields

$$d\tilde{\boldsymbol{\sigma}}' = \mathbf{C}^e : d\boldsymbol{\varepsilon} - d\lambda \, \mathbf{C}^e : \frac{\partial g}{\partial \tilde{\boldsymbol{\sigma}}'} \ . \qquad (3.224)$$

The two terms accounting for the hardening behaviour are rewritten as

$$B_1 = -\frac{\partial f}{\partial \boldsymbol{\rho}^s} : \frac{d\boldsymbol{\rho}^s}{d\lambda} \qquad \text{and} \qquad B_2 = -\frac{\partial f}{\partial \kappa^s} \frac{d\kappa^s}{d\lambda} \ . \qquad (3.225)$$

Substituting (3.224) and (3.225) into (3.223) and collecting the terms containing $d\lambda$ results in

$$df(\tilde{\boldsymbol{\sigma}}', \boldsymbol{\rho}^s, \kappa^s) = \frac{\partial f}{\partial \tilde{\boldsymbol{\sigma}}'} : \mathbf{C}^e : d\boldsymbol{\varepsilon} - d\lambda \left[\frac{\partial f}{\partial \tilde{\boldsymbol{\sigma}}'} : \mathbf{C}^e : \frac{\partial g}{\partial \tilde{\boldsymbol{\sigma}}'} + B_1 + B_2 \right] = 0 \ . \qquad (3.226)$$

The consistency parameter $d\lambda$ can then be determined from equation (3.226) as [Zienkiewicz(1991)]

$$d\lambda = \frac{\dfrac{\partial f}{\partial \tilde{\boldsymbol{\sigma}}'} : \mathbf{C}^e : d\boldsymbol{\varepsilon}}{\dfrac{\partial f}{\partial \tilde{\boldsymbol{\sigma}}'} : \mathbf{C}^e : \dfrac{\partial g}{\partial \tilde{\boldsymbol{\sigma}}'} + B_1 + B_2} \ . \qquad (3.227)$$

It should be noted that in (3.227) the two terms B_1 and B_2 still contain the consistency parameter $d\lambda$ (see (3.225)). However, when considering certain types of hardening, $d\lambda$ disappears in the equations (3.225) as shown in [Zienkiewicz(1991)].

Finally, the elastic-plastic material tensor is left to be defined since the stiffness of the material changes when plastic strains occur. A relationship similar to equation (3.214) may be written in incremental form as

$$d\tilde{\boldsymbol{\sigma}}' = \mathbf{C}^{ep} : d\boldsymbol{\varepsilon} \ , \qquad (3.228)$$

where $\mathbf{C}^{ep}$ denotes the elastic-plastic material tensor. Insertion of the consistency parameter $d\lambda$ (equation (3.227)) into the relationship for the effective averaged stresses as formulated in (3.224) yields

$$d\tilde{\boldsymbol{\sigma}}' = \mathbf{C}^e : d\boldsymbol{\varepsilon} - \frac{\dfrac{\partial f}{\partial \tilde{\boldsymbol{\sigma}}'} : \mathbf{C}^e \otimes \mathbf{C}^e : \dfrac{\partial g}{\partial \tilde{\boldsymbol{\sigma}}'}}{\dfrac{\partial f}{\partial \tilde{\boldsymbol{\sigma}}'} : \mathbf{C}^e : \dfrac{\partial g}{\partial \tilde{\boldsymbol{\sigma}}'} + B_1 + B_2} : d\boldsymbol{\varepsilon} \ . \qquad (3.229)$$

This equation can be written in a form similar to (3.228) as

$$d\tilde{\boldsymbol{\sigma}}' = \left[\mathbf{C}^e - \frac{\frac{\partial f}{\partial \tilde{\boldsymbol{\sigma}}'} : \mathbf{C}^e \otimes \mathbf{C}^e : \frac{\partial g}{\partial \tilde{\boldsymbol{\sigma}}'}}{\frac{\partial f}{\partial \tilde{\boldsymbol{\sigma}}'} : \mathbf{C}^e : \frac{\partial g}{\partial \tilde{\boldsymbol{\sigma}}'} + B_1 + B_2}\right] : d\boldsymbol{\varepsilon} . \qquad (3.230)$$

A comparison of the two relations (3.228) and (3.230) finally yields the definition of the elastic-plastic material tensor $\mathbf{C}^{ep}$ represented by the term in brackets in the latter equation:

$$\mathbf{C}^{ep} = \mathbf{C}^e - \frac{\frac{\partial f}{\partial \tilde{\boldsymbol{\sigma}}'} : \mathbf{C}^e \otimes \mathbf{C}^e : \frac{\partial g}{\partial \tilde{\boldsymbol{\sigma}}'}}{\frac{\partial f}{\partial \tilde{\boldsymbol{\sigma}}'} : \mathbf{C}^e : \frac{\partial g}{\partial \tilde{\boldsymbol{\sigma}}'} + B_1 + B_2} . \qquad (3.231)$$

It should be noted that for a non-hardening material, i.e. for elastic-perfectly plastic material behaviour, the terms B_1 and B_2 in the denominator of (3.231) vanish.

3.5.9 Drucker-Prager material model

In this section a particular material model for the soil skeleton is introduced in order to illustrate the derivations of the previous Section 3.5.8 dealing with elastic-plastic constitutive behaviour. The Drucker-Prager model described subsequently is characterised by a linear elastic-perfectly plastic material behaviour without any hardening. For convenience the derivation is performed using an associated flow rule. However, it would be straightforward to define a plastic potential function and take into account a flow rule formulated in terms of this plastic potential according to equation (3.222_2).

When dealing with isotropic material behaviour, it is convenient to consider the state of stress to consist of two components, namely hydrostatic and deviatoric stresses. This decomposition equally applies to effective and total stresses. For the effective averaged stresses $\tilde{\boldsymbol{\sigma}}'$, used to describe the constitutive behaviour of the soil skeleton, these two components are

$$\tilde{\boldsymbol{\sigma}}' = \tilde{\sigma}'^m \mathbf{I} + \tilde{\mathbf{s}}' = \frac{1}{3}\tilde{I}_1'^\sigma \mathbf{I} + \tilde{\mathbf{s}}' , \qquad (3.232)$$

with $\tilde{\sigma}'^m$, $\tilde{I}_1'^\sigma$ and $\tilde{\mathbf{s}}'$ denoting the mean stress, the first invariant of the stress tensor and the deviator stress, respectively. It should be noted that the quantities with $\tilde{}'$ are deduced from the effective averaged stress tensor $\tilde{\boldsymbol{\sigma}}'$.

A yield criterion defines the limits of elasticity under any possible combination of stress states. When written in terms of stress components, it is called a yield function and represents a surface in the n-dimensional stress space, separating the elastic region from an outer zone of impermissible stress states. In the principal stress space σ_1, σ_2, σ_3 the Drucker-Prager yield surface is defined by a right-circular cone with its axis equally inclined with respect to each of the coordinate directions and its apex in the tension octant. Using stress invariants, the Drucker-Prager yield function can be formulated as

$$f\left(\tilde{I}_1'^{\sigma}, \tilde{\mathbf{s}}'\right) = \|\tilde{\mathbf{s}}'\| + \frac{\mu}{\sqrt{3}} \tilde{I}_1'^{\sigma} - \sqrt{2}\, \tau_F = 0 \,, \tag{3.233}$$

where $\|\tilde{\mathbf{s}}'\|$ denotes the norm of the deviator stress and μ and τ_F are material parameters. The derivative of equation (3.233) with respect to the effective averaged stress tensor $\tilde{\boldsymbol{\sigma}}'$ yields the so-called consistency condition

$$df\left(\tilde{I}_1'^{\sigma}, \tilde{\mathbf{s}}'\right) = \frac{\partial f}{\partial \tilde{I}_1'^{\sigma}} d\tilde{I}_1'^{\sigma} + \frac{\partial f}{\partial \tilde{\mathbf{s}}'} : d\tilde{\mathbf{s}}' = 0 \,, \tag{3.234}$$

which is employed to determine the consistency parameter $d\lambda$.

For the Drucker-Prager criterion the derivative of the yield function with respect to the first stress invariant $\tilde{I}_1'^{\sigma}$ is obtained as

$$\frac{df}{d\tilde{I}_1'^{\sigma}} = \frac{\mu}{\sqrt{3}} \tag{3.235}$$

and the derivative with respect to the deviator stress $\tilde{\mathbf{s}}'$ as

$$\frac{df}{d\tilde{\mathbf{s}}'} = \frac{1}{2} \frac{1}{\sqrt{\tilde{\mathbf{s}}' : \tilde{\mathbf{s}}'}} 2\tilde{\mathbf{s}}' = \frac{\tilde{\mathbf{s}}'}{\|\tilde{\mathbf{s}}'\|} \,. \tag{3.236}$$

Additionally, the derivative with respect to $\tilde{\boldsymbol{\sigma}}'$ is needed for both the first invariant $\tilde{I}_1'^{\sigma}$ of the stress tensor and the deviator stress $\tilde{\mathbf{s}}'$:

$$\tilde{I}_1'^{\sigma} = \tilde{\boldsymbol{\sigma}}' : \mathbf{I} \quad \rightarrow \quad \frac{\partial \tilde{I}_1'^{\sigma}}{\partial \tilde{\boldsymbol{\sigma}}'} = \mathbf{I} \,, \tag{3.237}$$

$$\tilde{\mathbf{s}}' = \tilde{\boldsymbol{\sigma}}' - \frac{1}{3} \tilde{I}_1'^{\sigma} \mathbf{I} \quad \rightarrow \quad \frac{\partial \tilde{\mathbf{s}}'}{\partial \tilde{\boldsymbol{\sigma}}'} = \tilde{\mathbf{I}}'^{dev} \,. \tag{3.238}$$

In equation (3.238_2) the quantity $\tilde{\mathbf{I}}'^{dev}$ denotes the so-called deviatoric operator, a fourth-order tensor. Application of $\tilde{\mathbf{I}}'^{dev}$ to a second-order tensor yields the deviatoric part of the respective tensor.

For the case of an associated flow rule the infinitesimal increment of the plastic strains can be calculated according to equation (3.222$_1$) as

$$d\boldsymbol{\varepsilon}^p = d\lambda \frac{\partial f\left(\tilde{I}_1^{\prime\sigma}, \tilde{\mathbf{s}}'\right)}{\partial \tilde{\boldsymbol{\sigma}}'} = d\lambda \left(\frac{\partial f}{\partial \tilde{I}_1^{\prime\sigma}} \frac{\partial \tilde{I}_1^{\prime\sigma}}{\partial \tilde{\boldsymbol{\sigma}}'} + \frac{\partial f}{\partial \tilde{\mathbf{s}}'} : \frac{\partial \tilde{\mathbf{s}}'}{\partial \tilde{\boldsymbol{\sigma}}'} \right) =$$

$$= d\lambda \left(\frac{\mu}{\sqrt{3}} \mathbf{I} + \frac{\tilde{\mathbf{s}}'}{\|\tilde{\mathbf{s}}'\|} : \tilde{\mathbf{I}}'^{dev} \right) =$$

$$= d\lambda \left(\frac{\mu}{\sqrt{3}} \mathbf{I} + \frac{\tilde{\mathbf{s}}'}{\|\tilde{\mathbf{s}}'\|} \right), \quad (3.239)$$

where the relations (3.235) to (3.238) have been used. On the other hand, the infinitesimal plastic strain increment can be subdivided into a volumetric and a deviatoric part similar to the stresses (see equation (3.232)):

$$d\boldsymbol{\varepsilon}^p = d\varepsilon^{m,p} \mathbf{I} + d\mathbf{e}^p . \quad (3.240)$$

Comparing the relationship (3.240) with the last of the equations (3.239) yields expressions for the mean value and the deviatoric part of the plastic strain increment in terms of the consistency parameter $d\lambda$:

$$d\varepsilon^{m,p} = d\lambda \frac{\mu}{\sqrt{3}} \quad \text{and} \quad d\mathbf{e}^p = d\lambda \frac{\tilde{\mathbf{s}}'}{\|\tilde{\mathbf{s}}'\|} . \quad (3.241)$$

When making use of the decomposition of the elastic strains into a volumetric and a deviatoric part, $\varepsilon^{vol,e}$ and $\mathbf{e}^e$, where the superscript e indicates the elastic behaviour of the material, and employing the definitions for the bulk modulus K^s and the shear modulus G^s, (3.218) and (3.217), respectively, Hooke's law can be formulated as follows [Mang(2000)],

$$\tilde{\boldsymbol{\sigma}}' = K^s \varepsilon^{vol,e} \mathbf{I} + 2G^s \mathbf{e}^e . \quad (3.242)$$

A comparison of the equations (3.242) and (3.232) results in a relationship between the first invariants of the effective averaged stress tensor $\tilde{\boldsymbol{\sigma}}'$ and the strain tensor $\boldsymbol{\varepsilon}$:

$$K^s \varepsilon^{vol,e} \mathbf{I} = \frac{1}{3} \tilde{I}_1^{\prime\sigma} \mathbf{I} \quad \rightarrow \quad \tilde{I}_1^{\prime\sigma} = 3K^s \varepsilon^{vol,e} = 3K^s I_1^{\varepsilon,e} . \quad (3.243)$$

In incremental form (3.243$_2$) reads as

$$d\tilde{I}_1^{\prime\sigma} = 3K^s \left(dI_1^\varepsilon - dI_1^{\varepsilon,p} \right), \quad (3.244)$$

where the decomposition of the first strain invariant into an elastic and a plastic part has been used. By comparing the second terms of (3.242) and (3.232) one obtains a relationship between the effective averaged deviator stress $\tilde{\mathbf{s}}'$ and the deviator strain $\mathbf{e}^e$ as

$$d\tilde{\mathbf{s}}' = 2G^s \left(d\mathbf{e} - d\mathbf{e}^p \right) , \qquad (3.245)$$

where again an incremental form and the partitioning into elastic and plastic portions have been employed. Inserting the equations (3.241) into (3.244) and (3.245) and using the definition $dI_1^{\varepsilon,p} = 3\,d\varepsilon^{m,p}$ of the first invariant yields the subsequent two relations for the increment of the first invariant of the stress tensor,

$$d\tilde{I}_1'^{\sigma} = 3K^s \left(dI_1^{\varepsilon} - d\lambda\sqrt{3}\mu \right) , \qquad (3.246)$$

and for the increment of the deviator stress,

$$d\tilde{\mathbf{s}}' = 2G^s \left(d\mathbf{e} - d\lambda \frac{\tilde{\mathbf{s}}'}{\|\tilde{\mathbf{s}}'\|} \right) . \qquad (3.247)$$

Finally, the equations (3.235), (3.236), (3.246) and (3.247) can be substituted into the consistency condition (3.234) resulting in

$$\frac{\mu}{\sqrt{3}} 3K^s \left(dI_1^{\varepsilon} - d\lambda\sqrt{3}\mu \right) + \frac{\tilde{\mathbf{s}}'}{\|\tilde{\mathbf{s}}'\|} : 2G^s \left(d\mathbf{e} - d\lambda \frac{\tilde{\mathbf{s}}'}{\|\tilde{\mathbf{s}}'\|} \right) =$$

$$= \sqrt{3}K^s \mu\, dI_1^{\varepsilon} - 3K^s \mu^2 d\lambda + 2G^s \frac{\tilde{\mathbf{s}}'}{\|\tilde{\mathbf{s}}'\|} : d\mathbf{e} - 2G^s d\lambda = 0 . \qquad (3.248)$$

After collecting the terms containing $d\lambda$ one obtains

$$d\lambda \left(2G^s + 3K^s \mu^2 \right) = 2G^s \frac{\tilde{\mathbf{s}}'}{\|\tilde{\mathbf{s}}'\|} : d\mathbf{e} + \sqrt{3}K^s \mu\, dI_1^{\varepsilon} , \qquad (3.249)$$

and by introducing the relationship $dI_1^{\varepsilon} = d\boldsymbol{\varepsilon} : \mathbf{I}$ the consistency parameter $d\lambda$ can be expressed as

$$d\lambda = \frac{2G^s (\tilde{\mathbf{s}}'/\|\tilde{\mathbf{s}}'\|) + \sqrt{3}K^s \mu \mathbf{I}}{2G^s + 3K^s \mu^2} : d\boldsymbol{\varepsilon} . \qquad (3.250)$$

According to the general derivation presented in Section 3.5.8, the elastic-plastic material tensor $\mathbf{C}^{ep}$ remains to be determined. To this end, equation (3.242) is written in an incremental form, using once again the partitioning into elastic and plastic portions for the first strain invariant and the deviator strain:

$$d\tilde{\boldsymbol{\sigma}}' = K^s (dI_1^{\varepsilon} - dI_1^{\varepsilon,p})\mathbf{I} + 2G^s (d\mathbf{e} - d\mathbf{e}^p) . \qquad (3.251)$$

Inserting the definitions for the plastic parts (3.241) and collecting terms yields

$$d\tilde{\boldsymbol{\sigma}}' = K^s \left(dI_1^\varepsilon - d\lambda\sqrt{3}\mu \right) \mathbf{I} + 2G^s \left(d\mathbf{e} - d\lambda \frac{\tilde{\mathbf{s}}'}{\|\tilde{\mathbf{s}}'\|} \right) =$$

$$= K^s dI_1^\varepsilon \mathbf{I} - d\lambda\sqrt{3}K^s\mu \mathbf{I} + 2G^s d\mathbf{e} - d\lambda\, 2G^s \frac{\tilde{\mathbf{s}}'}{\|\tilde{\mathbf{s}}'\|} =$$

$$= K^s d\varepsilon^{vol} \mathbf{I} + 2G^s d\mathbf{e} - d\lambda \left(\sqrt{3}K^s\mu \mathbf{I} + 2G^s \frac{\tilde{\mathbf{s}}'}{\|\tilde{\mathbf{s}}'\|} \right). \quad (3.252)$$

Substituting the elasticity tensor (3.215), expressed in terms of K^s and G^s, and the consistency parameter (3.250) into this equation results in

$$d\tilde{\boldsymbol{\sigma}}' = \mathbf{C}^e : d\boldsymbol{\varepsilon} - \frac{2G^s(\tilde{\mathbf{s}}'/\|\tilde{\mathbf{s}}'\|) + \sqrt{3}K^s\mu \mathbf{I}}{2G^s + 3K^s\mu^2} \otimes \left(\sqrt{3}K^s\mu \mathbf{I} + 2G^s \frac{\tilde{\mathbf{s}}'}{\|\tilde{\mathbf{s}}'\|} \right) : d\boldsymbol{\varepsilon} =$$

$$= \left[\mathbf{C}^e - \frac{2G^s(\tilde{\mathbf{s}}'/\|\tilde{\mathbf{s}}'\|) + \sqrt{3}K^s\mu \mathbf{I}}{2G^s + 3K^s\mu^2} \otimes \left(\sqrt{3}K^s\mu \mathbf{I} + 2G^s \frac{\tilde{\mathbf{s}}'}{\|\tilde{\mathbf{s}}'\|} \right) \right] : d\boldsymbol{\varepsilon}. \quad (3.253)$$

Equation (3.253) immediately gives the elastic-plastic material tensor defined as the ratio $d\tilde{\boldsymbol{\sigma}}'/d\boldsymbol{\varepsilon}$,

$$\mathbf{C}^{ep} = \frac{d\tilde{\boldsymbol{\sigma}}'}{d\boldsymbol{\varepsilon}} = \mathbf{C}^e - \frac{2G^s(\tilde{\mathbf{s}}'/\|\tilde{\mathbf{s}}'\|) + \sqrt{3}K^s\mu \mathbf{I}}{2G^s + 3K^s\mu^2} \otimes \left(\sqrt{3}K^s\mu \mathbf{I} + 2G^s \frac{\tilde{\mathbf{s}}'}{\|\tilde{\mathbf{s}}'\|} \right). \quad (3.254)$$

According to (3.254), for plastic loading the elasticity tensor $\mathbf{C}^e$ has to be adjusted by a term depending on the current stress state yielding the elastic-plastic material tensor $\mathbf{C}^{ep}$. The elastic-plastic material tensor (3.254) is valid for a Drucker-Prager material without hardening in combination with an associated flow rule.

3.5.10 Equation of state for the water phase

The constitutive equation for the water phase in general provides a relationship between the density ρ^w, the hydrostatic stress p^w, and the temperature T of the fluid:

$$\rho^w = \rho^w(p^w, T). \quad (3.255)$$

Fernandez [Fernandez(1972)] proposed the following form of the equation of state,

$$\rho^w = \rho^{ow} \left[1 - \beta^w T - C^w(p^w - p^{ow}) \right], \quad (3.256)$$

which for isothermal conditions considered in the present model can be written as

$$\rho^w = \rho^{ow}\left[1 - C^w(p^w - p^{ow})\right] . \qquad (3.257)$$

In (3.257) the two quantities ρ^{ow} and p^{ow} are reference values for the density and for the hydrostatic stress in the water phase, respectively. p^{ow} is usually assumed to be equal to the atmospheric air pressure. C^w denotes the compressibility of the fluid phase which is inverse to the bulk modulus K^w,

$$C^w = \frac{1}{K^w} . \qquad (3.258)$$

Employing the definition (3.258) and considering an atmospheric reference pressure equal to zero, equation (3.257) can be formulated as

$$\frac{\rho^w}{\rho^{ow}} = 1 - \frac{1}{K^w} p^w . \qquad (3.259)$$

The negative sign on the right hand side of (3.259) reflects the definition of compressive stress components being negative quantities. The time derivative of equation (3.259), which is necessary for specifying the individual balance laws, can be calculated as

$$\frac{1}{\rho^{ow}}\frac{\partial \rho^w}{\partial t} = -\frac{1}{K^w}\frac{\partial p^w}{\partial t} . \qquad (3.260)$$

Considering the two constituents water and air, which are of particular interest in the present model, some important facts should be outlined concerning the compressibility of a mixture of these fluids filling the void space of a soil.

Since the water molecules form a lattice structure, air may dissolve in the water and fill these openings. The volume of dissolved air amounts to approximately 2.0 % of the total volume. However, this water lattice is relatively rigid and stable and thus the changes of the density of water due to the presence of dissolved air are rather small. Figure 3.7(a) shows the effect of dissolved air on the density of water for a certain range of temperatures [Fredlund(1993)]. Furthermore, the volume of dissolved air in water is essentially independent of water and air stresses which can be demonstrated using the ideal gas law and Henry's law. At a certain temperature the volume of dissolved air is a constant for different hydrostatic stress states. The ratio between the volume of dissolved gas in a liquid and the volume of the liquid is called the volumetric coefficient of solubility, h, which varies with temperature as shown in Figure 3.7(b) [Fredlund(1993)].

The compressibility is usually employed to relate the volume change of a fluid to the change in the hydrostatic stress. The isothermal compressibility C^w is defined as

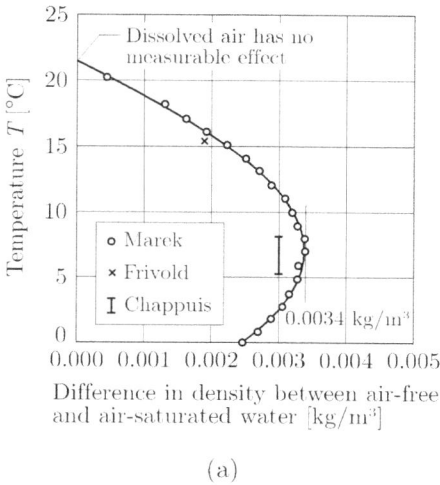 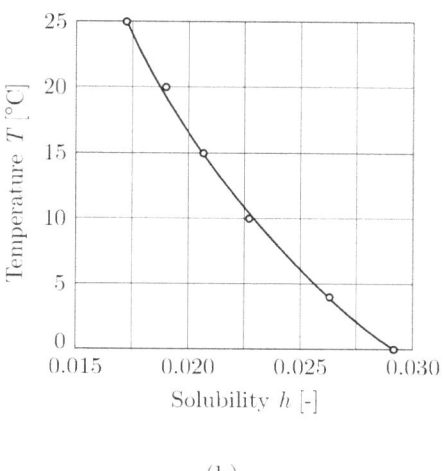

(a) (b)

Figure 3.7: (a) Effect of dissolved air on the density of water and (b) coefficient of solubility for a certain temperature range.

the volume change of a fixed mass with respect to a stress change per unit volume at a constant temperature:

$$C^w = \frac{1}{v^w}\frac{dv^w}{dp^w} . \tag{3.261}$$

In equation (3.261) v^w is the volume of water and dv^w/dp^w describes the volume change with respect to the hydrostatic stress change dp^w.

In an unsaturated soil the pore fluid consists of water, free air and air dissolved in water. Thus, the compressibility C^{wa} of the mixture is conceived to be constituted by three parts:

$$C^{wa} = S^w C^w B^w + S^a C^a B^a + h S^w C^a B^a . \tag{3.262}$$

The first and the second term of (3.262) represent the compressibility of the water and the free air, respectively. C^w and C^a denote the individual compressibilities of the fluids according to equation (3.261) and B^w and B^a are the pore water and pore air pressure parameters for isotropic loading, i.e. they describe the pressure change of the particular fluid phase with respect to the total stress change. The third term accounts for the dissolved air in water. The different components of the compressibility of a water-air mixture versus the degree of saturation are plotted in Figure 3.8 [Fredlund(1993)]. This figure indicates that the compressibility of a water-air mixture is predominantly

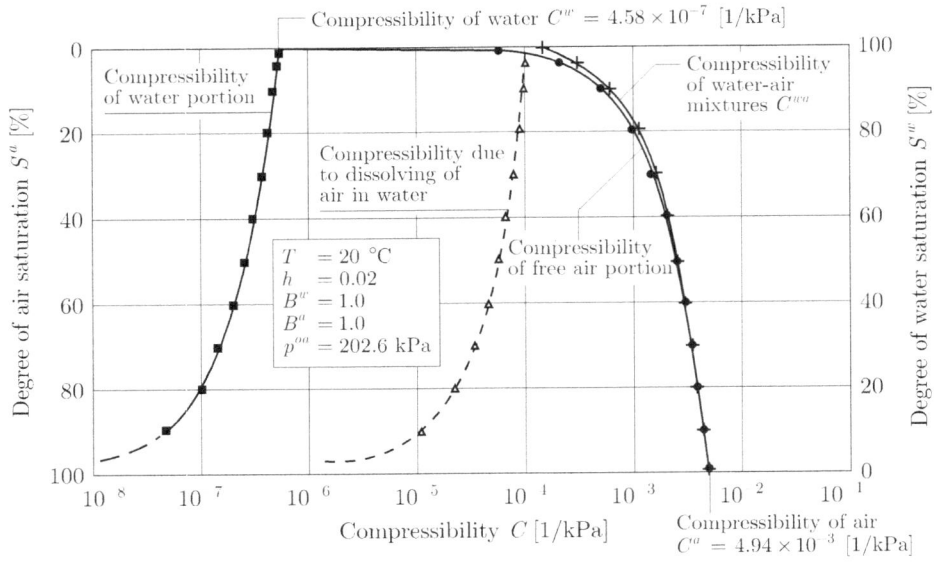

Figure 3.8: Components of the compressibility of a water-air mixture.

influenced by the compressibility of the free air portion. When the voids of the soil are completely filled with air, that is $S^w = 0.0$, the compressibility of the pore fluid is equal to the isothermal compressibility of air. At a fully saturated state, that is at $S^w = 1.0$, the pore fluid compressibility becomes equal to that of water. The compressibility of water is only of importance in a calculation when the soil is fully saturated. However, the inclusion of just 1.0 % of air in the soil is sufficient to significantly increase the pore fluid compressibility. The compressibility due to the solution of air in water is approximately two orders of magnitude larger than the one of water. Air dissolving in water significantly affects the compressibility of a water-air mixture when the free air volume becomes less than about 20.0 % of the volume of the voids.

Finally, Figure 3.9 shows a semi-logarithmic plot of the compressibility of a water-air mixture versus the degree of saturation for various initial absolute air pressures [Fredlund(1993)]. Basically, the effect of air solubility on the compressibility is the same for any initial air pressure. It should be mentioned that a consideration of the effect of air dissolving in water results in a non-smooth transition for the compressibility of a water-air mixture as saturation is reached, i.e. as S^w approaches one. A discontinuity occurs at the point where there is no more free air to be dissolved in the water, that is at

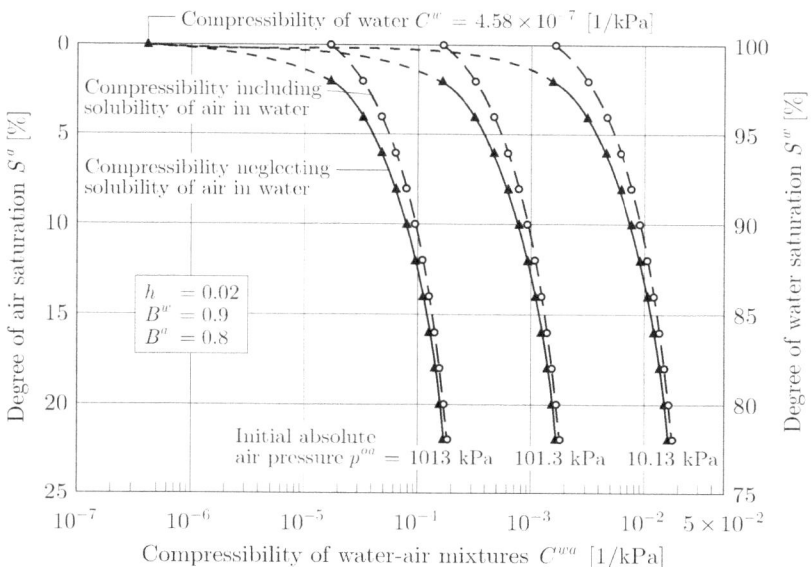

Figure 3.9: Effect of the solubility of air in water on the compressibility of a water-air mixture for different initial absolute air pressures.

$S^a = 0.0$. In this case the terms two and three of equation (3.262) have to be dropped, the second term because $S^a = 0.0$ and the third one due to the vanishing coefficient of solubility if there is no free air to be dissolved. As a consequence the compressibility C^{wa} abruptly decreases to the compressibility of water. On the other hand, if the free air does not have time to be dissolved in the water, the last term of equation (3.262) must be set to zero. Therefore, the compressibility of the pore fluid shows a smooth behaviour when the fully water saturated state is reached (see Figure 3.9).

3.5.11 Equation of state for the air phase

Similar to the equation of state for the water phase (confer Section 3.5.10), the constitutive equation for the air phase provides a relationship between the density ρ^a, the hydrostatic air stress p^a and the temperature T. Considering air as being an ideal gas and taking into account only isothermal conditions, the gas law reads as [Malvern(1969)]

$$-p^a v^a = R = const. \, , \tag{3.263}$$

with v^a describing the volume of air and R denoting the gas constant for the particular gas. The negative sign reflects the definition of compressive stress components being negative quantities. The density of the air phase is obtained by relating the mass to the volume of the fluid, that is

$$\rho^a = \frac{m^a}{v^a} \; . \tag{3.264}$$

Rewriting equation (3.264) with respect to v^a and inserting into (3.263) yields

$$\rho^a = -\frac{m^a}{R} p^a = -\overline{m}^a p^a \; , \tag{3.265}$$

which is the desired relationship between the fluid density and the hydrostatic fluid stress. It should be mentioned that $\overline{m}^a$ is a constant. The time derivative of equation (3.265) can then be calculated as

$$\frac{\partial \rho^a}{\partial t} = -\overline{m}^a \frac{\partial p^a}{\partial t} \; , \tag{3.266}$$

and is employed for specifying the individual balance laws.

3.5.12 Capillary stress versus degree of water saturation

The relationship between the matric suction and the degree of water saturation (matric suction versus degree of saturation curve) of a soil depends not only on the fluid properties but also on the structure of the porous medium. It describes the ability of the soil to store water in its pore spaces. Instead of the degree of water saturation the water content may be used for plotting the curve which is then called the soil-water characteristic curve. Due to the highly complex geometry of the pores in a soil an analytical derivation of this relationship on the microscale is impossible. Therefore, it is usually determined on the macroscale by means of laboratory tests. In the last decades much effort was spent by research workers from such different fields as biology, physics, petroleum engineering or soil mechanics for deriving analytical equations to approximate experimental data.

Experiments to determine the relationship between the matric suction s and the degree of saturation S^w are usually performed starting with a fully water saturated soil specimen subject to an excess air pressure. Figure 3.10(a) shows a qualitative plot of the type of curves which may be obtained from laboratory tests. By means of this figure the most important characteristics of such kind of experiments are explained subsequently. Departing from the state of a fully water saturated soil, that is $S^w = 1.0$ ($\equiv 100.0$ %) and $s = 0.0$, a gradual decrease of the applied excess hydrostatic air stress and thus also

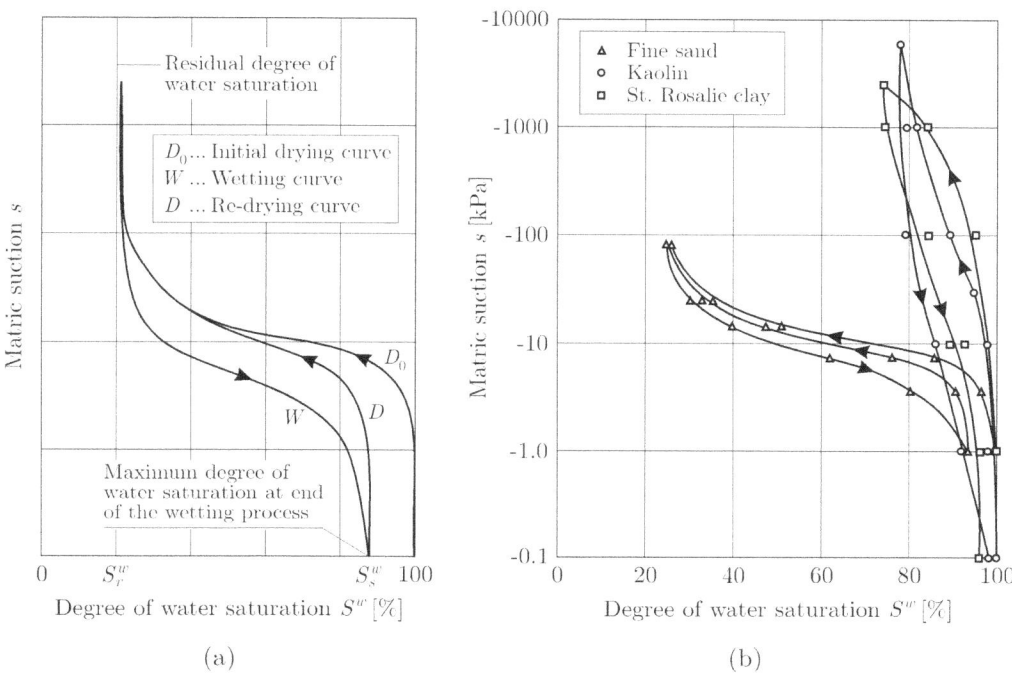

Figure 3.10: Matric suction versus degree of saturation curves: (a) General shape of the curves usually obtained from drying-wetting cycles and (b) experimental data for three widely different types of soil [Alonso(1987)].

of the matric suction (or the capillary stress), which is actually an increase in terms of absolute values since compressive stress components are defined as negative quantities (see Section 3.5.1), results in a displacement of the pore water. The degree of water saturation of the soil is reduced according to curve D_0 (initial drying curve) in Figure 3.10(a). Hence, the decrease in matric suction s is related to a decrease in the degree of water saturation S^w.

Following the very steep part of curve D_0, a characteristic property for an unsaturated soil is encountered when the degree of water saturation starts decreasing below the fully saturated state, namely the so-called air entry value p_b^c, also referred to as 'bubbling pressure' in ceramics engineering or 'displacement pressure' in petroleum engineering. It is related to the maximum pore size in a soil and describes the value which the matric suction has to reach before the air actually enters the water saturated pore spaces. It should be pointed out that air entry values are well defined in soils with uniform pore size distributions but are not clearly observed in soils having a broader pore size distribution. After the air entry value has been exceeded a further decrease of matric suction first yields a drainage (desorption) of the large pore spaces. The smaller pores are dewatered if the matric suction is reduced even more. Approaching the residual degree of water saturation S_r^w it is impossible to further withdraw water from the pores by decreasing s. The actual value of the residual degree of water saturation depends on the pore structure of the soil. The more irregular the particles, the greater is the residual degree of saturation.

Subsequent wetting of the specimen (imbibition) follows the wetting curve W (Figure 3.10(a)) until a maximum value of the degree of water saturation S_s^w is reached being perhaps somewhat smaller than 1.0 because of the inclusion of air bubbles. A repeated drainage process yields curve D approaching curve D_0 (Figure 3.10(a)). Thus, for a dewetting-wetting cycle the relationship between matric suction and the degree of water saturation exhibits hysteresis as Figure 3.10(a) indicates very clearly. In other words, different degrees of water saturation correspond to a certain matric suction value depending on the investigation of either a dewetting or a wetting procedure. For a wetting process the obtained degree of water saturation is smaller than for drainage.

Experimental data of draining-wetting cycles [Alonso(1987)] for three widely different types of soil, namely for a natural clay, a kaolin and a fine sand, are shown in Figure 3.10(b). While the sand is a rather incompressible material, the clays are highly compressible so that changes in suction cause large volume changes. The two clays lose full saturation in a suction range of 1.0 to 10.0 kPa but no definite air entry value can be specified. Despite the applied high suctions the degree of saturation is not reduced significantly. A much more definite air entry value is encountered for the fine sand which becomes fairly dry for suctions less than 100.0 kPa.

The hysteresis phenomenon may be compared to the fact that different heights of capillary rise are encountered in a capillary tube of radius r_c containing a bulb of radius r_b, $r_b > r_c$, at midheight of the capillary rise (Figure 3.11). The presence of the bulb prevents the water from rising up beyond the base of the bulb, i.e. a full development of the capillary height is not possible (Figure 3.11(b)). On the other hand, the capillary height can be fully developed if the bulb is filled by submerging it below the water surface and then raising it above the surface (Figure 3.11(c)).

In a soil the nonuniform pore size distribution can result in the above mentioned

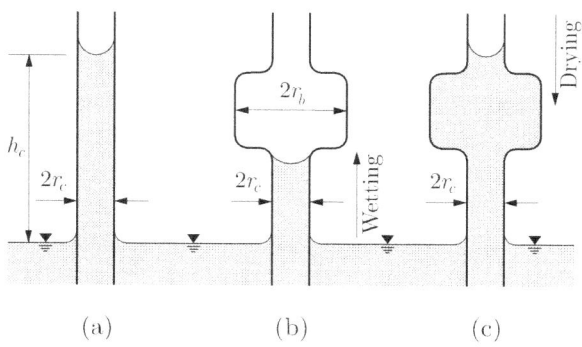

Figure 3.11: Radius effects on the capillary height.

hysteresis effect in the matric suction versus degree of saturation curve. Additionally, the contact angle at an advancing interface during a wetting process is different from that at a receding interface during a drying process (see Section 3.5.6). According to [Bear(1979)], the contact angle at an advancing interface is greater than the one at a receding interface. The above factors as well as the presence of entrapped air in the soil are considered to be the main causes for the hysteresis in the relationship between matric suction and the degree of water saturation (Figure 3.10). In rather coarse-textured soils this hysteresis effect is very small, it may already be neglected for coarse sands. The more fine-grained the soil, the bigger this effect becomes [Rodriguez(1993)].

As already mentioned, the degree of water saturation is commonly expressed as a function of matric suction and obtained experimental data are usually approximated by means of empirical relations. The matric suction versus degree of saturation curve can then be employed for a prediction of the coefficient of permeability; in particular, the relative permeability coefficient, which predominantly depends on the degree of water saturation, may be calculated. To this end, numerous semi-empirical equations for the coefficient of permeability have been derived using the matric suction versus degree of saturation curve (see Section 3.5.13) whereas other approaches are based on employing the soil-water characteristic curve [Fredlund(1993)]. In the literature several more or less sophisticated approximations for the relationship between matric suction s and the degree of water saturation S^w (or the water content) have been suggested, e.g., [Brooks(1964), Genuchten(1980), Genuchten(1985), Vogel(2001), Tuli(2001)]. Subsequently, two approximations for the matric suction (or capillary stress) versus degree of saturation curve are introduced which are used in Section 3.5.13 to derive equations for calculating the permeabilities of the soil with respect to the fluid phases water and air, respectively.

Taking into account only isothermal conditions and employing the more general expression of the capillary stress (equation (3.209)) instead of the matric suction, the relationship between the capillary stress and the degree of water saturation reads as

$$S^w = S^w(p^c) \,. \tag{3.267}$$

For the non-isothermal case the degree of water saturation would of course also depend on the temperature.

A relatively simple empirical equation was developed by Brooks and Corey [Brooks(1964)]. In their model only the drainage curve is used. Additionally, the soil skeleton is assumed to be incompressible. In terms of the capillary stress the proposed equation [Brooks(1964)] may be written as

$$S^w = S^w_r + (S^w_s - S^w_r)\left(\frac{p^c_b}{p^c}\right)^\lambda \,. \tag{3.268}$$

In (3.268) S^w_s denotes the maximum degree of water saturation. S^w_r is the residual degree of water saturation, defined as the degree of saturation at which a decrease (i.e. an increase in absolute values) of the capillary stress does not yield any significant changes in the degree of saturation. The air entry value p^c_b describes the value matric suction (or the capillary stress) must exceed before air can enter the pores of the soil and the pore size distribution index λ gives the dependence of the curve on the properties of the soil. The quantity λ is a small value for soils with a wide range of pore sizes while a more uniform distribution of pore sizes results in a larger value for the pore size distribution index λ.

By means of the so-called effective degree of saturation S_e, which is defined by combining the values for the current, the maximum and the residual degree of saturation,

$$S_e = \frac{S^w - S^w_r}{S^w_s - S^w_r} \,, \tag{3.269}$$

equation (3.268) can be reformulated as

$$S_e = \left(\frac{p^c_b}{p^c}\right)^\lambda \,. \tag{3.270}$$

Both equations (3.268) and (3.270) are only valid for capillary stresses less than the air entry value, i.e. for $p^c < p^c_b$ or absolute values $|p^c| > |p^c_b|$. A qualitative plot of equation (3.268) is shown in Figure 3.12(a). Mathematically speaking, the curve has the shape of a hyperbola with the two asymptotes being the residual degree of saturation S_r^w parallel to the ordinate and a zero capillary stress in the direction of the abscissa (Figure 3.12(a)). The parameter λ defines the curvature in the region of the intersection of both asymptotes. A continuous description of the behaviour near fully water saturated states is of course not possible with this approach. However, as mentioned in [Genuchten(1985)], this equation yields good agreement with experimental data for relatively coarse-textured soils with rather narrow pore size distributions, i.e. for large values of λ. According to the experimental data shown in Figure 3.12(b),

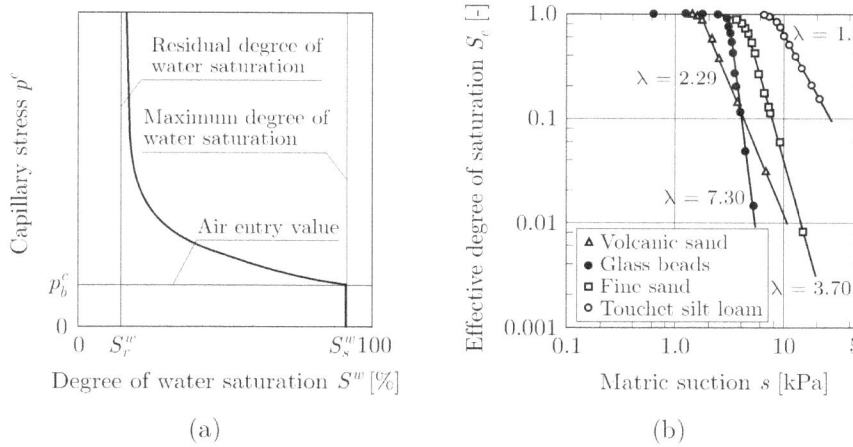

Figure 3.12: (a) Capillary stress versus degree of water saturation curve proposed by Brooks and Corey [Brooks(1964)] and (b) matric suction versus effective degree of saturation for different materials [Fredlund(1993)].

equation (3.270) can readily be visualised as follows: In a plot of matric suction (or capillary stress) versus the effective degree of saturation the intersection point between the sloping straight line and the saturation ordinate at $S_e = 1.0$ gives the air entry value of the soil. The pore size distribution index λ is defined as the negative slope of this line.

In order to achieve a better approximation of experimental data in the range of the fully saturated state, that is near $S^w = 1.0$ ($\equiv 100.0$ %), various improved equations have been suggested in the literature (compare the overview given in [Genuchten(1985)]). These equations are characterised by yielding continuously differentiable (smooth) S-shaped curves. Concerning the further use of the relationships for deriving equations to calculate the coefficients of permeability, the different approaches are more or less suitable. Subsequently, the model proposed by van Genuchten [Genuchten(1980), Genuchten(1985)] is described in some more detail. Compared to [Brooks(1964)] this approach is more general: Firstly, the S-shaped curves are better approximations of the experimentally obtained data (Figure 3.10(b)) and secondly, mathematically speaking, the equation is continuously differentiable and invertible.

The equation proposed by van Genuchten [Genuchten(1985)] may be formulated in a way similar to (3.268) as

$$S^w = S_r^w + (S_s^w - S_r^w)\left[1 + \left(\frac{p^c}{p_b^c}\right)^n\right]^{-m}, \qquad (3.271)$$

where m and n represent empirical parameters to fit the curve to experimental data. Figure 3.13(a) shows a plot of the qualitative shape of this curve; the ability to approximate test data of various types of soil such as sand, silt or clay is outlined in Figure 3.13(b). In general the parameters m and n are independent. However, van Genuchten also investigated various relations between these two constants and compared the performance

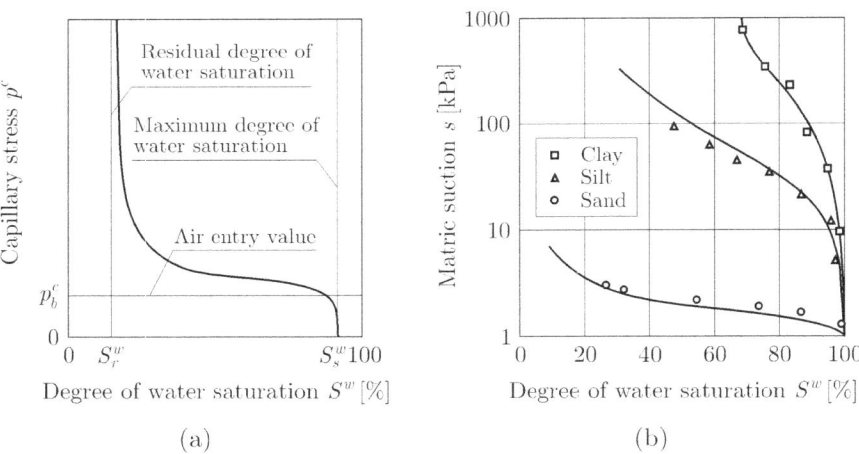

Figure 3.13: (a) Capillary stress versus degree of saturation curve proposed by van Genuchten [Genuchten(1985)] and (b) matric suction versus degree of saturation data (symbols) and their approximation by van Genuchten's relationship (solid lines).

of equation (3.271) for these restricted cases with experimental results. A relationship of $n = 1/(1-m)$ proved to be successful for many soils. The parameter m should be specified in the range of zero and one. To check his model, van Genuchten used an extensive collection of experimental data for a large number of different soils which in extracts is given in [Mualem(1976)].

Using the effective degree of saturation as defined in equation (3.269), the relationship of van Genuchten reads as

$$S_e = \left[1 + \left(\frac{p^c}{p_b^c}\right)^n\right]^{-m}. \tag{3.272}$$

The equations described in this section, in particular the form given in (3.272), may now be employed to derive empirical relations to calculate the coefficients of relative permeability with respect to the two fluid phases water and air which is dealt with in the next section. These relative permeability coefficients are then combined with the permeabilities for the fully saturated state to calculate the permeability for the unsaturated soil.

3.5.13 Permeability coefficients for the fluid phases

The experimental determination of the hydraulic conductivity of soils, i.e. the coefficient of permeability with respect to the water phase, is a difficult task, in particular when dealing with unsaturated soils. In the latter case this is due to the hydraulic conductivity being a function of the water content (or the degree of saturation) of the soil. For this reason many attempts have been made to express the permeability by simple parametric equations as a function of soil properties which are fairly easy to determine such as the water content or the degree of saturation. A brief review of the different approaches is presented in a recent paper by Brutsaert [Brutsaert(2000)].

For partially saturated soils the permeability coefficient with respect to an individual fluid phase may be calculated as the product of the permeability for the fully saturated state and the relative permeability coefficient for the particular fluid phase (confer Section 3.5.7). The latter is introduced in order to account for the partially saturated state and usually depends on the water content or the degree of saturation of the soil. One of the most widely used expressions is a power law where the relative permeability is calculated as some power of the effective degree of saturation (equation (3.269)). This power law has been derived on the basis of some widely different conceptual models of the pore geometry.

According to [Brutsaert(2000)], three types of models can be distinguished: (i) Uniform pore size models, where the porous medium is characterised by some equivalent uniform pore size and the variability of the pore sizes is not considered. (ii) Parallel models, where the pore system is assumed to be equivalent to a bundle of uniform capillary tubes of many different sizes. The distribution of the pore sizes is derived from the soil-water characteristic curve. Via the capillary stress the tensile stresses in the water phase are related to the water content or the degree of saturation and these tensile stresses, by equation (3.204) for the capillary rise, depend on the pore radius. (iii) Series-parallel models, which also start with a bundle of parallel tubes, each with a different but uniform size. These tubes are then cut normal to the direction of the flow with two resulting faces which, after some random rearrangement of the tubes, are joined again. This way the random variations of the pore sizes both in the plane normal to the direction of flow and along the direction of flow are taken into account. Once again the distribution of the pore sizes is derived from the relationship between capillary stress and degree of water saturation and the capillary rise equation [Brutsaert(2000)].

In the present formulation a model proposed by Mualem [Mualem(1974), Mualem(1976)] is employed to obtain closed-form analytical equations for the permeability coefficients of the soil with respect to the two fluid phases water and air. The derivations are based on the consideration of a combination of two cylindrical tubes in series with radii r_1 and r_2 and lengths l_1 and l_2. The lengths are assumed to be proportional to the radii, $l_1/l_2 = r_1/r_2$. The flow in each tube is regarded to obey Poiseuille's equation. According to [Mualem(1976)], the hydraulic conductivity is then found to vary with the product $r_1 r_2$ and the relative permeability coefficient can be calculated as

$$k^{rw} = S_e^p \frac{\int_{R_{min}}^{R} \int_{R_{min}}^{R} r_1 r_2 \, f(r_1) f(r_2) \, dr_1 dr_2}{\int_{R_{min}}^{R_{max}} \int_{R_{min}}^{R_{max}} r_1 r_2 \, f(r_1) f(r_2) \, dr_1 dr_2} , \qquad (3.273)$$

where both radii r_1 and r_2 are assumed to vary between R_{min} and R_{max}. In (3.273) the quantities R_{min}, R and R_{max} represent the minimum, an arbitrary and the maximum pore radius in a soil, respectively. R can be interpreted as a critical pore radius separating the voids filled with water (radii smaller than R) from the voids filled with air (radii greater than R). The product $a(r_1, r_2) = f(r_1) f(r_2) \, dr_1 dr_2$ is the probability of pores of radius r_1 encountering pores of radius r_2 at a certain location. $f(r_1)$ and $f(r_2)$ are pore water distribution functions yielding expressions of the type

$$\int_{R_{min}}^{R} f(r)\,dr = \theta(R) \quad \text{and} \quad \int_{R_{min}}^{R_{max}} f(r)\,dr = \theta_{sat}, \qquad (3.274)$$

where $\theta(R)$ and θ_{sat} denote the current water content and the water content at fully saturated conditions, respectively. The integrals (3.274) can be explained such that a pore with an arbitrary radius r is filled with water if r lies within the range of R_{min} and R, thus yielding a water content of the soil of $\theta(R)$ depending of course on the value of the upper limit R of the integral. If the integration covers all possible pore radii of the soil and r varies within R_{min} and R_{max} as assumed, the water content for a completely saturated soil θ_{sat} is obtained. The meaning of the coefficient S_e^p in equation (3.273) is twofold, it contains (i) a correction factor accounting for 'partial correlation' between the pores of the radii r_1 and r_2 at a given water content $\theta(R)$ and (ii) a tortuosity factor taking into account the eccentricity of the flow path [Mualem(1976)]. According to [Burdine(1953)], both the correlation and tortuosity factors are power functions of the water content. Equation (3.273) can then be reformulated as

$$k^{rw} = S_e^p \frac{\int_{R_{min}}^{R} r_1 f(r_1)\,dr_1 \int_{R_{min}}^{R} r_2 f(r_2)\,dr_2}{\int_{R_{min}}^{R_{max}} r_1 f(r_1)\,dr_1 \int_{R_{min}}^{R_{max}} r_2 f(r_2)\,dr_2} = S_e^p \left[\frac{\int_{R_{min}}^{R} r f(r)\,dr}{\int_{R_{min}}^{R_{max}} r f(r)\,dr} \right]^2. \qquad (3.275)$$

For water filled pores the relationship $f(r)\,dr = d\theta$ has to hold. According to equation (3.207), the pore radius is connected to the capillary stress. In a rather general form this relationship can be written as $r = C/p^c$ with C being a constant. Applying these two expressions to equation (3.275) and using the effective degree of saturation instead of the water content results in the following relationship for the relative permeability coefficient with respect to the water phase:

$$k^{rw} = S_e^p \left[\frac{\int_0^{S_e} (1/p^c)\,ds_e}{\int_0^1 (1/p^c)\,ds_e} \right]^2. \qquad (3.276)$$

It should be noted that the constant C cancels in (3.276). The capillary stress is of course a function of the effective degree of saturation, $p^c = p^c(S_e)$, as can be seen from the equations (3.270) and (3.272). To determine the power p Mualem investigated a number of forty-five different soils with data on drainage available in the literature from which he concluded a value of $p = 0.5$ to fit best for a great many of soils [Mualem(1976)].

For a further use of the equation (3.276) the capillary stress p^c as a function of the effective degree of saturation S_e is needed. The relationship (3.272) proposed by van Genuchten can be rewritten in an inverse form as

$$p^c(S_e) = \left(S_e^{-1/m} - 1\right)^{1/n} , \qquad (3.277)$$

where the air entry value p_b^c has been omitted because it cancels in equation (3.276) appearing both in the numerator and in the denominator. The inverse, $1/p^c(S_e)$, is then obtained as

$$\frac{1}{p^c(S_e)} = \left(\frac{1}{S_e^{-1/m} - 1}\right)^{1/n} = \left(\frac{S_e^{1/m}}{1 - S_e^{1/m}}\right)^{1/n} . \qquad (3.278)$$

Subsequently, to avoid confusion between the integrand and the limits of the integral, S_e is replaced by s_e in the integrand. Insertion of the relationship (3.278) into the integral in the numerator of equation (3.276) yields

$$\int_0^{S_e} \frac{1}{p^c(s_e)} \, ds_e = \int_0^{S_e} \left(\frac{s_e^{1/m}}{1 - s_e^{1/m}}\right)^{1/n} ds_e . \qquad (3.279)$$

For convenience the substitutions

$$s_e = y^m , \qquad ds_e = m \, y^{m-1} \, dy , \qquad y = s_e^{1/m} , \qquad (3.280)$$

may be introduced. Using the definitions (3.280) together with equation (3.279) results in the following form of the respective integral:

$$\int_0^{S_e} \frac{1}{p^c(s_e)} \, ds_e = \int_0^{S_e^{1/m}} \left(\frac{y}{1-y}\right)^{1/n} m \, y^{m-1} \, dy = \int_0^{S_e^{1/m}} m \, y^{m-1+1/n} (1-y)^{-1/n} \, dy . \qquad (3.281)$$

After inserting (3.278) into the integral in the denominator of equation (3.276) and employing again the definitions (3.280) one obtains

$$\int_0^1 \frac{1}{p^c(s_e)} \, ds_e = \int_0^1 \left(\frac{s_e^{1/m}}{1 - s_e^{1/m}}\right)^{1/n} ds_e = \int_0^1 m \, y^{m-1+1/n} (1-y)^{-1/n} \, dy . \qquad (3.282)$$

As already mentioned, in general the parameters m and n are independent. However, as reported in [Genuchten(1985)], the relationship $n = 1/(1 - m)$ proved to give good approximations for experimental data of a broad range of soils. Using this simplifying assumption in the form $1/n = (1-m)$, equation (3.281) can be formulated as

$$\int_0^{S_e} \frac{1}{p^c(s_e)} \, ds_e = m \int_0^{S_e^{1/m}} (1-y)^{m-1} \, dy = m \left[-\frac{1}{m}(1-y)^m \right]_0^{S_e^{1/m}} = 1 - \left(1 - S_e^{1/m}\right)^m . \quad (3.283)$$

Employing the above relationship between m and n to equation (3.282) yields a constant:

$$\int_0^1 \frac{1}{p^c(s_e)} \, ds_e = m \int_0^1 (1-y)^{m-1} \, dy = m \left[-\frac{1}{m}(1-y)^m \right]_0^1 = 1 . \quad (3.284)$$

Keeping in mind the equations (3.273) and (3.274$_2$), the latter result should be expected, representing the fully saturated soil ($S_e = 1$ if $S^w = S_s^w = 1.0$, confer definition (3.269)). Finally, the expressions (3.283) and (3.284) are inserted into the equation (3.276) for the coefficient of relative permeability with respect to the water phase resulting in

$$k^{rw} = \sqrt{S_e} \left[1 - \left(1 - S_e^{1/m}\right)^m \right]^2 . \quad (3.285)$$

To obtain the actual value of the permeability with respect to the water phase, the relative permeability coefficient (3.285) has to be multiplied by the coefficient of permeability for the fully saturated state (compare equation (3.213)).

The coefficient of relative permeability with respect to the air phase can be derived similar to the relative permeability for the water phase (3.276). However, the limits of the integral in the numerator have to be changed. In equation (3.276) this integral is evaluated within the range $[0, S_e]$, thus covering the water filled pores. For deriving the relative permeability with respect to the air phase this integral should catch the part of the void space filled with air, i.e. the counterpart of the integral in (3.276) is needed. Hence, for the relative air permeability the integral in the numerator should be evaluated from S_e to one. Similar considerations apply to the tortuosity factor. Therefore, the coefficient of relative permeability with respect to the air phase can be calculated as [Mualem(1976)]

$$k^{ra} = \sqrt{1 - S_e} \left[\frac{\int_{S_e}^1 (1/p^c) \, ds_e}{\int_0^1 (1/p^c) \, ds_e} \right]^2 . \quad (3.286)$$

When evaluating the integral in the numerator of (3.286), van Genuchten's equation in the form (3.278) has to be inserted into the integrand. Subsequent application of the definitions (3.280) and the relationship $1/n = (1-m)$ between the parameters m and n yields the following form for the numerator,

$$\int_{S_e}^{1} \frac{1}{p^c(s_e)} ds_e = m \int_{S_e^{1/m}}^{1} (1-y)^{m-1} dy = m \left[-\frac{1}{m}(1-y)^m \right]_{S_e^{1/m}}^{1} = \left(1 - S_e^{1/m}\right)^m , \quad (3.287)$$

and the denominator is calculated as in (3.284). Insertion into (3.286) finally results in the equation for the relative permeability coefficient with respect to the air phase:

$$k^{ra} = \sqrt{1 - S_e} \left(1 - S_e^{1/m}\right)^{2m} . \quad (3.288)$$

Similar to the water phase the permeability with respect to the air phase is obtained by multiplying the relative permeability coefficient (3.288) by the permeability for the completely dry state (equation (3.213)).

Experimental data for the relative permeabilities depending on the degree of saturation for a typical soil (sand) are shown in Figure 3.14(a) [Wyckoff(1936)]. Considering the two curves in Figure 3.14(a) for an increasing degree of saturation of the particular fluid phase also the corresponding relative permeability increases, i.e. the higher the degree of saturation, the higher the conductivity of the soil with respect to the fluid becomes. The inclination of the relative water permeability curve is relatively small for low degrees of water saturation and increases for higher values of saturation. This results from the fact that at low degrees of saturation the very small pores are filled with water first. The larger pores, which, hydraulically speaking, are much more efficient, are filled at the higher degrees of water saturation which explains the larger slope of the curve near the fully water saturated state.

On the other hand, for the relative air permeability the behaviour is somewhat contrary. Here, for low degrees of air saturation the large pores, which contribute to the airflow substantially, are filled first. Therefore, the slope of the relative air permeability curve is rather steep already for low degrees of air saturation. The small pores are filled only when the degree of air saturation reaches unity. Since they do not contribute to the airflow very much, the inclination of the curve decreases for high degrees of air saturation.

Figure 3.14(a) also indicates very clearly the existence of both a residual degree of water saturation and a residual degree of air saturation. A reduction of the degree of water saturation below the residual value is not possible by means of dewatering due to an applied air pressure, only evaporation or diffusion could yield lower degrees of water saturation. In the range below the residual degree of water saturation the water phase is no longer continuous which prevents any water flow. Within this range a change in the degree of water saturation is less important since the water only fills the very small pores which barely contribute to the airflow. Hence, the relative air permeability coefficient does not show any significant changes below the residual degree of water saturation.

The residual degree of air saturation may be encountered due to entrapped air bubbles at the end of a wetting process (confer Figure 3.10). A further reduction of the air saturation is possible due to compression or diffusion of the air. In the range below the residual air saturation the air phase is not continuous any longer and thus no airflow exists. However, because of the presence of air bubbles in large pores a change in the air saturation yields also a significant change in the relative coefficient of water permeability.

The range in between these two residual degrees of saturation is characterised by a continuous distribution of both fluid phases, water and air, and consequently by the existence of a two-phase flow. According to Kezdi [Kezdi(1976)], these residual degrees of saturation are negligible for non-cohesive materials. On the other hand, for cohesive

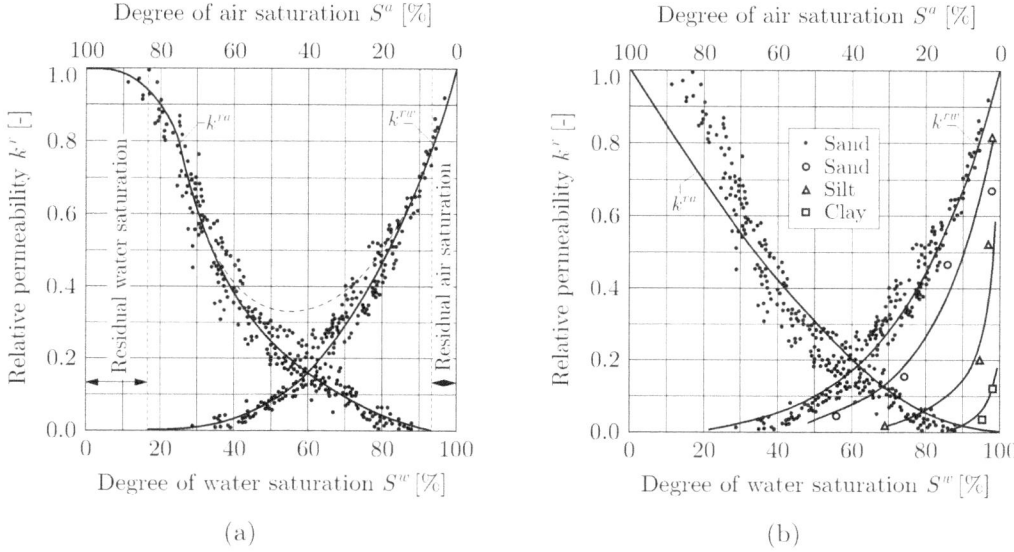

Figure 3.14: (a) Experimental data of the relative permeabilities for a sand [Wyckoff(1936)] and (b) approximation of experimental results by the empirical equations proposed by Mualem [Mualem(1976)].

soils they may become rather significant. Values of 10.0 % to 30.0 % for the water phase and of 10.0 % to 40.0 % for the air phase are reasonable depending on the type of soil under consideration.

Figure 3.14(b) shows plots of the relative permeability curves according to the equations (3.285) and (3.288) in comparison with experimental data [Wyckoff(1936), Mualem(1976)]. The empirical relations approximate the test results fairly well. Only for very high degrees of air saturation the empirical curve for the relative air permeability deviates somewhat more from the experimental data of the particular soil.

When considering the problem of dewatering of soils by means of compressed air, an initially completely water saturated domain exhibits drainage in the course of the process. Therefore, the maximum value of the degree of saturation S_s^w in the matric suction versus degree of saturation curve is specified as $S_s^w = 1.0$, or at least very close to one if a small amount of air bubbles is taken into account which might be present in the groundwater. As Figure 3.14(a) indicates, the air permeability is negligibly small within a certain range close to fully saturated conditions. However, when using a maximum degree of water saturation of $S_s^w = 1.0$ (as for the matric suction versus degree of saturation curve) in combination with the effective degree of saturation S_e and equation (3.288), compared to the water permeability the obtained air permeability at nearly saturated conditions is not negligible at all. The ratio of air to water permeability becomes the larger, the less permeable the soil is. Table 3.3 shows this ratio $(k^{oa}k^{ra})/(k^{ow}k^{rw})$ for a degree of saturation of $S^w = 0.99$ and a certain range of soils (different m values). The value of S_r^w has been chosen as $S_r^w = 0.0$. For a residual saturation greater than zero the ratio becomes even larger. For simplicity the air permeability for the completely dry state,

m	k^{ow}	k^{oa}	k^{rw} [-]	k^{ra} [-]	$(k^{oa}k^{ra})/(k^{ow}k^{rw})$
[-]	[m/s]	[m/s]	Eq. (3.285)	Eq. (3.288)	[-]
0.80	$1.00 \cdot 10^{-4}$	$7.44 \cdot 10^{-3}$	0.936188	0.000090	0.0072
0.65	$1.00 \cdot 10^{-5}$	$7.44 \cdot 10^{-4}$	0.867616	0.000438	0.0376
0.36	$1.00 \cdot 10^{-6}$	$7.44 \cdot 10^{-5}$	0.523898	0.007528	1.0691
0.17	$1.00 \cdot 10^{-7}$	$7.44 \cdot 10^{-6}$	0.147322	0.037848	19.1140

Table 3.3: Permeability ratio for $S^w = 0.99$ and different soils.

k^{oa}, is calculated from the water permeability k^{ow} using the dynamic viscosities of both fluids, $k^{oa} = k^{ow} \mu^w / \mu^a$, where a ratio of $\mu^w / \mu^a = 74.4$ has been employed which is valid for a temperature of $10.0\,^\circ$ C.

For the numerical simulation of a dewatering process usually initial conditions for both the hydrostatic water and air stresses have to be specified. These initial stresses increase with depth below the groundwater table and differ from each other in accordance to a certain degree of saturation as will be outlined in some more detail in Chapter 6. Insertion of these fluid stresses into Darcy's law yields a zero water velocity while the velocity of the air phase equals the air permeability. Since the air permeability at the almost water saturated state of the soil is not negligible when computed from equation (3.288), an airflow in the direction of the groundwater table occurs which, due to the coupled nature of the problem, also induces a water flow. However, both the water and airflow do not exist in reality.

To avoid these fluid flows resulting purely from the numerical simulation of the initial conditions, according to the work of Hochgürtel [Hochguertel(1998)], a critical degree of water saturation S_k^w is introduced which is then used to calculate the effective degree of saturation S_{ek}, entering the relative permeability equations (3.285) and (3.288). Thus, the effective degree of saturation employed for computing the relative permeability coefficients is defined as

$$S_{ek} = \frac{S^w - S_r^w}{S_k^w - S_r^w} \tag{3.289}$$

in analogy to equation (3.269). Inserting this relationship into (3.285) and (3.288) results in the relative permeability for the water phase reading as

$$k^{rw} = \sqrt{S_{ek}} \left[1 - \left(1 - S_{ek}^{1/m}\right)^m \right]^2, \tag{3.290}$$

and the relative permeability for the air phase being

$$k^{ra} = \sqrt{1 - S_{ek}} \left(1 - S_{ek}^{1/m}\right)^{2m}. \tag{3.291}$$

This critical value for the degree of saturation, S_k^w, can be interpreted as some sort of limit. If the degree of saturation S^w exceeds this limit S_k^w, no airflow should occur in the soil which is achieved in an approximate manner by choosing a very small value for the relative air permeability coefficient.

3.6 Field equations for the three-phase medium

3.6.1 Introduction

After the specification of a set of constitutive equations in the previous Section 3.5, the general forms of the various balance equations as derived in Section 3.4 may now be specialised for the individual phases by introducing the constitutive assumptions into these general balance laws. Therefore, this section is dedicated to the formulation of the mass and linear momentum balance equations for the different constituents of the model – the soil skeleton and the two fluid phases – as well as for the whole three-phase mixture. First, the mass balance equations for the soil skeleton and for the fluid phases are derived and second, the linear momentum balance equations for the soil skeleton and for the fluid phases as well as for the three-phase medium are given. The finite element formulation discussed in the subsequent chapter of this work will be based on these equations. For reasons of simplicity the bar, serving as the averaging symbol for the velocities and accelerations, is omitted in this section and in the remainder of the work because all quantities are referred to the macroscopic situation.

3.6.2 Mass balance for the soil skeleton

The general macroscopic mass balance equation (3.122) can be specialised for the soil skeleton ($\pi = s$) as

$$\frac{D^s[(1-n)\rho^s]}{Dt} + (1-n)\rho^s \operatorname{div} \mathbf{v}^s = 0 \;, \tag{3.292}$$

where the volume fraction for the solid phase, $\eta^s = (1-n)$, has been introduced as given in Table 3.1. The exchange term $\eta^\pi \rho^\pi e_\rho^\pi$ in (3.122), i.e. the exchange of mass with the other phases, is assumed to be zero for the solid phase. Performing the material time derivative of the first term in (3.292) and dividing the resulting equation by ρ^s yields

$$-\frac{D^s n}{Dt} + \frac{1-n}{\rho^s}\frac{D^s \rho^s}{Dt} + (1-n)v^s_{i,i} = 0 \;. \tag{3.293}$$

In (3.293) index notation has been used for the divergence term.

Since the solid phase is considered to be incompressible, which is a common assumption in soil mechanics (one must be careful in distinguishing between the solid phase and the soil skeleton, the latter is of course deformable), the material time derivative of the density of the solid phase in equation (3.293) has to vanish, that is

$$\frac{D^s \rho^s}{Dt} = 0 \, . \tag{3.294}$$

Because of the incompressible soil grains the divergence of the velocity of the solid phase, $v^s_{i,i}$, can be related to the rate of the volumetric strain of the soil skeleton as

$$v^s_{i,i} = \frac{\partial \dot{u}_i}{\partial x_i} = \dot{\varepsilon}_{ii} = \dot{\varepsilon}^{vol} \, . \tag{3.295}$$

Insertion of the equations (3.294) and (3.295) into (3.293) finally yields the mass balance equation for the soil skeleton:

$$\frac{D^s n}{Dt} = (1-n)\dot{\varepsilon}^{vol} \, . \tag{3.296}$$

Due to the assumed incompressibility of the solid phase the rate of the volumetric strain of the soil skeleton, $\dot{\varepsilon}^{vol}$, multiplied by the volume fraction $(1-n)$ of the solid phase is equivalent to the positive rate of the porosity.

3.6.3 Mass balance for the fluid phases

Specifying the general macroscopic mass balance equation (3.122) for a fluid phase with $\pi = f$ and using the volume fraction of the particular phase as given in Table 3.1, $\eta^f = nS^f$, yields

$$\frac{D^f(nS^f \rho^f)}{Dt} + nS^f \rho^f \operatorname{div} \mathbf{v}^f = 0 \, . \tag{3.297}$$

The right hand side of this equation is zero because any mass exchange for the fluid phases is neglected, e.g., the change of water into vapour is not taken into account which might be justified due to the assumption of isothermal conditions. Performing the material time derivative of the first term on the left hand side of equation (3.297) and dividing by $S^f \rho^f$ results in

$$\frac{D^f n}{Dt} + \frac{n}{S^f}\frac{D^f S^f}{Dt} + \frac{n}{\rho^f}\frac{D^f \rho^f}{Dt} + n \operatorname{div} \mathbf{v}^f = 0 \, . \tag{3.298}$$

For convenience the material time derivatives with respect to the fluid phase f are converted into material time derivatives with respect to the moving solid s by employing equation (3.13) with $\alpha = f$ and $\pi = s$. This yields

$$\frac{D^s n}{Dt} + \operatorname{grad} n \cdot \mathbf{v}^{fs} + \frac{n}{S^f}\frac{D^s S^f}{Dt} + \frac{n}{S^f}\operatorname{grad} S^f \cdot \mathbf{v}^{fs} +$$

$$+ \frac{n}{\rho^f}\frac{D^s \rho^f}{Dt} + \frac{n}{\rho^f}\operatorname{grad}\rho^f \cdot \mathbf{v}^{fs} + n \operatorname{div}\mathbf{v}^f = 0 \,, \tag{3.299}$$

where $\mathbf{v}^{fs}$ denotes the velocity of the fluid phase relative to the solid. Inherent in this terminology is the commonly employed assumption in soil mechanics of an incompressible solid phase, as previously stated in Section 3.5.8. In this case the velocities of the soil skeleton and of the solid phase may be regarded as equivalent. For a compressible solid phase the superscript s of the relative velocity $\mathbf{v}^{fs}$ should be avoided in order to prevent a misleading interpretation because the velocity of the soil skeleton is then constituted by two portions, namely one due to the deformation of the solid particles itself and a second one due to movements of the particles as rigid bodies.

Using the mass balance equation for the soil skeleton (3.296) together with (3.295) to rewrite the first term and employing index notation for the divergence term results in

$$(1-n)v_{i,i}^s + \frac{n}{S^f}\frac{D^s S^f}{Dt} + \frac{n}{\rho^f}\frac{D^s \rho^f}{Dt} + n\, v_{i,i}^f +$$

$$+ \operatorname{grad} n \cdot \mathbf{v}^{fs} + \frac{n}{S^f}\operatorname{grad} S^f \cdot \mathbf{v}^{fs} + \frac{n}{\rho^f}\operatorname{grad}\rho^f \cdot \mathbf{v}^{fs} = 0 \,. \tag{3.300}$$

The definition of the relative velocity (3.10) can be applied to the first and the fourth term of (3.300). Multiplying the respective equation by S^f yields

$$S^f v_{i,i}^s + n\frac{D^s S^f}{Dt} + \frac{nS^f}{\rho^f}\frac{D^s \rho^f}{Dt} + nS^f v_{i,i}^{fs} +$$

$$+ S^f \frac{\partial n}{\partial x_i}v_i^{fs} + n\frac{\partial S^f}{\partial x_i}v_i^{fs} + \frac{nS^f}{\rho^f}\frac{\partial \rho^f}{\partial x_i}v_i^{fs} = 0 \,. \tag{3.301}$$

It should be mentioned that neither the intrinsic velocity $\mathbf{v}^f$ of the fluid nor its relative velocity $\mathbf{v}^{fs}$ are measured in laboratory or field tests but rather the so-called artificial velocity $\tilde{\mathbf{v}}^{fs}$, i.e. the relative velocity averaged by the portion nS^f of the particular fluid phase. Thus, in the present context the artificial velocity can be defined as

$$\tilde{v}_i^{fs} = nS^f v_i^{fs} \qquad \text{or} \qquad v_i^{fs} = \frac{\tilde{v}_i^{fs}}{nS^f} \,. \tag{3.302}$$

Chapter 3. Mechanics of the coupled three-phase model

The divergence of the relative velocity $\mathbf{v}^{fs}$ can then be calculated as

$$v_{i,i}^{fs} = \frac{\partial}{\partial x_i}\left(\frac{\tilde{v}_i^{fs}}{nS^f}\right) = \frac{\tilde{v}_{i,i}^{fs}}{nS^f} - \frac{\tilde{v}_i^{fs}}{(nS^f)^2}\frac{\partial(nS^f)}{\partial x_i} , \qquad (3.303)$$

and hence for the fourth term of equation (3.301) one obtains

$$nS^f v_{i,i}^{fs} = \tilde{v}_{i,i}^{fs} - \frac{\tilde{v}_i^{fs}}{nS^f}\frac{\partial(nS^f)}{\partial x_i} . \qquad (3.304)$$

Performing the partial derivative in (3.304) and inserting the resulting equation into (3.301) yields

$$S^f v_{i,i}^s + n\frac{D^s S^f}{Dt} + \frac{nS^f}{\rho^f}\frac{D^s \rho^f}{Dt} + \tilde{v}_{i,i}^{fs} - \frac{\tilde{v}_i^{fs}}{n}\frac{\partial n}{\partial x_i} - \frac{\tilde{v}_i^{fs}}{S^f}\frac{\partial S^f}{\partial x_i} +$$
$$+ S^f \frac{\partial n}{\partial x_i}v_i^{fs} + n\frac{\partial S^f}{\partial x_i}v_i^{fs} + \frac{nS^f}{\rho^f}\frac{\partial \rho^f}{\partial x_i}v_i^{fs} = 0 . \qquad (3.305)$$

Finally, this equation may be simplified by means of the definition of the artificial velocity (3.302) – terms five and six then cancel out with terms seven and eight, respectively, yielding the mass balance equation for the fluid phase f, $f = w, a$:

$$S^f \dot{\varepsilon}^{vol} + n\frac{D^s S^f}{Dt} + \frac{nS^f}{\rho^f}\frac{D^s \rho^f}{Dt} + \operatorname{div} \tilde{\mathbf{v}}^{fs} + \frac{1}{\rho^f}\operatorname{grad} \rho^f \cdot \tilde{\mathbf{v}}^{fs} = 0 , \qquad (3.306)$$

where (3.295) and vector notation for divergence and gradient terms have been used.

3.6.4 Alternative derivation of the mass balance equations

Based on the assumption that the mass of an individual constituent π has to stay constant for the body under consideration, an alternative way of deriving the mass balance equations is introduced in this section. Any sources or sinks in the domain are excluded. The mass of the constituent π can then be determined as

$$m^\pi = \int_V \eta^\pi \rho^\pi \, dV = const. , \qquad (3.307)$$

where the volume fraction of the respective phase has been used according to definition (3.40). Since this mass m^π is considered as a constant in any configuration, its material time derivative has to vanish, that is

$$\frac{D^\pi m^\pi}{Dt} = 0 . \qquad (3.308)$$

Employing the definition for the material time derivative of a volume integral, as given in equation (3.14), yields for the derivative of (3.307)

$$\frac{D^\pi m^\pi}{Dt} = \int_V \left[\frac{D^\pi (\eta^\pi \rho^\pi)}{Dt} + \eta^\pi \rho^\pi \, \text{div} \, \mathbf{v}^\pi \right] dV = 0 \ . \tag{3.309}$$

The material time derivative in equation (3.309) is equal to zero whenever the condition of a vanishing integrand is satisfied, that is

$$\frac{D^\pi (\eta^\pi \rho^\pi)}{Dt} + \eta^\pi \rho^\pi \, \text{div} \, \mathbf{v}^\pi = 0 \ . \tag{3.310}$$

Finally, this equation can be specified for the individual phases, i.e. for the soil skeleton with $\pi = s$ and for the fluid phases with $\pi = f$. If the volume fractions according to Table 3.1 are used, the mass balance equation for the soil skeleton is obtained as

$$\frac{D^s [(1-n)\rho^s]}{Dt} + (1-n)\rho^s \, \text{div} \, \mathbf{v}^s = 0 \ , \tag{3.311}$$

which is equivalent to equation (3.292) of Section 3.6.2. For the fluid phases this specification yields

$$\frac{D^f (nS^f \rho^f)}{Dt} + nS^f \rho^f \, \text{div} \, \mathbf{v}^f = 0 \ , \tag{3.312}$$

which may be compared to equation (3.297) of Section 3.6.3.

3.6.5 Linear momentum balance for the soil skeleton

Specialisation of the general macroscopic linear momentum balance equation (3.162) for the soil skeleton with $\pi = s$ yields

$$\text{div}[(1-n)\boldsymbol{\sigma}^s] + (1-n)\rho^s(\mathbf{g} - \mathbf{a}^s) + (1-n)\rho^s \mathbf{t}^s = \mathbf{0} \ , \tag{3.313}$$

where the volume and area fractions given in Table 3.1 have been used. The intrinsic momentum supply term $\eta^s \rho^s \mathbf{e}^s_{\rho \hat{\mathbf{r}}^s}$ for the solid phase is assumed to be zero.

3.6.6 Linear momentum balance for the fluid phases

To obtain the linear momentum balance law for a fluid phase, the general macroscopic linear momentum balance equation (3.162) is specified with $\pi = f$,

$$\text{div}(nS^f \boldsymbol{\sigma}^f) + nS^f \rho^f (\mathbf{g} - \mathbf{a}^f) + nS^f \rho^f \mathbf{t}^f = \mathbf{0} \ . \tag{3.314}$$

The employed volume and area fractions have been given in Table 3.1 and again, as for the soil skeleton, the intrinsic momentum supply term is not taken into account. The definition of the hydrostatic stress tensor in the fluid phase (3.185) may now be inserted into equation (3.314) and for the exchange of momentum due to mechanical interactions, $nS^f \rho^f \, \mathbf{t}^f$, equation (3.210) can be employed which yields the following form of the balance law,

$$\operatorname{div}\left(nS^f p^f \, \mathbf{I}\right) + nS^f \rho^f \left(\mathbf{g} - \mathbf{a}^f\right) - p^f \operatorname{grad}\left(nS^f\right) - nS^f \, \mathbf{R}^f \cdot \mathbf{v}^{fs} = \mathbf{0} \; . \tag{3.315}$$

The first term in this equation containing the divergence of the hydrostatic fluid stress p^f can be calculated using

$$\operatorname{div}\left(nS^f p^f \, \mathbf{I}\right) = nS^f \operatorname{div}\left(p^f \, \mathbf{I}\right) + p^f \operatorname{grad}\left(nS^f\right) \; . \tag{3.316}$$

Applying the definition of the artificial velocity as given in (3.302_1) to the last term of (3.315) results in

$$nS^f \operatorname{div}\left(p^f \, \mathbf{I}\right) + p^f \operatorname{grad}\left(nS^f\right) + nS^f \rho^f \left(\mathbf{g} - \mathbf{a}^f\right) - p^f \operatorname{grad}\left(nS^f\right) - \mathbf{R}^f \cdot \tilde{\mathbf{v}}^{fs} = \mathbf{0} \; . \tag{3.317}$$

Obviously the second and the fourth term cancel and the remainder of the equation can be reordered to calculate the artificial velocity as

$$\tilde{\mathbf{v}}^{fs} = \left(\mathbf{R}^f\right)^{-1} \cdot \left[nS^f \operatorname{div}\left(p^f \, \mathbf{I}\right) + nS^f \rho^f \left(\mathbf{g} - \mathbf{a}^f\right)\right] \; . \tag{3.318}$$

For convenience index notation is used to reformulate the divergence term,

$$\operatorname{div}\left(p^f \, \mathbf{I}\right) = \left(p^f \delta_{ij}\right)_{,j} = p^f_{,j} \delta_{ij} = p^f_{,i} = \operatorname{grad} p^f \; . \tag{3.319}$$

For the inverse of the tensor $\mathbf{R}^f$ the definition given in (3.213) can be substituted into (3.318) which finally results in the following form of Darcy's law, sometimes also called generalised Darcy's law,

$$\tilde{\mathbf{v}}^{fs} = \frac{\mathbf{k}^{of} k^{rf}}{\rho^f g} \cdot \left[\operatorname{grad} p^f + \rho^f \left(\mathbf{g} - \mathbf{a}^f\right)\right] \; , \tag{3.320}$$

which is assumed to be valid for the transport of both water and air considering slow phenomena. In equation (3.320) the permeability tensor $\mathbf{k}^{of}$ is multiplied by a scalar factor k^{rf}, the so-called relative permeability coefficient (see Section 3.5.13), which is introduced in order to account for the flow of the two fluids water and air through the pores of the soil.

The familiar form of Darcy's law is obtained from (3.320) when dealing with quasi-static conditions ($\mathbf{a}^f$ can be neglected) and a completely water saturated soil (k^{rf} is equal to one):

$$\tilde{\mathbf{v}}^{fs} = \frac{\mathbf{k}^{of}}{\rho^f g} \cdot \left(\operatorname{grad} p^f + \rho^f \mathbf{g}\right) . \tag{3.321}$$

For soils which can be approximated as isotropic materials the permeability tensor $\mathbf{k}^{of}$ reduces to a scalar k^{of} [Fredlund(1993), Verruijt(1995)].

A rather general form of Darcy's law from the viewpoint of modern thermodynamics is discussed in a recent paper by Smith [Smith(2000)]. This approach elucidates the relationship between Darcy's artificial velocity and the Gibbs free energy of the pore fluid. The individual components contributing to Gibbs free energy for the case of fully and partially saturated soils, respectively, are explained.

3.6.7 Linear momentum balance for the mixture

Summation of the linear momentum balance equation for the soil skeleton (3.313) and of the linear momentum balance equation for the fluid phase (3.314), the latter for both the water and the air phase, yields the linear momentum balance for the three-phase mixture. According to the constraint (3.154), any exchange terms have to vanish in this sum. Therefore, the linear momentum balance for the three-phase medium can be formulated as

$$\operatorname{div}\left[(1-n)\boldsymbol{\sigma}^s + nS^w\boldsymbol{\sigma}^w + nS^a\boldsymbol{\sigma}^a\right] + \left[(1-n)\rho^s + nS^w\rho^w + nS^a\rho^a\right]\mathbf{g} -$$

$$- (1-n)\rho^s\mathbf{a}^s - nS^w\rho^w\mathbf{a}^w - nS^a\rho^a\mathbf{a}^a = \mathbf{0} . \tag{3.322}$$

Employing the definitions for the averaged stress tensor (3.190) and for the averaged density (3.184) of the three-phase mixture, equation (3.322) can be reduced to

$$\operatorname{div}\tilde{\boldsymbol{\sigma}} + \tilde{\rho}\mathbf{g} = (1-n)\rho^s\mathbf{a}^s + nS^w\rho^w\mathbf{a}^w + nS^a\rho^a\mathbf{a}^a . \tag{3.323}$$

When considering the special case of quasi-static conditions, the acceleration terms on the right hand side of (3.323) can be neglected, thus yielding the equilibrium equations

$$\operatorname{div}\tilde{\boldsymbol{\sigma}} + \tilde{\rho}\mathbf{g} = \mathbf{0} . \tag{3.324}$$

Compared to the conventional form of the equilibrium conditions (confer, e.g., [Malvern(1969)]) equation (3.324) only differs in the tilde added to the quantities $\boldsymbol{\sigma}$

and ρ which indicates that these quantities are the averaged stress tensor $\tilde{\boldsymbol{\sigma}}$ and the averaged density $\tilde{\rho}$ of the three-phase medium, respectively.

3.6.8 Alternative derivation of the equilibrium equations

Another possible approach to derive the equilibrium equations for the three-phase medium consists in applying the fundamental law of dynamics to an infinitesimal volume element of the three-phase medium (representative elementary volume). Consider an inertial reference frame where d'Alembert's principle is assumed to hold:

$$d\tilde{\mathbf{R}} + (-d\tilde{m}\,\tilde{\mathbf{a}}) = \mathbf{0} \ . \tag{3.325}$$

In equation (3.325) $d\tilde{\mathbf{R}}$ describes a vector of resulting forces exerted on the infinitesimal volume element under consideration, containing body forces, surface tractions and internal forces, and $(-d\tilde{m}\,\tilde{\mathbf{a}})$ denotes the inertia force. The tilde indicates that the particular quantities refer to the averaged volume element. Hence, in the formulation (3.325) the fundamental law of dynamics can be interpreted as equilibrium condition for a certain time t. In the inertia term $\tilde{\mathbf{a}}$ is the vector of acceleration. For the case of a three-phase medium the averaged volume element depicted in Figure 3.1(b) is employed as basis for the subsequent derivations. The mass of this volume element, $d\tilde{m}$, can be calculated as

$$d\tilde{m} = \tilde{\rho}\,dv \ , \tag{3.326}$$

with $\tilde{\rho}$ denoting the averaged density of the three-phase medium (confer equation (3.184) of Section 3.5.2).

For deriving the equilibrium equations a volume element bounded by surfaces normal to the coordinate axes x_1, x_2, x_3 of a Cartesian coordinate system is considered (Figure 3.15). When assuming edges of the length dx_1, dx_2, dx_3, the volume and the individual surface elements can be calculated as

$$dv = dx_1\,dx_2\,dx_3 \ , \quad da^{(1)} = dx_2\,dx_3 \ , \quad da^{(2)} = dx_1\,dx_3 \ , \quad da^{(3)} = dx_1\,dx_2 \ . \tag{3.327}$$

$da^{(i)}$ being a surface element normal to the respective coordinate axis x_i, $i = 1, 2, 3$. Each of the three pairs of parallel boundaries of the volume element is defined by two planes $x_i = const.$ and $x_i + dx_i = const.$, the tractions acting on these planes are $-\tilde{\mathbf{t}}^{(i)}(x_i)$ and $\tilde{\mathbf{t}}^{(i)}(x_i + dx_i)$ as depicted in Figure 3.15(a). The tilde again indicates that the respective quantities are averaged quantities of the three-phase medium. The sum of the forces being present on any two parallel surfaces of the volume element thus can be written as

$$\left[-\tilde{\mathbf{t}}^{(i)}(x_i) + \tilde{\mathbf{t}}^{(i)}(x_i+dx_i)\right]da^{(i)} \ , \quad i=1,2,3 \ . \tag{3.328}$$

When assuming the components of the stress tensor to be smooth and differentiable functions of space, the traction vector $\tilde{\mathbf{t}}^{(i)}(x_i+dx_i)$, which is acting on the surface element $x_i + dx_i = const.$, may be represented by a Taylor series in the neighbourhood of point x_i as follows:

$$\tilde{\mathbf{t}}^{(i)}(x_i+dx_i) = \tilde{\mathbf{t}}^{(i)}(x_i) + \frac{\partial \tilde{\mathbf{t}}^{(i)}(x_i)}{\partial x_i}dx_i + \cdots \ . \tag{3.329}$$

Terms of an order greater than one can be neglected in equation (3.329). Inserting the truncated Taylor series (3.329) into the sum of the forces (3.328) and using the respective

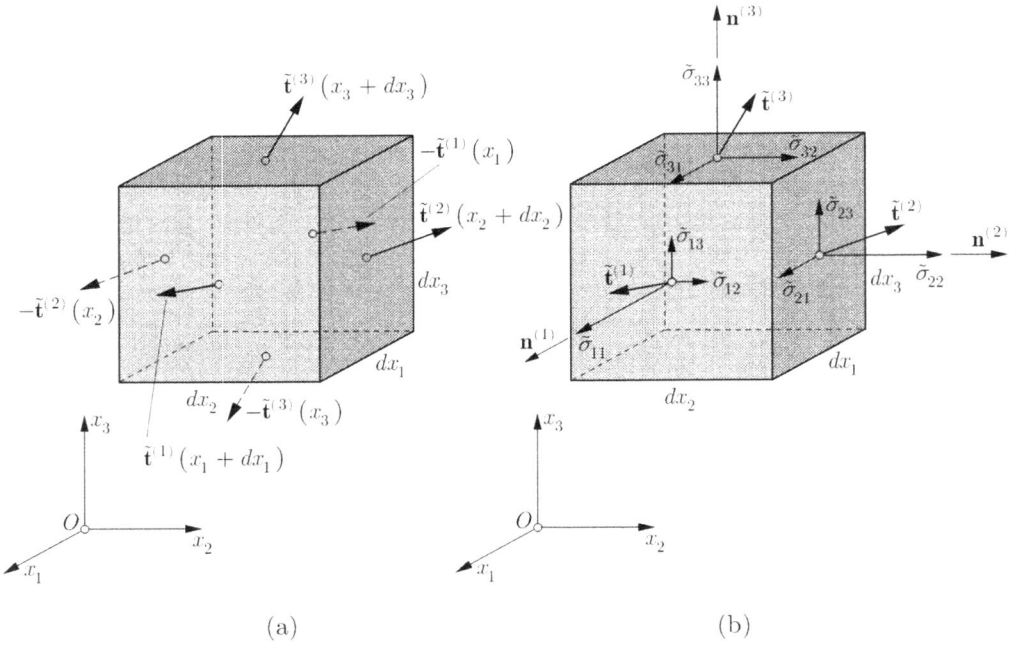

Figure 3.15: (a) Averaged stress vectors acting on the surfaces of an infinitesimal representative elementary volume (REV) and (b) decomposition of the stress vectors into the components of the averaged stress tensor.

relations (3.327) results in

$$\left[-\tilde{\mathbf{t}}^{(i)}(x_i) + \tilde{\mathbf{t}}^{(i)}(x_i) + \frac{\partial \tilde{\mathbf{t}}^{(i)}(x_i)}{\partial x_i} dx_i \right] da^{(i)} =$$

$$= \frac{\partial \tilde{\mathbf{t}}^{(i)}(x_i)}{\partial x_i} dx_i \, da^{(i)} = \frac{\partial \tilde{\mathbf{t}}^{(i)}(x_i)}{\partial x_i} dv \, , \quad i = 1, 2, 3 \, . \qquad (3.330)$$

Formulating equation (3.330) for the three coordinate directions x_1, x_2, x_3, multiplying the body force vector $\tilde{\mathbf{b}}$ by the representative elementary volume dv and inserting into the vector of the resulting forces, $d\tilde{\mathbf{R}}$, yields

$$d\tilde{\mathbf{R}} = \left(\frac{\partial \tilde{\mathbf{t}}^{(1)}}{\partial x_1} + \frac{\partial \tilde{\mathbf{t}}^{(2)}}{\partial x_2} + \frac{\partial \tilde{\mathbf{t}}^{(3)}}{\partial x_3} + \tilde{\mathbf{b}} \right) dv \, . \qquad (3.331)$$

Using (3.331) and the definition of the mass of the volume element (3.326) in combination with d'Alembert's principle (3.325) and dividing the resulting equation by the volume element dv, one obtains the following form of the equilibrium equations:

$$\frac{\partial \tilde{\mathbf{t}}^{(1)}}{\partial x_1} + \frac{\partial \tilde{\mathbf{t}}^{(2)}}{\partial x_2} + \frac{\partial \tilde{\mathbf{t}}^{(3)}}{\partial x_3} + \tilde{\mathbf{b}} - \tilde{\rho} \tilde{\mathbf{a}} = \mathbf{0} \, . \qquad (3.332)$$

The body forces for the volume element can be calculated as $\tilde{\mathbf{b}} = \tilde{\rho} \mathbf{g}$ with $\mathbf{g}$ being the vector of gravitational acceleration. For quasi-static conditions, i.e. when neglecting the acceleration term $\tilde{\rho} \tilde{\mathbf{a}}$ in (3.332), the equilibrium equations can be formulated as

$$\frac{\partial \tilde{\mathbf{t}}^{(1)}}{\partial x_1} + \frac{\partial \tilde{\mathbf{t}}^{(2)}}{\partial x_2} + \frac{\partial \tilde{\mathbf{t}}^{(3)}}{\partial x_3} + \tilde{\rho} \mathbf{g} = \mathbf{0} \, . \qquad (3.333)$$

When considering the volume element shown in Figure 3.15(b), each of the traction vectors can be written as linear combination of the stress components acting on the particular surface element, i.e. one normal stress component in the direction of the coordinate axis normal to the surface and two shear stress components lying in the particular plane and acting in the directions of the other two coordinate axes:

$$\tilde{\mathbf{t}}^{(1)} = \tilde{\sigma}_{11} \mathbf{e}_1 + \tilde{\sigma}_{12} \mathbf{e}_2 + \tilde{\sigma}_{13} \mathbf{e}_3 \, ,$$

$$\tilde{\mathbf{t}}^{(2)} = \tilde{\sigma}_{21} \mathbf{e}_1 + \tilde{\sigma}_{22} \mathbf{e}_2 + \tilde{\sigma}_{23} \mathbf{e}_3 \, , \qquad (3.334)$$

$$\tilde{\mathbf{t}}^{(3)} = \tilde{\sigma}_{31} \mathbf{e}_1 + \tilde{\sigma}_{32} \mathbf{e}_2 + \tilde{\sigma}_{33} \mathbf{e}_3 \, .$$

Insertion of (3.334) into (3.333) finally yields the three equilibrium equations for quasi-static conditions, expressed in terms of the stress components,

$$\frac{\partial \tilde{\sigma}_{11}}{\partial x_1} + \frac{\partial \tilde{\sigma}_{21}}{\partial x_2} + \frac{\partial \tilde{\sigma}_{31}}{\partial x_3} + \tilde{\rho} g_1 = 0 \; ,$$

$$\frac{\partial \tilde{\sigma}_{12}}{\partial x_1} + \frac{\partial \tilde{\sigma}_{22}}{\partial x_2} + \frac{\partial \tilde{\sigma}_{32}}{\partial x_3} + \tilde{\rho} g_2 = 0 \; , \tag{3.335}$$

$$\frac{\partial \tilde{\sigma}_{13}}{\partial x_1} + \frac{\partial \tilde{\sigma}_{23}}{\partial x_2} + \frac{\partial \tilde{\sigma}_{33}}{\partial x_3} + \tilde{\rho} g_3 = 0 \; .$$

Employing index notation these three equations can be written in a short form as

$$\tilde{\sigma}_{ji,j} + \tilde{\rho} g_i = 0 \; , \tag{3.336}$$

which is equivalent to the vector equation (3.324) given in Section 3.6.7.

3.7 Special cases included in the three-phase formulation

3.7.1 Introduction

The theoretical framework of the three-phase formulation presented in this chapter is applicable to a broad range of problems (confer also Chapter 2). The particular model considering the fluid phases water and air (or compressed air) allows for a description of various phenomena usually encountered in the field of geotechnical engineering. However, not only the complete model as proposed in this chapter may be useful, e.g., for describing the process of dewatering of soils by means of compressed air, but also different special cases, which form certain subsets of the full three-phase formulation, can be employed to tackle various tasks in geomechanics. In particular, a two-phase formulation for dewatering of soils under atmospheric conditions and a two-phase formulation for saturated conditions (consolidation) are contained in the complete model. Additionally, the uncoupled approach can be deduced. Of course the familiar equation for drained conditions may be obtained as well. Each of the subsequent sections is dedicated to one of these special cases, starting with the more complex model of dewatering under atmospheric conditions and proceeding towards the formulation for drained analyses.

3.7.2 Dewatering under atmospheric conditions

When assuming the hydrostatic stress in the air phase to be equal to the atmospheric pressure (sometimes the term 'passive air assumption' is used), a model for dewatering of soils under atmospheric conditions is obtained. In this case any changes in the hydrostatic air stresses are neglected. Consequently, the mass balance equation for the air phase can be omitted and the original three-phase model is reduced to a two-phase formulation. However, a description of unsaturated soil behaviour is still possible, i.e. the degree of water saturation may vary during the calculation. Subsequently, the equations derived in the previous sections are simplified for this special case.

Since the air stresses are considered to be atmospheric, the capillary stress (equation (3.209)) is equal to the negative hydrostatic water stress which also governs the degree of water saturation:

$$S^w = S^w(-p^w) \,. \tag{3.337}$$

The mass balance equation (3.306) can be specialised for the water phase, $f = w$, because no mass balance for the air phase has to be taken into account:

$$S^w \dot{\varepsilon}^{vol} + n \frac{D^s S^w}{Dt} + \frac{nS^w}{\rho^w} \frac{D^s \rho^w}{Dt} + \operatorname{div} \tilde{\mathbf{v}}^{ws} + \frac{1}{\rho^w} \operatorname{grad} \rho^w \cdot \tilde{\mathbf{v}}^{ws} = 0 \,. \tag{3.338}$$

In a similar manner Darcy's law (3.320) may be written for the water phase,

$$\tilde{\mathbf{v}}^{ws} = \frac{\mathbf{k}^{ow} k^{rw}}{\rho^w g} \cdot [\operatorname{grad} p^w + \rho^w(\mathbf{g} - \mathbf{a}^w)] \,. \tag{3.339}$$

The linear momentum balance equation (3.323) of the three-phase mixture for this special case reduces to

$$\operatorname{div} \tilde{\boldsymbol{\sigma}} + \tilde{\rho} \mathbf{g} = (1-n)\rho^s \mathbf{a}^s + nS^w \rho^w \mathbf{a}^w \,. \tag{3.340}$$

It should be mentioned that in (3.340) the quantity $\tilde{\rho}$ still contains a portion due to the density of the air phase, i.e. the definition (3.184) of the averaged density for the three-phase medium is valid. For the averaged stress tensor $\tilde{\boldsymbol{\sigma}}$, however, the hydrostatic air stress $\boldsymbol{\sigma}^a$ is zero and thus equation (3.190) changes to

$$\tilde{\boldsymbol{\sigma}} = (1-n)\boldsymbol{\sigma}^s + nS^w \boldsymbol{\sigma}^w \,. \tag{3.341}$$

When considering quasi-static conditions, further simplifications of the equations (3.338) to (3.340) are possible. The convective term in the mass balance law (3.338) may be dropped which results in

$$S^w \dot{\tilde{\varepsilon}}^{vol} + n\frac{D^s S^w}{Dt} + \frac{nS^w}{\rho^w}\frac{D^s \rho^w}{Dt} + \text{div}\,\tilde{\mathbf{v}}^{ws} = 0 \;. \tag{3.342}$$

In the equations (3.339) and (3.340) the acceleration terms can be neglected which yields Darcy's law in the form

$$\tilde{\mathbf{v}}^{ws} = \frac{\mathbf{k}^{ow} k^{rw}}{\rho^w g} \cdot (\text{grad}\,p^w + \rho^w \mathbf{g}) \tag{3.343}$$

and the linear momentum balance equation

$$\text{div}\,\tilde{\boldsymbol{\sigma}} + \tilde{\rho}\mathbf{g} = \mathbf{0} \;. \tag{3.344}$$

As a practical example for this special case one may think of a dewatering process which is solely driven by gravity, i.e. the water flows in the pores of the soil because of its dead weight.

3.7.3 Consolidation

For a water saturated soil under atmospheric conditions the consolidation theory is obtained. Similar to the case of dewatering under atmospheric conditions the hydrostatic air stress is equal to the atmospheric pressure and thus the mass balance equation for the air phase is not necessary. Consequently, this special case is a two-phase formulation as well.

The degree of water saturation is now constant during the analysis which means that the soil stays fully saturated according to the specified initial conditions. Therefore, instead of the equations (3.267) or (3.337) the relations

$$S^w = 1 = const. \quad \text{and} \quad S^a = 0 = const. \tag{3.345}$$

have to hold. The second of the equations (3.345) of course results from the saturation condition, i.e. from $S^w + S^a = 1$. For a fully saturated state of the soil the coefficient of permeability with respect to the particular fluid phase may be considered as a constant and hence for the relative permeability coefficient

$$k^{rw} = 1 = const. \tag{3.346}$$

is valid when dealing with consolidation problems.

Taking into account the special assumptions (3.345) and (3.346) the equations for the two-phase formulation given in Section 3.7.2 can be further simplified. Since the degree of water saturation is a constant (confer (3.345$_1$)), its time derivative is zero and consequently the second term in (3.338) vanishes. Thus, insertion of (3.345$_1$) yields the following form of the mass balance equation for the fluid phase,

$$\dot{\varepsilon}^{vol} + \frac{n}{\rho^w}\frac{D^s\rho^w}{Dt} + \operatorname{div}\tilde{\mathbf{v}}^{ws} + \frac{1}{\rho^w}\operatorname{grad}\rho^w \cdot \tilde{\mathbf{v}}^{ws} = 0 \ . \tag{3.347}$$

For a soil skeleton completely saturated with one fluid phase, e.g., water, Darcy's law can be written as

$$\tilde{\mathbf{v}}^{ws} = \frac{\mathbf{k}^{ow}}{\rho^w g} \cdot [\operatorname{grad} p^w + \rho^w(\mathbf{g} - \mathbf{a}^w)] \ , \tag{3.348}$$

where the relative permeability coefficient according to (3.346) has been used. For a degree of water saturation of $S^w = 1$ the linear momentum balance equation is obtained from (3.340) as

$$\operatorname{div}\tilde{\boldsymbol{\sigma}} + \tilde{\rho}\mathbf{g} = (1-n)\rho^s\mathbf{a}^s + n\rho^w\mathbf{a}^w \ . \tag{3.349}$$

The definitions for the averaged stress tensor $\tilde{\boldsymbol{\sigma}}$ and the averaged density $\tilde{\rho}$ of the three-phase mixture have to be adjusted for this special case as well. Applying the relations (3.345) to equation (3.341) yields the averaged stress tensor for consolidation problems,

$$\tilde{\boldsymbol{\sigma}} = (1-n)\boldsymbol{\sigma}^s + n\boldsymbol{\sigma}^w \ , \tag{3.350}$$

the averaged density for a two-phase medium results from (3.184),

$$\tilde{\rho} = (1-n)\rho^s + n\rho^w \ . \tag{3.351}$$

When dealing with quasi-static conditions, some additional simplifications of the equations (3.347) to (3.349) are possible. The convective term in the mass balance equation (3.347) can be omitted,

$$\dot{\varepsilon}^{vol} + \frac{n}{\rho^w}\frac{D^s\rho^w}{Dt} + \operatorname{div}\tilde{\mathbf{v}}^{ws} = 0 \ . \tag{3.352}$$

Furthermore, any acceleration terms in Darcy's law and in the linear momentum balance equation can be dropped, resulting in the subsequent form of Darcy's law,

$$\tilde{\mathbf{v}}^{ws} = \frac{\mathbf{k}^{ow}}{\rho^w g} \cdot (\operatorname{grad} p^w + \rho^w\mathbf{g}) \ , \tag{3.353}$$

and in the linear momentum balance

$$\operatorname{div} \tilde{\boldsymbol{\sigma}} + \tilde{\rho}\mathbf{g} = \mathbf{0} \ . \tag{3.354}$$

The compressibility of the fluid phase may be taken into account. However, in conventional consolidation analyses, dealing with the deformation of a porous material resulting from the flow of pore water, the compressibility of the fluid is usually neglected. In such cases the second term in the equations (3.347) or (3.352), which, by the constitutive law (3.260) for the fluid phase, is related to the bulk modulus und thus to the compressibility of the fluid, can be omitted yielding a mass balance equation for the water phase as

$$\dot{\varepsilon}^{vol} + \operatorname{div} \tilde{\mathbf{v}}^{ws} = 0 \ , \tag{3.355}$$

or, using the divergence of the velocity of the soil skeleton, as

$$\operatorname{div} \mathbf{v}^s + \operatorname{div} \tilde{\mathbf{v}}^{ws} = 0 \ . \tag{3.356}$$

Mass balance equations of the form (3.355) or (3.356) or similar versions may be found in texts dealing with consolidation analyses, compare, e.g., [Buchmaier(1985), Verruijt(1995), Smith(1998)].

3.7.4 Uncoupled approach

This section is dedicated to some remarks concerning the uncoupled approach for multi-phase problems which is quite popular in engineering practice. It is characterised by treating the flow of the fluids in the soil in a first step (assuming a rigid soil skeleton) and determining the deformations of the soil in a second step. Consequently, any interactions between the fluid flow in the soil and the deformations of the soil are neglected in this simplified approach. However, the influence of these coupling effects on the obtained solution depends on the type of problem under consideration and may be more or less significant.

Subsequently, the uncoupled approach is deduced from the coupled three-phase formulation described in the previous sections. As a starting point for the derivations the linear momentum balance equation for the soil skeleton (3.313), as given in Section 3.6.5, is used. In the last term of this equation $\mathbf{t}^s$ accounts for mechanical interactions between the solid phase and the other phases present in the domain. Thus, the term $\tilde{\mathbf{f}}^s = (1-n)\rho^s \mathbf{t}^s$ can be considered as an averaged interaction force of the soil skeleton when keeping in mind that $(1-n)$ is the portion of the solid phase within the representative elementary volume. The linear momentum balance equation can then be written as

$$\text{div}[(1-n)\boldsymbol{\sigma}^s] + (1-n)\rho^s(\mathbf{g}-\mathbf{a}^s) + \tilde{\mathbf{f}}^s = \mathbf{0} \ . \tag{3.357}$$

In equation (3.357) the stress tensor in the solid phase $\boldsymbol{\sigma}^s$ can be expressed by means of the effective averaged stresses in the soil skeleton and the hydrostatic stresses in the fluid phases water and air. To this end, the definition of the effective averaged stress tensor as given in equation (3.201) is employed in combination with the hydrostatic stress of the two fluid phases exerted on the solid phase, (3.192), reading as

$$\tilde{\boldsymbol{\sigma}}' = (1-n)\boldsymbol{\sigma}^s - (1-n)S^w p^w \mathbf{I} - (1-n)S^a p^a \mathbf{I} \ . \tag{3.358}$$

Calculation of the divergence of $\tilde{\boldsymbol{\sigma}}'$ in (3.358) and a reordering of terms results in

$$\text{div}[(1-n)\boldsymbol{\sigma}^s] = \text{div}\,\tilde{\boldsymbol{\sigma}}' + (1-n)S^w \text{grad}\,p^w + (1-n)S^a \text{grad}\,p^a \ . \tag{3.359}$$

This expression can now be substituted into equation (3.357) and the subsequent form of the linear momentum balance law for the soil skeleton is obtained:

$$\text{div}\,\tilde{\boldsymbol{\sigma}}' + (1-n)S^w \text{grad}\,p^w + (1-n)S^a \text{grad}\,p^a + (1-n)\rho^s(\mathbf{g}-\mathbf{a}^s) + \tilde{\mathbf{f}}^s = \mathbf{0} \ . \tag{3.360}$$

Similar to the linear momentum balance for the soil skeleton, the linear momentum balance equation (3.314) for the fluid phase f contains an averaged interaction force $\tilde{\mathbf{f}}^f = nS^f \rho^f \, \mathbf{t}^f$ (nS^f is the volume fraction of the respective fluid phase) and thus reads as

$$\text{div}\left(nS^f \boldsymbol{\sigma}^f\right) + nS^f \rho^f (\mathbf{g}-\mathbf{a}^f) + \tilde{\mathbf{f}}^f = \mathbf{0} \ . \tag{3.361}$$

The divergence term can be reformulated as

$$\text{div}\left(nS^f \boldsymbol{\sigma}^f\right) = \text{div}\left(nS^f p^f \mathbf{I}\right) = nS^f \text{grad}\,p^f + p^f \text{grad}\left(nS^f\right) \ , \tag{3.362}$$

where the definition of the stress tensor in the fluid phase, as given in (3.185), has been employed. Inserting the relationship (3.362) into the linear momentum balance equation (3.361) yields

$$nS^f \text{grad}\,p^f + p^f \text{grad}\left(nS^f\right) + nS^f \rho^f (\mathbf{g}-\mathbf{a}^f) + \tilde{\mathbf{f}}^f = \mathbf{0} \ . \tag{3.363}$$

From equation (3.363) the interaction forces $\tilde{\mathbf{f}}^f$ can be deduced for the two fluid phases water ($f=w$) and air ($f=a$):

$$\tilde{\mathbf{f}}^w = -nS^w \operatorname{grad} p^w - p^w \operatorname{grad}(nS^w) - nS^w \rho^w (\mathbf{g} - \mathbf{a}^w) ,$$
$$\tilde{\mathbf{f}}^a = -nS^a \operatorname{grad} p^a - p^a \operatorname{grad}(nS^a) - nS^a \rho^a (\mathbf{g} - \mathbf{a}^a) .$$
(3.364)

According to the constraint (3.154), the interaction forces have to sum up to zero which, for the three-phase medium, means that

$$\tilde{\mathbf{f}}^s + \tilde{\mathbf{f}}^w + \tilde{\mathbf{f}}^a = \mathbf{0} \quad \text{or} \quad \tilde{\mathbf{f}}^s = -\tilde{\mathbf{f}}^w - \tilde{\mathbf{f}}^a ,$$
(3.365)

when neglecting other interaction terms present in (3.154). By substituting (3.364) into (3.365$_2$) the interaction force for the soil skeleton $\tilde{\mathbf{f}}^s$ can be expressed as

$$\tilde{\mathbf{f}}^s = nS^w \operatorname{grad} p^w + p^w \operatorname{grad}(nS^w) + nS^w \rho^w (\mathbf{g} - \mathbf{a}^w) +$$
$$+ nS^a \operatorname{grad} p^a + p^a \operatorname{grad}(nS^a) + nS^a \rho^a (\mathbf{g} - \mathbf{a}^a) .$$
(3.366)

The generalised Darcy's law for the fluid phase f (equation (3.320)) can be employed to rewrite the gradient of the hydrostatic fluid stress, $\operatorname{grad} p^f$, as

$$\operatorname{grad} p^f = \frac{\rho^f g}{k^{rf}} (\mathbf{k}^{of})^{-1} \cdot \tilde{\mathbf{v}}^{fs} - \rho^f \mathbf{g} + \rho^f \mathbf{a}^f .$$
(3.367)

This relationship has to be specified for the water phase with $f = w$ and for the air phase with $f = a$:

$$\operatorname{grad} p^w = \frac{\rho^w g}{k^{rw}} (\mathbf{k}^{ow})^{-1} \cdot \tilde{\mathbf{v}}^{ws} - \rho^w \mathbf{g} + \rho^w \mathbf{a}^w ,$$
$$\operatorname{grad} p^a = \frac{\rho^a g}{k^{ra}} (\mathbf{k}^{oa})^{-1} \cdot \tilde{\mathbf{v}}^{as} - \rho^a \mathbf{g} + \rho^a \mathbf{a}^a .$$
(3.368)

Substituting the two equations (3.368) into the interaction force for the soil skeleton (relationship (3.366)) yields

$$\tilde{\mathbf{f}}^s = nS^w \frac{\rho^w g}{k^{rw}} (\mathbf{k}^{ow})^{-1} \cdot \tilde{\mathbf{v}}^{ws} - nS^w \rho^w \mathbf{g} + nS^w \rho^w \mathbf{a}^w +$$
$$+ p^w \operatorname{grad}(nS^w) + nS^w \rho^w \mathbf{g} - nS^w \rho^w \mathbf{a}^w +$$
$$+ nS^a \frac{\rho^a g}{k^{ra}} (\mathbf{k}^{oa})^{-1} \cdot \tilde{\mathbf{v}}^{as} - nS^a \rho^a \mathbf{g} + nS^a \rho^a \mathbf{a}^a +$$
$$+ p^a \operatorname{grad}(nS^a) + nS^a \rho^a \mathbf{g} - nS^a \rho^a \mathbf{a}^a .$$
(3.369)

Since most of the terms in (3.369) cancel, this equation can be reduced to

$$\tilde{\mathbf{f}}^s = nS^w \frac{\rho^w g}{k^{rw}} (\mathbf{k}^{ow})^{-1} \cdot \tilde{\mathbf{v}}^{ws} + p^w \mathrm{grad}(nS^w) +$$

$$+ nS^a \frac{\rho^a g}{k^{ra}} (\mathbf{k}^{oa})^{-1} \cdot \tilde{\mathbf{v}}^{as} + p^a \mathrm{grad}(nS^a) \ . \tag{3.370}$$

Inserting the expressions (3.368) for the gradient of the hydrostatic water and air stresses, respectively, and the relationship (3.370) for the interaction force of the soil skeleton into the linear momentum balance law for the soil skeleton, as given in (3.360), results in

$$\mathrm{div}\, \tilde{\boldsymbol{\sigma}}' + (1-n)S^w \frac{\rho^w g}{k^{rw}} (\mathbf{k}^{ow})^{-1} \cdot \tilde{\mathbf{v}}^{ws} - (1-n)S^w \rho^w \mathbf{g} + (1-n)S^w \rho^w \mathbf{a}^w +$$

$$+ (1-n)S^a \frac{\rho^a g}{k^{ra}} (\mathbf{k}^{oa})^{-1} \cdot \tilde{\mathbf{v}}^{as} - (1-n)S^a \rho^a \mathbf{g} + (1-n)S^a \rho^a \mathbf{a}^a +$$

$$+ (1-n)\rho^s(\mathbf{g} - \mathbf{a}^s) + nS^w \frac{\rho^w g}{k^{rw}} (\mathbf{k}^{ow})^{-1} \cdot \tilde{\mathbf{v}}^{ws} + p^w \mathrm{grad}(nS^w) +$$

$$+ nS^a \frac{\rho^a g}{k^{ra}} (\mathbf{k}^{oa})^{-1} \cdot \tilde{\mathbf{v}}^{as} + p^a \mathrm{grad}(nS^a) = \mathbf{0} \ , \tag{3.371}$$

and after collecting terms the following form is obtained:

$$\mathrm{div}\, \tilde{\boldsymbol{\sigma}}' + S^w \frac{\rho^w g}{k^{rw}} (\mathbf{k}^{ow})^{-1} \cdot \tilde{\mathbf{v}}^{ws} + p^w \mathrm{grad}(nS^w) + S^a \frac{\rho^a g}{k^{ra}} (\mathbf{k}^{oa})^{-1} \cdot \tilde{\mathbf{v}}^{as} +$$

$$+ p^a \mathrm{grad}(nS^a) + (1-n)(\rho^s - S^w \rho^w - S^a \rho^a)\mathbf{g} -$$

$$- (1-n)(\rho^s \mathbf{a}^s - S^w \rho^w \mathbf{a}^w - S^a \rho^a \mathbf{a}^a) = \mathbf{0} \ . \tag{3.372}$$

This general form of the linear momentum balance equation for the soil skeleton can be simplified in order to obtain certain special cases contained in the formulation. When dealing with quasi-static conditions, the acceleration terms in (3.372) can be neglected, that is $\mathbf{a}^s = \mathbf{a}^w = \mathbf{a}^a = \mathbf{0}$, yielding

$$\mathrm{div}\, \tilde{\boldsymbol{\sigma}}' + S^w \frac{\rho^w g}{k^{rw}} (\mathbf{k}^{ow})^{-1} \cdot \tilde{\mathbf{v}}^{ws} + p^w \mathrm{grad}(nS^w) + S^a \frac{\rho^a g}{k^{ra}} (\mathbf{k}^{oa})^{-1} \cdot \tilde{\mathbf{v}}^{as} +$$

$$+ p^a \mathrm{grad}(nS^a) + (1-n)(\rho^s - S^w \rho^w - S^a \rho^a)\mathbf{g} = \mathbf{0} \ . \tag{3.373}$$

Equation (3.373) may be interpreted as follows. The first term denotes the divergence of the effective averaged stress tensor. The last term expresses the weight of the soil skeleton due to the presence of the two fluid phases. The remaining terms represent the drag forces arising from the flow of the fluids filling the voids of the soil skeleton. Hence, when considering a partially saturated porous medium, the drag forces $\tilde{\mathbf{f}}^d$ are constituted by three different kinds of gradients:

$$\tilde{\mathbf{f}}^d = S^w \frac{\rho^w g}{k^{rw}} (\mathbf{k}^{ow})^{-1} \cdot \tilde{\mathbf{v}}^{ws} + S^a \frac{\rho^a g}{k^{ra}} (\mathbf{k}^{oa})^{-1} \cdot \tilde{\mathbf{v}}^{as} +$$
$$+ n(p^w \operatorname{grad} S^w + p^a \operatorname{grad} S^a) + (S^w p^w + S^a p^a) \operatorname{grad} n , \qquad (3.374)$$

where the first two terms are the hydraulic gradients of the two fluids, the third term contains the spatial variation of the two degrees of saturation and the fourth term represents the gradient of the porosity. Thus, the flow of the fluids exerts forces onto the soil skeleton due to spatial variations of the hydraulic head and the degree of saturation of the fluids as well as of the porosity of the soil skeleton.

For a porosity constant within the domain equation (3.374) reduces to

$$\tilde{\mathbf{f}}^d = S^w \frac{\rho^w g}{k^{rw}} (\mathbf{k}^{ow})^{-1} \cdot \tilde{\mathbf{v}}^{ws} + S^a \frac{\rho^a g}{k^{ra}} (\mathbf{k}^{oa})^{-1} \cdot \tilde{\mathbf{v}}^{as} + n(p^w \operatorname{grad} S^w + p^a \operatorname{grad} S^a) . \quad (3.375)$$

In this case drag forces arise from spatial gradients of the hydraulic head and the degree of saturation of the individual fluid phases, respectively.

Finally, this equation is further simplified for a completely water saturated porous medium, where $S^w = 1$, $S^a = 0$ and $k^{rw} = 1$ are valid. If the degree of saturation is constant, the gradient vanishes and consequently $\tilde{\mathbf{f}}^d$ is obtained as

$$\tilde{\mathbf{f}}^d = \rho^w g (\mathbf{k}^{ow})^{-1} \cdot \tilde{\mathbf{v}}^{ws} . \qquad (3.376)$$

For the completely water saturated medium drag forces arise only due to the hydraulic gradient of the water phase. The linear momentum balance equation for the soil skeleton in this case can be written as

$$\operatorname{div} \tilde{\boldsymbol{\sigma}}' + \rho^w g (\mathbf{k}^{ow})^{-1} \cdot \tilde{\mathbf{v}}^{ws} + (1 - n)(\rho^s - \rho^w) \mathbf{g} = \mathbf{0} , \qquad (3.377)$$

where the first term represents the divergence of the effective averaged stress tensor and the last term describes the buoyant weight of the soil skeleton.

Hence, in an uncoupled approach, after dealing with the flow of the fluids in a rigid soil skeleton yielding the drag forces (3.374), (3.375) or (3.376) depending on the problem under consideration, the deformations of the soil skeleton are calculated from the linear momentum balance equations (3.372), (3.373) or (3.377) taking into account the drag forces and the weight of the soil skeleton due to the presence of fluid phases (e.g., buoyant weight in the case of a fully water saturated soil skeleton).

3.7.5 Drained conditions

To obtain the special case for drained conditions, there are no fluids to be considered and thus the fluid stresses and consequently also the mass balance equations for the fluids can be omitted. Only the equilibrium equations are necessary in this case, relating the internal forces of the soil skeleton to any present external forces consisting of body forces and surface tractions.

For the case of zero water stresses ($\boldsymbol{\sigma}^w = \mathbf{0}$) and for a dry soil, that is for $\rho^w = 0$ and $\mathbf{a}^w = \mathbf{0}$, with a porosity n the linear momentum balance equation (3.349) can be further simplified yielding the relationship

$$\operatorname{div} \tilde{\boldsymbol{\sigma}} + \tilde{\rho} \mathbf{g} = (1-n) \rho^s \mathbf{a}^s \ . \tag{3.378}$$

Considering the above assumptions, the definitions for the averaged stress tensor (3.350) and the averaged density (3.351) reduce to

$$\tilde{\boldsymbol{\sigma}} = (1-n) \boldsymbol{\sigma}^s \quad \text{and} \quad \tilde{\rho} = (1-n) \rho^s \ . \tag{3.379}$$

Insertion of (3.379) into (3.378) results in the linear momentum balance equation for a dry soil with a porosity n,

$$\operatorname{div}[(1-n) \boldsymbol{\sigma}^s] + (1-n) \rho^s \mathbf{g} = (1-n) \rho^s \mathbf{a}^s \ . \tag{3.380}$$

When imagining a material with a very small value for the porosity (or a vanishing porosity in the limiting case), the equilibrium equations for a single-phase material are obtained as

$$\operatorname{div} \boldsymbol{\sigma}^s + \rho^s \mathbf{g} = \rho^s \mathbf{a}^s \ , \tag{3.381}$$

which for quasi-static conditions ($\mathbf{a}^s = \mathbf{0}$) reduce to

$$\operatorname{div} \boldsymbol{\sigma}^s + \rho^s \mathbf{g} = \mathbf{0} \ . \tag{3.382}$$

Chapter 4

Numerical formulation for the three-phase model

4.1 Introduction

After the mathematical derivation of the basic equations for the three-phase model in the previous chapter, this part is devoted to the numerical solution of the set of governing equations. To this end, a finite element approach is chosen in the present work. According to standard procedures, several steps are necessary for the solution of partial differential equations (PDEs) by means of the Finite Element Method (FEM).

One usually emanates from a so-called weak formulation of the basic equations which therefore is described first in the subsequent sections. This weak formulation is obtained by weighting the respective equations with appropriate test functions and then integrating over the domain under consideration. Alternatively, for certain classes of problems a variational formulation may serve as basis for the finite element treatment of a system of PDEs.

After introducing the weak formulation of the basic equations, consisting of the mass balances for the individual phases and of the linear momentum balance for the three-phase mixture, the finite element framework requires a subdivision of the particular domain into finite elements. Hence, before dealing with the derivation of the spatially discretised equations, the choice of the primary variables of the model and their approximation for the numerical solution are discussed briefly.

Since the equations for the particular three-phase model contain time derivatives of the unknowns, an additional numerical integration in the time domain has to be performed. Hence, a survey of the various time integration schemes is presented which is followed by a description of the particular method employed in the current formulation.

Finally, after performing the integration in the time domain a non-linear coupled system of equations is obtained which has to be solved iteratively for the incremental nodal values of the unknowns. To this end, two different approaches are proposed concluding this chapter.

4.2 Weak formulation of the basic equations

4.2.1 Introduction

Considering the partial differential equations presented in Chapter 3, a solution satisfying these equations would be called a 'strong', or 'local', solution [Hughes(1987)]. However, in general the Finite Element Method only allows for an approximation of such a strong solution, i.e. the equations are not satisfied exactly at each point of the analysed domain. Usually PDEs are solved by means of the FEM on the basis of so-called 'weak' formulations of the governing partial differential equations of the model.

Emanating from a partial differential equation, the corresponding weak formulation is obtained by multiplying the respective equation by an appropriate 'weight' or 'test' function and subsequent integrating over the domain under consideration. The general idea of weak formulations is that they allow for a solution with lower (or weaker) smoothness constraints. However, according to [Celia(1992)], the test functions should be chosen sufficiently smooth to guarantee a meaningful integration by parts.

Since the exact solution is only approximated, a residuum remains for each equation. This residual value, multiplied by a weighting function, is required to vanish in the integral over the volume of the domain under consideration. This integral constraint for the residuum is less stringent than the vanishing residuum for the strong solution.

For certain types of problems finite element solutions of PDEs may be based on so-called 'variational' formulations of the governing equations. A fundamental variational method is the principle of virtual work (or the principle of virtual displacements) [Malvern(1969)]. The principle of virtual displacements may be regarded as an alternative way of expressing the equilibrium conditions.

In the subsequent sections weak formulations of the linear momentum balance for the three-phase mixture and of the mass balance equations for the different constituents are derived after formally stating the local problem. These weak equations serve as a basis for the discretised formulation presented hereafter. In order to further simplify the model some additional restrictions are introduced. Within the framework presented here only slow phenomena in the soil are investigated. Therefore, two simplifications of the three-phase model are possible: (i) Material time derivatives are equivalent to partial time derivatives (implying any convective terms to be dropped) and (ii) acceleration terms (inertia effects) can be neglected, i.e. only quasi-static conditions are taken into account.

4.2.2 Formal statement of the problem

Before deriving the weak formulations of the governing equations of the three-phase model the mathematical problem should be stated properly. Therefore, this section is devoted to the specification of the partial differential equations together with the appropriate boundary and initial conditions.

Let n_{sd} (= 1, 2 or 3) denote the number of space dimensions of the problem under consideration. Let $V \subset \mathbb{R}^{n_{sd}}$ be an open set with a piecewise smooth boundary A. Assume in addition that A admits the decompositions

$$A = \overline{A_{\mathbf{u}} \cup A_{\mathbf{u}}^q} = \overline{A_w \cup A_w^q} = \overline{A_a \cup A_a^q} \,,$$

$$A_{\mathbf{u}} \cap A_{\mathbf{u}}^q = \emptyset \qquad A_w \cap A_w^q = \emptyset \qquad A_a \cap A_a^q = \emptyset \,, \tag{4.1}$$

where $A_{\mathbf{u}}$, $A_{\mathbf{u}}^q$, A_w, A_w^q, A_a and A_a^q are open sets in A. $A_{\mathbf{u}}$, A_w and A_a are the parts of the boundary A with *Dirichlet* boundary conditions for the soil skeleton, the water phase and the air phase, respectively, being prescribed displacements or fluid pressures. $A_{\mathbf{u}}^q$, A_w^q and A_a^q denote the parts of A with specified *Neumann* boundary conditions which are either prescribed surface tractions for the soil skeleton or prescribed surface flows for the fluid phases.

Furthermore, $\cup$ is the set union symbol. Thus, for instance, $A_{\mathbf{u}} \cup A_{\mathbf{u}}^q$ means the set of all points contained in either $A_{\mathbf{u}}$ or $A_{\mathbf{u}}^q$. $\cap$ is the set intersection symbol. Hence, e.g., $A_{\mathbf{u}} \cap A_{\mathbf{u}}^q$ denotes the set of all points contained in both $A_{\mathbf{u}}$ and $A_{\mathbf{u}}^q$. The empty set is described by $\emptyset$. For example, $A_{\mathbf{u}} \cap A_{\mathbf{u}}^q = \emptyset$ means that there is no point contained in both $A_{\mathbf{u}}$ and $A_{\mathbf{u}}^q$, that is $A_{\mathbf{u}}$ and $A_{\mathbf{u}}^q$ do not intersect or overlap. A bar above the set indicates set closure, i.e. the union of the set with its boundary. For instance, for a two-dimensional case ($n_{sd} = 2$) each of the parts $A_\mathbf{u}$ or $A_\mathbf{u}^q$ of the boundary A is a curve except its end points (open set) whereas by the closure of the set, $\overline{A_\mathbf{u} \cup A_\mathbf{u}^q}$, the curve together with its end points is understood [Hughes(1987)]. Considerations similar to the ones given above for $A_\mathbf{u}$ and $A_\mathbf{u}^q$ also apply to the parts A_w, A_w^q, A_a and A_a^q of the boundary A.

Employing the assumptions mentioned in Section 4.2.1 to the 'strong' form of the linear momentum balance equation (3.323) and making use of the partitioning of the total averaged stress tensor according to (3.199) yields

$$\text{div }\tilde{\boldsymbol{\sigma}}' + \text{grad}\,(S^w p^w) + \text{grad}\,(S^a p^a) + \tilde{\rho}\mathbf{g} = \mathbf{0}\ , \tag{4.2}$$

where the acceleration terms on the right hand side have been omitted due to the assumed quasi-static conditions. Equation (4.2) may also be referred to as 'equation of motion' or 'equilibrium equation' for the three-phase medium. Applying the same simplifying assumptions to the mass balance equation for the fluid phase (3.306) results in

$$S^f \text{div }\dot{\mathbf{u}} + n\dot{S}^f + \frac{nS^f}{\rho^f}\dot{\rho}^f + \text{div }\tilde{\mathbf{v}}^{fs} = 0\ , \tag{4.3}$$

where the rate of the volumetric strain, $\dot{\varepsilon}^{vol}$, has been replaced by the divergence of the rate of the soil displacements, div $\dot{\mathbf{u}}$, the material time derivatives have been substituted by partial time derivatives and the convective term in (3.306) has been neglected. For the rate of the degree of saturation and the rate of the density of the fluid the respective constitutive equations can be inserted. For the two fluid phases water and air these relations are given as

$$\frac{\dot{\rho}^w}{\rho^w} = -\frac{\dot{p}^w}{K^w} \quad \text{and} \quad \frac{\dot{\rho}^a}{\rho^a} = \frac{\dot{p}^a}{p^a}\ ,$$
$$\dot{S}^w = -\dot{S}^a = \frac{dS^w}{dp^c}\dot{p}^c = \frac{dS^w}{dp^c}(\dot{p}^a - \dot{p}^w)\ . \tag{4.4}$$

Equation (4.4$_1$) for the water phase follows from (3.260), the relationship (4.4$_2$) for the air phase is obtained from (3.265) and (3.266) and finally the rate of the degrees of saturation (4.4$_3$) is expressed by means of (3.267).

Employing the constitutive equations (4.4) together with the mass balance law (4.3), the latter specialised for both the water and air phases, and introducing the constitutive

rate equations for the soil skeleton, obtained from (3.228), the *strong form of the initial value/boundary value problem (I/BVP)* for the three-phase model can be stated:
Find the displacement field $\mathbf{u}(\mathbf{x}, t)$ and the hydrostatic water and air stress fields, $p^w(\mathbf{x}, t)$ and $p^a(\mathbf{x}, t)$, such that

$$\operatorname{div} \tilde{\boldsymbol{\sigma}}' + \operatorname{grad}(S^w p^w) + \operatorname{grad}(S^a p^a) + \tilde{\rho}\mathbf{g} = \mathbf{0} \,, \tag{4.5}$$

$$S^w \operatorname{div} \dot{\mathbf{u}} + n\frac{dS^w}{dp^c}(\dot{p}^a - \dot{p}^w) - \frac{nS^w}{K^w}\dot{p}^w + \operatorname{div} \tilde{\mathbf{v}}^{ws} = 0 \,, \tag{4.6}$$

$$S^a \operatorname{div} \dot{\mathbf{u}} - n\frac{dS^w}{dp^c}(\dot{p}^a - \dot{p}^w) - \frac{nS^a}{K^a}\dot{p}^a + \operatorname{div} \tilde{\mathbf{v}}^{as} = 0 \,, \tag{4.7}$$

$$\dot{\boldsymbol{\sigma}}' = \mathbf{C}^{ep} : \dot{\boldsymbol{\varepsilon}} \,, \tag{4.8}$$

are satisfied in the domain under consideration together with the boundary conditions

$$\begin{aligned}
\mathbf{u} &= \bar{\mathbf{u}} && \text{on} \quad A_{\mathbf{u}} \,, & \tilde{\boldsymbol{\sigma}} \cdot \mathbf{n} &= \bar{\mathbf{t}} && \text{on} \quad A_{\mathbf{u}}^q \,, \\
p^w &= \bar{p}^w && \text{on} \quad A_w \,, & \tilde{\mathbf{v}}^{ws} \cdot \mathbf{n} &= \bar{q}^w && \text{on} \quad A_w^q \,, \\
p^a &= \bar{p}^a && \text{on} \quad A_a \,, & \tilde{\mathbf{v}}^{as} \cdot \mathbf{n} &= \bar{q}^a && \text{on} \quad A_a^q \,,
\end{aligned} \tag{4.9}$$

and the initial conditions

$$\begin{aligned}
\mathbf{u}(\mathbf{x}, 0) &= \mathbf{u}_0(\mathbf{x}) \,, \\
p^w(\mathbf{x}, 0) &= p_0^w(\mathbf{x}) \,, \\
p^a(\mathbf{x}, 0) &= p_0^a(\mathbf{x}) \,.
\end{aligned} \tag{4.10}$$

It should be mentioned that the bulk modulus K^a for the air phase in the balance law (4.7) is equal to the negative hydrostatic air stress, $K^a = -p^a$. For convenience in the notation, K^a has been used in (4.7), yielding identical structures of both mass balance equations (4.6) and (4.7). In the set of equations (4.9) the quantities $\bar{\mathbf{u}}$, $\bar{p}^w$ and $\bar{p}^a$ denote prescribed values for the displacements of the soil skeleton and for the hydrostatic stress in the water and air phases at the boundary of the considered domain. $\bar{\mathbf{t}}$ is a vector of prescribed surface tractions and $\bar{q}^w$ and $\bar{q}^a$ are the prescribed water flow and airflow through the surface of the domain. In the equations (4.10) the quantities $\mathbf{u}_0(\mathbf{x})$, $p_0^w(\mathbf{x})$ and $p_0^a(\mathbf{x})$ represent initial values of the soil displacements and the hydrostatic stresses in the two fluid phases, respectively.

However, the strong form of the initial value/boundary value problem discussed in this section is not suited for a numerical solution by means of the Finite Element Method. For this latter purpose it proves more convenient to consider the weak form of the I/BVP. Therefore, weak forms of the individual equations (4.5) to (4.7) are derived in the subsequent sections to arrive at the final weak formulation of the problem.

4.2.3 Weak form of the linear momentum balance equation

To derive a weak formulation of the linear momentum balance law for the three-phase mixture, the relationship (4.5) is used as a starting point. This equation is weighted by a test function $\delta\mathbf{u}$ (also called weighting function) and then integrated over the volume V of the domain under consideration yielding the following weak formulation of the linear momentum balance law:

$$G(\mathbf{u}, p^w, p^a, \delta\mathbf{u}) = \int_V \delta\mathbf{u} \cdot \left[\operatorname{div} \tilde{\boldsymbol{\sigma}}' + \operatorname{grad}(S^w p^w) + \operatorname{grad}(S^a p^a) + \tilde{\rho}\mathbf{g}\right] dV . \quad (4.11)$$

The test function $\delta\mathbf{u}$ in (4.11) can be interpreted as a virtual displacement field of the soil skeleton.

Green's formula (also known as Green's first identity) [Eriksson(1996)] may be employed to rewrite each of the first three terms in the integral in equation (4.11). For the first term, containing the divergence of the effective averaged stress tensor, this identity reads as

$$\int_V \delta\mathbf{u} \cdot \operatorname{div} \tilde{\boldsymbol{\sigma}}' \, dV = \int_A \delta\mathbf{u} \cdot \tilde{\boldsymbol{\sigma}}' \cdot \mathbf{n} \, dA - \int_V \operatorname{grad} \delta\mathbf{u} : \tilde{\boldsymbol{\sigma}}' \, dV . \quad (4.12)$$

The second term, consisting of the gradient of the hydrostatic stress in the water phase, may be reformulated as

$$\int_V \delta\mathbf{u} \cdot \operatorname{grad}(S^w p^w) \, dV = \int_A \delta\mathbf{u} \cdot (S^w p^w \mathbf{I}) \cdot \mathbf{n} \, dA - \int_V S^w p^w \operatorname{div} \delta\mathbf{u} \, dV , \quad (4.13)$$

and for the third integral, containing the gradient of the hydrostatic air stress, one obtains in an analogous manner

$$\int_V \delta\mathbf{u} \cdot \operatorname{grad}(S^a p^a) \, dV = \int_A \delta\mathbf{u} \cdot (S^a p^a \mathbf{I}) \cdot \mathbf{n} \, dA - \int_V S^a p^a \operatorname{div} \delta\mathbf{u} \, dV . \quad (4.14)$$

Insertion of the equations (4.12) to (4.14) into the weak form of the linear momentum balance (4.11) yields

$$G(\mathbf{u}, p^w, p^a, \delta\mathbf{u}) = \int_A \delta\mathbf{u} \cdot \tilde{\boldsymbol{\sigma}}' \cdot \mathbf{n}\, dA - \int_V \operatorname{grad} \delta\mathbf{u} : \tilde{\boldsymbol{\sigma}}'\, dV +$$

$$+ \int_A \delta\mathbf{u} \cdot (S^w p^w\, \mathbf{I}) \cdot \mathbf{n}\, dA - \int_V S^w p^w \operatorname{div} \delta\mathbf{u}\, dV +$$

$$+ \int_A \delta\mathbf{u} \cdot (S^a p^a\, \mathbf{I}) \cdot \mathbf{n}\, dA - \int_V S^a p^a \operatorname{div} \delta\mathbf{u}\, dV +$$

$$+ \int_V \tilde{\rho}\, \delta\mathbf{u} \cdot \mathbf{g}\, dV \ . \tag{4.15}$$

The three integrals over the surface A can now be collected which results in

$$G(\mathbf{u}, p^w, p^a, \delta\mathbf{u}) = \int_A \delta\mathbf{u} \cdot \left(\tilde{\boldsymbol{\sigma}}' + S^w p^w\, \mathbf{I} + S^a p^a\, \mathbf{I} \right) \cdot \mathbf{n}\, dA -$$

$$- \int_V \operatorname{grad} \delta\mathbf{u} : \tilde{\boldsymbol{\sigma}}'\, dV - \int_V S^w p^w \operatorname{div} \delta\mathbf{u}\, dV - \int_V S^a p^a \operatorname{div} \delta\mathbf{u}\, dV +$$

$$+ \int_V \tilde{\rho}\, \delta\mathbf{u} \cdot \mathbf{g}\, dV \ . \tag{4.16}$$

According to (3.199), the term in brackets in the integral over the surface A represents the total averaged stress tensor $\tilde{\boldsymbol{\sigma}}$ of the three-phase mixture. Employing the definition of the Neumann boundary conditions (see set of equations (4.9)), $\tilde{\boldsymbol{\sigma}} \cdot \mathbf{n} = \bar{\mathbf{t}}$, the *weak formulation of the linear momentum balance equation* is finally obtained as

$$G(\mathbf{u}, p^w, p^a, \delta\mathbf{u}) = - \int_V \operatorname{grad} \delta\mathbf{u} : \tilde{\boldsymbol{\sigma}}'\, dV - \int_V S^w p^w \operatorname{div} \delta\mathbf{u}\, dV - \int_V S^a p^a \operatorname{div} \delta\mathbf{u}\, dV +$$

$$+ \int_V \tilde{\rho}\, \delta\mathbf{u} \cdot \mathbf{g}\, dV + \int_{A_\mathbf{u}^q} \delta\mathbf{u} \cdot \bar{\mathbf{t}}\, dA \ . \tag{4.17}$$

For the last term of (4.17) an integration over the total boundary A of the domain is not necessary since surface tractions are only specified on the portion $A_\mathbf{u}^q$ of A. Hence, the integration domain has been reduced to the part $A_{\mathbf{u}}^q$ of the total surface which exhibits Neumann boundary conditions. Equation (4.17) can be considered as an extension of the principle of virtual displacements to the case of a partially saturated soil which is indicated by the second and the third integral containing the hydrostatic stresses in the two fluid phases water and air. For a single-phase material only the first and the last two terms of (4.17) are necessary to express the weak form of equilibrium of the body.

4.2.4 Weak form of the mass balance equation for the water phase

For the derivation of the weak formulation of the mass balance equation for the water phase one proceeds from (4.6). This equation is weighted by a test function δp^w and then integrated over the volume V of the domain:

$$H^w(\mathbf{u}, p^w, p^a, \delta p^w) = \int_V \delta p^w \left[S^w \operatorname{div} \dot{\mathbf{u}} + n \frac{dS^w}{dp^c}(\dot{p}^a - \dot{p}^w) - \frac{nS^w}{K^w}\dot{p}^w + \operatorname{div} \tilde{\mathbf{v}}^{ws} \right] dV .$$
(4.18)

The test function δp^w in (4.18) can be interpreted as a field of virtual hydrostatic water stresses.

Green's formula may be used to rewrite the integral containing the divergence of the artificial velocity $\tilde{\mathbf{v}}^{ws}$ as

$$\int_V \delta p^w \operatorname{div} \tilde{\mathbf{v}}^{ws} \, dV = \int_A \delta p^w \, \tilde{\mathbf{v}}^{ws} \cdot \mathbf{n} \, dA - \int_V \operatorname{grad} \delta p^w \cdot \tilde{\mathbf{v}}^{ws} \, dV .$$
(4.19)

Insertion of (4.19) into (4.18) results in the following weak formulation of the mass balance equation for the water phase,

$$H^w(\mathbf{u}, p^w, p^a, \delta p^w) = \int_V \delta p^w \, S^w \operatorname{div} \dot{\mathbf{u}} \, dV + \int_V \delta p^w \, n \frac{dS^w}{dp^c}(\dot{p}^a - \dot{p}^w) \, dV$$
$$- \int_V \delta p^w \frac{nS^w}{K^w}\dot{p}^w \, dV + \int_A \delta p^w \, \tilde{\mathbf{v}}^{ws} \cdot \mathbf{n} \, dA$$
$$- \int_V \operatorname{grad} \delta p^w \cdot \tilde{\mathbf{v}}^{ws} \, dV .$$
(4.20)

The appropriate Neumann boundary condition (equations (4.9)), $\tilde{\mathbf{v}}^{ws} \cdot \mathbf{n} = \bar{q}^w$, can be substituted into the integral over the surface A. Darcy's law, as given in (3.320) for the general case of a fluid phase f possessing an acceleration $\mathbf{a}^f$, is used in the last integral of (4.20) over the volume V which yields the following *weak formulation of the mass balance equation for the water phase*:

$$H^w(\mathbf{u}, p^w, p^a, \delta p^w) = \int_V \delta p^w \, S^w \text{div}\, \dot{\mathbf{u}}\, dV + \int_V \delta p^w \, n \frac{dS^w}{dp^c}(\dot{p}^a - \dot{p}^w)\, dV -$$

$$- \int_V \delta p^w \frac{nS^w}{K^w}\dot{p}^w\, dV - \int_V \text{grad}\, \delta p^w \cdot \frac{\mathbf{k}^{ow}k^{rw}}{\rho^w g} \cdot \text{grad}\, p^w\, dV -$$

$$- \int_V \text{grad}\, \delta p^w \cdot \frac{\mathbf{k}^{ow}k^{rw}}{g} \cdot \mathbf{g}\, dV + \int_{A_w^q} \delta p^w \, \bar{q}^w\, dA \ . \quad (4.21)$$

In accordance with the assumptions described in Section 4.2.1 the acceleration $\mathbf{a}^f$ of the water phase has been neglected in Darcy's law (thus accounting for quasi-static conditions). The integration over the surface A in equation (4.21) again has been reduced to the portion A_w^q of the total surface where the Neumann boundary conditions are prescribed.

4.2.5 Weak form of the mass balance equation for the air phase

Starting with equation (4.7), the weak formulation of the mass balance law for the air phase is derived similar to the mass balance equation for the water phase. Hence, (4.7) is weighted by a test function δp^a, which can be interpreted as a virtual hydrostatic air stress field, and integrated over the volume V of the domain yielding

$$H^a(\mathbf{u}, p^w, p^a, \delta p^a) = \int_V \delta p^a \left[S^a \text{div}\,\dot{\mathbf{u}} - n\frac{dS^w}{dp^c}(\dot{p}^a - \dot{p}^w) - \frac{nS^a}{K^a}\dot{p}^a + \text{div}\,\tilde{\mathbf{v}}^{as} \right] dV \ . \quad (4.22)$$

Using Green's formula and inserting both the Neumann boundary condition and Darcy's law for the air phase (equation (3.320) with $f = a$ and $\mathbf{a}^f = \mathbf{0}$), as done for the water phase in Section 4.2.4, the *weak formulation of the mass balance equation for the air phase* is finally obtained as

$$H^a(\mathbf{u}, p^w, p^a, \delta p^a) = \int_V \delta p^a \, S^a \operatorname{div} \dot{\mathbf{u}} \, dV - \int_V \delta p^a \, n \frac{dS^w}{dp^c} (\dot{p}^a - \dot{p}^w) \, dV -$$

$$- \int_V \delta p^a \frac{n S^a}{K^a} \dot{p}^a \, dV - \int_V \operatorname{grad} \delta p^a \cdot \frac{\mathbf{k}^{oa} k^{ra}}{\rho^a g} \cdot \operatorname{grad} p^a \, dV -$$

$$- \int_V \operatorname{grad} \delta p^a \cdot \frac{\mathbf{k}^{oa} k^{ra}}{g} \cdot \mathbf{g} \, dV + \int_{A_a^q} \delta p^a \, \overline{q}^a \, dA \,, \qquad (4.23)$$

where the integration of the surface flow $\overline{q}^a$ is performed only over the part A_a^q of A with prescribed Neumann boundary conditions.

4.2.6 Summary of the weak formulation

After the formal statement of the problem has been made (confer Section 4.2.2) and the weak forms of the governing equations have been derived, the definition of the required solution spaces opens this summarising section on the weak formulation of the three-phase problem. Two different sets of spaces should be defined, one containing the candidate, or trial, solutions for the unknown functions $\mathbf{u}(\mathbf{x}, t)$, $p^w(\mathbf{x}, t)$ and $p^a(\mathbf{x}, t)$ (called *space of configurations*), and a second one composed of the corresponding weighting or test functions $\delta \mathbf{u}(\mathbf{x})$, $\delta p^w(\mathbf{x})$ and $\delta p^a(\mathbf{x})$ (termed *space of variations*), respectively [Hughes(1987)].

Let

$$\mathcal{C}_{\mathbf{u}} := \left\{ \mathbf{u}(\mathbf{x}, t) : V \times T \mapsto \mathbb{R}^3 \,\middle|\, u_i \in H^1, \, \mathbf{u} = \overline{\mathbf{u}} \; \forall \mathbf{x} \in A_{\mathbf{u}} \right\} , \qquad (4.24)$$

$$\mathcal{C}_{pw} := \left\{ p^w(\mathbf{x}, t) : V \times T \mapsto \mathbb{R} \,\middle|\, p^w \in H^1, \, p^w = \overline{p}^w \; \forall \mathbf{x} \in A_w \right\} , \qquad (4.25)$$

$$\mathcal{C}_{pa} := \left\{ p^a(\mathbf{x}, t) : V \times T \mapsto \mathbb{R} \,\middle|\, p^a \in H^1, \, p^a = \overline{p}^a \; \forall \mathbf{x} \in A_a \right\} , \qquad (4.26)$$

be the spaces of configurations, containing the solutions for the unknown displacements of the soil skeleton $\mathbf{u}(\mathbf{x}, t)$, the hydrostatic stress in the water phase $p^w(\mathbf{x}, t)$ and the hydrostatic stress in the air phase $p^a(\mathbf{x}, t)$. The possible solutions are contained in the space H^1 and satisfy the Dirichlet boundary conditions as can be recognised from the definitions (4.24) to (4.26).

Let
$$\mathcal{V}_{\mathbf{u}} := \left\{ \delta\mathbf{u}(\mathbf{x}) : V \mapsto \mathbb{R}^3 \,\middle|\, \delta u_i \in H^1, \delta\mathbf{u} = \mathbf{0} \quad \forall \mathbf{x} \in A_u \right\}, \tag{4.27}$$

$$\mathcal{V}_{pw} := \left\{ \delta p^w(\mathbf{x}) : V \mapsto \mathbb{R} \,\middle|\, \delta p^w \in H^1, \delta p^w = 0 \quad \forall \mathbf{x} \in A_w \right\}, \tag{4.28}$$

$$\mathcal{V}_{pa} := \left\{ \delta p^a(\mathbf{x}) : V \mapsto \mathbb{R} \,\middle|\, \delta p^a \in H^1, \delta p^a = 0 \quad \forall \mathbf{x} \in A_a \right\}. \tag{4.29}$$

be the corresponding spaces of weighting functions, satisfying the homogeneous counterpart of the Dirichlet boundary conditions.

In the above defined spaces H^1 denotes the Sobolev space of functions of degree one,
$$H^1(V) := \left\{ f \,\middle|\, f \in L_2 ; \frac{\partial f}{\partial x_i} \in L_2 \right\} \quad \text{with} \quad L_2(V) := \left\{ f \,\middle|\, \int_V f^2 \, dV < \infty \right\}. \tag{4.30}$$

In general, H^k denotes a Sobolev space of functions with k square-integrable derivatives. Thus, in H^1 the first derivatives are required to be square-integrable, the latter is expressed by the second of the definitions (4.30). The square-integrability is necessary to preclude certain singularities in the trial solutions. The space L_2 is called Hilbert space of square-integrable functions. A Hilbert space is a vector space with a scalar product that is complete, which means that any Cauchy sequence in the space is convergent.

Finally, by means of (4.17), (4.21) and (4.23) the *weak form of the initial value/boundary value problem (I/BVP)* for the three-phase model can be stated as: Find $\mathbf{u}(\mathbf{x},t) \in \mathcal{C}_{\mathbf{u}}$, $p^w(\mathbf{x},t) \in \mathcal{C}_{pw}$ and $p^a(\mathbf{x},t) \in \mathcal{C}_{pa}$ such that

$$G(\mathbf{u}, p^w, p^a, \delta\mathbf{u}) = -\int_V \operatorname{grad} \delta\mathbf{u} : \tilde{\boldsymbol{\sigma}}' \, dV - \int_V S^w p^w \operatorname{div} \delta\mathbf{u} \, dV - \int_V S^a p^a \operatorname{div} \delta\mathbf{u} \, dV +$$
$$+ \int_V \tilde{\rho}\, \delta\mathbf{u} \cdot \mathbf{g} \, dV + \int_{A_{\mathbf{u}}^q} \delta\mathbf{u} \cdot \bar{\mathbf{t}} \, dA = 0, \tag{4.31}$$

$$H^w(\mathbf{u}, p^w, p^a, \delta p^w) = \int_V \delta p^w \, S^w \, \dot{\boldsymbol{\varepsilon}} : \mathbf{I} \, dV - \int_V \delta p^w \left(\frac{n S^w}{K^w} + n \frac{dS^w}{dp^c} \right) \dot{p}^w \, dV +$$
$$+ \int_V \delta p^w \, n \frac{dS^w}{dp^c} \dot{p}^a \, dV - \int_V \operatorname{grad} \delta p^w \cdot \frac{\mathbf{k}^{ow} k^{rw}}{\rho^w g} \cdot \operatorname{grad} p^w \, dV -$$
$$- \int_V \operatorname{grad} \delta p^w \cdot \frac{\mathbf{k}^{ow} k^{rw}}{g} \cdot \mathbf{g} \, dV + \int_{A_w^q} \delta p^w \, \bar{q}^w \, dA = 0, \tag{4.32}$$

$$H^a(\mathbf{u}, p^w, p^a, \delta p^a) = \int_V \delta p^a \, S^a \, \dot{\boldsymbol{\varepsilon}} : \mathbf{I} \, dV - \int_V \delta p^a \left(\frac{nS^a}{K^a} + n\frac{dS^w}{dp^c} \right) \dot{p}^a \, dV +$$

$$+ \int_V \delta p^a \, n\frac{dS^w}{dp^c} \dot{p}^w \, dV - \int_V \operatorname{grad} \delta p^a \cdot \frac{\mathbf{k}^{oa} k^{ra}}{\rho^a g} \cdot \operatorname{grad} p^a \, dV -$$

$$- \int_V \operatorname{grad} \delta p^a \cdot \frac{\mathbf{k}^{oa} k^{ra}}{g} \cdot \mathbf{g} \, dV + \int_{A_a^q} \delta p^a \, \overline{q}^a \, dA = 0 \; . \tag{4.33}$$

for any $\delta \mathbf{u} \in \mathcal{V}_\mathbf{u}$, $\delta p^w \in \mathcal{V}_{pw}$ and $\delta p^a \in \mathcal{V}_{pa}$. In (4.32) and (4.33) use of the identity $\operatorname{div} \dot{\mathbf{u}} = \dot{\boldsymbol{\varepsilon}} : \mathbf{I}$ has been made. The equations (4.31) to (4.33) may now be used as basis for deriving the discretised formulation of the problem in the next section.

4.3 Spatial discretisation of the weak formulation

4.3.1 Introduction

The initial value/boundary value problem defined by the set of equations (4.31) to (4.33) can be solved exactly only for very simple cases. Most practical problems, however, require some form of approximation which in the present context is achieved by means of the Finite Element Method.

To obtain a solution for the presented three-phase formulation, two major steps have to be distinguished. First, a spatial discretisation of the problem has to be performed, i.e. the domain under consideration is subdivided into a certain number of finite regions, the so-called 'finite elements'. The unknowns of the problem are approximated within each of these finite elements in terms of a set of shape functions and the respective nodal values. Second, since there are also time derivatives of the unknowns involved in the formulation, an additional numerical integration in the time domain has to be performed. While this section is devoted to the first step, the second step is treated in the next section.

At the outset of this section the basic principles of the spatial approximation of the variables within a finite element are discussed. Subsequently, the discretised forms of the linear momentum balance equation and the mass balance equations are derived. Since it is quite common in a finite element formulation for multi-phase problems to choose different approximations for the different kinds of variables, i.e. for the displacements of the soil skeleton and the hydrostatic stresses in the fluid phases, a separate section deals with the proper choice of finite elements. It should be mentioned that in the subsequent sections matrix notation is applied. Dots indicating certain matrix or vector operations are omitted for the sake of brevity.

4.3.2 Variables of the model

As indicated in Section 4.3.1, the Finite Element Method requires a subdivision of the continuum into a certain number of finite elements. Hence, the domain with the volume V is discretised by non-overlapping elements of volume V_e. Associated with this discretisation are NP_u and NP_p nodal points with displacement and with hydrostatic stress degrees of freedom, respectively. It should be mentioned that in general NP_u and NP_p need not be equal.

The choice of the displacements of the soil skeleton and the hydrostatic stresses in the fluid phases water and air as primary variables is considered to be most convenient for the current approach, i.e. for the kinds of problems investigated. Of course it is not the only possible choice. In the literature, several different models exist: An approach which is also based on the displacements and the two fluid stresses has been used by Alonso and co-workers [Alonso(1995)]. Other models rely on the displacements, the capillary stress and one of the individual fluid stresses, either the stress in the water phase [Schrefler(1990)] or the stress in the air phase [Gawin(1995)]. Instead of the water stresses the degree of water saturation may be chosen as degree of freedom as, e.g., in [Li(1990)] or [Li(1992)]. A comparison of the different models concerning their advantages and drawbacks is presented in [Schrefler(1991)].

Subsequently, the approximations of the primary variables within a single finite element e are specified. For any arbitrary point within the finite element the displacements of the soil skeleton $\mathbf{u}_e$ are given in terms of a matrix of suitable shape functions $\mathbf{N_u}$ and the respective vector of the nodal values of the displacements $\mathbf{U}_e$. The same interpolation is employed for the virtual displacements $\delta\mathbf{u}_e$ (Galerkin-type approximation). In mathematical terms these relations are formulated as [Zienkiewicz(1989)]

$$\mathbf{u}_e = \mathbf{N_u}\,\mathbf{U}_e \quad \text{and} \quad \delta\mathbf{u}_e = \mathbf{N_u}\,\delta\mathbf{U}_e \,. \tag{4.34}$$

In (4.34) the matrix of shape functions $\mathbf{N_u}$ has the form $\mathbf{N_u} = \left[\mathbf{I}\,N_u^{(i)} \vdots \mathbf{I}\,N_u^{(j)} \vdots \ldots\right]$ with $\mathbf{I}$ denoting the unit matrix and $N_u^{(i)}, N_u^{(j)}, \ldots$ being the individual shape functions. The relationship between the strains $\boldsymbol{\varepsilon}_e$ in the soil skeleton and the nodal values of the displacements reads as

$$\varepsilon_e = \mathbf{B_u}\,\mathbf{U}_e \quad \text{and} \quad \delta\varepsilon_e = \mathbf{B_u}\,\delta\mathbf{U}_e\,, \tag{4.35}$$

where the matrix $\mathbf{B_u}$ contains the partial derivatives of the shape functions $\mathbf{N_u}$ with respect to the coordinate directions. It should be mentioned that the relationship (4.35_2) is only valid if a geometrically linear theory is employed.

The hydrostatic and the virtual hydrostatic stresses in the water phase, p_e^w and δp_e^w, are approximated at any arbitrary point within the finite element e by means of a vector of suitable shape functions $\mathbf{N}_p$ and the respective nodal values of the hydrostatic and the virtual hydrostatic water stresses, $\mathbf{P}_e^w$ and $\delta\mathbf{P}_e^w$, as

$$p_e^w = \mathbf{N}_p\,\mathbf{P}_e^w \quad \text{and} \quad \delta p_e^w = \mathbf{N}_p\,\delta\mathbf{P}_e^w\,. \tag{4.36}$$

Again, according to (4.36), the same interpolations are chosen for the hydrostatic and the virtual hydrostatic water stresses. Contrary to the displacements the hydrostatic fluid stress is a scalar quantity and therefore the matrix of shape functions degenerates to a vector, $\mathbf{N}_p = \lfloor N_p^{(i)}\ \vdots\ N_p^{(j)}\ \vdots\ \ldots \rfloor$. The gradients of the hydrostatic and the virtual hydrostatic stresses in the water phase are derived from (4.36) as

$$\boldsymbol{\nabla} p_e^w = \mathbf{B}_p\,\mathbf{P}_e^w \quad \text{and} \quad \boldsymbol{\nabla}\delta p_e^w = \mathbf{B}_p\,\delta\mathbf{P}_e^w\,, \tag{4.37}$$

where $\mathbf{B}_p = \boldsymbol{\nabla}\mathbf{N}_p$ contains the partial derivatives of the shape functions with respect to the coordinate directions.

Similar to the hydrostatic water stresses, the hydrostatic and the virtual hydrostatic stresses in the air phase, p_e^a and δp_e^a, are interpolated by means of a vector of shape functions $\mathbf{N}_p$ and the respective vectors of nodal values, $\mathbf{P}_e^a$ and $\delta\mathbf{P}_e^a$, as

$$p_e^a = \mathbf{N}_p\,\mathbf{P}_e^a \quad \text{and} \quad \delta p_e^a = \mathbf{N}_p\,\delta\mathbf{P}_e^a\,. \tag{4.38}$$

The gradients of the hydrostatic and the virtual hydrostatic air stresses, $\boldsymbol{\nabla}p_e^a$ and $\boldsymbol{\nabla}\delta p_e^a$, are obtained as

$$\boldsymbol{\nabla}p_e^a = \mathbf{B}_p\,\mathbf{P}_e^a \quad \text{and} \quad \boldsymbol{\nabla}\delta p_e^a = \mathbf{B}_p\,\delta\mathbf{P}_e^a\,, \tag{4.39}$$

where $\mathbf{B}_p$ again contains the partial derivatives of the shape functions $\mathbf{N}_p$.

The approximations (4.34) to (4.39) for the displacements of the soil skeleton and for the hydrostatic stresses in the fluid phases water and air can now be inserted into the weak formulation of the governing equations to obtain a discretised form of these

Chapter 4. Numerical formulation for the three-phase model 177

equations. First of all, for simplicity the domain under consideration is assumed to be discretised by only one finite element e. Since the assembly of the individual elements follows standard procedures [Zienkiewicz(1989)] it will be mentioned just briefly later on.

4.3.3 Discretised linear momentum balance equation

The weak formulation of the linear momentum balance equation (4.31) can be used to derive a discretised form of the respective equation. Insertion of the approximations for the primary variables as given in (4.34) to (4.39) yields the *discretised equilibrium equation*

$$-(\delta \mathbf{U}_e)^T \int_{V_e} \mathbf{B_u}^T \tilde{\boldsymbol{\sigma}}' \, dV_e -$$

$$- (\delta \mathbf{U}_e)^T \int_{V_e} \mathbf{B_u}^T \mathbf{1} \, S^w \, \mathbf{N}_p \, dV_e \, \mathbf{P}_e^w - (\delta \mathbf{U}_e)^T \int_{V_e} \mathbf{B_u}^T \mathbf{1} \, S^a \, \mathbf{N}_p \, dV_e \, \mathbf{P}_e^a +$$

$$+ (\delta \mathbf{U}_e)^T \int_{V_e} \mathbf{N_u}^T \tilde{\mathbf{b}} \, dV_e + (\delta \mathbf{U}_e)^T \int_{(A_\mathbf{u}^q)_e} \mathbf{N_u}^T \bar{\mathbf{t}} \, dA_e = 0 \; , \qquad (4.40)$$

where for the divergence of the virtual displacements $\delta \mathbf{u}$ occurring in (4.31) the identity $\mathrm{div}\,\delta \mathbf{u} = \delta \boldsymbol{\varepsilon} : \mathbf{I}$ has been used. Thus, both operators grad and div can be expressed by means of the matrix $\mathbf{B_u}$ which contains the derivatives of the shape functions with respect to the coordinate directions. Additionally, it should be mentioned that the product $\mathbf{B_u}^T \mathbf{1}$, present in the second and the third term of (4.40), denotes the divergence. The integration in equation (4.40) is now performed over the volume V_e and the surface part $(A_\mathbf{u}^q)_e$ of the finite element e, respectively. $\mathbf{1}$ denotes a vector comparable to the Kronecker delta, $\mathbf{1} = \lfloor 1, 1, 1, 0, 0, 0 \rfloor^T$, and $\tilde{\mathbf{b}}$ is the vector of the body forces, $\tilde{\mathbf{b}} = \tilde{\rho}\,\mathbf{g}$.

4.3.4 Discretised water mass balance equation

The discretised formulation of the mass balance equation for the water phase is developed from its weak counterpart (4.32) by inserting the approximations for the unknown displacements of the soil skeleton and the hydrostatic stresses in the two fluid phases given in the relations (4.34) to (4.39). This results in the subsequent form of the *discretised water mass balance equation:*

$$(\delta \mathbf{P}_e^w)^T \int_{V_e} \mathbf{N}_p^T S^w \mathbf{1}^T \mathbf{B_u} \, dV_e \, \dot{\mathbf{U}}_e - (\delta \mathbf{P}_e^w)^T \int_{V_e} \mathbf{N}_p^T \left(\frac{nS^w}{K^w} + n \frac{dS^w}{dp^c} \right) \mathbf{N}_p \, dV_e \, \dot{\mathbf{P}}_e^w +$$

$$+ (\delta \mathbf{P}_e^w)^T \int_{V_e} \mathbf{N}_p^T n \frac{dS^w}{dp^c} \mathbf{N}_p \, dV_e \, \dot{\mathbf{P}}_e^a - (\delta \mathbf{P}_e^w)^T \int_{V_e} \mathbf{B}_p^T \frac{\mathbf{k}^{ow} k^{rw}}{\rho^w g} \mathbf{B}_p \, dV_e \, \mathbf{P}_e^w -$$

$$- (\delta \mathbf{P}_e^w)^T \int_{V_e} \mathbf{B}_p^T \frac{\mathbf{k}^{ow} k^{rw}}{g} \mathbf{g} \, dV_e + (\delta \mathbf{P}_e^w)^T \int_{(A_w^q)_e} \mathbf{N}_p^T \bar{q}^w \, dA_e = 0 \, , \qquad (4.41)$$

where $(A_w^q)_e$ is the part of the surface A_e of the finite element e with a prescribed surface flow $\bar{q}^w$. The gradient operators apparent in the weak formulation (4.32) are expressed by means of the matrix $\mathbf{B}_p$.

4.3.5 Discretised air mass balance equation

The discretised formulation of the mass balance equation for the air phase is derived from the respective weak counterpart (4.33). Again the approximations of the primary variables (4.34) to (4.39) are used and the integration is performed over the volume V_e and the part $(A_a^q)_e$ of the surface with a prescribed airflow $\bar{q}^a$, respectively. Hence, the *discretised air mass balance equation* reads as

$$(\delta \mathbf{P}_e^a)^T \int_{V_e} \mathbf{N}_p^T S^a \mathbf{1}^T \mathbf{B_u} \, dV_e \, \dot{\mathbf{U}}_e - (\delta \mathbf{P}_e^a)^T \int_{V_e} \mathbf{N}_p^T \left(\frac{nS^a}{K^a} + n \frac{dS^w}{dp^c} \right) \mathbf{N}_p \, dV_e \, \dot{\mathbf{P}}_e^a +$$

$$+ (\delta \mathbf{P}_e^a)^T \int_{V_e} \mathbf{N}_p^T n \frac{dS^w}{dp^c} \mathbf{N}_p \, dV_e \, \dot{\mathbf{P}}_e^w - (\delta \mathbf{P}_e^a)^T \int_{V_e} \mathbf{B}_p^T \frac{\mathbf{k}^{oa} k^{ra}}{\rho^a g} \mathbf{B}_p \, dV_e \, \mathbf{P}_e^a -$$

$$- (\delta \mathbf{P}_e^a)^T \int_{V_e} \mathbf{B}_p^T \frac{\mathbf{k}^{oa} k^{ra}}{g} \mathbf{g} \, dV_e + (\delta \mathbf{P}_e^a)^T \int_{(A_a^q)_e} \mathbf{N}_p^T \bar{q}^a \, dA_e = 0 \, . \qquad (4.42)$$

4.3.6 System of spatially discretised equations

Since in the case of a domain discretised by a single finite element e the equations (4.40) to (4.42) have to hold for arbitrary virtual displacements $\delta \mathbf{U}_e$ and for arbitrary virtual hydrostatic fluid stresses $\delta\mathbf{P}_e^w$ and $\delta\mathbf{P}_e^a$, respectively, the set of equations in a more compact form reads as

$$\mathbf{f}_e^{in}(\mathbf{U}_e) + \mathbf{C}_e^{sw}\mathbf{P}_e^w + \mathbf{C}_e^{sa}\mathbf{P}_e^a = \mathbf{f}_e^{ex},$$

$$(\mathbf{C}_e^{sw})^T\dot{\mathbf{U}}_e - \mathbf{Q}_e^{ww}\dot{\mathbf{P}}_e^w - \mathbf{H}_e^{ww}\mathbf{P}_e^w + \mathbf{C}_e^{wa}\dot{\mathbf{P}}_e^a = \dot{\mathbf{f}}_e^w, \quad (4.43)$$

$$(\mathbf{C}_e^{sa})^T\dot{\mathbf{U}}_e + \mathbf{C}_e^{wa}\dot{\mathbf{P}}_e^w - \mathbf{Q}_e^{aa}\dot{\mathbf{P}}_e^a - \mathbf{H}_e^{aa}\mathbf{P}_e^a = \dot{\mathbf{f}}_e^a.$$

The following abbreviations have been used in the above set of governing equations for the linear momentum balance equation of the three-phase system (4.43_1),

$$\mathbf{f}_e^{in} = \int_{V_e} \mathbf{B}_\mathbf{u}^T \tilde{\boldsymbol{\sigma}}' \, dV_e, \quad \mathbf{f}_e^{ex} = \int_{V_e} \mathbf{N}_\mathbf{u}^T \tilde{\mathbf{b}} \, dV_e + \int_{(A_\mathbf{u}^q)_e} \mathbf{N}_\mathbf{u}^T \bar{\mathbf{t}} \, dA_e,$$

$$\mathbf{C}_e^{sw} = \int_{V_e} \mathbf{B}_\mathbf{u}^T \mathbf{1} S^w \mathbf{N}_p \, dV_e, \quad \mathbf{C}_e^{sa} = \int_{V_e} \mathbf{B}_\mathbf{u}^T \mathbf{1} S^a \mathbf{N}_p \, dV_e,$$
(4.44)

for the mass balance equation of the water phase (4.43_2),

$$\mathbf{Q}_e^{ww} = \int_{V_e} \mathbf{N}_p^T \left(\frac{nS^w}{K^w} + n\frac{dS^w}{dp^c}\right) \mathbf{N}_p \, dV_e, \quad \mathbf{C}_e^{wa} = \int_{V_e} \mathbf{N}_p^T n\frac{dS^w}{dp^c} \mathbf{N}_p \, dV_e,$$

$$\mathbf{H}_e^{ww} = \int_{V_e} \mathbf{B}_p^T \frac{k^{ow}k^{rw}}{\rho^w g} \mathbf{B}_p \, dV_e, \quad \dot{\mathbf{f}}_e^w = \int_{V_e} \mathbf{B}_p^T \frac{k^{ow}k^{rw}}{g} \mathbf{g} \, dV_e - \int_{(A_w^q)_e} \mathbf{N}_p^T \bar{q}^w \, dA_e,$$
(4.45)

and for the mass balance equation of the air phase (4.43_3),

$$\mathbf{Q}_e^{aa} = \int_{V_e} \mathbf{N}_p^T \left(\frac{nS^a}{K^a} + n\frac{dS^w}{dp^c}\right) \mathbf{N}_p \, dV_e,$$

$$\mathbf{H}_e^{aa} = \int_{V_e} \mathbf{B}_p^T \frac{k^{oa}k^{ra}}{\rho^a g} \mathbf{B}_p \, dV_e, \quad \dot{\mathbf{f}}_e^a = \int_{V_e} \mathbf{B}_p^T \frac{k^{oa}k^{ra}}{g} \mathbf{g} \, dV_e - \int_{(A_a^q)_e} \mathbf{N}_p^T \bar{q}^a \, dA_e.$$
(4.46)

In the above system of equations (4.43) $\mathbf{U}_e$, $\mathbf{P}_e^w$ and $\mathbf{P}_e^a$ denote the vectors of the nodal values for the displacements of the soil skeleton and for the hydrostatic stresses in the fluid phases water and air. $\mathbf{f}_e^{in}$ and $\mathbf{f}_e^{ex}$ are the vectors of the internal forces of the soil skeleton and of the external forces, consisting of body forces and surface tractions, respectively.

The matrices $\mathbf{C}_e^{sw}$, $\mathbf{C}_e^{sa}$ and $\mathbf{C}_e^{wa}$ (coupling matrices) describe the coupling between the individual constituents, i.e. between the soil skeleton and the water phase, between the soil skeleton and the air phase and between the two fluid phases water and air. $\mathbf{Q}_e^{ww}$ and $\mathbf{Q}_e^{aa}$ (compressibility matrices) contain the compressibility of the particular fluid (note that for a compressible solid phase they would also contain the compressibility of the solid phase) and the constitutive relationship between the degree of water saturation and the capillary stress. $\mathbf{H}_e^{ww}$ and $\mathbf{H}_e^{aa}$ are the permeability matrices, expressing the permeability of the soil with respect to the water and the air phase, and the vectors $\mathbf{\dot{f}}_e^w$ and $\mathbf{\dot{f}}_e^a$ are associated with the flow of the respective fluid due to gravity and the flow through the surface of the domain under consideration.

The discretised form (4.43) of the governing equations is valid for a domain discretised by a single finite element e. However, in general a domain is subdivided into a certain number of elements and therefore, the contributions of the individual finite elements have to be assembled to represent the equations describing the behaviour of the complete body. The assembly of the individual element matrices and vectors follows standard procedures which can be found in the literature, e.g., [Zienkiewicz(1989)]. Hence, any details are omitted here. The set of equations for a domain discretised by more than one finite element is obtained from (4.43) as

$$\mathbf{f}^{in}(\mathbf{U}) + \mathbf{C}^{sw}\mathbf{P}^w + \mathbf{C}^{sa}\mathbf{P}^a = \mathbf{f}^{ex},$$

$$(\mathbf{C}^{sw})^T \mathbf{\dot{U}} - \mathbf{Q}^{ww}\mathbf{\dot{P}}^w - \mathbf{H}^{ww}\mathbf{P}^w + \mathbf{C}^{wa}\mathbf{\dot{P}}^a = \mathbf{\dot{f}}^w, \quad (4.47)$$

$$(\mathbf{C}^{sa})^T \mathbf{\dot{U}} + \mathbf{C}^{wa}\mathbf{\dot{P}}^w - \mathbf{Q}^{aa}\mathbf{\dot{P}}^a - \mathbf{H}^{aa}\mathbf{P}^a = \mathbf{\dot{f}}^a,$$

where the matrices and vectors in these equations result from assembling the element matrices and vectors in (4.44) to (4.46) to global matrices and vectors.

The matrices and vectors in these equations depend on the nodal displacements $\mathbf{U}$ of the soil skeleton and on the nodal hydrostatic stresses in the two fluid phases water and air, $\mathbf{P}^w$ and $\mathbf{P}^a$. The vector of the internal forces of the soil skeleton, $\mathbf{f}^{in}$, is a function of the nodal displacements $\mathbf{U}$. The coupling matrices $\mathbf{C}^{sw}$, $\mathbf{C}^{sa}$ and $\mathbf{C}^{wa}$ as well as the compressibility matrices $\mathbf{Q}^{ww}$ and $\mathbf{Q}^{aa}$ depend on the degree of saturation S^w or S^a of the respective fluid phase. Because of the saturation condition, that is $S^w + S^a = 1.0$, the coupling matrices and the compressibility matrices may be considered as functions of the degree of water saturation. However, the degree of water saturation depends on the capillary stress (equation (3.267)) which is a function of the hydrostatic stresses in the individual fluid phases water and air (equation (3.209)). Consequently, the coupling matrices and the compressibility matrices depend on the nodal values of the hydrostatic stresses in the fluid phases, $\mathbf{P}^w$ and $\mathbf{P}^a$. The permeability matrices $\mathbf{H}^{ww}$ and $\mathbf{H}^{aa}$ and the flow vectors $\dot{\mathbf{f}}^w$ and $\dot{\mathbf{f}}^a$ contain the coefficients of permeability of the respective fluid phase and are thus, according to the equations (3.285) and (3.288), functions of the degree of water saturation as well. Hence, they also depend on the nodal values of the hydrostatic stresses $\mathbf{P}^w$ and $\mathbf{P}^a$.

To obtain the special cases of the formulation discussed in Section 3.7, the following assumptions have to be introduced into the general set of equations.

For the case of dewatering under atmospheric conditions (Section 3.7.2) the hydrostatic stress in the air phase (exceeding the atmospheric air pressure) is equal to zero, that is $p^a = 0$, and hence also $\delta p^a = 0$. Consequently, the mass balance equation for the air phase can be omitted as well as the coupling terms between the air phase and the other two constituents solid and water. This yields the subsequent reduced set of equations

$$\begin{aligned} \mathbf{f}^{in}(\mathbf{U}) + \mathbf{C}^{sw}\,\mathbf{P}^w &= \mathbf{f}^{ex}\;, \\ (\mathbf{C}^{sw})^T\,\dot{\mathbf{U}} - \mathbf{Q}^{ww}\,\dot{\mathbf{P}}^w - \mathbf{H}^{ww}\,\mathbf{P}^w &= \dot{\mathbf{f}}^w\;. \end{aligned} \quad (4.48)$$

In (4.48) the matrices $\mathbf{C}^{sw}$, $\mathbf{Q}^{ww}$ and $\mathbf{H}^{ww}$ and the vector $\dot{\mathbf{f}}^w$ are still functions of the degree of water saturation S^w and thus, via the capillary stress, of the nodal values of the hydrostatic water stress $\mathbf{P}^w$.

For the special case of consolidation (Section 3.7.3) the degree of saturation is constant, that is $S^w = 1.0$, and consequently also the matrices in (4.48) have constant coefficients. The derivative of the degree of water saturation in the compressibility matrix $\mathbf{Q}^{ww}$ (first of the equations (4.45)) is equal to zero. If the fluid phase is assumed to be incompressible, i.e. $K^w \to \infty$, which is a common assumption for consolidation analyses, then the compressibility matrix vanishes completely and the system of equations describing the consolidation problem reads as

$$\begin{aligned} \mathbf{f}^{in}(\mathbf{U}) + \mathbf{C}^{sw}\,\mathbf{P}^w &= \mathbf{f}^{ex}\;, \\ (\mathbf{C}^{sw})^T\,\dot{\mathbf{U}} - \mathbf{H}^{ww}\,\mathbf{P}^w &= \dot{\mathbf{f}}^w\;. \end{aligned} \quad (4.49)$$

A set of equations similar to (4.49) may be found in the literature on the finite element treatment of the consolidation phenomenon [Buchmaier(1985), Britto(1987),

Smith(1998)]. In classical Biot consolidation even the term $\dot{\mathbf{f}}^w$ on the right hand side of (4.49) is neglected [Smith(1998)], i.e. no flow due to the unit weight of water and no prescribed surface flows are taken into account.

For drained conditions the fluid phases are not taken into account and consequently the respective mass balance equations can be omitted. In this case only the equilibrium between the internal forces of the soil skeleton, $\mathbf{f}^{in}$, and any external forces $\mathbf{f}^{ex}$, consisting of body forces and surface tractions, has to be satisfied, that is

$$\mathbf{f}^{in}(\mathbf{U}) = \mathbf{f}^{ex} \ . \tag{4.50}$$

This is the familiar form of the equilibrium equations encountered in finite element analyses of deformable solids [Zienkiewicz(1989), Zienkiewicz(1991)].

4.4 Numerical integration in the time domain

4.4.1 Introduction

Compared to the spatial discretisation discussed in the previous section, the time domain is fundamentally different in that the time dimension is open ended and information can only propagate forward. Hence, approximations of time-dependent behaviour must be inherently one-sided, always looking to the past for information to predict the future. Consequently, problems involving the time dimension are always initial value problems in time, meaning that current solutions are predicted based on information specified for a certain initial point of time, say, $t = 0$. Numerical algorithms that propagate information forward in time are called 'time-marching' or 'time-stepping' algorithms.

For a numerical treatment the time domain is subdivided into finite time increments Δt and the current time instant t_{n+1} is related to the previous time instant t_n by the relationship $t_{n+1} = t_n + \Delta t$. The solution at t_{n+1} is obtained from the respective solutions at previous time instants by so-called 'recurrence relations'. Within the current work solutions for the space-time problem are gained by first performing a semi-discretisation in space followed by a discretisation in the time domain. Therefore, the starting point in this section will be the set of semi-discrete equations (4.47) which still contains time derivatives of the unknowns.

A variety of different time integration schemes are available in the literature [Hughes(1987), Zienkiewicz(1991)]. Time-dependent problems may be classified into either first-order or second-order problems according to the appearance of only first or first and second time derivatives of the unknowns. The integration methods are divided into 'single-step' recurrence relations and 'multistep' schemes, depending on the number of previous time steps used for the prediction of the current solution. Consequently, not only the amount of storage demand but also the computational efficiency are closely related to the chosen method, in particular when dealing with (i) non-linear problems which demand an iterative solution procedure and/or (ii) domains with a large number of degrees of freedom.

After presenting a brief survey of the various time integration schemes for first-order and second-order ordinary differential equations, the approach employed in the current model is introduced. The approximations for the unknowns of the formulation are defined which may then be inserted into the semi-discrete set of equations (4.47) presented in the previous section.

4.4.2 Brief survey of the various time integration schemes

After spatial discretisation the obtained equations are either first-order or second-order ordinary differential equations depending on the type of problem to be investigated. First-order differential equations usually arise from problems like, e.g., heat transfer, seepage, fluid flow or electric conduction and are characterised by containing only first time derivatives of the unknowns. On the other hand, second-order differential equations comprise also second time derivatives of the variables. They are encountered in dynamic analyses, e.g., in structural dynamics or earthquake analysis, taking into account inertia and damping effects.

For the numerical integration in the time domain of both first-order and second-order differential equations either single-step or multistep recurrence schemes can be applied. A broad variety of methods is available in the respective literature, e.g., [Hughes(1987), Zienkiewicz(1991)]. The common characteristic of all the single-step algorithms is that they only use the information of the previous time step t_n for calculating the current solution at t_{n+1}. However, there exist explicit and implicit members of this family of solution schemes, each exhibiting different accuracy, stability and convergence properties. Since only information of one previous time step is necessary, these algorithms are rather economical with respect to storage demand. Additionally, they are very suited for the use of varying time step sizes. When considering multistep algorithms the prediction of the current solution is based on information of several previous points of time, as inherent in the terminology. It is possible to derive solution procedures in which the variables at time t_{n+1} are only related to values of the solution at t_n, t_{n-1}, t_{n-2}, etcetera, without explicitly introducing any derivatives and assuming an equal interval Δt between each set. However, these algorithms are in general less convenient to use than the single-step processes as they do not permit an easy change of the time step magnitude [Zienkiewicz(1991)].

Concerning the spatial discretisation of a space-time problem, it is usually truncation error analysis that is employed to investigate if a certain approximation is acceptable, i.e. the truncation error must decrease to zero as the mesh refinement increases (consistency). Mathematically speaking, the consistency condition may be formulated as

$$\lim_{\Delta h \to 0} e_t = 0 , \qquad (4.51)$$

where e_t denotes the truncation error and Δh is a typical measure of the element size. For time-marching algorithms, on the other hand, truncation error and consistency are not the only important properties but stability of the time-stepping scheme is also critical. Stability can be formally defined by the constraint that the difference between the numerical and the analytical solution must remain bounded for all time. Thus, for any point of time t_n the subsequent relationship should be satisfied:

$$\|\mathbf{Y}_n - \mathbf{Y}(t_n)\| < c . \qquad (4.52)$$

In (4.52) $\mathbf{Y}_n$ denotes the numerical approximation of the exact solution $\mathbf{Y}(t_n)$ and c is a finite-valued constant. However, a necessary condition for any numerical procedure is that it should produce approximations that become arbitrarily close to the exact solution as the spatial and temporal steps decrease, i.e. that numerical errors can be controlled by sufficiently fine discretisation. A numerical approximation to a differential equation that reproduces the exact solution upon continuous spatial and temporal mesh refinement is said to be convergent. The mathematical statement of this criterion reads as

$$\lim_{\substack{\Delta h \to 0 \\ \Delta t \to 0}} \|\mathbf{Y}_n - \mathbf{Y}(t_n)\| = 0 . \qquad (4.53)$$

Convergence can be investigated by means of the Lax equivalence theorem which states that consistent and stable approximations are also convergent. Thus, convergence may be shown by truncation error analysis (consistency) and by stability analysis [Celia(1992)].

Probably the most extensively investigated methods for solving first-order ordinary differential equations are the single-step generalised trapezoidal family of methods

[Hughes(1983), Borja(1991a)], primarily because they are self-starting and simple to implement. Additionally, their storage demands are rather economical. Three of the most well-known members of this family are mentioned subsequently: (i) The Euler forward scheme is an explicit method; it is only conditionally stable and first-order accurate in time. (ii) The trapezoidal rule (Crank-Nicolson scheme) is an implicit method; it possesses the main advantages of being unconditionally stable and second-order accurate. (iii) The implicit Euler backward scheme is unconditionally stable as well but only first-order accurate. It is known that the Euler backward method can damp high-frequency components of the system, which enter into the solution because of the numerical approximation, but at the expense of accuracy (only first-order). On the other hand, the Crank-Nicolson scheme is second-order accurate but lacks the numerical dissipation of the Euler backward method. However, both algorithms are implicit and unconditionally stable and also compatible with the implicit stress point algorithms widely used in computational plasticity [Borja(1991a)]. A member of this family is applied in the current three-phase model and is thus discussed in more detail in the next section.

One-step recurrence schemes can also be derived by means of a weighted residual (finite element) approach [Zienkiewicz(1991)]. As usual within the finite element framework, the solution is assumed to vary polynomial in the interval (time interval Δt in this case). In [Zienkiewicz(1991)] approximations of the unknowns both with the lowest (linear) polynomials as well as with polynomials of an arbitrary degree p are presented. These approximations can then be (i) inserted into the semi-discrete set of equations, (ii) multiplied with a weighting function and (iii) integrated over the time interval Δt which gives the required recurrence algorithm. However, a suitable choice of weighting functions may yield formulae identical to the generalised trapezoidal method [Borja(1991a)].

Alternatively, collocation schemes using truncated Taylor series can be applied for the solution of the semi-discrete equations in the time domain. Satisfaction of the respective equations may be considered, e.g., only at the endpoints of the time interval Δt (collocation). The unknowns and their derivatives are approximated separately by truncated Taylor series. These approximations are then substituted into the semi-discrete equations which yields a solution for the highest-order derivative of the unknowns forming the basis of the recurrence scheme. Unfortunately, the obtained process may not be self-starting. A rather popular and widely used member of this family of methods is the well-known Newmark algorithm [Zienkiewicz(1991)]. It should be mentioned that the latter two general procedures can of course be applied to both first-order and second-order ordinary differential equations.

In addition to the one-step methods mentioned above, the so-called 'one-step, multistage' methods are widely employed. Here algorithms are obtained from the use of two or more stages in the calculation from one time level to the next. The family of the Runge-Kutta methods is a well-known member of the one-step, multistage algorithms. Recently, a complex-time-step algorithm based on the Newmark method has been proposed in [Fung(1998)] and [Fung(1999)] for second-order and first-order differential equations, respectively. This algorithm combines the two desirable properties for a time-marching method, namely unconditional stability together with a high order of accuracy and controllable numerical dissipation in the high frequency range. Numerical results are evaluated independently at a certain number of stages (may be complex) within the time step under consideration and then combined linearly to give higher-order accurate results at the end of the time step. The desirable order of accuracy of an algorithm determines the number of necessary intermediate solutions. It should be mentioned that this algorithm does not require systems of nested equations to be solved as in the implicit Runge-Kutta methods.

The previously mentioned schemes may be regarded as being somewhat similar to the so-called multistep methods which predict the current solution on the basis of a certain number of preceding time steps. However, in this case not the solutions of any sub-steps within the current time increment are employed but rather the solutions from previous converged points of time. A general $(q+2)$-step method uses the time levels $n+1, n, n-1, \ldots, n-q$. Hence, the current solution is obtained from a number of temporally earlier solutions which implies significant shortcomings. First, these methods require information at more than one previous time level to proceed to the solution at a new level. Therefore, start-up is complicated by the fact that for the first step information is known at only one level and thus a special start-up algorithm is needed. Second, this class of algorithms is of course more storage consuming as a number of solution sets has to be kept. Finally, a change in the magnitude of the time step is more cumbersome to realise than for the single-step methods. However, much classical work on stability and accuracy refers to such multistep algorithms [Zienkiewicz(1991)]. Typical members of this family are, e.g., the Adams-Bashforth, Adams-Moulton and Gear's methods. A family of k-step, k-order linear multistep methods based on the backward differentiation formulae has been used to solve non-linear consolidation problems in [Borja(1991a)]. The advantages of these methods are a high (k) order of achievable accuracy, high-frequency numerical dissipation and the possibility of variable step sizes. However, the method is not self-starting and more than one solution has to be kept in storage.

Finally, some remarks should be made concerning non-linear problems. In ordinary differential equations for solid mechanics problems non-linearities may occur due to non-linear material behaviour or due to non-linear geometric behaviour because of large deformations. In all cases a proper choice of the time-stepping procedure is of primary importance since the solution process at each recurrent step between t_n and t_{n+1} requires iteration. Clearly, it is not at all desirable to use many iteration steps for each time step. If convergence does not occur for a small number of iteration steps, it might be advantageous to reduce the size of the time increment.

4.4.3 Approximation of the variables in the current model

As indicated in the previous section, the generalised trapezoidal method is employed for the numerical integration of the semi-discrete set of equations (4.47) in the time domain. Being a one-step method, the numerical implementation of the algorithm is rather simple and storage demand is economical. In particular, the latter is considered advantageous since the three-phase formulation is storage consuming anyway due to its always non-linear character. Additionally, variable time step sizes are easy to handle in a one-step method.

Employing the *generalised trapezoidal method* an approximation of the unknowns at time t_{n+1} can be calculated as

$$\mathbf{Y}_{n+1} = \mathbf{Y}_n + \Delta t_{n+1} \dot{\mathbf{Y}}_{n+\alpha} \tag{4.54}$$

and of the first time derivatives of the unknowns as

$$\dot{\mathbf{Y}}_{n+\alpha} = (1-\alpha)\dot{\mathbf{Y}}_n + \alpha \dot{\mathbf{Y}}_{n+1} . \tag{4.55}$$

In the equations (4.54) and (4.55) $\mathbf{Y}_n$ and $\mathbf{Y}_{n+1}$ denote the vectors of unknowns at the points of time t_n and t_{n+1}, respectively. The dotted quantities are the corresponding time derivatives. Δt_{n+1} describes the current time step size and α is a given parameter, $0 \leq \alpha \leq 1$. An alternative and more convenient form of (4.54) is obtained after insertion of equation (4.55),

$$\mathbf{Y}_{n+1} = \mathbf{Y}_n + \Delta t_{n+1} \left[(1-\alpha)\dot{\mathbf{Y}}_n + \alpha \dot{\mathbf{Y}}_{n+1}\right] . \tag{4.56}$$

The time derivative of the vector of unknowns at time t_{n+1} can then be formulated by restructuring (4.56) as

$$\dot{\mathbf{Y}}_{n+1} = \frac{1}{\alpha \Delta t_{n+1}} \left[\mathbf{Y}_{n+1} - \mathbf{Y}_n - \Delta t_{n+1}(1-\alpha)\dot{\mathbf{Y}}_n \right] . \tag{4.57}$$

The approximations for the unknowns (4.56) and their time derivatives (4.57) may be inserted into the semi-discrete system of equations (4.47) in different ways. Two possible approaches are discussed, e.g., in [Borja(1991a)] with respect to consolidation problems. The first, so-called '**v**-form' is characterised by eliminating the unknowns from the spatially discretised set of equations, meaning that only the time derivatives of the vector of unknowns remain in the equations. Consequently, the system is solved for the first time derivatives of the unknowns and then the vector of the unknowns is updated. The second, so-called '**d**-form' is the complete opposite, i.e. the time derivatives are eliminated and the system of equations is solved for the unknowns. It should be remarked explanatory that the above terminology stems from 'classical' finite element analysis where the unknowns are the displacements (might be called **d**) and their time derivatives are the velocities (usually denoted with **v**). As shown in [Borja(1991a)], employing the '**d**-form' enables a complete elimination of the time derivatives of the unknowns from the set of semi-discrete equations. Furthermore, symmetric matrices may be obtained for certain cases which is not possible for the '**v**-form'. In addition to potential savings associated with matrix symmetry, the elimination of any time derivatives of the unknowns from the solution automatically results in a reduction of required computer storage.

In the current model the second approach is chosen in order to gain the above mentioned advantages. Therefore, the equations (4.56) and (4.57) are reformulated to include the difference $\Delta \mathbf{Y}_{n+1} = \mathbf{Y}_{n+1} - \mathbf{Y}_n$ between the solutions at time t_{n+1} and time t_n. This yields the following form for the time derivative of the vector of unknowns,

$$\dot{\mathbf{Y}}_{n+1} = \frac{\Delta \mathbf{Y}_{n+1}}{\alpha \Delta t_{n+1}} - \frac{1-\alpha}{\alpha} \dot{\mathbf{Y}}_n . \tag{4.58}$$

Insertion of (4.58) into (4.56) results in

$$\mathbf{Y}_{n+1} = \mathbf{Y}_n + \Delta t_{n+1} \left[(1-\alpha)\dot{\mathbf{Y}}_n + \frac{\Delta \mathbf{Y}_{n+1}}{\Delta t_{n+1}} - (1-\alpha)\dot{\mathbf{Y}}_n \right] = \mathbf{Y}_n + \Delta \mathbf{Y}_{n+1} \tag{4.59}$$

concerning the approximation of the unknowns. Both equations (4.58) and (4.59) now contain the desired increment $\Delta \mathbf{Y}_{n+1}$ of the unknowns corresponding to the time increment $\Delta t_{n+1} = t_{n+1} - t_n$.

Different values can be chosen for the parameter α, within the given range $0 \leq \alpha \leq 1$, resulting in different kinds of solution schemes. Setting $\alpha = 0$ in (4.56) yields the explicit Euler forward method, $\alpha = 0.5$ results in the implicit Crank-Nicolson scheme and $\alpha = 1$ gives the implicit Euler backward method. The two implicit methods seem to be advantageous since they are unconditionally stable and compatible with implicit stress point algorithms used in computational plasticity. Although the Crank-Nicolson scheme is of second-order accuracy, the first-order Euler backward method is chosen for the present model because of its high-frequency damping possibilities. Employing the equations (4.59) and (4.58), the latter with a factor α equal to one, for the Euler backward method the vector of the unknowns and its time derivative at time t_{n+1} can be expressed as

$$\mathbf{Y}_{n+1} = \mathbf{Y}_n + \Delta \mathbf{Y}_{n+1} \quad \text{and} \quad \dot{\mathbf{Y}}_{n+1} = \frac{\Delta \mathbf{Y}_{n+1}}{\Delta t_{n+1}} . \quad (4.60)$$

In the semi-discrete set of equations (4.47), the unknowns are the displacements $\mathbf{U}$ of the soil skeleton and the hydrostatic stresses in the fluid phases water and air, $\mathbf{P}^w$ and $\mathbf{P}^a$. Therefore, the general approximations (4.60) have to be specialised for the displacements of the soil skeleton as

$$\mathbf{U}_{n+1} = \mathbf{U}_n + \Delta \mathbf{U}_{n+1} \quad \text{and} \quad \dot{\mathbf{U}}_{n+1} = \frac{\Delta \mathbf{U}_{n+1}}{\Delta t_{n+1}} , \quad (4.61)$$

for the hydrostatic stresses in the water phase as

$$\mathbf{P}^w_{n+1} = \mathbf{P}^w_n + \Delta \mathbf{P}^w_{n+1} \quad \text{and} \quad \dot{\mathbf{P}}^w_{n+1} = \frac{\Delta \mathbf{P}^w_{n+1}}{\Delta t_{n+1}} , \quad (4.62)$$

and for the hydrostatic stresses in the air phase as

$$\mathbf{P}^a_{n+1} = \mathbf{P}^a_n + \Delta \mathbf{P}^a_{n+1} \quad \text{and} \quad \dot{\mathbf{P}}^a_{n+1} = \frac{\Delta \mathbf{P}^a_{n+1}}{\Delta t_{n+1}} . \quad (4.63)$$

The approximations (4.61) to (4.63) for the unknowns and for their time derivatives can now be inserted into the semi-discrete system of equations (4.47) which is then solved for the incremental nodal values of the unknowns, $\Delta \mathbf{U}_{n+1}$, $\Delta \mathbf{P}^w_{n+1}$ and $\Delta \mathbf{P}^a_{n+1}$. This final part, however, is dealt with in the next section.

4.5 Solution of the final coupled set of equations

4.5.1 Introduction

After the explanation of the spatial and temporal discretisation procedures in the previous sections the solution of the finally obtained coupled system of equations remains to be treated. To this end, the approximations of the unknowns in the time domain introduced in Section 4.4 are inserted into the spatially discretised set of equations (4.47). As already mentioned, the coupled system of equations is always non-linear, even for the case of linear elastic material behaviour of the soil skeleton. To obtain a solution of the final coupled set of equations, two different types of iterative solution procedures are introduced. The two methods presented here are (i) the fixed-point iteration procedure which has the advantage of a simple implementation, and (ii) the Newton-Raphson iteration procedure which is more costly in programming due to the necessary derivatives but possesses superior convergence properties.

To conclude this chapter on the numerical solution of the coupled three-phase formulation, some remarks on a suitable choice of finite elements are made. Since usually different orders of interpolation are employed for the different kinds of the degrees of freedom, i.e. soil displacements and hydrostatic fluid stresses, in a multi-phase formulation, possible meaningful combinations of approximations for the variables are explained.

The obtained system of equations for the complete discretised domain is a linear algebraic system with a number of n_{eq} equations which is solved by means of a direct solution procedure of the Gaussian type.

4.5.2 Finally obtained coupled system of equations

After the spatial discretisation of the governing equations, as derived in Section 4.3, the semi-discrete set of equations (4.47) is obtained. For the point of time t_{n+1} it can be formulated as follows:

$$\mathbf{f}^{in}_{n+1}(\mathbf{U}_{n+1}) + \mathbf{C}^{sw}_{n+1}\mathbf{P}^{w}_{n+1} + \mathbf{C}^{sa}_{n+1}\mathbf{P}^{a}_{n+1} = \mathbf{f}^{ex}_{n+1},$$

$$(\mathbf{C}^{sw}_{n+1})^T \dot{\mathbf{U}}_{n+1} - \mathbf{Q}^{ww}_{n+1}\dot{\mathbf{P}}^{w}_{n+1} - \mathbf{H}^{ww}_{n+1}\mathbf{P}^{w}_{n+1} + \mathbf{C}^{wa}_{n+1}\dot{\mathbf{P}}^{a}_{n+1} = \dot{\mathbf{f}}^{w}_{n+1}, \quad (4.64)$$

$$(\mathbf{C}^{sa}_{n+1})^T \dot{\mathbf{U}}_{n+1} + \mathbf{C}^{wa}_{n+1}\dot{\mathbf{P}}^{w}_{n+1} - \mathbf{Q}^{aa}_{n+1}\dot{\mathbf{P}}^{a}_{n+1} - \mathbf{H}^{aa}_{n+1}\mathbf{P}^{a}_{n+1} = \dot{\mathbf{f}}^{a}_{n+1}.$$

The unknowns are the displacements of the soil skeleton and their time derivatives, $\mathbf{U}_{n+1}$ and $\dot{\mathbf{U}}_{n+1}$, the hydrostatic stresses in the water phase and their time derivatives, $\mathbf{P}^w_{n+1}$ and $\dot{\mathbf{P}}^w_{n+1}$, and the hydrostatic stresses in the air phase together with their time derivatives, $\mathbf{P}^a_{n+1}$ and $\dot{\mathbf{P}}^a_{n+1}$. For these unknowns the approximations given in the equations (4.61) to (4.63) can be inserted, relating the unknown solution at time t_{n+1} to the known solution at time t_n which yields

$$\mathbf{f}^{in}_{n+1}(\mathbf{U}_{n+1}) + \mathbf{C}^{sw}_{n+1}\left(\mathbf{P}^w_n + \Delta\mathbf{P}^w_{n+1}\right) + \mathbf{C}^{sa}_{n+1}\left(\mathbf{P}^a_n + \Delta\mathbf{P}^a_{n+1}\right) = \mathbf{f}^{ex}_{n+1},$$

$$\left(\mathbf{C}^{sw}_{n+1}\right)^T \frac{\Delta\mathbf{U}_{n+1}}{\Delta t_{n+1}} - \mathbf{Q}^{ww}_{n+1}\frac{\Delta\mathbf{P}^w_{n+1}}{\Delta t_{n+1}} - \mathbf{H}^{ww}_{n+1}\left(\mathbf{P}^w_n + \Delta\mathbf{P}^w_{n+1}\right) + \mathbf{C}^{wa}_{n+1}\frac{\Delta\mathbf{P}^a_{n+1}}{\Delta t_{n+1}} = \dot{\mathbf{f}}^w_{n+1},$$

$$\left(\mathbf{C}^{sa}_{n+1}\right)^T \frac{\Delta\mathbf{U}_{n+1}}{\Delta t_{n+1}} + \mathbf{C}^{wa}_{n+1}\frac{\Delta\mathbf{P}^w_{n+1}}{\Delta t_{n+1}} - \mathbf{Q}^{aa}_{n+1}\frac{\Delta\mathbf{P}^a_{n+1}}{\Delta t_{n+1}} - \mathbf{H}^{aa}_{n+1}\left(\mathbf{P}^a_n + \Delta\mathbf{P}^a_{n+1}\right) = \dot{\mathbf{f}}^a_{n+1}.$$

(4.65)

Multiplication of the second and the third equation in (4.65) by the time increment Δt_{n+1} yields

$$\begin{bmatrix} \mathbf{0} & \mathbf{C}^{sw}_{n+1} & \mathbf{C}^{sa}_{n+1} \\ \left(\mathbf{C}^{sw}_{n+1}\right)^T & -\mathbf{Q}^{ww}_{n+1} - \Delta t_{n+1}\mathbf{H}^{ww}_{n+1} & \mathbf{C}^{wa}_{n+1} \\ \left(\mathbf{C}^{sa}_{n+1}\right)^T & \mathbf{C}^{wa}_{n+1} & -\mathbf{Q}^{aa}_{n+1} - \Delta t_{n+1}\mathbf{H}^{aa}_{n+1} \end{bmatrix} \begin{Bmatrix} \Delta\mathbf{U}_{n+1} \\ \Delta\mathbf{P}^w_{n+1} \\ \Delta\mathbf{P}^a_{n+1} \end{Bmatrix} =$$

$$= \begin{Bmatrix} \mathbf{f}^{ex}_{n+1} - \mathbf{f}^{in}_{n+1}(\mathbf{U}_{n+1}) - \mathbf{C}^{sw}_{n+1}\mathbf{P}^w_n - \mathbf{C}^{sa}_{n+1}\mathbf{P}^a_n \\ \Delta t_{n+1}\left(\dot{\mathbf{f}}^w_{n+1} + \mathbf{H}^{ww}_{n+1}\mathbf{P}^w_n\right) \\ \Delta t_{n+1}\left(\dot{\mathbf{f}}^a_{n+1} + \mathbf{H}^{aa}_{n+1}\mathbf{P}^a_n\right) \end{Bmatrix}, \quad (4.66)$$

which represents the most general form of the coupled set of equations for the three-phase soil model. To obtain the special cases included in the formulation (confer derivations of Section 3.7), individual terms have to be dropped in (4.66) according to the spatially discretised equations (4.48) to (4.50).

However, the vector on the right hand side of (4.66) still contains the vector of the internal forces of the soil skeleton at time t_{n+1} which is a function of the unknown displacements. Similar to the approximation of the unknowns given in (4.61) to (4.63), the vector of internal forces may be split into two parts, namely the known portion $\mathbf{f}_n^{in}$ at time instant t_n and the unknown increment $\Delta\mathbf{f}_{n+1}^{in}$ due to the time step Δt_{n+1},

$$\mathbf{f}_{n+1}^{in} = \mathbf{f}_n^{in} + \Delta\mathbf{f}_{n+1}^{in} . \qquad (4.67)$$

Assuming linear elastic material behaviour of the soil skeleton, the incremental part of the internal forces $\Delta\mathbf{f}_{n+1}^{in}$ can be expressed in terms of the stiffness matrix $\mathbf{K}$ of the soil skeleton and of the incremental values for the displacements $\Delta\mathbf{U}_{n+1}$ of the soil skeleton as $\Delta\mathbf{f}_{n+1}^{in} = \mathbf{K}\,\Delta\mathbf{U}_{n+1}$.

Inserting (4.67) into (4.66) yields the following system of equations for the three-phase model with linear elastic behaviour of the soil skeleton,

$$\begin{bmatrix} \mathbf{K} & \mathbf{C}_{n+1}^{sw} & \mathbf{C}_{n+1}^{sa} \\ \left(\mathbf{C}_{n+1}^{sw}\right)^T & -\mathbf{Q}_{n+1}^{ww} - \Delta t_{n+1}\,\mathbf{H}_{n+1}^{ww} & \mathbf{C}_{n+1}^{wa} \\ \left(\mathbf{C}_{n+1}^{sa}\right)^T & \mathbf{C}_{n+1}^{wa} & -\mathbf{Q}_{n+1}^{aa} - \Delta t_{n+1}\,\mathbf{H}_{n+1}^{aa} \end{bmatrix} \begin{Bmatrix} \Delta\mathbf{U}_{n+1} \\ \Delta\mathbf{P}_{n+1}^{w} \\ \Delta\mathbf{P}_{n+1}^{a} \end{Bmatrix} = $$

$$= \begin{Bmatrix} \mathbf{f}_{n+1}^{ex} - \mathbf{f}_n^{in} - \mathbf{C}_{n+1}^{sw}\mathbf{P}_n^w - \mathbf{C}_{n+1}^{sa}\mathbf{P}_n^a \\ \Delta t_{n+1}\left(\dot{\mathbf{f}}_{n+1}^{w} + \mathbf{H}_{n+1}^{ww}\mathbf{P}_n^w\right) \\ \Delta t_{n+1}\left(\dot{\mathbf{f}}_{n+1}^{a} + \mathbf{H}_{n+1}^{aa}\mathbf{P}_n^a\right) \end{Bmatrix}. \qquad (4.68)$$

It should be mentioned that, although (4.68) refers to the special case of linear elastic behaviour of the soil skeleton, the equation is still non-linear. This follows from the dependence of the matrices $\mathbf{C}$, $\mathbf{Q}$ and $\mathbf{H}$ and of the vectors $\dot{\mathbf{f}}$ on the nodal values of the fluid stresses, $\mathbf{P}^w$ and $\mathbf{P}^a$, because $\mathbf{C}$, $\mathbf{Q}$, $\mathbf{H}$ and $\dot{\mathbf{f}}$ are functions either of the degree of water saturation or of the permeability coefficients (confer equations (4.44) to (4.46) containing the definition of these matrices/vectors).

Equation (4.68) can be extended to take into account non-linear material behaviour of the soil skeleton by evaluating the soil stresses and, consequently, the stiffness matrix $\mathbf{K}$ and the vector of the internal forces $\mathbf{f}^{in}$ according to the employed constitutive model for the soil skeleton.

Finally, the matrix equation (4.68) may be written in a more concise form as

$$\mathbf{A}(\mathbf{Y}_{n+1})\,\Delta\mathbf{Y}_{n+1} = \mathbf{f}(\mathbf{Y}_{n+1})\,, \qquad (4.69)$$

where $\Delta\mathbf{Y}_{n+1} = \left[\Delta\mathbf{U}_{n+1}, \Delta\mathbf{P}^w_{n+1}, \Delta\mathbf{P}^a_{n+1}\right]^T$ represents the vector of the incremental nodal values of the unknowns for which equation (4.69) has to be solved. A solution can be obtained by means of an iterative procedure. Subsequently, two different solution algorithms proposed in the literature, e.g., in [Kreyszig(1993)], are discussed on the basis of the general equation (4.69). The method introduced first is the so-called 'fixed-point' iteration which has the advantage of simple implementation. The second approach is the Newton method. Concerning the implementation procedure the latter algorithm is more costly, however, for the benefit of improved convergence.

4.5.3 Fixed-point iteration

The system of equations (4.69) in general has a non-linear character, i.e. both the coefficient matrix $\mathbf{A}(\mathbf{Y}_{n+1})$ on the left hand side and the vector $\mathbf{f}(\mathbf{Y}_{n+1})$ on the right hand side of the equation depend on the current values of the unknowns. Consequently, an iterative method has to be used for solving the equations. Employing such a procedure one usually starts from an initial guess and computes, step by step, improved approximations of the unknown solution.

To this end, equation (4.69) can be rewritten in a more convenient form as

$$\Delta\mathbf{Y}^i_{n+1} = \left[\mathbf{A}\left(\mathbf{Y}^{i-1}_{n+1}\right)\right]^{-1}\mathbf{f}\left(\mathbf{Y}^{i-1}_{n+1}\right)\,, \qquad (4.70)$$

where $\Delta\mathbf{Y}^i_{n+1}$ is the current value of the approximation of the solution increment for the nodal values and i denotes the iteration step. A solution of (4.70) is also called a fixed point, motivating the designation 'fixed-point iteration' [Kreyszig(1993)]. The coefficients of $\mathbf{A}(\mathbf{Y}_{n+1})$ and $\mathbf{f}(\mathbf{Y}_{n+1})$ are evaluated on the basis of the approximation to the solution $\mathbf{Y}^{i-1}_{n+1}$, obtained in the previous iteration step as $\mathbf{Y}^{i-1}_{n+1} = \mathbf{Y}_n + \Delta\mathbf{Y}^{i-1}_{n+1}$. After solving (4.70) for the new solution increment $\Delta\mathbf{Y}^i_{n+1}$ the current values of the unknowns can be computed as

$$\mathbf{Y}^i_{n+1} = \mathbf{Y}_n + \Delta\mathbf{Y}^i_{n+1}\,. \qquad (4.71)$$

As this equation indicates, the current solution $\mathbf{Y}^i_{n+1}$ at time instant t_{n+1} is always based on the known solution at time instant t_n to which the latest 'guess' $\Delta\mathbf{Y}^i_{n+1}$, i.e. the incremental value of the unknowns according to the evaluation of $\mathbf{A}(\mathbf{Y}_{n+1})$ and $\mathbf{f}(\mathbf{Y}_{n+1})$ using the previous solution $\mathbf{Y}_{n+1}^{i-1}$, is added. If the approximation to the solution becomes sufficiently close to the exact solution, or in other words, if the approximation to the solution only changes within a specified (small) value of tolerance, the iteration procedure is terminated.

4.5.4 Newton method

An alternative iterative procedure to solve (4.69) is the Newton method. This approach is rather popular because of its excellent convergence properties. For convenience (4.69) may be formulated as

$$\begin{aligned}\boldsymbol{\Psi}(\mathbf{Y}_{n+1}) &= \mathbf{A}(\mathbf{Y}_{n+1})\,\Delta\mathbf{Y}_{n+1} - \mathbf{f}(\mathbf{Y}_{n+1}) \\ &= \mathbf{A}(\mathbf{Y}_{n+1})\,(\mathbf{Y}_{n+1} - \mathbf{Y}_n) - \mathbf{f}(\mathbf{Y}_{n+1}) = \mathbf{0}\,.\end{aligned} \quad (4.72)$$

Hence, the solution $\mathbf{Y}_{n+1}$ is obtained by enforcing $\boldsymbol{\Psi}(\mathbf{Y}_{n+1}) = \mathbf{0}$. If one thinks of a given function in conventional 2D-space, the underlying idea of Newton's method is the approximation of the graph of this function by suitable tangents. This basic concept can of course be generalised and applied to equation (4.72). Thus, the linear approximation of $\boldsymbol{\Psi}(\mathbf{Y}_{n+1})$ in the vicinity of a known approximate solution $\mathbf{Y}_{n+1}^{i-1}$ is obtained as

$$\boldsymbol{\Psi}(\mathbf{Y}_{n+1}^{i}) \approx \boldsymbol{\Psi}(\mathbf{Y}_{n+1}^{i-1}) + \frac{\partial \boldsymbol{\Psi}(\mathbf{Y}_{n+1}^{i-1})}{\partial \mathbf{Y}_{n+1}^{i-1}}\,\Delta\mathbf{Y}_{i,n+1}\,, \quad (4.73)$$

where i again denotes the iteration step. Performing the partial derivative in the second term of (4.73) results in

$$\boldsymbol{\Psi}(\mathbf{Y}_{n+1}^{i}) \approx \boldsymbol{\Psi}(\mathbf{Y}_{n+1}^{i-1}) + \left[\frac{\partial \mathbf{A}(\mathbf{Y}_{n+1}^{i-1})}{\partial \mathbf{Y}_{n+1}^{i-1}}\,(\mathbf{Y}_{n+1}^{i-1} - \mathbf{Y}_n) + \right. \\ \left. + \mathbf{A}(\mathbf{Y}_{n+1}^{i-1}) - \frac{\partial \mathbf{f}(\mathbf{Y}_{n+1}^{i-1})}{\partial \mathbf{Y}_{n+1}^{i-1}}\right]\Delta\mathbf{Y}_{i,n+1}\,. \quad (4.74)$$

In the above equations $\boldsymbol{\Psi}(\mathbf{Y}_{n+1}^{i})$ can be interpreted as a residual value which remains due to the applied approximation of the exact solution. Consequently, the condition

$$\boldsymbol{\Psi}(\mathbf{Y}_{n+1}^{i}) = \mathbf{0} \quad (4.75)$$

is employed to determine an improved approximation for the unknown incremental nodal values $\Delta \mathbf{Y}_{n+1}^i$. Restructuring of (4.74) yields

$$\Delta \mathbf{Y}_{i,n+1} = \left[\frac{\partial \mathbf{A}(\mathbf{Y}_{n+1}^{i-1})}{\partial \mathbf{Y}_{n+1}^{i-1}} \left(\mathbf{Y}_{n+1}^{i-1} - \mathbf{Y}_n \right) + \mathbf{A}(\mathbf{Y}_{n+1}^{i-1}) - \frac{\partial \mathbf{f}(\mathbf{Y}_{n+1}^{i-1})}{\partial \mathbf{Y}_{n+1}^{i-1}} \right]^{-1}$$
$$\left[\mathbf{f}(\mathbf{Y}_{n+1}^{i-1}) - \mathbf{A}(\mathbf{Y}_{n+1}^{i-1}) \left(\mathbf{Y}_{n+1}^{i-1} - \mathbf{Y}_n \right) \right] , \quad (4.76)$$

where equation (4.72) has been inserted for $\mathbf{\Psi}(\mathbf{Y}_{n+1}^{i-1})$. After computing $\Delta \mathbf{Y}_{i,n+1}$ the updated solution for the incremental nodal values of the unknowns is obtained as

$$\Delta \mathbf{Y}_{n+1}^i = \Delta \mathbf{Y}_{n+1}^{i-1} + \Delta \mathbf{Y}_{i,n+1} , \quad (4.77)$$

with

$$\Delta \mathbf{Y}_{n+1}^{i-1} = \mathbf{Y}_{n+1}^{i-1} - \mathbf{Y}_n \quad \text{and} \quad \Delta \mathbf{Y}_{n+1}^0 = \mathbf{0} . \quad (4.78)$$

Thus, in iteration step i the increment $\Delta \mathbf{Y}_{n+1}^i$ is added to the converged values of the unknowns $\mathbf{Y}_n$ at time instant t_n.

When comparing the two equations (4.70) and (4.77) together with (4.76) for the calculation of the current incremental nodal values of the unknowns, it becomes quite obvious that Newton's method is more costly considering the implementation procedure because of the required derivatives of $\mathbf{A}$ and $\mathbf{f}$ with respect to the unknowns. However, the benefit is a rapid convergence. The fundamental difference in the updating procedure for the incremental nodal values of the unknowns (see equations (4.70) and (4.77)) is given by the fact that in the Newton method the contributions $\Delta \mathbf{Y}_{i,n+1}$ from the individual iteration steps are accumulated to the incremental solution $\Delta \mathbf{Y}_{n+1}^i$ for time t_{n+1}, while in the fixed-point iteration procedure the latter is determined repeatedly.

4.5.5 Suitable choice of finite elements

The finite elements chosen for all the example problems discussed in the subsequent chapters are based on the parametric concept. These kinds of elements allow for a reasonable approximation of arbitrarily shaped boundaries of a domain. The fundamental idea of parametric finite elements consists of mapping elements of simple geometry, e.g., triangles, rectangles or cubes, to elements of a more complicated shape by means of a coordinate transformation procedure. The former are usually referred to as 'parent' elements while the latter are called 'distorted' elements. Care has to be taken to ensure that the coordinate transformation is unique for any arbitrary point located within the element. To formulate the transformation between the natural coordinates of the parent element and the Cartesian coordinates of the geometrically distorted element, it is common and very convenient to use shape functions identical to those chosen for the interpolation of the unknown variables [Zienkiewicz(1989)]. The respective elements are denoted as isoparametric elements.

In the finite element treatment of multi-phase problems usually different orders of interpolation are applied to the different kinds of variables, i.e. for the displacement degrees of freedom and the hydrostatic fluid stress degrees of freedom. This is necessary in particular when dealing with the undrained limit state where the permeability and often also the compressibility matrices are set to zero, that is $\mathbf{Q} = \mathbf{0}$ and $\mathbf{H} = \mathbf{0}$. Consequently, zero diagonal terms appear in the equation (4.68). The matrices for such a limiting case are similar to those encountered in the solution of problems of incompressible elasticity or fluid mechanics. In these studies limitations are placed on the approximating functions $\mathbf{N_u}$ (shape functions for displacements) and $\mathbf{N}_p$ (shape functions for hydrostatic stresses) if the Babuska-Brezzi conditions are to be satisfied [Zienkiewicz(1990a)]. A selection of elements satisfying these criteria for the undrained state is shown in Figure 4.1.

However, if this limiting (undrained) condition is never imposed, a wide choice of elements is available. Because of the presence of first-order spatial derivatives in the differential operators, C_0-continuous interpolation functions should be used [Zienkiewicz(1989)]. In most of the examples presented in the subsequent chapters finite elements with

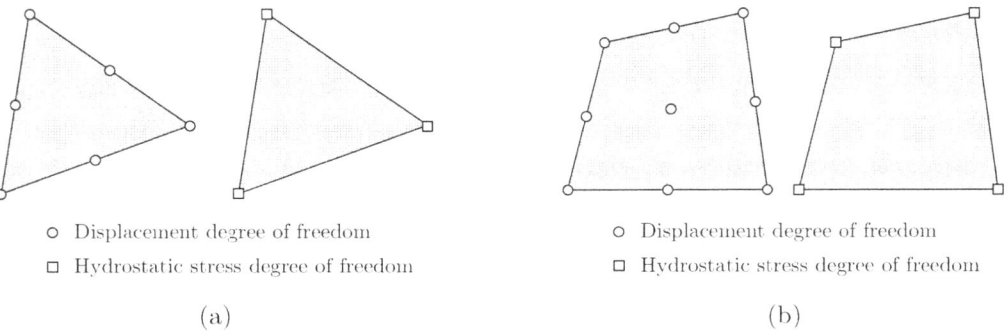

Figure 4.1: Elements employed in multi-phase flow analyses: (a) triangular and (b) quadrilateral elements with the appropriate number of degrees of freedom.

quadratic interpolation for the displacements of the soil skeleton and bilinear interpolation for the hydrostatic stresses in the fluid phases are used.

4.6 Stability, consistency, accuracy

4.6.1 Introduction

The primary requirement for the solution algorithms presented in the previous sections is their *convergence*, i.e. the approximate solution should approach the exact solution as element size and time step size decrease toward zero (equation (4.53)). To investigate the convergence of an algorithm, two additional notions have to be considered, namely *stability* and *consistency*. As can be shown [Hughes(1987)], once stability and consistency are verified, convergence of the algorithm is guaranteed. Furthermore, one should be concerned with the *accuracy* of an algorithm, i.e. with the rate of convergence of the method.

The subsequent three sections are dedicated to brief discussions on stability, consistency and accuracy, respectively. These considerations are based on respective investigations proposed in the literature, in particular on the descriptions given in [Hughes(1987)] and on papers dealing with the convergence of algorithms applied to non-linear consolidation problems [Borja(1991a), Turska(1993), Turska(1994)].

4.6.2 Stability

The set of equations (4.47) obtained after the spatial discretisation of the three-phase problem can be reformulated in matrix notation as

$$\begin{bmatrix} 0 & 0 & 0 \\ (\mathbf{C}^{sw})^T & -\mathbf{Q}^{ww} & \mathbf{C}^{wa} \\ (\mathbf{C}^{sa})^T & \mathbf{C}^{wa} & -\mathbf{Q}^{aa} \end{bmatrix} \begin{Bmatrix} \dot{\mathbf{U}} \\ \dot{\mathbf{P}}^w \\ \dot{\mathbf{P}}^a \end{Bmatrix} + \begin{bmatrix} 0 & \mathbf{C}^{sw} & \mathbf{C}^{sa} \\ 0 & -\mathbf{H}^{ww} & 0 \\ 0 & 0 & -\mathbf{H}^{aa} \end{bmatrix} \begin{Bmatrix} \mathbf{U} \\ \mathbf{P}^w \\ \mathbf{P}^a \end{Bmatrix} = \begin{Bmatrix} \mathbf{f}^{ex} - \mathbf{f}^{in} \\ \dot{\mathbf{f}}^w \\ \dot{\mathbf{f}}^a \end{Bmatrix} , \qquad (4.79)$$

or in a much more convenient short form as

$$\mathbf{A}_1 \dot{\mathbf{Y}} + \mathbf{A}_2 \mathbf{Y} = \mathbf{f} . \qquad (4.80)$$

The time domain approximations (generalised trapezoidal method) of the vector of the unknowns $\mathbf{Y}$ and its time derivative $\dot{\mathbf{Y}}$ have been introduced in the equations (4.54) and (4.55). For the point of time $t_{n+\alpha}$ these relations can be restructured as

$$\dot{\mathbf{Y}}_{n+\alpha} = \frac{\mathbf{Y}_{n+1} - \mathbf{Y}_n}{\Delta t} \quad \text{and} \quad \mathbf{Y}_{n+\alpha} = (1-\alpha)\mathbf{Y}_n + \alpha\mathbf{Y}_{n+1}, \qquad (4.81)$$

where for the sake of simplicity the time increment size Δt has been assumed constant, that is $\Delta t_n = \Delta t_{n+1} = \Delta t$. Equation (4.80) can be written for the time $t_{n+\alpha}$ and both approximations (4.81) may be inserted for the unknowns $\mathbf{Y}_{n+\alpha}$ and $\dot{\mathbf{Y}}_{n+\alpha}$. This yields

$$\mathbf{A}_1 \left[\frac{\mathbf{Y}_{n+1} - \mathbf{Y}_n}{\Delta t} \right] + \mathbf{A}_2 [(1-\alpha)\mathbf{Y}_n + \alpha\mathbf{Y}_{n+1}] = \mathbf{f}_{n+\alpha}, \qquad (4.82)$$

and after collecting the terms containing $\mathbf{Y}_{n+1}$ and $\mathbf{Y}_n$, respectively, the equation reads as

$$[\mathbf{A}_1 + \alpha\Delta t\, \mathbf{A}_2]\, \mathbf{Y}_{n+1} - [\mathbf{A}_1 - (1-\alpha)\Delta t\, \mathbf{A}_2]\, \mathbf{Y}_n = \Delta t\, \mathbf{f}_{n+\alpha}. \qquad (4.83)$$

The most crucial point of this equation, concerning its further application to investigate certain numerical properties, is the singularity of the matrices $\mathbf{A}_1$ and $\mathbf{A}_2$. However, the vector of the internal forces $\mathbf{f}^{in}$ may be rewritten to partly overcome this difficulty. $\mathbf{f}^{in}$ is given in equation (4.44_1). For the effective averaged stress tensor $\tilde{\boldsymbol{\sigma}}'$ the constitutive relationship for the linear elastic soil skeleton (3.214) can be substituted into (4.44_1) together with equation (4.35_1) for the strains of the soil skeleton. This procedure results in a relationship between the vector of the internal forces and the displacements of the soil skeleton, $\mathbf{f}^{in} = \mathbf{K}\mathbf{U}$, where $\mathbf{K}$ denotes the linear elastic stiffness matrix of the soil skeleton. Consequently, instead of using $\mathbf{f}^{in}$ on the right hand side of (4.79), the term $\mathbf{K}\mathbf{U}$ can be inserted on the left hand side of this equation, thus overcoming the singularity of the matrix $\mathbf{A}_2$ because the first diagonal term of $\mathbf{A}_2$ in this case becomes non-zero. Another consequence of this substitution is that the sum $\mathbf{A}_1 + \mathbf{A}_2$, and in particular $[\mathbf{A}_1 + \alpha\Delta t\, \mathbf{A}_2]$, is no longer singular provided $\alpha \neq 0$. Similar considerations can be applied if the behaviour of the soil skeleton is not assumed to be linear elastic but rather elastic-plastic. Concerning the zero diagonal term in $\mathbf{A}_1$, instead of the first of the three equations (4.47), i.e. the equilibrium equations for the mixture, a time derivative of this equation may be used (compare [Lewis(1998)]). In this case, time derivatives of the displacements of the soil skeleton and of the hydrostatic fluid stresses are contained in the equilibrium equations and consequently the first row of $\mathbf{A}_1$ becomes non-zero. Thus, also the sum $\mathbf{A}_1 + \mathbf{A}_2$, or $[\mathbf{A}_1 + \alpha\Delta t\, \mathbf{A}_2]$, is non-zero.

Keeping these considerations in mind, it is possible to proceed in the derivations. According to [Hughes(1987)], the temporally discretised equation (4.83) can be decomposed into a number n_{eq} of uncoupled scalar equations applying a modal reduction technique. The essential property used in reducing (4.83) to a single-degree-of-freedom form (SDOF form) is the orthogonality of the eigenvectors of the associated eigenvalue problem. The modally decomposed form of equation (4.83) then reads as

$$[1 + \alpha \Delta t \lambda_{ni}] y_{n+1} - [1 - (1-\alpha) \Delta t \lambda_{ni}] y_n = \Delta t f_{n+\alpha} , \qquad (4.84)$$

where $\lambda_{ni} \in \mathbb{R}_+$ is the i-th-mode eigenvalue of the system in the time interval $[t_n, t_{n+1}]$. To investigate the stability behaviour, it is sufficient to consider the homogeneous model equation

$$[1 + \alpha \Delta t \lambda_{ni}] y_{n+1} = [1 - (1-\alpha) \Delta t \lambda_{ni}] y_n . \qquad (4.85)$$

Noting that the term in square brackets on the left hand side of (4.85) is greater than zero, that is $[1 + \alpha \Delta t \lambda_{ni}] > 0$, for all allowable values of the parameters, equation (4.85) can be written in a more concise form as

$$y_{n+1} = A_n y_n , \qquad (4.86)$$

where

$$A_n = \frac{1 - (1-\alpha) \Delta t \lambda_{ni}}{1 + \alpha \Delta t \lambda_{ni}} \qquad (4.87)$$

denotes the so-called 'amplification factor'.

The numerical solution is then sought according to the following stability requirements:

$$\begin{aligned} |y_{n+1}| < |y_n| \quad &\text{for} \quad \lambda_{ni} > 0 \\ y_{n+1} = y_n \quad &\text{for} \quad \lambda_{ni} = 0 . \end{aligned} \qquad (4.88)$$

The second of the equations (4.88) follows directly from the definition of the amplification factor A_n (confer equation (4.87)), since for a zero eigenvalue λ_{ni} the factor A_n becomes equal to one. The first condition of (4.88) is equivalent to insisting on an absolute value of the amplification factor less than one, that is $|A_n| < 1$. Recursive evaluation of the relationship (4.86) yields

$$y_n = A_{n-1} y_{n-1} , \qquad y_{n-1} = A_{n-2} y_{n-2} , \quad \ldots \qquad (4.89)$$

and after substitution of the set (4.89) into equation (4.86) one obtains

$$y_{n+1} = \prod_{j=0}^{n} A_j \, y_0 \,. \tag{4.90}$$

In this equation $\prod_{j=0}^{n}$ denotes the product of all the amplification factors A_0 to A_n which arises due to the successive substitution. Finally, the stability requirement according to (4.88$_1$) and (4.90) can be formulated as

$$\left| \prod_{j=0}^{n} A_j \right| < 1 \,, \tag{4.91}$$

which is achieved for any time increment Δt if $\alpha \geq 1/2$ (unconditional stability). For a value of $\alpha < 1/2$ restrictions are imposed on the magnitude of the allowable time step size (conditional stability). In this case, the stability condition must hold for all modes, i.e. for all λ_{ni}, in the system. Consequently, the greatest λ_{ni}, namely $\lambda_{n_{eq}}$, imposes the most stringent restriction upon the time step which might be a severe constraint in a large system of equations, i.e. for $n_{eq} \gg 1$.

4.6.3 Consistency

According to the Lax equivalence theorem [Celia(1992)], for the investigation of the convergence behaviour of an algorithm to solve a space-time problem not only stability but also consistency have to be proved. However, once stability and consistency are verified, convergence is guaranteed [Hughes(1987)]. Therefore, this section is devoted to the consistency of the algorithm.

From equation (4.83), taking into account the relationship between the vector of the internal forces and the displacements of the soil skeleton, the temporally discrete model problem may be written as

$$\mathbf{Y}_{n+1} - [\mathbf{A}_1 + \alpha \Delta t \, \mathbf{A}_2]^{-1} [\mathbf{A}_1 - (1-\alpha) \Delta t \, \mathbf{A}_2] \, \mathbf{Y}_n - \Delta t [\mathbf{A}_1 + \alpha \Delta t \, \mathbf{A}_2]^{-1} \mathbf{f}_{n+\alpha} = \mathbf{0} \,. \tag{4.92}$$

It should be mentioned that, as in the preceding Section 4.6.2, for the sake of simplicity the time step Δt is assumed constant. If the approximate solutions $\mathbf{Y}_n$ and $\mathbf{Y}_{n+1}$ on the left hand side of equation (4.92) are replaced by the corresponding exact solutions $\mathbf{Y}(t_n)$

and $\mathbf{Y}(t_{n+1})$, the following expression is obtained,

$$\mathbf{Y}(t_{n+1}) - [\mathbf{A}_1 + \alpha \Delta t\, \mathbf{A}_2]^{-1}[\mathbf{A}_1 - (1-\alpha)\Delta t\, \mathbf{A}_2]\,\mathbf{Y}(t_n) - $$
$$- \Delta t [\mathbf{A}_1 + \alpha \Delta t\, \mathbf{A}_2]^{-1}\mathbf{f}_{n+\alpha} = \Delta t\, \boldsymbol{\tau}(t_n)\,, \qquad (4.93)$$

where $\boldsymbol{\tau}(t_n)$ is called the local truncation error. If $|\boldsymbol{\tau}(t_n)| \leq c\,\Delta t^k$, for all $t \in [0,T]$, with c being a constant independent of Δt and $k > 0$, the algorithm (4.92) is said to be consistent. The parameter k is called the order of accuracy or the rate of convergence. Subtracting (4.93) from (4.92) yields the following error equation

$$\mathbf{e}(t_{n+1}) - [\mathbf{A}_1 + \alpha \Delta t\, \mathbf{A}_2]^{-1}[\mathbf{A}_1 - (1-\alpha)\Delta t\, \mathbf{A}_2]\,\mathbf{e}(t_n) = -\Delta t\, \boldsymbol{\tau}(t_n)\,, \qquad (4.94)$$

which may be written in a more concise form as

$$\mathbf{e}(t_{n+1}) = \mathbf{M}_n^A\,\mathbf{e}(t_n) - \Delta t\, \boldsymbol{\tau}(t_n)\,, \qquad (4.95)$$

with

$$\mathbf{M}_n^A = [\mathbf{A}_1 + \alpha \Delta t\, \mathbf{A}_2]^{-1}[\mathbf{A}_1 - (1-\alpha)\Delta t\, \mathbf{A}_2]\,, \qquad (4.96)$$

a matrix similar to the amplification factor defined in the previous section (see equation (4.87)). $\mathbf{M}_n^A$ is calculated using the solution at time t_n. Recursive evaluation of (4.95) for all the previous time steps gives

$$\mathbf{e}(t_n) = \mathbf{M}_{n-1}^A\,\mathbf{e}(t_{n-1}) - \Delta t\, \boldsymbol{\tau}(t_{n-1})\,,$$
$$\mathbf{e}(t_{n-1}) = \mathbf{M}_{n-2}^A\,\mathbf{e}(t_{n-2}) - \Delta t\, \boldsymbol{\tau}(t_{n-2})\,, \quad \ldots \qquad (4.97)$$

These relations (4.97) can be inserted into (4.95) by successive substitution which results in the error $\mathbf{e}(t_{n+1})$ as

$$\begin{aligned}\mathbf{e}(t_{n+1}) &= \mathbf{M}_n^A\,\mathbf{M}_{n-1}^A\,\mathbf{M}_{n-2}^A \ldots \mathbf{M}_0^A\,\mathbf{e}(t_0) - \\ &\quad - \Delta t \Big[\mathbf{M}_n^A\,\mathbf{M}_{n-1}^A\,\mathbf{M}_{n-2}^A \ldots \mathbf{M}_1^A\,\boldsymbol{\tau}(t_0) + \ldots \\ &\quad + \mathbf{M}_n^A\,\mathbf{M}_{n-1}^A\,\boldsymbol{\tau}(t_{n-2}) + \mathbf{M}_n^A\,\boldsymbol{\tau}(t_{n-1}) + \boldsymbol{\tau}(t_n)\Big] = \\ &= \left(\prod_{j=0}^{n}\mathbf{M}_j^A\right)\mathbf{e}(t_0) - \Delta t \sum_{j=0}^{n}\left(\prod_{j}\mathbf{M}_j^A\right)\boldsymbol{\tau}(t_{n-j})\,, \qquad (4.98)\end{aligned}$$

where $\prod_j \mathbf{M}_j^A$ denotes a product of j terms according to equation (4.98$_1$). The first term on the right hand side of (4.98$_2$) vanishes since no error is assumed to occur for the initial value, that is

$$\mathbf{e}(t_0) = \mathbf{Y}_0 - \mathbf{Y}(t_0) = \mathbf{0} \ . \tag{4.99}$$

Taking absolute values and evaluating (4.98$_2$) at time instant t_n instead of t_{n+1} yields

$$|\mathbf{e}(t_n)| = \Delta t \left| \sum_{j=0}^{n-1} \left(\prod_j \mathbf{M}_j^A \right) \boldsymbol{\tau}(t_{n-1-j}) \right| \leq \Delta t \sum_{j=0}^{n-1} \left| \prod_j \mathbf{M}_j^A \right| \left| \boldsymbol{\tau}(t_{n-1-j}) \right| \leq$$

$$\leq \Delta t \sum_{j=0}^{n-1} \left| \boldsymbol{\tau}(t_{n-1-j}) \right| \leq t_n \max \left| \boldsymbol{\tau}(t) \right| \leq t_n \, c \, (\Delta t)^k \ . \tag{4.100}$$

The derivations in (4.100) should be explained as follows: For convenience absolute values are used as mentioned. The Schwartz inequality may be applied to show that in the first line of (4.100) the first term on the right hand side is less than or equal to the second term where now two separate absolute values appear. Since the stability condition in the sense of (4.91) is assumed to be valid, that is $\left| \prod_j \mathbf{M}_j^A \right| < 1$, the first term in the second line of (4.100) results which itself is less than or equal to the time t_n multiplied by the maximum value of the local truncation error $\boldsymbol{\tau}(t)$. Finally, the last term in the second line is obtained because the truncation error may be expressed in terms of a constant and the time increment to the power k. From relating the left hand side of (4.100) to the first term in the second line of this equation one recognises that $\mathbf{e}(t_n)$ decays to zero as the time step size decreases, that is $\mathbf{e}(t_n) \to 0$ as $\Delta t \to 0$. Or in other words, the approximate solution converges to the exact solution with a decreasing Δt, meaning that $\mathbf{Y}_n \to \mathbf{Y}(t_n)$ as $\Delta t \to 0$. Furthermore, the last term in the second line gives the rate of convergence k, that is $\mathbf{e}(t_n) = \mathbf{O}(\Delta t^k)$.

4.6.4 Accuracy

To conclude the considerations on the convergence of the algorithm, the accuracy of the one-step generalised trapezoidal family of methods is investigated. The following simplified system of differential equations is considered,

$$\dot{\mathbf{Y}}^* = \mathbf{f}^*(\mathbf{Y}, t) \ , \quad \mathbf{Y}^*(0) = \mathbf{Y}_0^* \ , \quad \mathbf{Y}^* \in \mathbb{R}^{n_{eq}} \ , \quad t \geq 0 \ . \tag{4.101}$$

In (4.101) $\mathbb{R}^{n_{eq}}$ denotes a space with the eigenvectors as basis, meaning that any element in $\mathbb{R}^{n_{eq}}$ can be written as a linear combination of the eigenvectors. The set of equations

(4.80) can be put in the form (4.101) by subtracting $\mathbf{A}_2 \mathbf{Y}$ on both sides and subsequent multiplication by $\mathbf{A}_1^{-1}$. Of course the remarks on the singularity of $\mathbf{A}_1$ made in Section 4.6.2 have to be kept in mind. The solution of (4.101) based on the generalised trapezoidal method is given as

$$\mathbf{Y}^*_{n+1} = \mathbf{Y}^*_n + \Delta t \left[\alpha \, \mathbf{f}^*_{n+1} + (1-\alpha) \, \mathbf{f}^*_n \right] . \quad (4.102)$$

According to [Borja(1991a)], an operator $\mathbf{L}(\mathbf{Y}^*(t_{n+1}))$ may be defined as

$$\mathbf{L}(\mathbf{Y}^*(t_{n+1})) = -\mathbf{Y}^*(t_{n+1}) + \mathbf{Y}^*(t_n) + \Delta t \left[\alpha \, \mathbf{f}^*(t_{n+1}) + (1-\alpha) \, \mathbf{f}^*(t_n) \right] . \quad (4.103)$$

The terms $\mathbf{Y}^*(t_n)$ and $\mathbf{f}^*(t_n)$ in (4.103) can be expanded in Taylor series with remainders about time t_{n+1}:

$$\begin{aligned}
\mathbf{Y}^*(t_n) &= \mathbf{Y}^*(t_{n+1}) + (-\Delta t) \, \dot{\mathbf{Y}}^*(t_{n+1}) + \frac{(-\Delta t)^2}{2!} \ddot{\mathbf{Y}}^*(t_{n+1}) + \\
&\quad + \frac{(-\Delta t)^3}{3!} \dddot{\mathbf{Y}}^*(t_{n+1}) + \mathbf{O}(\Delta t^4) \\
\mathbf{f}^*(t_n) &= \mathbf{f}^*(t_{n+1}) + (-\Delta t) \, \dot{\mathbf{f}}^*(t_{n+1}) + \frac{(-\Delta t)^2}{2!} \ddot{\mathbf{f}}^*(t_{n+1}) + \\
&\quad + \frac{(-\Delta t)^3}{3!} \dddot{\mathbf{f}}^*(t_{n+1}) + \mathbf{O}(\Delta t^4) .
\end{aligned} \quad (4.104)$$

Inserting these relations (4.104) into equation (4.103) and using the model equation (4.101$_1$) to eliminate the terms $\mathbf{f}^*$ in (4.104$_2$) results in

$$\begin{aligned}
\mathbf{L}(\mathbf{Y}^*(t_{n+1})) &= -\mathbf{Y}^*(t_{n+1}) + \mathbf{Y}^*(t_{n+1}) - \Delta t \, \dot{\mathbf{Y}}^*(t_{n+1}) + \tfrac{1}{2} \Delta t^2 \, \ddot{\mathbf{Y}}^*(t_{n+1}) - \\
&\quad - \tfrac{1}{6} \Delta t^3 \, \dddot{\mathbf{Y}}^*(t_{n+1}) + \mathbf{O}(\Delta t^4) + \alpha \Delta t \, \dot{\mathbf{Y}}^*(t_{n+1}) + (1-\alpha)\Delta t \, \dot{\mathbf{Y}}^*(t_{n+1}) - \\
&\quad - (1-\alpha)\Delta t^2 \, \ddot{\mathbf{Y}}^*(t_{n+1}) + \tfrac{1}{2}(1-\alpha)\Delta t^3 \, \dddot{\mathbf{Y}}^*(t_{n+1}) + \mathbf{O}(\Delta t^4) = \\
&= \tfrac{1}{2} \Delta t^2 \, \ddot{\mathbf{Y}}^*(t_{n+1}) - (1-\alpha)\Delta t^2 \, \ddot{\mathbf{Y}}^*(t_{n+1}) - \\
&\quad - \tfrac{1}{6}\Delta t^3 \, \dddot{\mathbf{Y}}^*(t_{n+1}) + \tfrac{1}{2}(1-\alpha)\Delta t^3 \, \dddot{\mathbf{Y}}^*(t_{n+1}) + \mathbf{O}(\Delta t^4) = \\
&= \left(-\tfrac{1}{2} + \alpha\right)\Delta t^2 \, \ddot{\mathbf{Y}}^*(t_{n+1}) + \left(\tfrac{1}{3} - \tfrac{1}{2}\alpha\right)\Delta t^3 \, \dddot{\mathbf{Y}}^*(t_{n+1}) + \mathbf{O}(\Delta t^4) = \\
&= \sum_{q=0}^{k+1} C_q \, \Delta t^q \, \mathbf{Y}^{*(q)}(t_{n+1}) + \mathbf{O}(\Delta t^{k+2}) ,
\end{aligned} \quad (4.105)$$

which finally gives a rather concise and convenient expression for determining the order of accuracy of a certain algorithm. $\mathbf{Y}^{*(q)}$ denotes the q-th time derivative of the unknowns $\mathbf{Y}^*$ and k is the order of accuracy of the particular method. The first non-zero coefficient C_{k+1} in the last line of (4.105) is called the truncation error coefficient. For $\alpha \neq 1/2$ one obtains

$$C_0 = C_1 = 0 , \qquad C_2 = \alpha - \tfrac{1}{2} \neq 0 ,$$

$$C_{k+1} = C_2 \quad \rightarrow \quad k = 1 . \tag{4.106}$$

Hence, the algorithms characterised by $\alpha \neq 1/2$, i.e. both the Euler forward and the Euler backward methods, are first-order accurate. For $\alpha = 1/2$ equation (4.105) yields

$$C_0 = C_1 = 0 , \qquad C_2 = \tfrac{1}{2} - \tfrac{1}{2} = 0 , \qquad C_3 = \tfrac{1}{12} \neq 0 ,$$

$$C_{k+1} = C_3 \quad \rightarrow \quad k = 2 , \tag{4.107}$$

meaning that the first non-zero truncation error coefficient is C_3. Consequently, the Crank-Nicolson scheme, characterised by $\alpha = 1/2$, is second-order accurate.

Chapter 5

Verification of the three-phase formulation

5.1 Introduction

After the derivation of the governing equations of a multi-phase model as well as their specification for the case of a three-phase formulation, the three phases being the incompressible solid phase constituting the deformable soil skeleton and the two compressible barotropic fluid phases water and air, in Chapter 3 and the subsequent explanation of the corresponding finite element framework in Chapter 4, Chapters 5 and 6 are devoted to the discussion of a series of example problems employed for the validation of the finite element code on the one hand and for the verification of the physical model and its applicability to a broad range of tasks encountered in the realm of geotechnical engineering on the other hand. As Section 3.7 indicates, certain simpler special cases are included in the complete three-phase formulation. To the latter class of problems belong geotechnical phenomena such as consolidation or the dewatering of soils under atmospheric conditions for the air phase, whereas the complete three-phase model is necessary when dealing for instance with dewatering of soils by means of compressed air.

For the implementation of the three-phase formulation into a finite element programme [Carter(1995)] a step by step procedure seems to be rather promising. Hence, starting from drained conditions, that is single-phase material calculations, the extension to a consolidation model is performed in a first step. In this case both the degree of saturation and the coefficient of permeability with respect to the fluid phase are constant (confer Section 3.7.3) which yields matrices in the system of equations containing only constant entries.

In a second step the degree of saturation as well as the permeability coefficient are allowed to vary with progress of time (see Section 3.7.2), entailing that the matrices are no longer constant, and therefore, an iterative solution procedure is required. This special case enables to model the process of dewatering of soils under atmospheric conditions. It should be mentioned that although the degree of water saturation may change the hydrostatic stress in the air phase is kept at atmospheric conditions (passive air assumption). Consequently, both consolidation and dewatering under atmospheric conditions are two-phase formulations with the hydrostatic air stresses (exceeding the atmospheric stresses) and thus also the mass balance equation of the air phase being neglected.

Finally, in the third step the hydrostatic stress in the air phase is taken into account requiring the complete three-phase formulation. This model is used when dealing, e.g., with dewatering of soils by means of compressed air.

The current chapter is devoted to the discussion of problems used for the verification of the described three-phase formulation. Starting with rather simple consolidation examples which can be solved by means of a two-phase model, the complexity of the problems under consideration is increased step by step. The individual sections of the chapter deal with the following examples: (i) First, the simplest case of one-dimensional consolidation is discussed. The so-called Terzaghi problem is solved numerically and the results are compared with the analytical solution originally derived by Terzaghi [Terzaghi(1925)]. (ii) Next, two-dimensional consolidation is dealt with. To this end, strip and circular footings are analysed and the results once again can be checked by means of analytical solutions available in the literature [Booker(1974)]. (iii) A laboratory test conducted at the Institute for Soil Mechanics and Foundation Engineering at Graz University of Technology dealing with the flow of compressed air through dry soil [Kammerer(2000)] can also be simulated numerically by means of a two-phase formulation. However, in contrast to the above mentioned consolidation problems, where usually an incompressible water phase is assumed, in this example the compressibility of the fluid phase (compressed air) has to be taken into account. The numerical results are compared with data obtained from the experiment. (iv) A two-phase simulation of an experiment well documented in the literature [Schrefler(1988), Lewis(1998)] is presented next. In this laboratory test [Liakopoulos(1965)] the dewatering (drainage) of a column of sand under atmospheric conditions, that is, the dewatering process is solely driven by gravity, is investigated. Numerical results are compared with test data [Liakopoulos(1965)] as well as calculations presented in the literature [Schrefler(1988)]. (v) A three-phase simulation of the latter experiment provides valuable insight into the differences obtained from the two-phase and the three-phase model, respectively.

5.2 Terzaghi's problem – one-dimensional consolidation

5.2.1 Description of the problem

Volume changes in soils are induced by stresses resulting from a variety of sources: mechanical loading by application or removal of surcharges, desiccation, water supply to the soil by rain or a changing groundwater table, etcetera. All of the above physical circumstances yield decreases or increases in volume, and, in terms of deformations, they result in a subsidence and settlement of the soil or in heaving. The subsidence of soils due to a volume decrease is known in geotechnical engineering as consolidation, while the heaving of soils, a volume increase, is commonly termed swelling. The former phenomenon (one-dimensional case) is considered in the current section.

Soils such as sand or clay consist of small particles and the void space between the particles may be filled with water. In geomechanics this kind of system is denoted a saturated porous medium. The deformations of such a material not only depend on the stiffness of the porous body but also on the behaviour of the fluid in the pores. In particular, the flow of the pore fluid influences the deformations of the soil. If the permeability of the porous material is small, the pore fluid needs considerable time to be expelled from the soil and thus hinders the deformations substantially. On the basis of the theory of consolidation a simultaneous investigation of the deformation of the porous medium and the flow of the pore fluid is possible.

Terzaghi's theory of consolidation, originally developed by Terzaghi in 1925 [Terzaghi(1925)] for the one-dimensional case and later extended to three dimensions by Biot [Biot(1941)], relies on the following assumptions with respect to the constitutive equations and the initial and boundary conditions: (i) The soil is considered as homogeneous or quasi-homogeneous. (ii) The pores of the soil are completely filled with water. (iii) Both the solid particles and the water phase are incompressible, thus the densities of these constituents are constant. (iv) The processes occurring during consolidation are slow enough so that the velocity and the velocity gradient of the solid phase are negligible. (v) The pore water pressure is considered to be the direct cause of consolidation. (vi) Darcy's law is valid for the flow of the fluid through the pores and the coefficient of permeability is assumed to be constant in this law. (vii) The relationship between the change in void ratio and the change in pore water pressure is assumed to be linear. (viii) A soil layer of thickness $2h$ and extending to infinity in the other two coordinate directions, lying in between two permeable but not deformable layers (Figure 5.1), consolidates due to a pore pressure gradient. Since the consolidating soil can drain only towards the two permeable layers, the drainage paths and thus all processes in the soil occur exclusively in the vertical direction. At the top and bottom of the consolidating soil complete drainage prevails – the pore pressure remains zero at these boundaries, and in this respect this assumption is a boundary condition. (ix) The ground surface is loaded at time $t = 0$ by a uniform load of magnitude q_0 (Figure 5.1) which is applied suddenly and then remains constant in time.

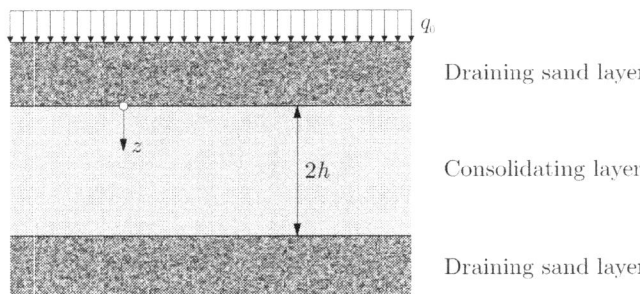

Figure 5.1: Physical model of Terzaghi's problem.

In connection with the above described assumptions underlying Terzaghi's consolidation theory it should be mentioned that the correlation between theory and experimental data, stemming either from laboratory or field, is only approximate. Discrepancies between the measured settlements and those predicted by the theory indicate that the latter only provides a good first approximation of the time-displacement curve. The part of the volume changes accounted for by the theory is referred to as the primary consolidation while the second part, not explained by it, is called secondary consolidation. Consequently, the theory is in reasonable agreement only with part of measured time-displacement curves, beyond this part the theoretical curve flattens, becoming asymptotic to a horizontal line, while the measurements increase for a long period of time even under very small excess pore pressures [Klausner(1991)].

5.2.2 Analytical solution

When investigating the flow of an incompressible fluid phase through a fully saturated porous medium, the mass balance equation as given in (3.355) has to hold. Furthermore, if Darcy's law is assumed to describe the fluid velocity, relationship (3.353) is valid.

Hence, for the case of one-dimensional consolidation considering only flow paths in the vertical (z-) direction, the mass balance equation (3.355) can be written as

$$\frac{\partial \varepsilon^{vol}}{\partial t} + \frac{\partial \tilde{v}_z^{ws}}{\partial z} = 0 , \tag{5.1}$$

and Darcy's law (3.353) reduces to

$$\tilde{v}_z^{ws} = \frac{k_z^{ow}}{\rho^w g} \left(\frac{\partial p^w}{\partial z} + \rho^w g \right) . \tag{5.2}$$

Substituting (5.2) into (5.1) and neglecting the spatial gradient of the hydraulic conductivity (second term of (5.2)) yields

$$\frac{\partial \varepsilon^{vol}}{\partial t} + \frac{\partial}{\partial z} \left(\frac{k_z^{ow}}{\rho^w g} \frac{\partial p^w}{\partial z} \right) = 0 , \tag{5.3}$$

which is a relationship between two basic variables, namely the volumetric strain ε^{vol} of the soil skeleton and the pore water pressure p^w.

In order to complete the system of equations the deformations of the soil skeleton have to be considered. In general this involves three types of equations: equilibrium, compatibility and stress-strain relations. From a mathematical point of view it would be most convenient if a second equation relating the volumetric strain to the pore pressure could be found.

According to assumption (viii) of Section 5.2.1 (infinite horizontal dimensions of the consolidating soil layer), when considering one-dimensional consolidation there are no horizontal deformations and strains, that is

$$\varepsilon_{xx} = \varepsilon_{yy} = 0 . \tag{5.4}$$

Thus, a volume change is only obtained from the vertical strain,

$$\varepsilon^{vol} = \varepsilon_{xx} + \varepsilon_{yy} + \varepsilon_{zz} = \varepsilon_{zz} \ . \tag{5.5}$$

By means of the Lamé parameters λ^s and G^s the linear elastic Hooke's law for the soil skeleton can be written as [Mang(2000)]

$$\tilde{\sigma}'_{ij} = \lambda^s \varepsilon^{vol} \delta_{ij} + 2G^s \varepsilon_{ij} \ , \tag{5.6}$$

relating the effective averaged stress tensor $\tilde{\sigma}'_{ij}$ to the strains ε_{ij} of the soil skeleton. Expressing (5.6) for the z-direction and using (5.5) the relationship

$$\tilde{\sigma}'_{zz} = \lambda^s \varepsilon^{vol} + 2G^s \varepsilon_{zz} = (\lambda^s + 2G^s)\varepsilon^{vol} \tag{5.7}$$

is obtained between the vertical effective averaged stress and the volumetric strain. According to (3.198), for a fully saturated porous medium $\tilde{\sigma}'_{zz}$ can be written as

$$\tilde{\sigma}'_{zz} = \tilde{\sigma}_{zz} - p^w \ , \tag{5.8}$$

that is, as the difference between total averaged vertical stress $\tilde{\sigma}_{zz}$ and water pressure p^w. Substituting (5.8) into (5.7) yields

$$\tilde{\sigma}_{zz} - p^w = (\lambda^s + 2G^s)\varepsilon^{vol} \tag{5.9}$$

and its time derivative reading as

$$\frac{\partial \varepsilon^{vol}}{\partial t} = \frac{1}{\lambda^s + 2G^s} \left(\frac{\partial \tilde{\sigma}_{zz}}{\partial t} - \frac{\partial p^w}{\partial t} \right) \ . \tag{5.10}$$

This is the second relationship between ε^{vol} and p^w that was sought for.

Substituting (5.10) into (5.3) yields the subsequent relatively simple diffusion-type differential equation for the pore water pressure p^w,

$$\frac{1}{\lambda^s + 2G^s} \frac{\partial p^w}{\partial t} = \frac{1}{\lambda^s + 2G^s} \frac{\partial \tilde{\sigma}_{zz}}{\partial t} + \frac{k_z^{ow}}{\rho^w g} \frac{\partial^2 p^w}{\partial z^2} \ . \tag{5.11}$$

In (5.11) the first term on the right hand side is given from the loading conditions. Additionally, it has been assumed that the ratio $k_z^{ow}/(\rho^w g)$ is a constant. For the solution of a particular problem equation (5.11) has to be considered in combination with a set of initial and boundary conditions.

The first term on the right hand side of (5.11) represents the loading rate which is very large (approaching infinity) at the moment of loading and zero afterwards. In order to study the behaviour at the time of loading one may integrate (5.11) over a short time interval Δt and then assume that $\Delta t \to 0$ yielding

$$\frac{1}{\lambda^s + 2G^s} \Delta p^w = \frac{1}{\lambda^s + 2G^s} \Delta \tilde{\sigma}_{zz} \quad \to \quad \Delta p^w = \Delta \tilde{\sigma}_{zz} \ . \tag{5.12}$$

Hence, if at time $t = 0$ the vertical stress suddenly increases from 0 to q_0 (due to a uniform external load of such magnitude), the pore water pressure will increase to the value p_0^w such that

$$p_0^w = q_0 \ , \tag{5.13}$$

indicating that, for an incompressible fluid phase, the initial pore water pressure equals the external load. Thus, due to the fact that the water is incompressible and no water has yet drained out of the soil, initially the water carries all the load. With reference to equation (5.13) it should be emphasised that only excess pore water pressures, i.e. pore water pressures in excess of the hydrostatic water stresses (increasing linearly with depth according to the unit weight of water), are considered in combination with the consolidation phenomenon. At time $t = 0$ the loading q_0 causes an initial excess pore water pressure to arise which completely dissipates in the course of the consolidation process.

Since the load is constant in time (see assumption (ix) in Section 5.2.1), after its application the term $\partial \tilde{\sigma}_{zz}/\partial t$ is equal to zero. Equation (5.11) then reduces to [Verruijt(1995)]

$$\frac{\partial p^w}{\partial t} = c_v \frac{\partial^2 p^w}{\partial z^2} \ , \tag{5.14}$$

where c_v is the so-called consolidation coefficient, a parameter of paramount importance when dealing with consolidation analyses,

$$c_v = \frac{k_z^{ow}(\lambda^s + 2G^s)}{\rho^w g} \ . \tag{5.15}$$

For a linear elastic soil skeleton and an incompressible fluid phase the consolidation coefficient depends on the hydraulic conductivity, on the unit weight of the fluid and on the elastic properties of the soil skeleton.

Solving the Terzaghi problem the following initial and boundary conditions have to be taken into account: At time $t = 0$ the soil layer is subjected to a uniform load of magnitude q_0. Thus, the initial condition is given as

$$t = 0: \quad p^w = p_0^w = q_0 \,. \tag{5.16}$$

Since the consolidating soil is embedded between two draining materials, the upper and lower boundaries of this soil layer are fully drained such that along these boundaries the excess pore water pressure p^w remains zero yielding boundary conditions of the form

$$z = 0: \quad p^w = 0 \,,$$
$$z = 2h: \quad p^w = 0 \,. \tag{5.17}$$

The mathematical problem to be solved is now completely determined by the equations (5.14) to (5.17).

The solution of the problem can be obtained by using the mathematical tools supplied by the theory of partial differential equations, for instance by the method of separation of variables or, even more conveniently, by the Laplace transform method [Kreyszig(1993)]. The Laplace transform is particularly useful for problems in which the variables are defined in a semi-infinite domain, say for $0 \leq t < \infty$, where t may, e.g., be the time and $t = 0$ indicates the initial value of time. The Laplace transform of a given function $f(t)$ is defined as [Kreyszig(1993)]

$$F(s) = \mathcal{L}(f) = \int_0^\infty f(t) \exp(-st)\, dt \,, \tag{5.18}$$

where s is a parameter which is assumed to be sufficiently large for the integral to exist (that is, the integral has some finite value). By the integration over the time domain, for various values of s, the function $f(t)$ is transformed into a function $F(s)$. When considering the Laplace transform of the time derivative of a function $f(t)$, partial integration can be used to find

$$\int_0^\infty \frac{df(t)}{dt} \exp(-st)\, dt = sF(s) - f(0) = s\mathcal{L}(f) - f(0) \,. \tag{5.19}$$

Thus, differentiation with respect to time is transformed into a multiplication by s, apart from the subtraction of the initial value $f(0)$.

The basic principle of Laplace transform is to multiply the differential equation, in this case (5.14), by the factor $\exp(-st)$ and then integrate the resulting equation in time from $t = 0$ to $t = \infty$ [Verruijt(1995)],

$$\int_0^\infty \frac{\partial p^w}{\partial t} \exp(-st)\, dt = c_v \frac{\partial^2}{\partial z^2} \int_0^\infty p^w \exp(-st)\, dt \ . \tag{5.20}$$

Applying the definitions (5.18) and (5.19) to the pore pressure, that is $f(t) = p^w(t)$, and making use of the initial condition (5.16) yields

$$s\mathcal{L}(p^w) - p_0^w = c_v \frac{\partial^2 \mathcal{L}(p^w)}{\partial z^2} \ . \tag{5.21}$$

The partial differential equation has now been reduced to an ordinary differential equation with the general solution

$$\mathcal{L}(p^w) = \frac{p_0^w}{s} + A \exp\left(z\sqrt{s/c_v}\right) + B \exp\left(-z\sqrt{s/c_v}\right) \ . \tag{5.22}$$

In (5.22) A and B are integration constants that may depend on the Laplace transform parameter s. They can be determined by means of the boundary conditions (5.17). The final result for the transformed pore water pressure is obtained as

$$\frac{\mathcal{L}(p^w)}{p_0^w} = \frac{1}{s} - \frac{\cosh\left[(h-z)\sqrt{s/c_v}\right]}{s \cosh\left[h\sqrt{s/c_v}\right]} \ . \tag{5.23}$$

The inverse transform of this expression may be determined by application of Heaviside's expansion theorem. This theorem is suitable for functions that can be written as a quotient of two polynomials,

$$Y(s) = \frac{F(s)}{G(s)} = \frac{a_1}{s - s_1} + \frac{a_2}{s - s_2} + \cdots + \frac{a_n}{s - s_n} \ , \tag{5.24}$$

where $s_1, \ldots, s_n$ are the single zeros of the function $G(s)$ and the coefficients a_i are evaluated by $a_i = F(s_i)/G'(s_i)$. The inverse transform of a function of the form given in (5.24), where the order of the denominator $G(s)$ should be higher than that of the numerator $F(s)$, consists of a sum of terms, one for each of the zeros of the denominator $G(s)$,

$$f(t) = \sum_{i=1}^n \frac{F(s_i)}{G'(s_i)} \exp(s_i t) \ . \tag{5.25}$$

According to the relationship (5.23) the functions $F(s)$, $G(s)$ and $G'(s)$ for this particular case can be written as

$$F(s) = \cosh\left[(h-z)\sqrt{s/c_v}\right],$$

$$G(s) = s \cosh\left[h\sqrt{s/c_v}\right], \qquad (5.26)$$

$$G'(s) = \cosh\left[h\sqrt{s/c_v}\right] + \frac{h\sqrt{s/c_v}}{2}\sinh\left[h\sqrt{s/c_v}\right],$$

and the pore water pressure p^w is then obtained by a sum of terms of the form

$$\frac{F(s_j)}{G'(s_j)}\exp(s_j t). \qquad (5.27)$$

The zeros of the function $G(s)$ in (5.26) are $s = 0$ and the zeros of the function $\cosh\left[h\sqrt{s/c_v}\right]$ which are

$$s = s_j = -(2j-1)^2 \frac{\pi^2 c_v}{4h^2}, \quad j = 1, 2, \ldots. \qquad (5.28)$$

By insertion of (5.28) the argument $h\sqrt{s/c_v}$ of the cosh-function for the latter case becomes $h\sqrt{s/c_v} = i(2j-1)\pi/2$, where i is the imaginary unit, $i = \sqrt{-1}$.

As can easily be seen, for the first zero, $s = 0$, the values of the numerator $F(s)$ and the derivative of the denominator $G(s)$ are both 1, thus yielding the coefficient $a_1 = F(0)/G'(0) = 1/1 = 1$, so that the contribution of this zero cancels with the first term on the right hand side of equation (5.23). Consequently, terms of the form (5.27) have to be summarised only for the zeros s_j of the cosh-function given in (5.28). The value of the numerator $F(s)$ for $s = s_j$ is obtained as

$$F(s_j) = \cosh\left[\frac{h-z}{h}i(2j-1)\frac{\pi}{2}\right] = \cos\left[\frac{h-z}{h}(2j-1)\frac{\pi}{2}\right], \qquad (5.29)$$

where the relationship $\cosh iz = \cos z$ has been used. The value of the derivative of the denominator $G(s)$ for $s = s_j$ is

$$G'(s_j) = \frac{\pi}{4}i(2j-1)\sinh\left[i(2j-1)\frac{\pi}{2}\right] =$$

$$= -\frac{\pi}{4}(2j-1)\sin\left[(2j-1)\frac{\pi}{2}\right] = -\frac{\pi}{4}(2j-1)(-1)^{j-1}, \qquad (5.30)$$

where $\sinh iz = i \sin z$ has been employed. Inserting both expressions (5.29) and (5.30) into (5.27) and summarising yields the final result for the pore water pressure p^w [Verruijt(1995)],

$$\frac{p^w}{p_0^w} = \frac{4}{\pi} \sum_{j=1}^{\infty} \left\{ \frac{(-1)^{j-1}}{(2j-1)} \cos\left[\frac{\pi}{2}(2j-1)\frac{h-z}{h}\right] \exp\left[-(2j-1)^2 \frac{\pi^2}{4} \frac{c_v t}{h^2}\right] \right\}. \tag{5.31}$$

Equation (5.31) is used to calculate the dissipation of the excess pore water pressure analytically.

Concerning the settlements of the soil surface during the consolidation process an analytical solution is available as well. The progress of the settlement in time can be obtained by means of the solution (5.31) when taking into account that the strain is determined by the effective stress (confer equations (5.7) and (5.8)),

$$\varepsilon^{vol} = \varepsilon_{zz} = \frac{1}{\lambda^s + 2G^s}(\tilde{\sigma}_{zz} - p^w). \tag{5.32}$$

The settlement w can then be calculated as the integral of this vertical strain ε_{zz} over the height $2h$ of the sample,

$$w = -\int_0^{2h} \varepsilon_{zz}\, dz = -\frac{1}{\lambda^s + 2G^s} \int_0^{2h} (\tilde{\sigma}_{zz} - p^w)\, dz =$$

$$= \frac{1}{\lambda^s + 2G^s} q_0 2h - \frac{1}{\lambda^s + 2G^s} \int_0^{2h} p^w\, dz. \tag{5.33}$$

The first term on the right hand side of (5.33) denotes the final settlement which will be reached after the complete dissipation of the excess pore water pressure,

$$w_\infty = \frac{2hq_0}{\lambda^s + 2G^s}. \tag{5.34}$$

Immediately after the application of the surface load q_0 the pore pressure is given by $p_0^w = q_0$ (see equation (5.16)) which means that (according to (5.33)) the instantaneous settlement at the moment of loading is

$$w_0 = 0. \tag{5.35}$$

Equation (5.35), however, is only valid for an incompressible fluid phase as presupposed.

In order to describe the settlement as a function of time it is convenient to introduce the degree of consolidation U, defined as

$$U = \frac{w - w_0}{w_\infty - w_0} = \frac{w}{w_\infty} \,. \tag{5.36}$$

This quantity will vary between 0 (at the moment of loading) and 1 (after consolidation is completed). By means of (5.33) to (5.35) it is found that the degree of consolidation is related to the pore water pressure by

$$U = \frac{1}{2h} \int_0^{2h} \frac{p_0^w - p^w}{p_0^w} \, dz \,. \tag{5.37}$$

Using the solution (5.31) for the pore pressure distribution the final expression for the degree of consolidation as a function of time reads as [Verruijt(1995)]

$$U = 1 - \frac{8}{\pi^2} \sum_{j=1}^{\infty} \frac{1}{(2j-1)^2} \exp\left[-(2j-1)^2 \frac{\pi^2}{4} \frac{c_v t}{h^2}\right] \,. \tag{5.38}$$

For $t \to \infty$ the degree of consolidation becomes indeed 1 while for $t = 0$ it is 0 because then the terms in the infinite sum add up to $\pi^2/8$.

Theoretically speaking, the consolidation process is finished if $t \to \infty$. However, for all practical purposes it can be considered as completed once the argument of the exponential function in the first term of the sum is about 4 or 5. This will be the case for a dimensionless time parameter of [Verruijt(1995)]

$$\frac{c_v t}{h^2} \approx 2 \,. \tag{5.39}$$

This very useful relationship enables to estimate the duration of the consolidation process as

$$t \approx \frac{2h^2}{c_v} = \frac{2h^2 \rho^w g}{k_z^{ow}(\lambda^s + 2G^s)} = \frac{2h^2 \rho^w g (1+\nu)(1-2\nu)}{k_z^{ow} E(1-\nu)} \,, \tag{5.40}$$

when inserting the definition (5.15) for the consolidation coefficient. It can also be employed to evaluate the influence of the various parameters on the consolidation process, e.g., the effect of a change of permeability or drainage length.

5.2.3 Numerical model

According to Figure 5.1, Terzaghi's problem deals with the consolidation of a soil layer with a height of $2h$ which is allowed to drain at the top and bottom boundaries. Hence, in the middle of the consolidating layer at $z = h$ a horizontal symmetry plane is encountered where the pore pressure gradient necessarily vanishes, $\partial p^w/\partial z = 0$. This symmetry is taken advantage of in the numerical simulation of the problem where only the upper half of the consolidating layer is modelled, that is a soil column of 1.0 m height.

The domain is discretised by one column of plane strain finite elements; however, three different meshes are employed for the simulation as depicted in Figure 5.2. The first and simplest mesh (Figure 5.2(a)) consists of ten equally-sized finite elements of a height of 0.10 m. For the second and the third simulation graded meshes are used, i.e. the element height is reduced in the direction of the draining boundary or in other words the higher the pore pressure gradient becomes the smaller the element size. The second mesh (Figure 5.2(b)), as well as the first one, is composed of ten finite elements where the element height now varies between 0.40 m and 0.01 m. Consequently, a comparison of the first and the second simulation shows the improvement of the results by just choosing a mesh more suitable for the particular problem under consideration, keeping the same number of elements and therefore degrees of freedom. Finally, a third mesh

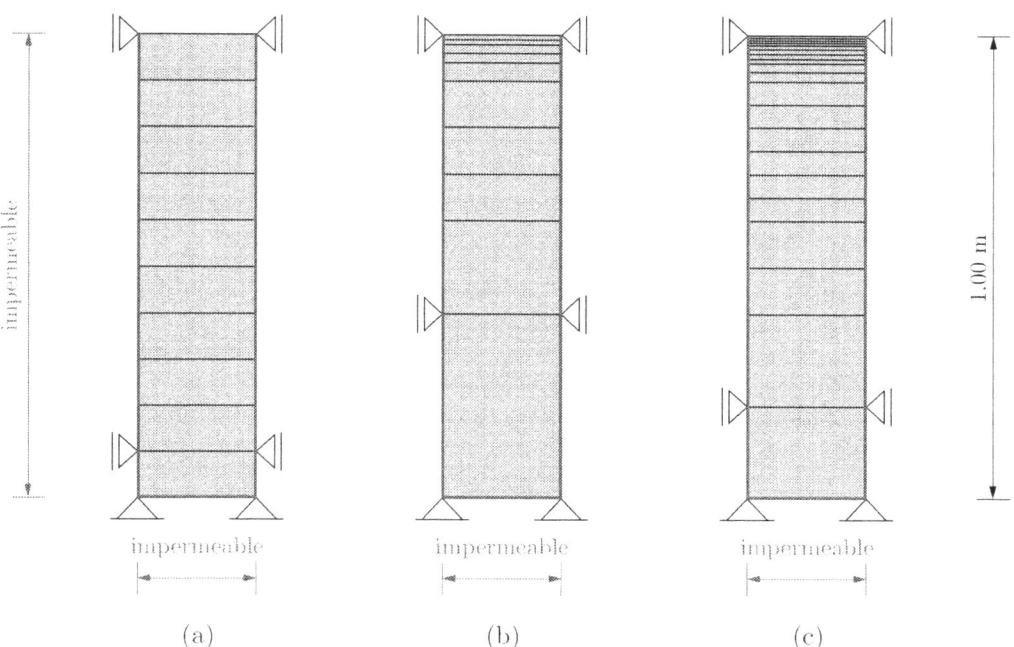

Figure 5.2: Different meshes employed to simulate Terzaghi's problem.

(Figure 5.2(c)) is used to analyse the example. This mesh is obtained by consistent refinement of the second mesh, that is, every element of the second mesh is split into two new elements in the vertical direction. The third mesh thus consists of twenty finite elements with heights between 0.20 m and 0.005 m.

Concerning the individual finite elements, a node of the element possesses a maximum number of three degrees of freedom (dof's), namely the two displacement components in the horizontal and vertical direction and the pore water pressure degree of freedom. As widely used in the numerical treatment of consolidation problems, different orders of interpolation are employed for the physically different degrees of freedom: quadratic interpolation is used for the displacements while bilinear interpolation is employed for the pressure degrees of freedom (confer Section 4.5.5). Hence, eight or nine nodes with displacement degrees of freedom are attached to each finite element, whereas pore pressure degrees of freedom are assigned to the four corner nodes of each element. The results presented in the next section have been obtained with eight displacement nodes per element. However, a calculation with nine displacement nodes per element using the first mesh yielded exactly the same results.

The boundary conditions for both the solid and fluid phases are shown in Figure 5.2. With respect to the displacements the domain is fully constrained at the bottom boundary and horizontally constrained at the vertical boundaries. For the water phase both the vertical and bottom surfaces of the domain are assumed to be impermeable, i.e. the fluid velocity normal to these boundaries is zero and no flow of pore water through the walls is permitted. At the top boundary of the domain the excess pore water pressure is constrained to zero and thus a permeable boundary is simulated.

The initial conditions are such that at time $t = 0$ a surface load of constant magnitude $q_0 = 10.0$ kPa is applied suddenly. Since the fluid phase is assumed to be incompressible, this yields an excess pore water pressure at time $t = 0$ of $p_0^w = q_0$ (see equation (5.16)) in the whole domain.

The material properties for both phases are summarised in Table 5.1. The values are chosen from the literature. Since the soil skeleton is considered to be linear elastic, only the two parameters Young's modulus and Poisson's ratio are necessary to describe its mechanical behaviour. For the flow process the permeability of the soil skeleton with respect to the water phase and the intrinsic density of the fluid are given in Table 5.1. Additionally, the gravitational acceleration is required.

To estimate the duration of the consolidation process, i.e. the time span after which

Parameter	Symbol	Unit	Value
Young's modulus	E	kPa	5000.0
Poisson's ratio	ν	-	0.35
Permeability water	k^{ow}	m/s	$1.0 \cdot 10^{-5}$
Density water	ρ^w	t/m^3	1.00
Gravitational acceleration	g	m/s^2	9.81

Table 5.1: Material properties for Terzaghi's problem.

the settlements are completed and any excess pore pressures have vanished again, equation (5.40) is very useful. According to this relationship, for given material properties the consolidation time can be calculated approximately. When inserting the parameters of Table 5.1 into equation (5.40) one obtains

$$t \approx \frac{2 \cdot 1.0 \cdot 1.0 \cdot 9.81 \cdot 1.35 \cdot 0.3}{1.0 \cdot 10^{-5} \cdot 5000.0 \cdot 0.65} \approx 245 \text{ sec}.$$

Hence, the 2.0 m high consolidating soil layer with the properties given in Table 5.1 needs approximately 245.0 seconds until the settlements are completed. In the numerical model a time span of 400.0 seconds is simulated. Within this interval the time increments are increased gradually, starting with a minimum time increment of 0.001 seconds which changes step by step to 10.0 seconds in the course of the calculation.

In any transient problem the spatial discretisation and the time step are related to the extent that, for a given element size, time steps smaller than a certain value yield no useful information (oscillatory results). This coupling of the spatial and temporal approximations is always most obvious at the start of the analysis, that is, immediately after prescribed changes in the boundary values. Therefore, the choice of the initial time step is an important issue in these kinds of problems. As the governing equations are parabolic, the initial solution (immediately after the sudden change in load) is a local, 'skin effect', solution [Vermeer(1981)]. With a finite element mesh of reasonable size for modelling the solution at a later time, when the changes in pore pressure have diffused into the bulk of the soil, this initial solution will be described poorly. The smaller the initial time step is chosen, the more pronounced this problem becomes. Thus, a simple criterion for the initial time step is suggested in [Vermeer(1981)] as

$$\Delta t \geq \frac{(\Delta h)^2}{6 c_v} = \frac{\rho^w g \,(1+\nu)(1-2\nu)}{6 \, k_z^{ow} \, E(1-\nu)} (\Delta h)^2 \,, \tag{5.41}$$

where Δh is the characteristic element size near the disturbance, i.e. in the vicinity of the draining surface in the particular case. For a mesh size of $\Delta h = 0.005$ m (this condition is met by the third mesh) and the material properties given in Table 5.1 the critical time increment size is obtained as $\Delta t \geq 0.00051$ sec. In the present calculations the minimum time increment thus is chosen to be 0.001 seconds.

5.2.4 Results of the numerical simulation

This section is devoted to a discussion of the numerical results obtained from calculations using the three different meshes shown in Figure 5.2. The subsequent Figures 5.3 to 5.5 depict the development of the normalised pore water pressure, i.e. the pore water pressure p^w related to the initial pore water pressure p_0^w, the latter being equal to the applied surface load, versus the normalised height, i.e. the vertical coordinate z related to the total height h of the column. All pressures in this section are considered to be pore water pressures in excess of the hydrostatic pore water pressures present in the soil prior to loading and after the end of the consolidation process.

Equally-sized finite element mesh

A first calculation is performed with the equally-sized finite element mesh shown in Figure 5.2(a). In the vicinity of the draining boundary this discretisation is relatively coarse, at least too coarse when considering an initial time step of 0.001 seconds as specified. To avoid oscillations an initial time step according to equation (5.41) of $\Delta t \geq 0.204$ sec would be appropriate. However, with this choice no information for earlier points of time would be available. As can be seen from Figure 5.3, for points of time smaller than 0.204 seconds oscillations in the pore water pressure are obtained which are the more pronounced the closer the time moves to the instant of loading. For example, 0.001 seconds after the application of the load the excess pore water pressure predicted by the finite element model is more than 25 % too large. This 'overshoot' decreases with increasing time, at 0.1 seconds after loading the difference between the analytical and the numerical solution is already less than 10 % and at $t = 1.0$ sec no oscillations can be recognised at all. Due to the coupled treatment of the problem these oscillations can also be observed in the soil displacements. However, at these early stages of the consolidation process the settlements are negligibly small anyway. For points of time equal to or greater than $t = 1.0$ sec the finite element results (dots) agree reasonably well with the analytical solution (solid lines) in Figure 5.3. A consideration of one individual point of time ($t \geq 1.0$ sec) indicates that the spatial discretisation is sufficiently fine to capture the behaviour of the excess pore water pressure along the height of the soil column. Since for different points of time the numerical results match the analytical solution perfectly well, the discretisation in the time domain seems to be suitable as well for all $t \geq 1.0$ sec. After a time span of 400.0 seconds the excess pore water pressures completely disappear in the whole domain. After about 250 seconds, which is predicted to be the end of the consolidation process by equation (5.40), the maximum excess pore water pressure in the domain (occurring of course in the symmetry plane) is still about 0.1 kPa, which is approximately 1 % of the initial excess pore water pressure. However, this result might be satisfactorily accurate for any practical purpose.

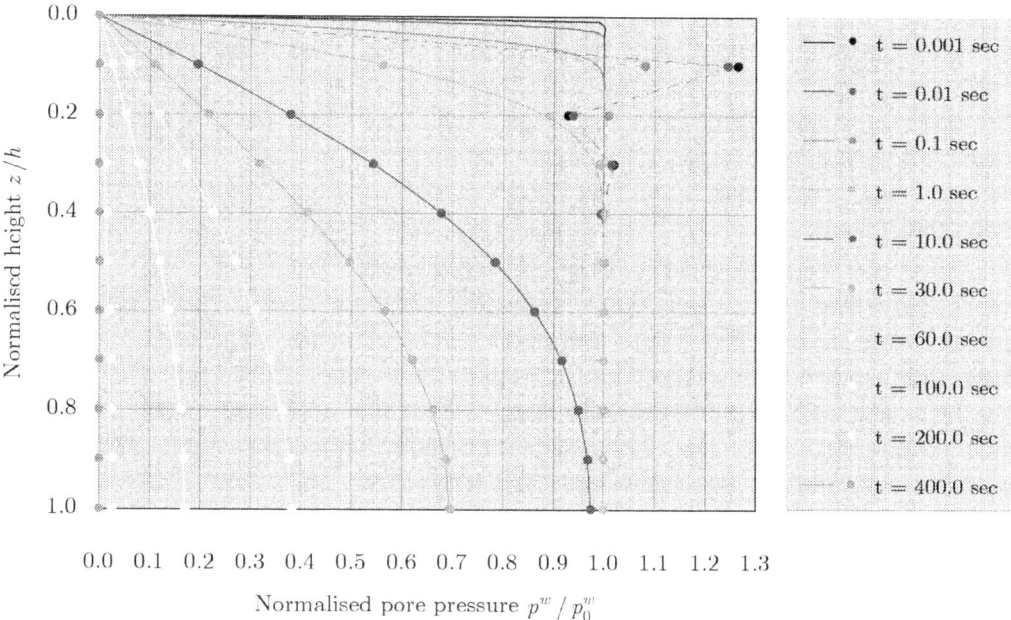

Figure 5.3: Evolution of the excess pore water pressure using the equidistant mesh (Figure 5.2(a)): analytical solution (solid lines) and numerical results (dots).

Graded finite element mesh

For the second mesh (Figure 5.2(b)) the same number of finite elements as for the first mesh is used. However, the element height is now decreasing in the direction of the draining boundary of the column. This simple improvement of the employed discretisation yields significantly better results as shown in Figure 5.4. Once again, the normalised pore water pressure, p^w/p_0^w, is plotted versus the normalised height, z/h, in this figure. Now the oscillations of the finite element solution are considerably smaller, a maximum of about 10 % is obtained, and restricted to a shorter period of time after the application of the surface load. At 0.01 seconds the oscillations almost disappear and no oscillations at all are recognised after $t = 0.1$ sec. However, for the current deciding element size of 0.01 m the critical time increment is obtained as $\Delta t \geq 0.00204$ sec. Thus, the chosen increment of 0.001 seconds still yields poor results at the very early stages after loading. The deviation of the numerical results (dots) from the analytical solution (solid line)

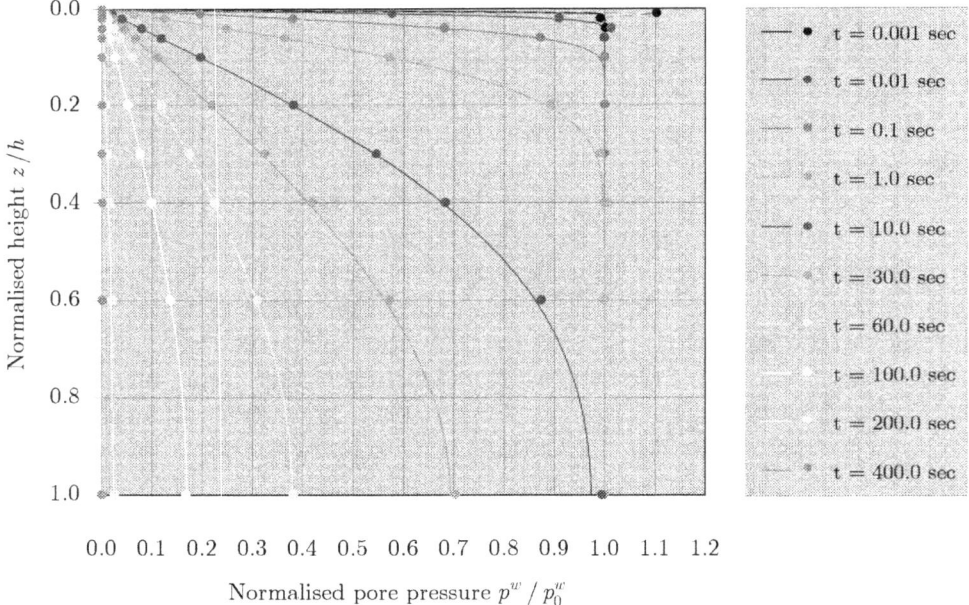

Figure 5.4: Evolution of the excess pore water pressure using the graded mesh (Figure 5.2(b)): analytical solution (solid lines) and numerical results (dots).

near the symmetry plane at $z/h = 1.0$, which is most pronounced at $t = 10.0$ sec, is due to the large element size in this region.

Fine graded finite element mesh

The third mesh (Figure 5.2(c)), obtained by consistent refinement of the second mesh, is also graded in the direction of the draining boundary. Figure 5.5 again shows the normalised pore water pressure, p^w/p_0^w, versus the normalised height, z/h. The element height of 0.005 m in the vicinity of the top surface now meets the requirement (5.41) for the minimum time increment. Consequently, no oscillations in the finite element solution are encountered and the numerical results match the analytical solution perfectly well over the whole time span of 400.0 seconds and for the entire domain. However, if one would be interested in results prior to $t = 0.001$ sec, then, keeping the current mesh size unchanged, oscillations could occur once again. Additionally, it should be mentioned

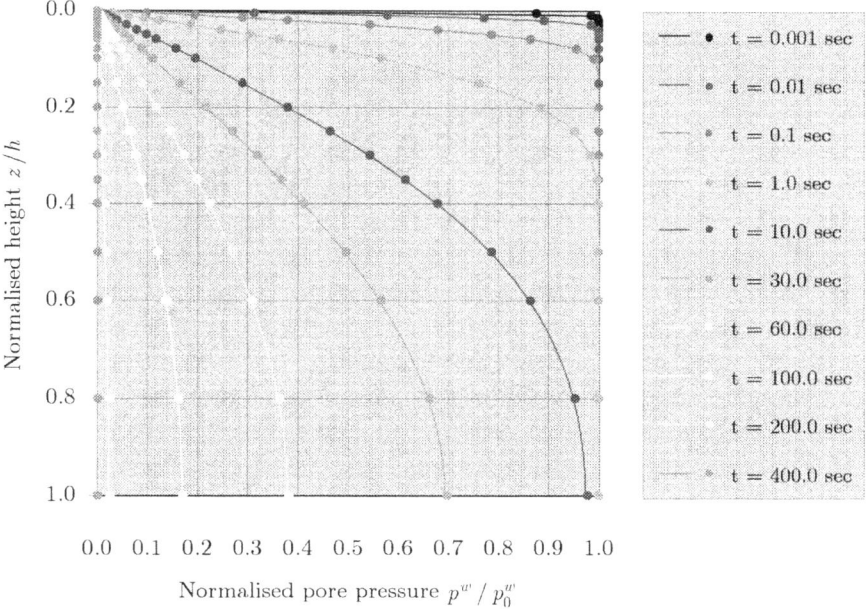

Figure 5.5: Evolution of the excess pore water pressure using the fine graded mesh (Figure 5.2(c)): analytical solution (solid lines) and numerical results (dots).

that in this solution it was not intended to optimise the time stepping procedure at all.

The above discussion suggests that poor numerical results at the very early stage of an analysis can be avoided (i) for a given finite element mesh by increasing the initial time increment – of course paying with a loss of information at the very beginning of the analysis or (ii) for a given initial time step size by decreasing the element height near the draining boundary, yielding perhaps a larger number of elements and thus more computer cost. However, quite big finite elements can be used in regions of the domain where pore pressure gradients are small.

Soil settlements versus time

Finally, Figure 5.6 shows the development of the soil settlements versus time. The degree of consolidation, i.e. the current settlement related to the final settlement (confer equation (5.36)), is plotted versus the dimensionless time parameter $c_v t/h^2$, where for the time axis a logarithmic scale proves to be very convenient. The four analytical

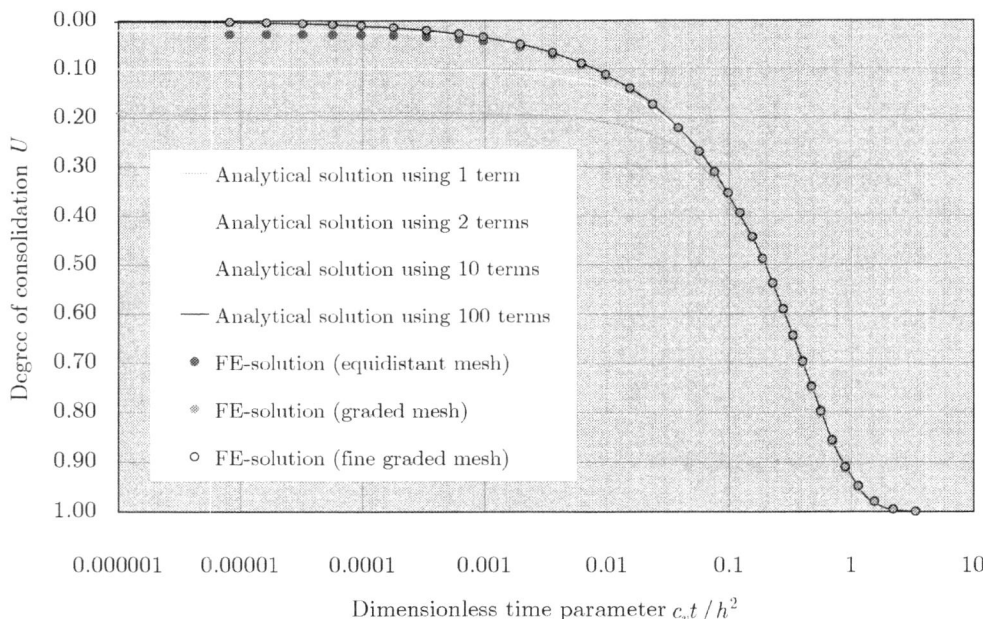

Figure 5.6: Evolution of the degree of consolidation for different meshes: analytical solution (solid lines) and numerical results (dots).

solutions (solid lines) should outline that quite a large number of terms in the sum of equation (5.38) is necessary to give a perfect analytical result (especially at early stages of the consolidation process), however, the first ten terms in the sum yield an acceptable approximation. Using less terms entails an overestimation of the degree of consolidation in the early period of time after loading. It should be mentioned that also for the analytical calculation of the pore water pressures (equation (5.31)) a rather large number of terms is required, especially, once again, for the very early points of time. After some seconds, however, ten terms yield a good approximation to the solution. The numerical results obtained by means of the equidistant mesh (Figure 5.2(a)) deviate somewhat from the analytical solution at the early stages of time, the predicted degree of consolidation is too large. For both graded meshes on the contrary the numerical results agree perfectly well with the analytical solution.

The maximum soil settlement computed by means of the finite element programme is obtained as $w = 1.246$ mm for all three meshes after about 400.0 seconds. The final settlement calculated from equation (5.34), using the parameters E and ν instead of Lamé's constants λ^s and G^s, is given as

$$w_\infty = \frac{2hq_0(1+\nu)(1-2\nu)}{E(1-\nu)} = \frac{2 \cdot 1.0 \cdot 10.0 \cdot 1.35 \cdot 0.3}{5000.0 \cdot 0.65} \equiv 2.492 \text{ mm},$$

which is exactly twice the computed finite element result due to the employed symmetry of the problem in the finite element solution. After the approximate duration of the consolidation process of 245.0 seconds, calculated according to (5.40), the finite element programme yields a maximum settlement of $w = 1.238$ mm which is about 99.4 % of the final value. Therefore, equation (5.40) is of course excellently suited for estimating the duration of the consolidation process in any practical purpose.

5.3 Footing on a water saturated soil layer

5.3.1 Description of the problem

Most consolidation problems of practical interest are two- or three-dimensional, so that the one-dimensional solution provided by Terzaghi's consolidation theory is useful only as indicator of settlement magnitudes and rates. Thus, a two-dimensional consolidation analysis is discussed in the current section. In engineering practice this phenomenon may be encountered, e.g., in connection with footings placed on water saturated soil layers.

A soil layer of thickness h and extending to infinity in the other two coordinate directions, resting on a rough rigid base and subject to a uniform surface loading q_0 is considered (Figure 5.7). With respect to the constitutive equations and the initial and boundary conditions basically the same assumptions as for the one-dimensional case (Terzaghi problem) are valid. The soil consists of a homogeneous isotropic perfectly elastic skeleton, the voids are completely filled (saturated) with water. Thus, it is assumed that the strains of the medium are related to the effective stresses by Hooke's law. Furthermore, for the flow of the fluid through the pores of the soil Darcy's law is valid. Both the solid particles and the fluid phase are considered to be incompressible. The rigid base, which the consolidating layer rests on, is assumed to be impermeable and therefore the soil is allowed to drain only towards the top permeable boundary (Figure 5.7). The ground surface is loaded at time $t = 0$ by a uniform load of magnitude q_0 over a given width of $2b$. The loading is applied suddenly and then remains constant in time.

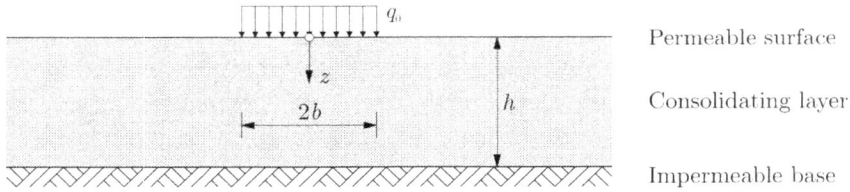

Figure 5.7: Footing on a consolidating soil layer.

The examples investigated in this section deal with a soil layer loaded either by a strip or a circular footing. Considering the above mentioned assumptions, in particular the linear elastic soil skeleton and the validity of Darcy's law, analytical solutions (e.g., [Booker(1974)]) are available, predicting the settlement history of a partially loaded layer of soil, which can be employed for the verification of numerical models [Booker(1982a), Booker(1982b)].

5.3.2 Analytical solution

The analytical solution given in [Booker(1974)] is used for verification purposes in the current section and should therefore be introduced subsequently in brief. The geometry of the problem investigated by Booker is shown schematically in Figure 5.7. The coordinate system is chosen such that the positive z-axis points downwards and the x- and y-axes lie in the plane of the ground surface, that is, the coordinate $z = 0$ denotes the upper boundary of the consolidating soil layer and the coordinate $z = h$ the lower one.

For the mathematical statement of the problem the governing equations are the equilibrium equations and the mass balance equation which in [Booker(1974)] are used in the form

$$G^s \nabla^2 \mathbf{u} - (\lambda^s + G^s) \boldsymbol{\nabla} \varepsilon^{vol} = \boldsymbol{\nabla} p^w \tag{5.42}$$

and

$$\frac{\partial \varepsilon^{vol}}{\partial t} = \frac{k^{ow}}{\rho^w g} \nabla^2 p^w \,, \tag{5.43}$$

where $\varepsilon^{vol} = -\boldsymbol{\nabla} \cdot \mathbf{u}$ denotes the volumetric strain of the soil skeleton and $\mathbf{u}$ is the vector of displacements with components u, v, w. Concerning the fluid phase (water), p^w denotes its excess pressure, ρ^w is the intrinsic density and k^{ow} the permeability of the soil skeleton with respect to the water phase. Compression is taken as positive for stresses/pressures in the above equations. It should be mentioned that the equations (5.42) and (5.43) can be derived from (3.354) and (3.355) when using Hooke's law for the soil skeleton as given in (5.6), i.e. written in terms of the Lamé constants λ^s and G^s.

In addition to the governing equations (5.42) and (5.43) two sets of boundary conditions have to be specified, namely for both boundaries of the consolidating soil layer. The upper surface is assumed to be permeable and part of it is subject to prescribed normal tractions:

$$\begin{aligned} \tilde{\sigma}'_{zz} &= q(x,y) & \text{for} \quad z = 0 \,, \\ \tilde{\sigma}'_{xz} &= \tilde{\sigma}'_{yz} = 0 & \text{for} \quad z = 0 \,, \\ p^w &= 0 & \text{for} \quad z = 0 \,. \end{aligned} \tag{5.44}$$

The lower boundary completely adheres to the rigid base and will be assumed to be impermeable, thus

$$\begin{aligned} \mathbf{u} &= \mathbf{0} & \text{for} \quad z = h \,, \\ \frac{\partial p^w}{\partial z} &= 0 & \text{for} \quad z = h \,. \end{aligned} \tag{5.45}$$

The pore water is considered to be incompressible, it thus follows that initially, when the load is applied, there can be no volume change, that is

$$\varepsilon^{vol} = 0 \qquad \text{for} \quad t = 0^+ \,. \tag{5.46}$$

The required solution is the solution of the equations (5.42) and (5.43) satisfying the boundary conditions (5.44) and (5.45) and the initial condition (5.46).

It is possible to deduce the initial and the final settlements from analytical solutions based on the theory of elasticity [Poulos(1991)]. At the beginning of the consolidation process the material is incompressible and it can be shown that the displacements are identical to those of an incompressible medium with a shear modulus G^s. At the end of consolidation it is clear that all excess pore pressures will have dissipated and thus as $t \to \infty$ the solution will reduce to that for a purely elastic material with the elastic constants λ^s and G^s. For instance, for the centre of a circular footing the initial settlement can be calculated from

$$w_0 = \frac{2b\, q_0\, I_w}{E} = \frac{2b\, q_0\, I_w}{2G^s(1+\nu)} = \frac{2}{3}\frac{bq_0}{G^s} 0.23 = 0.153 \frac{bq_0}{G^s} \,, \tag{5.47}$$

where the relationship (3.217) between Young's modulus E and shear modulus G^s is employed, $\nu = 0.50$ is assumed for the incompressible material and the value of $I_w = 0.23$ is extrapolated from values given in [Poulos(1991)]. The final settlement for the centre of a circular footing is obtained from (5.47) by inserting, e.g., $\nu = 0.00$ and $I_w = 0.485$

$$w_\infty = \frac{2b\, q_0\, I_w}{E} = \frac{2b\, q_0\, I_w}{2G^s(1+\nu)} = 0.485 \frac{bq_0}{G^s} \,. \tag{5.48}$$

The governing equations (5.42) and (5.43) are solved in [Booker(1974)] by applying a double Fourier transform, followed by a Laplace transform. However, any details of the solution procedure are omitted in the present work and the interested reader is referred to the particular paper. The expressions obtained from this solution procedure are then evaluated in [Booker(1974)] for three different cases, namely for a strip, circle and square, each loaded uniformly. The results of these three cases are given for a range of values of h/b (0.2, 0.5, 1.0, 2.0, 5.0) and Poisson's ratio ν (0.00, 0.25, 0.48). Two parameters are plotted versus a logarithmic time scale, either a related settlement $G^s w(0,t)/bq_0$ or the degree of consolidation $U = (w - w_0)/(w_\infty - w_0)$, where w denotes the deformation at the centre of the footing while w_0 and w_∞ describe the initial and final values, respectively. For most parts, however, the latter parameter is plotted versus time. Since both diagrams are available in [Booker(1974)] only for the case $h/b = 1.0$ and $\nu = 0.00$, these values are used in the numerical analyses although they might not be considered as the most realistic ones when it comes to real civil engineering problems.

5.3.3 Numerical model

When taking a look at Figure 5.7, the symmetry of the problem is easily encountered. The vertical symmetry plane runs through the centre of the footing and thus only one half of the domain has to be considered in the numerical model.

The finite element mesh chosen for the calculations is shown in Figure 5.8. As this figure indicates, a consolidating soil layer with a height of 8.0 m is investigated. In the horizontal direction a region with a width of 40.0 m is taken into account for the analysis. The loaded area is twice as wide as the depth of the soil layer under consideration yielding a value of h/b equal to 1.0. In the region of the loading and near the draining upper boundary smaller element sizes are used while a rather coarse discretisation is chosen for the rest of the domain. However, despite the small number of forty elements the obtained results are fairly accurate for all variables as a calculation with a refined mesh proved. Moreover, the good agreement between the numerical results and the analytical solution suggests that the used mesh is adequate – at least for the overall displacement response examined, as will be seen in the next section.

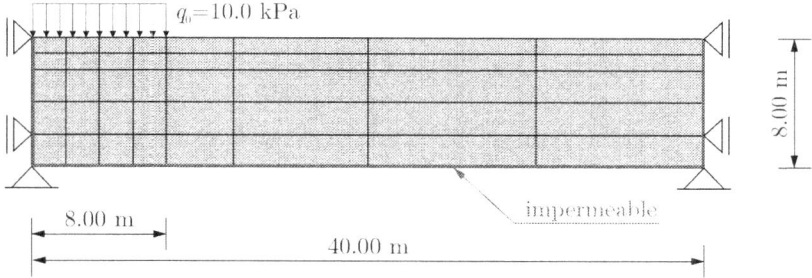

Figure 5.8: Finite element mesh employed for the footing example.

Similar to the one-dimensional consolidation case, a node of a finite element possesses a maximum number of three degrees of freedom, the two displacement components in the horizontal and vertical directions and the pore water pressure. Different interpolation is employed once again for soil displacements and fluid pressures. Eight or nine nodes are used for the displacements while pressure degrees of freedom are assigned to only four nodes per element.

Figure 5.8 also exhibits the boundary conditions determining the problem. With respect to the displacements the domain is fully constrained at the lower boundary and horizontally constrained at the vertical boundaries. Concerning the fluid phase both vertical boundaries and the base of the modelled region are impermeable, thus any fluid velocity components normal to these boundaries have to vanish. The ground surface is assumed to be completely permeable and therefore, the pore water pressure has to remain zero (or atmospheric). These boundary conditions exactly correspond to those specified in connection with the analytical solution (confer equations (5.44) and (5.45)). In engineering practice, however, such conditions might be used if a gravel layer is placed underneath the footing (loaded area) so that the water can also drain off under the loading. If the footing directly rests on the consolidating layer, an impermeable boundary should be assumed in the region of the footing.

As specified initial condition at the time $t = 0$ a surface load of constant magnitude $q_0 = 10.0$ kPa is applied suddenly over a width of 8.0 m, starting from the left vertical boundary of the domain. Prior to the consolidation process a geostatic stress state is calculated and stored as a set of history variables in all Gauss points. In the course of the calculation any arising stresses in the soil skeleton and in the fluid phase due to the applied surface load are added to these initial values. The geostatic stresses in both phases are calculated as a function of depth below the ground surface, using the unit weight of the respective constituent, and therefore increase linearly with depth.

The material properties (chosen from the literature) used for the analysis are summarised in Table 5.2. To describe the linear elastic constitutive behaviour of the soil skeleton, the two parameters Young's modulus and Poisson's ratio are necessary. Pois-

Parameter	Symbol	Unit	Value
Young's modulus	E	kPa	5000.0
Poisson's ratio	ν	-	0.00
Porosity	n	-	0.50
Permeability water	k^{ow}	m/d	0.1255251
Density solid	ρ^s	t/m^3	2.70
Density water	ρ^w	t/m^3	1.00
Gravitational acceleration	g	m/s^2	9.81

Table 5.2: Material properties for the footing example.

son's ratio is chosen to be zero in agreement with the diagrams available for the analytical solution. The permeability of the soil skeleton with respect to the water phase is calculated from the condition $c_v/h^2 = 1.0$ using the elastic properties of Table 5.2 and the height of the consolidating soil layer of 8.0 m. Furthermore, the intrinsic densities of both constituents solid and water as well as the gravitational acceleration are required for the analysis.

In order to estimate the duration of the consolidation process, equation (5.40) has been introduced in Section 5.2.2. Although the consolidation is not purely one-dimensional in the case under consideration, this relationship might be useful for a first approximation of the overall consolidation time. Inserting the parameters given in Table 5.2 yields

$$t \approx \frac{2 \cdot 8.0^2 \cdot 1.00 \cdot 9.81 \cdot 1.00 \cdot 1.00}{0.1255251 \cdot 5000.0 \cdot 1.00} \approx 2.0 \text{ days},$$

which is in good accordance with the diagrams shown in the next section. In the numerical simulation a time span of 10.0 days is simulated. The chosen initial time step size of 0.00005 day is increased to a value of 0.1 day in the course of the calculation.

5.3.4 Results of the numerical simulation

This section is devoted to the discussion of some results of the numerical simulation of the footing example. All calculations are performed with the mesh shown in Figure 5.8, no mesh convergence studies are presented for this example. However, three points have been investigated in some more detail: (i) the consequences of the choice of a certain number of degrees of freedom per element, (ii) the compressibility or incompressibility of the water phase and (iii) the type of boundary condition in the region of the loading. Results for both a strip and a circular footing are presented, using either plane strain or axisymmetric finite elements. It should be mentioned that only a part of the mesh (covering the first 20.0 m next to the symmetry plane) is plotted in the subsequent figures.

Analysis of a strip footing

First, the analysis of a strip footing is performed with the mesh of Figure 5.8. Each element possesses eight nodes with displacement degrees of freedom, pressure degrees of freedom are assigned to the four corner nodes of the element. The boundary conditions at the ground surface are in accordance with the analytical solution, i.e. also the loaded area is assumed to be completely permeable in the current simulation. The Figures 5.9 to 5.11 show results of this analysis.

Figure 5.9(a) shows the evolution of the excess pore water pressure after the application of the surface load. This pressure is in excess of the hydrostatic pressure distribution specified as initial condition which increases linearly from a value of zero at the ground surface to a maximum value of $p^w = 78.453$ kPa at the base of the consolidating layer. Immediately after loading the maximum value of the excess pore water pressure is about as high as the applied load, that is 10.0 kPa, which is due to the assumed incompressibility of the water phase. The slightly higher value in the first plot of Figure 5.9(a) results from some kind of pressure oscillation which might occur if the initial time step is chosen too large (see also Section 5.2). Thus, at the very beginning of the consolidation process the overburden is mainly carried by the water phase, a minor part is taken on by shear stresses. The maximum value of the excess water pressure is to be found in some distance (approximately 1.5 m) below the loaded area since at the ground surface the pore pressure is prescribed to be zero according to the specified boundary conditions (confer the equations (5.44)). At later stages of the consolidation process the maximum values of the excess pore water pressure occur in the vicinity of the bottom left corner of the domain (see plots two to four of Figure 5.9(a)). With increasing time these excess pore pressures decay. After about 1.0 day the excess water pressures have almost dissipated (the maximum value is already smaller than 10 % of the loading), as the last plot of Figure 5.9(a) indicates, and after 2.0 days the consolidation process is finally completed. Any excess water pressures have vanished and the initial hydrostatic pressure distribution is prevailing in the water phase.

The deformations shown in Figure 5.9(b) increase with proceeding time. The heave of the ground surface encountered at the beginning of the consolidation process disappears with time. Furthermore, the four plots of Figure 5.9(b) indicate, that the deformations are very much restricted to the vicinity of the loaded area (which is the first four columns of elements of the mesh), almost no displacements occur outside the distance of another 8.0 m from the edge of the loaded area.

Figure 5.10 shows the development of the effective vertical stresses during the consolidation process. Figure 5.10(a) indicates the initially applied geostatic stress state, a stress state increasing linearly with depth below the ground surface. By means of the material parameters given in Table 5.2 the maximum effective vertical stress can be

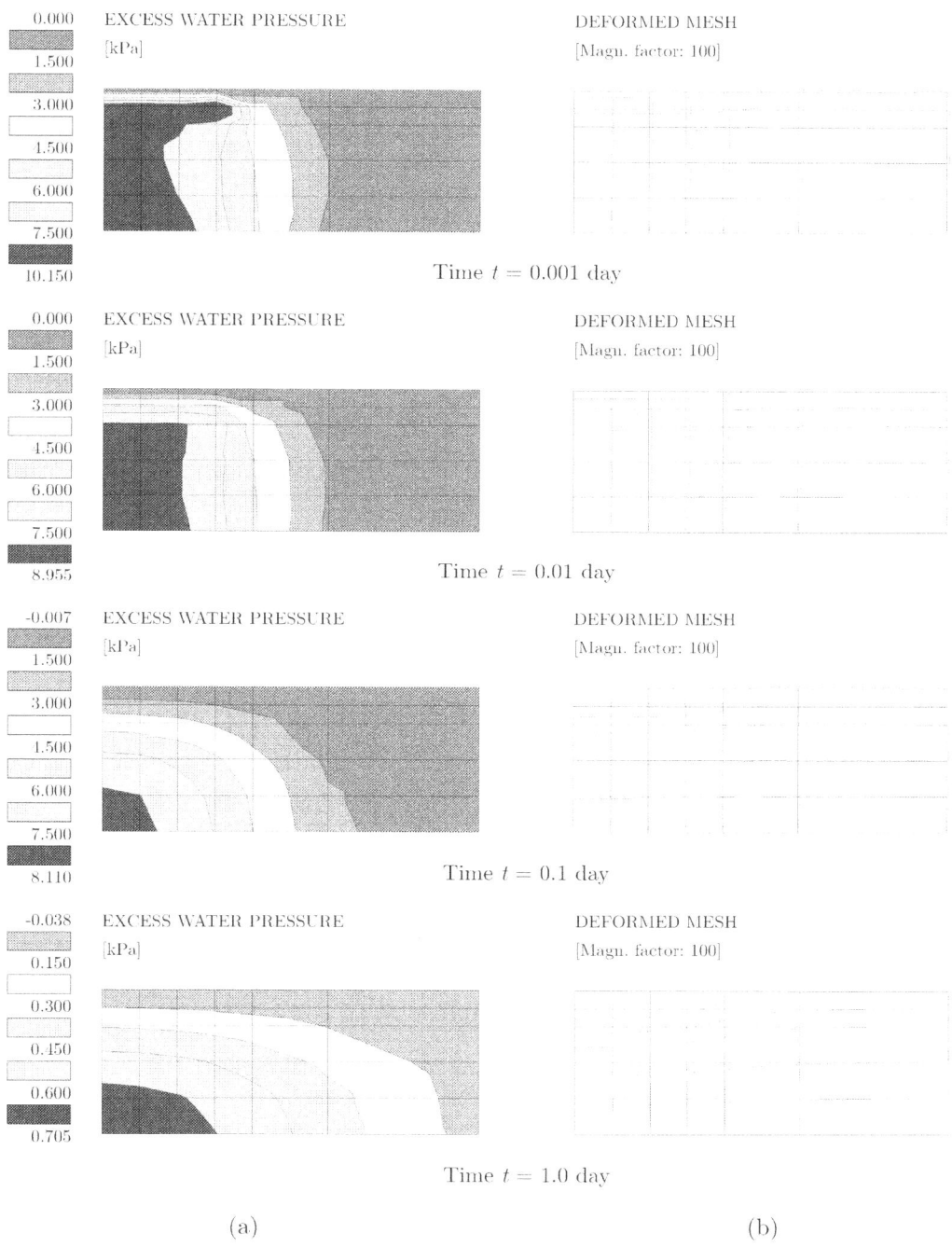

Figure 5.9: Evolution of (a) excess pore water pressure and (b) soil skeleton displacements during the consolidation period.

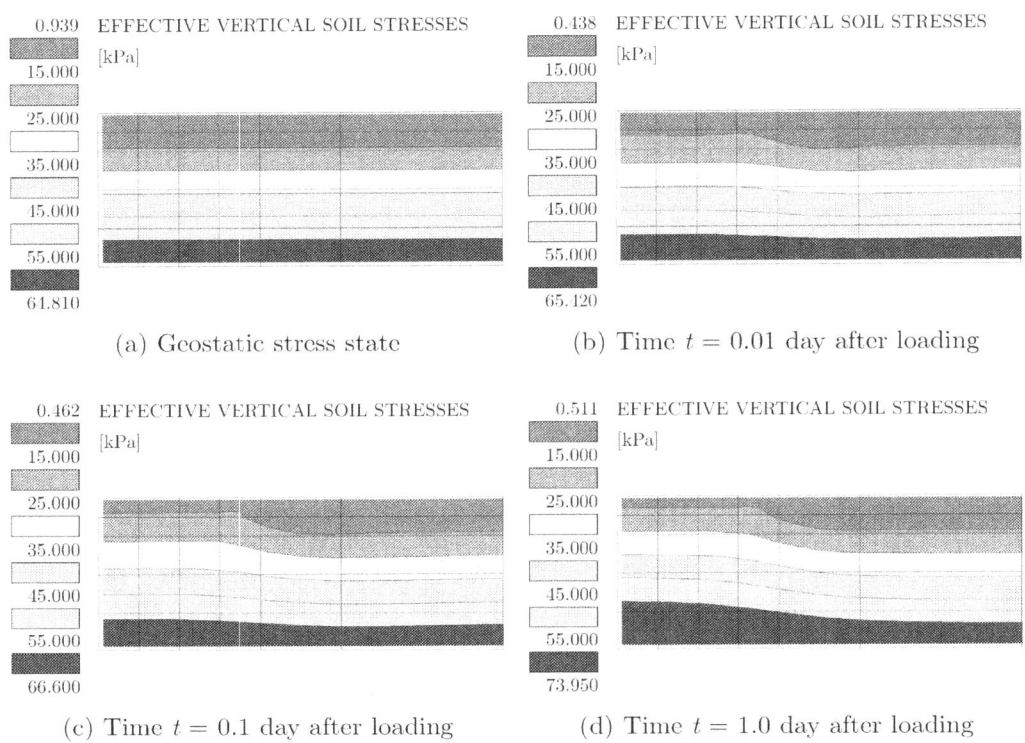

Figure 5.10: Evolution of the effective vertical stresses in the soil skeleton during the consolidation period.

calculated from the total vertical stress,

$$\tilde{\sigma}_{zz} = [(1 - 0.50) \cdot 2.70 + 0.50 \cdot 1.00] \cdot 9.81 \cdot 8.00 = 145.138 \text{ kPa} ,$$

by subtracting the hydrostatic fluid stress as

$$\tilde{\sigma}'_{zz} = \tilde{\sigma}_{zz} - p^w = 145.138 - 78.453 = 66.685 \text{ kPa} .$$

The maximum value given in Figure 5.10(a) is of course the stress referred to the lowest Gauss point next to the symmetry plane.

Immediately after loading the footing is primarily carried by the water phase, resulting in an increase of the total hydrostatic water stresses or an occurrence of excess pore water pressures in the vicinity of the applied load. This is due to the fact that the water phase is assumed to be incompressible and the permeability of the soil skeleton with respect to the water is rather low so that the water cannot be expelled from the pores sufficiently fast. However, with proceeding time the water manages to drain off and the load is transferred gradually from the fluid phase to the soil skeleton, yielding an increase of the effective soil stresses in the region underneath the loading (Figures 5.10(b) to 5.10(d)).

In the right part of the domain the geostatic stress state remains unchanged which indicates that the modelled domain is chosen adequately large. Figure 5.10(d) shows the distribution of the effective vertical stresses at steady state conditions. As can be seen, the maximum effective vertical stress has been increased by the amount of the applied load and all the load is now carried by the soil skeleton (since any excess water pressures have dissipated as Figure 5.9(a) indicates). It should be mentioned that the local change of the stresses next to the end of the loading can be described the better the finer the discretisation in this region is.

Finally, the numerical results are compared with the analytical solution of Booker [Booker(1974)] in Figure 5.11. The latter is given by the black solid line while the numerical results are indicated by the blue dots (FE-solution (8/4)). Figure 5.11(a) depicts the 'related' settlement $G^s w/bq_0$ versus a logarithmic time scale. A very close

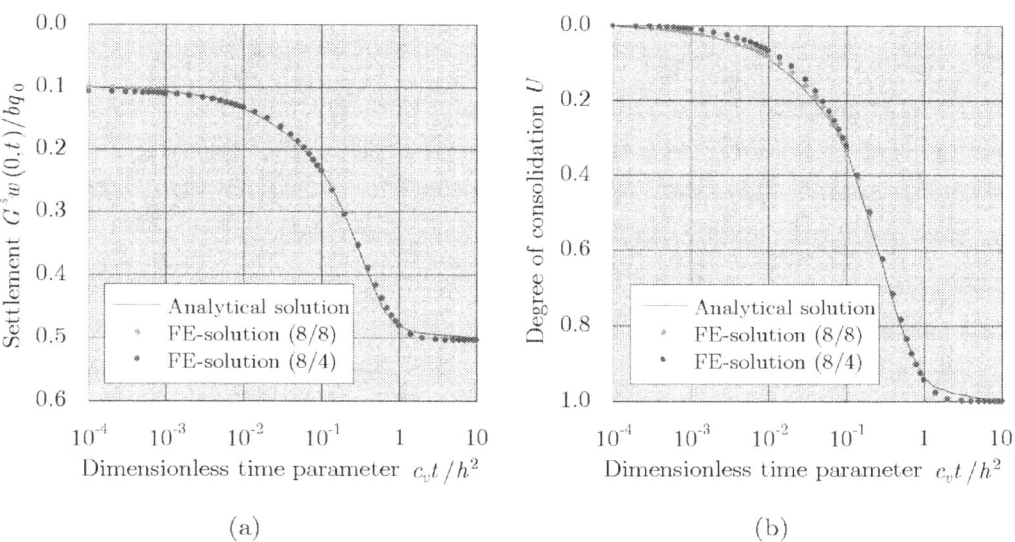

Figure 5.11: Time history of vertical deformation of the central point under the load: (a) settlement and (b) degree of consolidation.

agreement between the numerical results (blue dots) and the analytical solution (solid line) can be recognised for the whole period of time, only at the very beginning of the consolidation process the numerically predicted deformations are slightly too large. Since the factor c_v/h^2 is equal to one, the horizontal axis directly displays the consolidation time. As can be seen, the calculated time of 2.0 days (confer Section 5.3.3) corresponds to the prediction of the curves of this diagram.

Figure 5.11(b) shows the time-dependent development of the degree of consolidation, the latter being defined as $U = (w - w_0)/(w_\infty - w_0)$. The numerical results obtained for the degree of consolidation agree with the analytical solution fairly well, only in the vicinity of time 10^{-2} a somewhat more pronounced deviation is encountered. As these figures indicate, the settlement of the footing is predicted quite well although a rather coarse mesh is used for the numerical analysis.

Choice of degrees of freedom

As mentioned, various choices are possible for the number of degrees of freedom per finite element. For the results presented up to now elements with eight displacement and four pressure degrees of freedom are used. Figure 5.11 also contains numerical results (red diamonds) obtained with elements possessing eight pressure nodes. Considering Figure 5.11(a) some minor deviations of the numerical results (red diamonds) from the analytical solution (black solid line) are encountered at the beginning of the consolidation process, the former computations are slightly underestimating the settlements. For the remaining period of time, however, the accordance with the analytical solution is perfectly well. When employing elements with nine displacement and four pressure degrees of freedom, the results coincide with the blue dots in Figure 5.11(a).

When taking a look at the degree of consolidation plotted in Figure 5.11(b), the calculation with eight pressure degrees of freedom yields an even slightly better agreement with the analytical solution in the early stage of the consolidation process than the four pressure node computation. For later points of time, however, both numerical solutions almost coincide.

For a comparison of numerical results obtained with elements possessing either eight or nine displacement degrees of freedom in combination with four pressure nodes, some significant numbers concerning the maximum settlement w_{max} and the maximum effective vertical stress of the soil skeleton $\tilde{\sigma}'_{zz,max}$ are given in Table 5.3 for three points of time during the consolidation period. As can be seen, the differences between both

Solution increment	Time	Settlement w_{max}	Stress $\tilde{\sigma}'_{zz.max}$
Inc. 20 with 8/4	0.001 day	0.003544	65.12
Inc. 20 with 9/4		0.003546	65.12
Difference in [%]		0.06	0.00
Inc. 200 with 8/4	0.1 day	0.007520	66.60
Inc. 200 with 9/4		0.007521	66.60
Difference in [%]		0.01	0.00
Inc. 380 with 8/4	10.0 days	0.016100	74.66
Inc. 380 with 9/4		0.016100	74.66
Difference in [%]		0.00	0.00

Table 5.3: Results of calculations using elements with eight (8/4) or nine (9/4) displacement and four pressure degrees of freedom.

calculations are very small for the deformations (a maximum of 0.06 %) and negligible for the maximum effective vertical stresses, at least within the range of the given accuracy equal values for the stresses are obtained.

Compressibility of the fluid phase

A few words on the compressibility or incompressibility of the fluid phase. If the compressibility of the water phase is taken into account, a reasonable value for the bulk modulus of the water phase can be chosen according to Figure 3.8. The bulk modulus of pure water amounts to $2.0 \cdot 10^6$ kPa. If the groundwater contains approximately 1 % dissolved air (for example air bubbles), the compressibility of water increases substantially. In this case a bulk modulus of about 1000.0 kPa can be assumed for the water-air mixture. Subsequently, results of an analysis taking into account a bulk modulus for the water phase of $K^w = 1000.0$ kPa are presented.

If the water phase is assumed to be incompressible, the applied load is mainly carried by the fluid phase at the start of the consolidation process. In the course of the consolidation period water is expelled from the voids and the complete loading is gradually transferred to the soil skeleton. If the water phase is considered to be compressible, the arising excess pore water pressure immediately after loading is smaller, meaning that less load is carried by the water phase. For instance, in the present calculation after a time span of 0.001 day the maximum value of the excess water pressure only amounts to about 40 % of the respective value obtained for an incompressible fluid phase. Consequently, the soil skeleton is loaded to a higher extent at the beginning of the consolidation period, resulting in larger effective stresses and larger displacements.

Figure 5.12(a) contains a plot of the time history of the settlements. For comparison the results of Figure 5.11(a) are included in the diagram (analytical solution and two numerical results), the calculation using a compressible fluid phase is printed in orange coloured triangles (FE-solution (cpr)). Hence, the results plotted with blue dots and orange triangles only differ in the assumed compressibility of the water phase. As can be seen, the deformations at the start are almost four times as large when taking into account a compressible fluid in the pores. Furthermore, Figure 5.12(a) indicates that the consolidation process takes more time to complete. Only after about 10.0 days the final value of the settlement is reached which is five times longer than for the incompressible fluid phase. However, at steady state conditions the values of the maximum deformations are absolutely the same, no matter of the behaviour of the fluid phase. Figure 5.12(b)

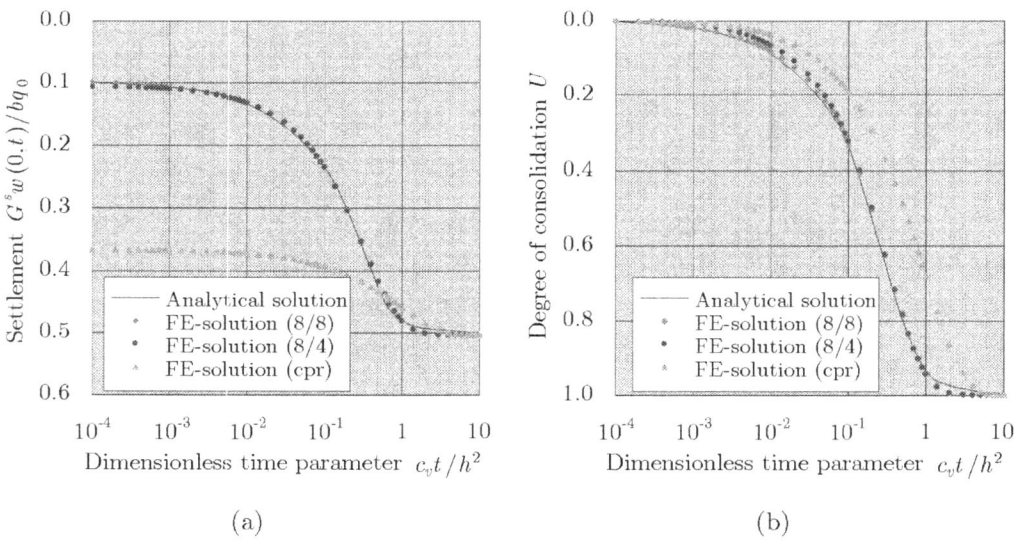

Figure 5.12: Time history of vertical deformation of the central point under the load (fluid phase compressible): (a) settlement and (b) degree of consolidation.

shows the degree of consolidation versus time. This figure also clearly outlines that the consolidation process is much slower when considering a compressible fluid phase instead of assuming incompressibility.

Boundary conditions in the loaded area

Next, the boundary conditions in the loaded area are investigated. According to the analytical solution, the region of loading is assumed to be completely permeable as the remainder of the ground surface. However, it might be argued that this assumption is not justified in many real problems since the footing may rest directly (without a draining gravel layer underneath) on the consolidating soil layer. In this case the loaded area should rather be specified as impermeable. Therefore, in a further calculation the boundary conditions at part of the ground surface are modified in this respect and results are compared with the ones obtained for the permeable surface (Figures 5.9 and 5.11).

Figure 5.13 shows the evolution of both the excess pore water pressure and the soil deformations for a part of the domain taking into account the modified boundary conditions. This figure should be compared with Figure 5.9. Since the pore water pressure is not constrained to zero at the loaded part of the ground surface any more, its maximum value in the early stage of the consolidation process occurs directly under the footing (see first plot of Figure 5.13(a)). In the course of the consolidation period, however, the excess water pressure assumes a rather uniformly distributed shape over the height of the layer which is due to the wide loaded area (8.0 m) and the comparatively small height (8.0 m) of the soil layer. Such a shape of the excess water pressure contours was of course prevented in the previous analysis (Figure 5.9(a)) by the boundary conditions specified in the region of the footing.

The last two plots of Figure 5.13(a) indicate that the consolidation process now takes much longer using the modified boundary conditions. After 1.0 day the excess pore pressures have decayed by 50 % and only after 10.0 days they have dissipated to sufficiently small values. However, one should remember that the loaded area takes 20 % of the total ground surface which is now impermeable and therefore prolongs the consolidation time substantially.

When the deformations at corresponding points of time are compared in the Figures 5.13(b) and 5.9(b), for the impermeable boundary they are of course smaller since the consolidation process takes more time to complete. However, what both figures have in common is that the deformations predominantly occur in the area underneath the

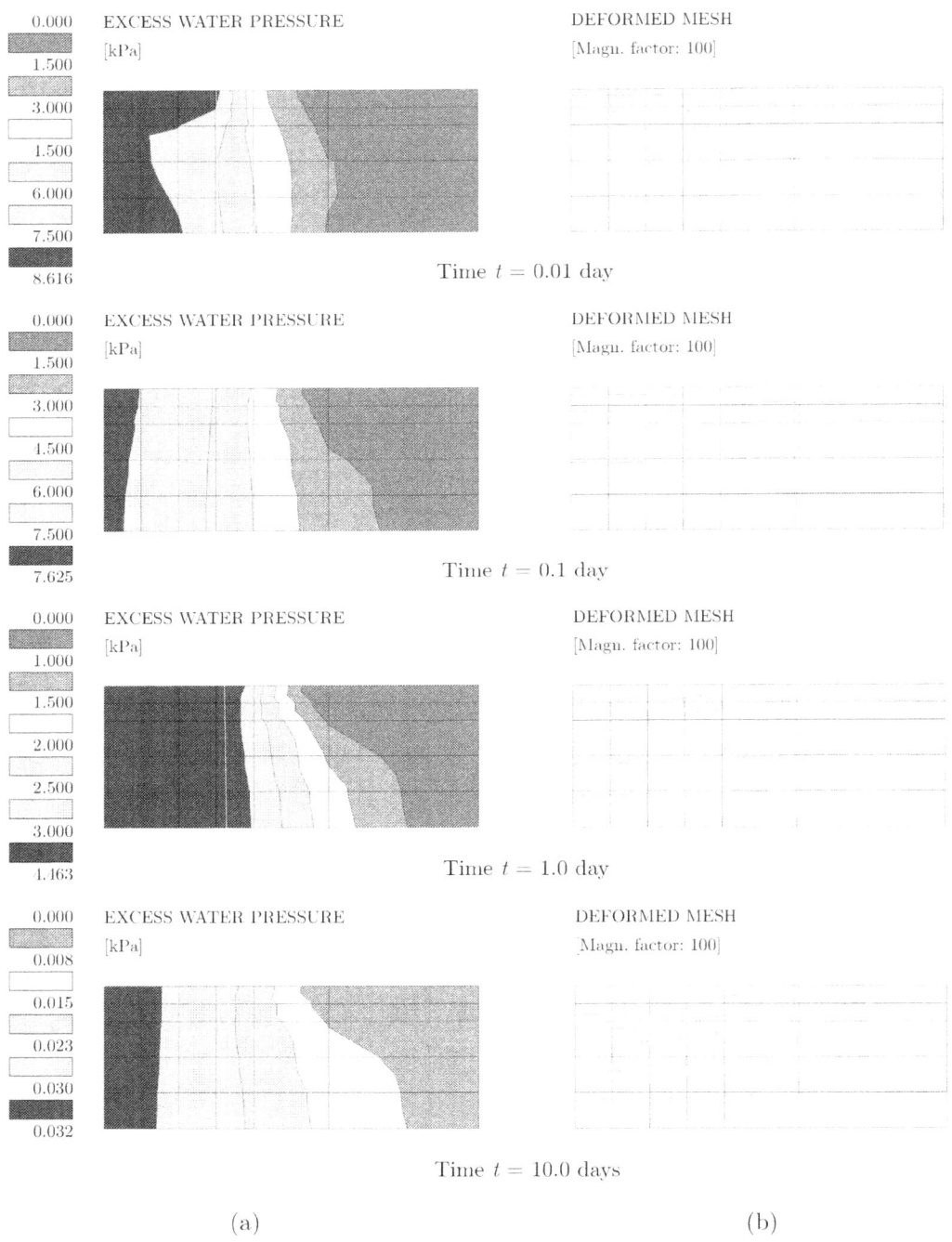

Figure 5.13: Evolution of (a) excess water pressure and (b) displacements during consolidation (impermeable boundary below loading).

loading. The settlements at the end of the consolidation process are of course identical to the ones obtained with a permeable boundary, that means the same effective stresses, hydrostatic fluid stresses and displacements are present at steady state conditions.

Figure 5.14 shows the evolution of both the maximum 'related' settlement and the degree of consolidation with time for the case of the impermeable boundary in the loaded area. The numerical results obtained with these modified boundary conditions are plotted in green squares (FE-solution (ib)). As both figures indicate very clearly, the shapes of the curves are identical as expected, however, there is a shift in time. The consolidation process for this case takes much longer than for the permeable loading area, even somewhat more than 10.0 days.

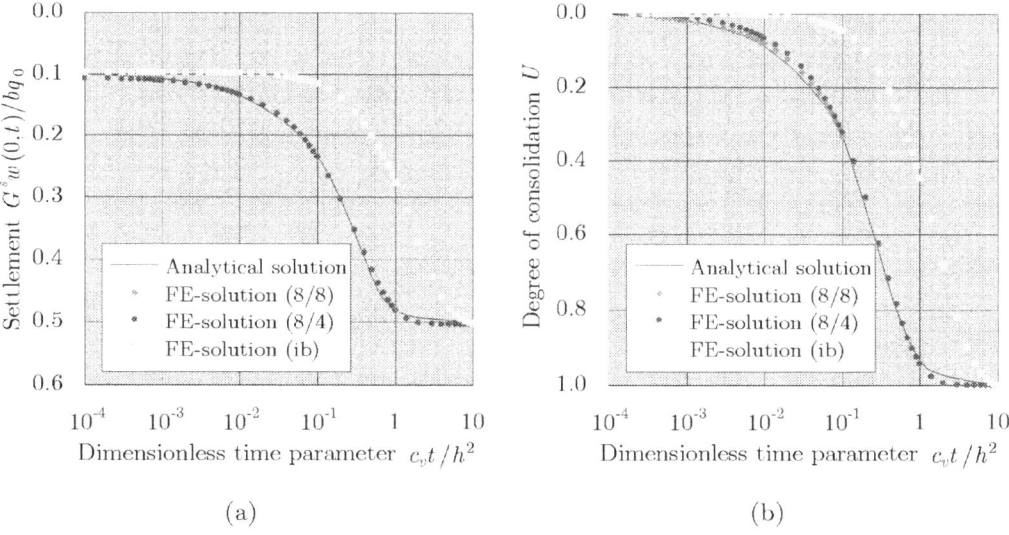

Figure 5.14: Time history of vertical deformation of the central point under the loading (impermeable boundary below the loading): (a) settlement and (b) degree of consolidation.

Analysis of a circular footing

Finally, some results for a circular footing are presented. The mesh given in Figure 5.8 can be used in combination with axisymmetric finite elements to simulate the consolidation of a soil layer loaded by a circular footing. Since there exists an analytical solution for this case [Booker(1974)], this example can be used to check the performance of the axisymmetric finite element.

Once again the time history of the settlement of the central point under the footing is considered. Figure 5.15 shows plots of the 'related' settlement and the degree of consolidation versus a logarithmic time scale for the case of a circular footing resting on a consolidating soil layer. Both Figures 5.15(a) and 5.15(b) contain the analytical solution of Booker (black solid line) and two numerical results stemming from calculations performed with elements possessing eight displacement degrees of freedom and either eight (red diamonds, FE-solution (8/8)) or four (blue dots, FE-solution (8/4)) pressure degrees of freedom. The agreement between analytical solution and numerical results is comparable to the plane strain case. The overall accordance between the results is very good, only in the time span between 10^{-2} and 10^{-1} the differences are somewhat more pronounced. Concerning the degree of consolidation the results obtained with eight pressure nodes match the analytical curve slightly better in the first stage of consolidation. The results from a calculation using nine displacement and four pressure nodes per element coincide with the blue dots in the Figures 5.15(a) and 5.15(b).

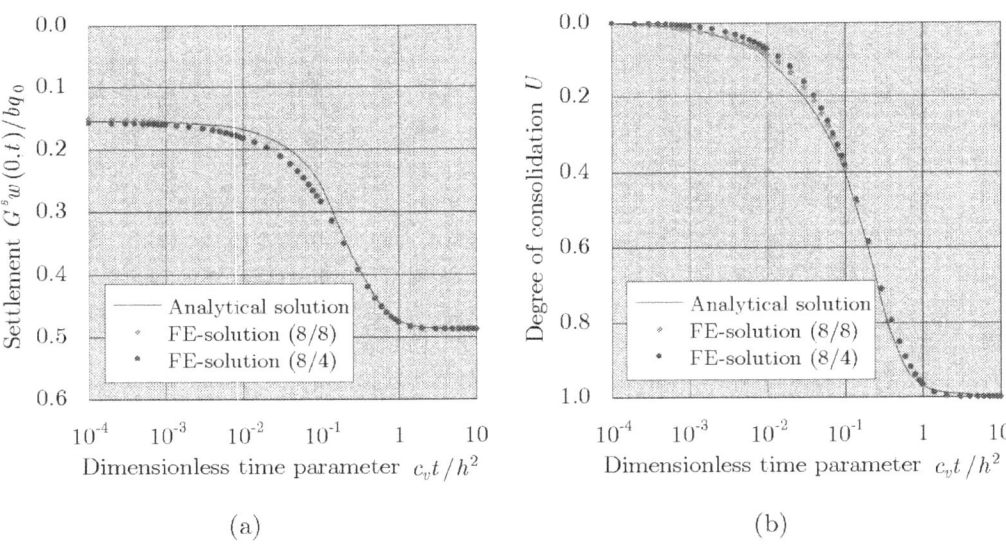

Figure 5.15: Time history of vertical deformation of the central point under the loading (circular footing): (a) settlement and (b) degree of consolidation.

It should be mentioned that the initial and final settlements of Figure 5.15(a) can also be calculated from solutions based on the theory of elasticity. The values computed in the equations (5.47) and (5.48) coincide with the initial and final points of the curve plotted in Figure 5.15(a) when multiplying the equations by the constant factor G^s/bq_0.

5.4 Flow of compressed air through dry sand

5.4.1 Description of the problem

To investigate the air loss through cracks in the shotcrete lining of a tunnel and the flow of compressed air in the adjacent soil, a number of different laboratory tests have been conducted at the Institute for Soil Mechanics and Foundation Engineering at Graz University of Technology [Kammerer(2000)]. These tests were part of a research programme within the framework of the *Austrian Joint Research Initiative* on *Numerical Simulation in Tunnelling* [Beer(1999)]. In a first set of experiments the loss of compressed air through a crack of defined width (tested widths lied between 0.08 mm and 1.04 mm), generated over the entire thickness of a shotcrete slab, has been studied. These tests were then extended to a system consisting of a shotcrete element overlaid by a soil layer of 1.0 m height. In the latter set of experiments (soil-shotcrete system) two different types of soil were investigated, namely a slightly gravelly, slightly silty sand and a slightly sandy, slightly clayey silt. Both soils were tested for completely dry as well as partially saturated conditions. In the current section the flow of compressed air through a dry sand is simulated numerically.

Figure 5.16 shows the set-up of this particular laboratory test. A steel container filled with dry sand is placed on top of a shotcrete element containing a crack of defined width. The soil is assembled in the box in 11 layers of 7.0 to 10.0 cm thickness. To avoid horizontal deformations of the walls of the steel container and consequently of the soil specimen, horizontal steel frames are used as reinforcement (not shown in the figure). The overburden above the crown of the tunnel is simulated in the experiment by a constant surface load of 250.0 kPa applied on top of the soil using a grid system of girders and beams which is vertically prestressed and fixed by four spindle rods. Thus, any heave of the soil is prevented. The bottom of the shotcrete slab is subjected to an excess air pressure causing compressed air to flow through the crack as well as the adjacent dry sand. In the experiment the following quantities are measured: (i) the applied excess air

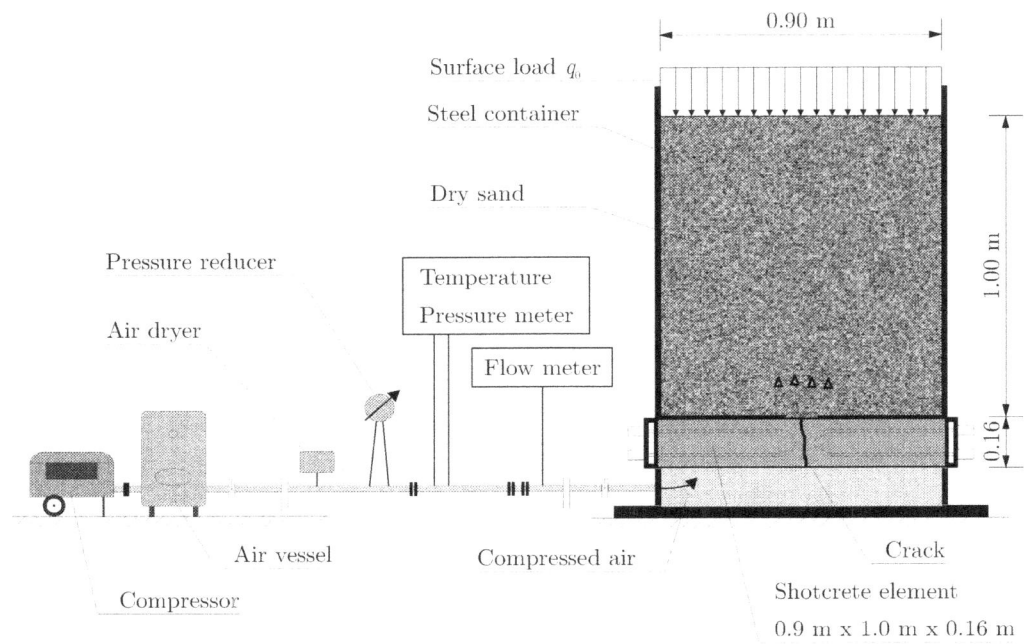

Figure 5.16: Set-up of the laboratory test conducted at Graz University of Technology [Kammerer(2000)].

pressure (air pressure in excess of the atmospheric pressure), (ii) the excess air pressure at a number of selected points in the soil specimen and (iii) the air loss through the crack in the shotcrete element.

In the experiment an excess air pressure of 2.0 bar was applied in steps of 0.1 bar at the lower boundary of the shotcrete element. However, this pressure was mainly reduced along the crack of the shotcrete slab, the amount of reduction of course depending on the crack width. Unfortunately, no pressure measurement was conducted right next to the upper end of the crack, only a few centimeters above the shotcrete element the air pressure in the soil was recorded. However, the overall air pressures measured in the soil were rather low [Kammerer(2000)]. According to [Kammerer(2000)], a steady state of the air pressure distribution in the dry sand was obtained for each pressure level almost immediately after the application of the respective pressure increment. Hence, the excess air pressure was increased by 0.1 bar after 1.0 minute, which was necessary for recording the measurements, yielding a total time of 20.0 minutes for the experiment to be performed.

5.4.2 Numerical model

In the numerical simulation the evolution of the excess air pressure in the soil specimen is studied. Similar to the examples discussed in Sections 5.2 and 5.3, a two-phase formulation is sufficient to simulate the problem. However, in contrast to the previously analysed examples in this case the compressibility of the fluid phase (compressed air) has to be taken into account.

Figure 5.17 shows the plane strain finite element meshes which are employed for obtaining the results presented in the next section. The first (rather coarse) mesh, depicted in Figure 5.17(a), consists of 90 elements. Most of the elements have a size of 9.0 times 10.0 cm, two columns of larger elements are located near the left and right vertical boundaries. Mesh two (Figure 5.17(b)) is assembled of 400 elements having dimensions of 4.5 times 5.0 cm. The plots shown in the subsequent section are based on calculations performed with this mesh. However, for the given convergence study two additional meshes are taken into account which are obtained by consistent refinement of mesh two. Consequently, the third mesh, shown in Figure 5.17(c), consists of 1600 elements and the finest mesh of 6400 elements.

Since a two-phase formulation is sufficient to describe the problem, a node of a finite element possesses a maximum number of three degrees of freedom, i.e. two displacements and one fluid pressure, similar to the consolidation examples of the previous

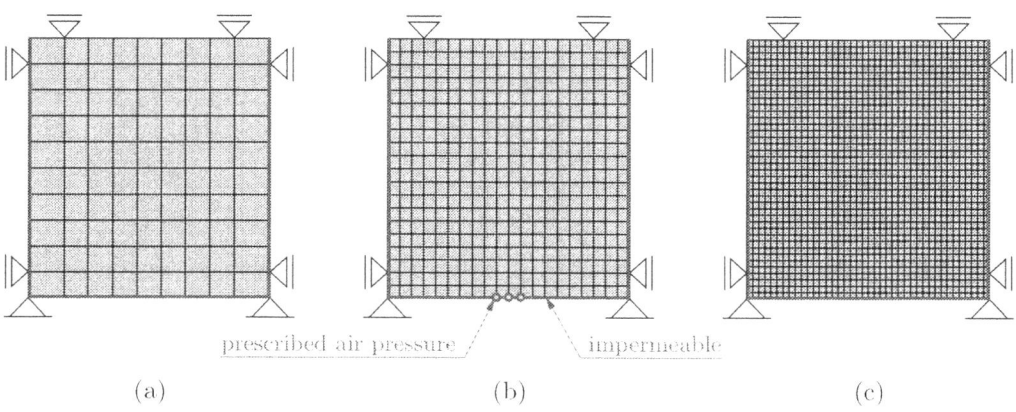

Figure 5.17: Three finite element meshes employed for the numerical simulation of the laboratory test conducted in Graz.

sections. Different interpolation is employed once again for displacements and pressures. Displacement degrees of freedom are assigned to eight nodes of each finite element, pressure degrees of freedom are considered for the four corner nodes. This type of element might be termed as isoparametric with respect to the displacements and superparametric with respect to the fluid pressures. Other combinations for the degrees of freedom per element are not investigated in the current example.

Figure 5.17 also illustrates the boundary conditions for both the solid and the fluid phase. With respect to the displacements the domain is fully constrained at the bottom, horizontally constrained at the vertical boundaries and vertically constrained at the upper boundary. Considering the fluid phase (compressed air), the vertical and bottom boundaries are assumed to be impermeable, atmospheric (i.e. zero) air pressure prevails along the upper boundary of the discretised region.

The overburden applied in the experiment on top of the soil specimen is accounted for in the numerical simulation in the geostatic stress state, i.e. the initial stresses in the soil skeleton are specified such that an effective vertical stress of 250.0 kPa is prescribed at the upper boundary of the soil skeleton which increases with depth according to the unit weights of both constituents. It should be mentioned that, compared to the rather high value of the surface load, any changes in the effective stresses due to the applied excess air pressure are of minor importance.

At the start of the experiment the excess air pressure is applied along the middle two elements at the bottom boundary of the soil specimen (see Figure 5.17(b)). Therefore, this excess air pressure represents a prescribed boundary condition for the fluid phase (compressed air) which either stays constant or linearly increases in the course of the analysis. In the numerical simulation of the laboratory test under consideration the applied excess air pressure is increased linearly from zero to its maximum value. On the other hand, for the investigation of the consolidation time of just one pressure increment, the fluid boundary condition is kept constant over the simulated period of time. The adjustment of the air pressure throughout the soil specimen to these prescribed boundary conditions is then investigated in the numerical simulation of the experiment.

In the calculation a maximum excess air pressure of 8.5 kPa (85.0 mbar) is applied at the bottom boundary of the soil specimen. However, this pressure is not increased step by step as in the laboratory test but by a linear ramp which might be justified since at each pressure level steady state conditions are obtained almost immediately after loading [Kammerer(2000)]. The simulated duration of the experiment amounts to 20.0 minutes.

To prove the fact of a negligible consolidation time, one pressure increment (0.425 kPa) is simulated numerically and the time span until attaining a steady state is investigated.

The material parameters summarised in the subsequent Table 5.4 are obtained as follows: Young's modulus E of the sand is determined by taking into account the overburden of the soil specimen. Data of confined compression tests serve as basis to identify the constrained modulus E_s (reciprocal of the coefficient of volume change) as a function of the effective vertical stress which is then used to calculate Young's modulus as $E = E_s(1+\nu)(1-2\nu)/(1-\nu)$. Poisson's ratio ν is computed from the coefficient of earth pressure at rest, $\nu = K_0/(1+K_0)$. Since the changes of the soil stresses during the

Parameter	Symbol	Unit	Value
Young's modulus	E	kPa	24000.0
Poisson's ratio	ν	-	0.23
Porosity	n	-	0.45
Permeability air	k^{oa}	m/s	$6.50 \cdot 10^{-6}$
Bulk modulus air	K^a	kPa	100.0
Density solid	ρ^s	t/m^3	2.71
Density air	ρ^a	t/m^3	$1.295 \cdot 10^{-3}$
Coefficient of earth pressure	K_0	-	0.30
Gravitational acceleration	g	m/s^2	9.81

Table 5.4: Material properties for the airflow test on a dry sand.

experiment are small compared to the initial soil stresses due to the applied overburden, linear elastic material behaviour of the soil skeleton seems to be an adequate assumption for the numerical simulation. The values for the porosity n and the solid density ρ^s are known from the experiment [Kammerer(2000)]. The coefficient of earth pressure at rest, K_0, is calculated from $K_0 = 1 - \sin\varphi'$, where the friction angle φ' of the soil is obtained from laboratory tests. The permeability with respect to the air phase, k^{oa}, is calculated from the intrinsic permeability k^i of the soil (confer equation (3.212)) determined from the experimentally obtained air pressure distribution [Kammerer(2000)]. The bulk modulus K^a at atmospheric conditions and the density ρ^a of the air phase as well as the gravitational acceleration are chosen from the literature.

5.4.3 Results of the numerical simulation

The results of the numerical simulation presented in this section are mainly based on calculations using the finite element mesh shown in Figure 5.17(b). As can be seen subsequently, the results obtained by means of this mesh are in satisfactory agreement with experimental data. First, the evolution of individual variables in the course of the simulation is presented, indicating various stages of the experiment. In a second analysis the time necessary to reach a steady state for one pressure increment is investigated. Finally, results of the mesh convergence study are given by comparing the air pressures and the vertical soil displacements at the end of the experiment obtained with the different meshes (Figure 5.17).

Simulation of the complete experiment

Figure 5.18 shows the development of the excess air pressure, i.e. the air pressure in excess of the atmospheric pressure, in the course of the experiment. Since the excess air pressure applied at the lower boundary of the domain increases with time (ramp loading), of course also the air pressure in the specimen increases. The appearance of the flow field in the soil is twofold: while it is pronouncedly two-dimensional in the region of the air injection, it is practically one-dimensional in the upper half of the specimen and in particular near the top boundary of the domain. This one-dimensional flow field in

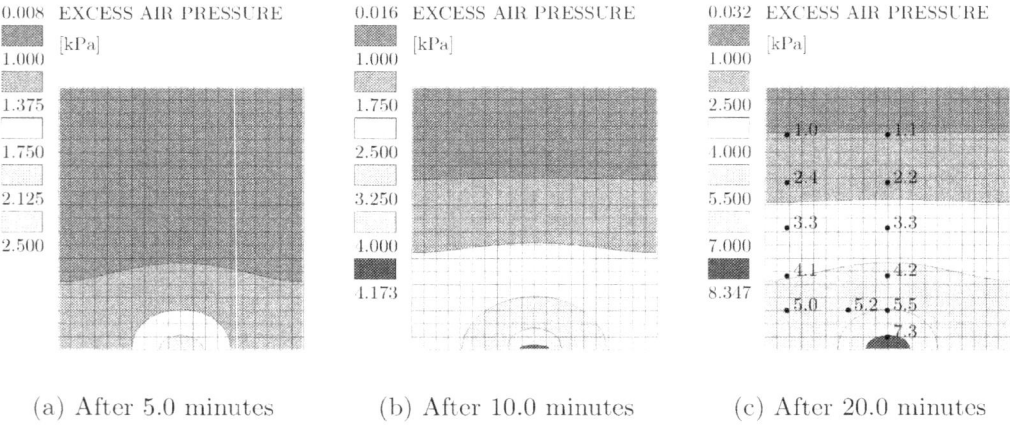

Figure 5.18: Evolution of the excess air pressure in dry sand specimen.

the upper region was used in the experiment to calculate the intrinsic permeability of the sand. Figure 5.18 depicts the calculated air pressure distribution in the soil 5.0, 10.0 and 20.0 minutes after the application of the excess air pressure (start of the laboratory test). As can be seen, the distribution of the air pressure is rather uniform in most parts of the specimen apart from the air injection region. A comparison of the numerical results after 20.0 minutes (Figure 5.18(c)) with available experimental data (dots with numbers) shows very good agreement.

When considering a section in the vertical symmetry plane, the development of the excess air pressure versus the height of the specimen is plotted in Figure 5.19. At the bottom boundary of the domain the increase of the applied excess air pressure with time to the maximum value of 8.5 kPa can be recognised while at the upper boundary the air pressure is constrained to zero (atmospheric pressure) and thus a permeable surface is simulated. In the course of the experiment the air pressure in the whole domain gradually increases. In the symmetry plane the pressure profiles exhibit a parabolic shape. The numerical results (blue solid line) for the final stage of the laboratory test (20.0 minutes)

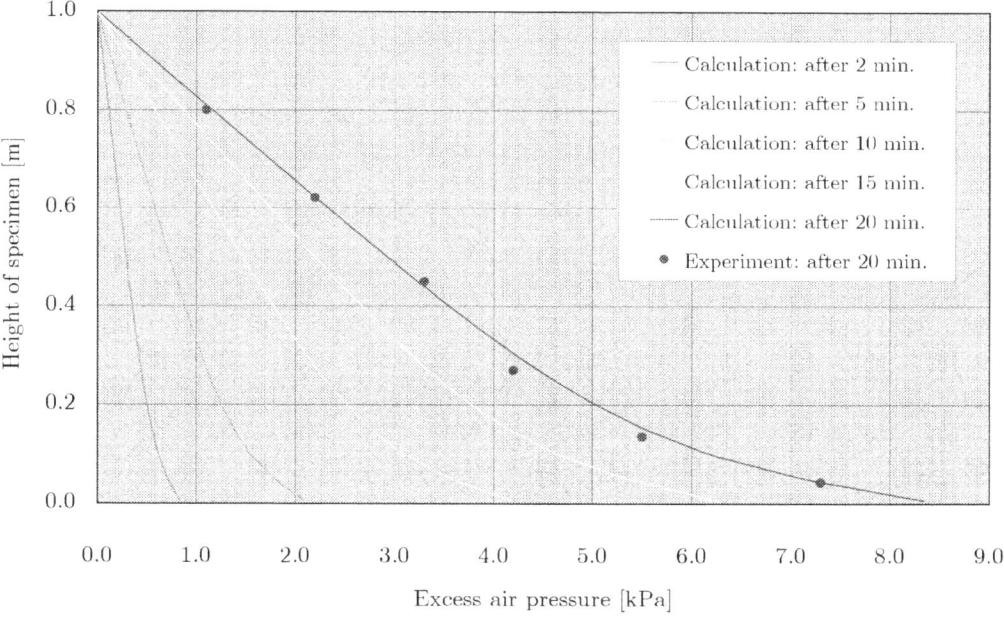

Figure 5.19: Evolution of the excess air pressure considering a section in the symmetry plane.

agree perfectly well with the experimental data (blue dots), as this figure outlines again.

In addition to the pore pressure the soil displacements are chosen as primary variables of the model. Therefore, results of the computed transient displacements of the soil skeleton are shown for two instants of time in Figure 5.20. Expectedly, the airflow induces upwardly directed displacements of the soil. Due to the applied vertical constraint at the top surface of the specimen (see Figure 5.17), the maximum values of the displacements occur at midheight of the domain. Considering the development with time the deformations increase with the applied excess air pressure. However, the absolute values of the displacements are very small, a maximum value of about $2.5 \cdot 10^{-5}$ m is encountered after 20.0 minutes.

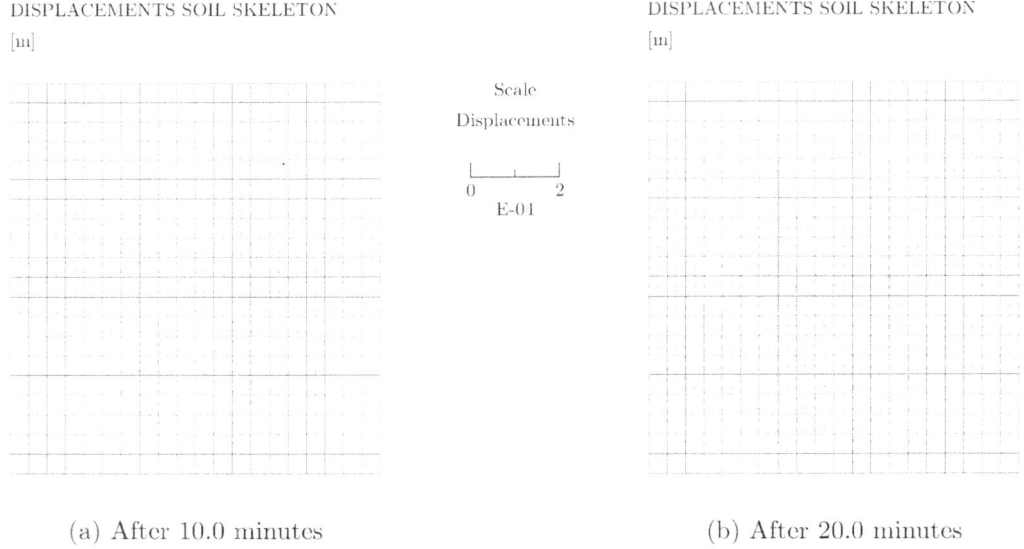

(a) After 10.0 minutes

(b) After 20.0 minutes

Figure 5.20: Evolution of the displacements of the soil skeleton.

The development of the effective vertical stresses after the application of the excess air pressure at the bottom boundary is shown in Figure 5.21. The initial geostatic stress state is specified as follows: at the top of the soil specimen an effective vertical stress of 250.0 kPa is prescribed, simulating the overburden applied in the experiment, which linearly increases with depth to a maximum value at the bottom boundary according to the unit weight of the soil. When taking a look at Figure 5.21(a), only minor deviations from the initial stress state are encountered. According to equation (3.198), the effective stresses decrease due to fluid pressure increases in the domain if the total stresses are assumed to be constant. A decrease of the effective vertical stresses is obvious in the lower half of the specimen, most pronounced changes occurring in the region of air injection (confer Figure 5.21(c)). In the upper part of the modelled domain the effective vertical stresses slightly increase due to the constrained top boundary and the upwardly directed displacements (see Figure 5.20) which induce additional compressive stresses in this area of the soil. As already mentioned, compared to the surface load of 250.0 kPa the changes in the effective stresses due to the applied excess air pressure are rather small. With respect to the initial stress state an increase of the minimum effective vertical stress of 1.1 % and a decrease of the maximum effective vertical stress of 0.8 % is encountered in the course of the simulated time span of 20.0 minutes.

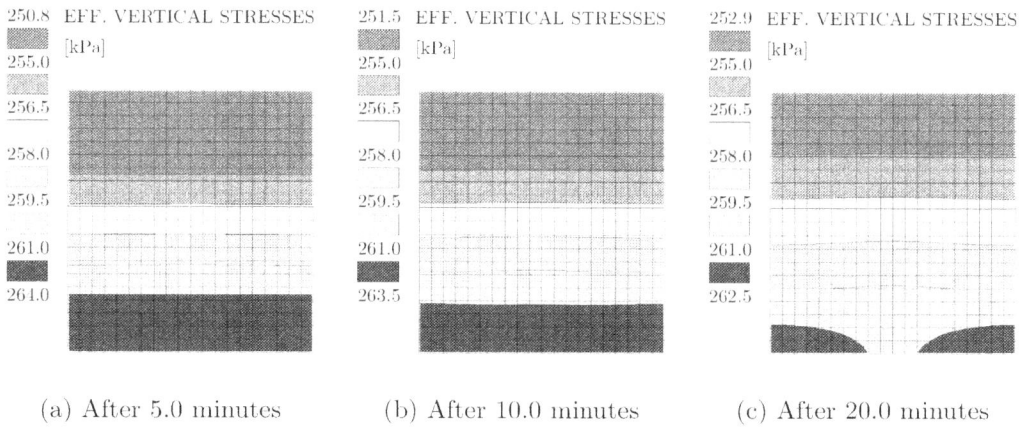

(a) After 5.0 minutes   (b) After 10.0 minutes   (c) After 20.0 minutes

Figure 5.21: Evolution of the effective vertical stresses in the soil skeleton.

Figure 5.22 shows the development of the velocity of the air phase for two instants of time after the application of the excess air pressure. The fluid velocities are plotted in Figure 5.22 in the central Gauss point of each finite element. Since the velocity depends on the pressure gradient of the fluid phase (confer, e.g., Darcy's law in the form (3.353)), it increases with time in the whole domain as the pressure gradient does (see Figure 5.19). The maximum value of the pressure gradient occurs of course in the region of air injection, inducing the highest fluid velocity in this area. With increasing distance from the air injection the velocity of the fluid phase decreases, it is, however, not zero at the upper boundary. In approximately the upper two thirds of the domain the airflow

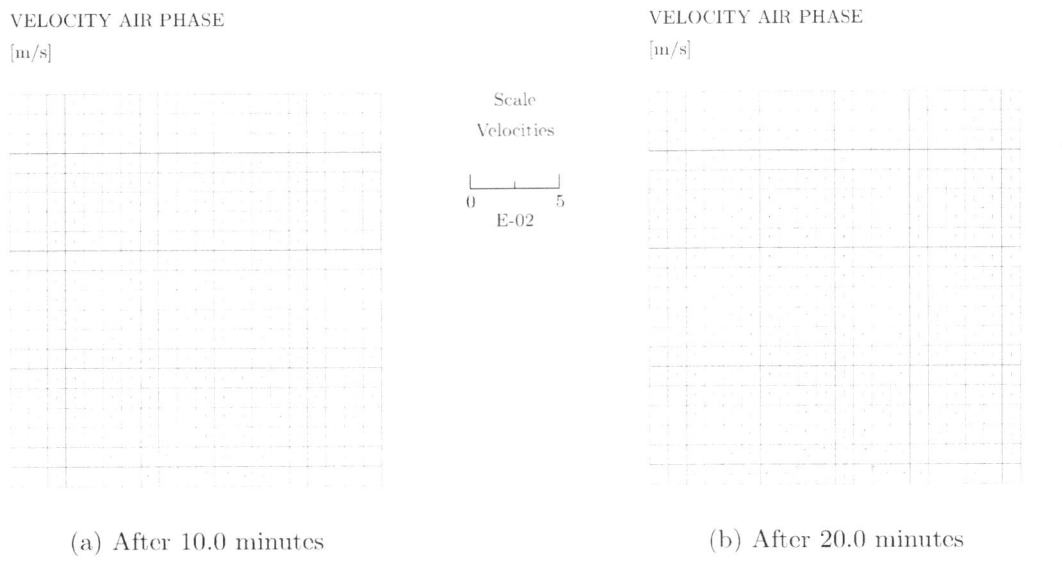

(a) After 10.0 minutes (b) After 20.0 minutes

Figure 5.22: Evolution of the velocity of the air phase.

takes place in vertical or almost vertical direction which means that the flow field in this part of the specimen is practically one-dimensional. A pronounced two-dimensional flow field is only encountered in the vicinity of the air injection region at the lower boundary of the domain.

Simulation of one pressure increment

In the experiment the excess air pressure was applied in steps of 0.1 bar, keeping the pressure constant at each level for 1.0 minute. However, according to [Kammerer(2000)], a steady state was obtained at each pressure level almost immediately after application of the pressure increment. The subsequent Figure 5.23 results from the numerical simulation of one suddenly applied pressure increment (size of 0.425 kPa) which is performed to study the time span necessary to reach a steady state. The excess air pressure evolution in the specimen is shown for three instants of time. After 0.5 seconds still a zero excess air pressure prevails in most of the domain (see Figure 5.23(a)). With increasing time the air pressure propagates over the whole soil specimen. For all points of time the distribution of the air pressure in most parts of the domain is quite uniform. Figure 5.23(c) finally shows the computed air pressure in the soil after steady state conditions are attained.

Chapter 5. Verification of the three-phase formulation

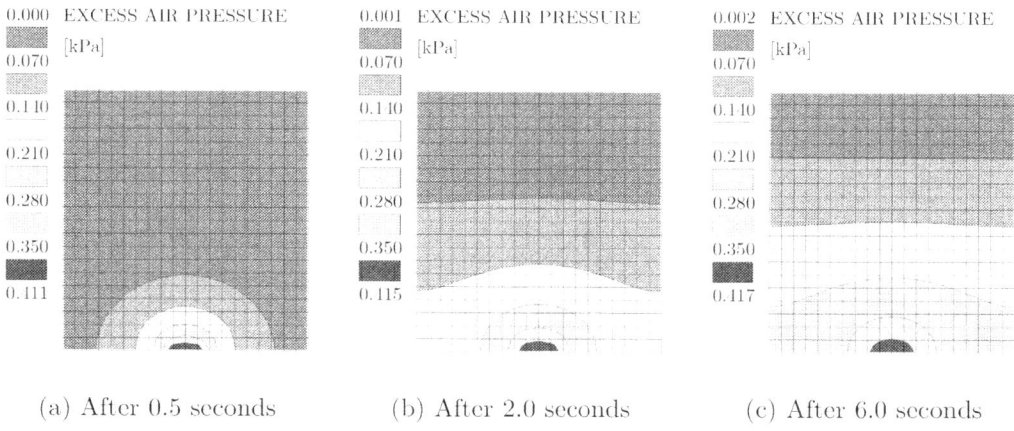

(a) After 0.5 seconds (b) After 2.0 seconds (c) After 6.0 seconds

Figure 5.23: Evolution of the excess air pressure within one load increment.

As this figure indicates, the time span to reach steady state conditions is quite short, only 6.0 seconds are necessary, which is in accordance to the observations made in the experiment.

Mesh convergence study

Finally, some results of the mesh convergence study, performed with the meshes depicted in Figure 5.17, are presented. The distribution of the primary variables in a vertical section in the symmetry plane obtained at the end of the experiment (after 20.0 minutes) is investigated. Figure 5.24 shows profiles of both the excess air pressure and the vertical soil displacements versus the height of the specimen stemming from calculations using the different finite element meshes. As both Figures 5.24(a) and 5.24(b) indicate, the results of all four employed meshes lie very close together. A maximum difference of about 5 % is encountered in the vertical displacements (Figure 5.24(b)). The results of the meshes three and four, mesh three shown in Figure 5.17(c) and mesh four obtained by consistent refinement, virtually coincide for both primary variables air pressure and vertical displacements. Figure 5.24(a) depicts that for the excess air pressure the results from all meshes are in rather good accordance with the experimental data. Figure 5.24(b) outlines that the actual displacements are very small and the maximum value occurs at midheight of the specimen due to the vertical constraints prescribed on the top and bottom boundaries as mentioned previously. A similar behaviour of these variables,

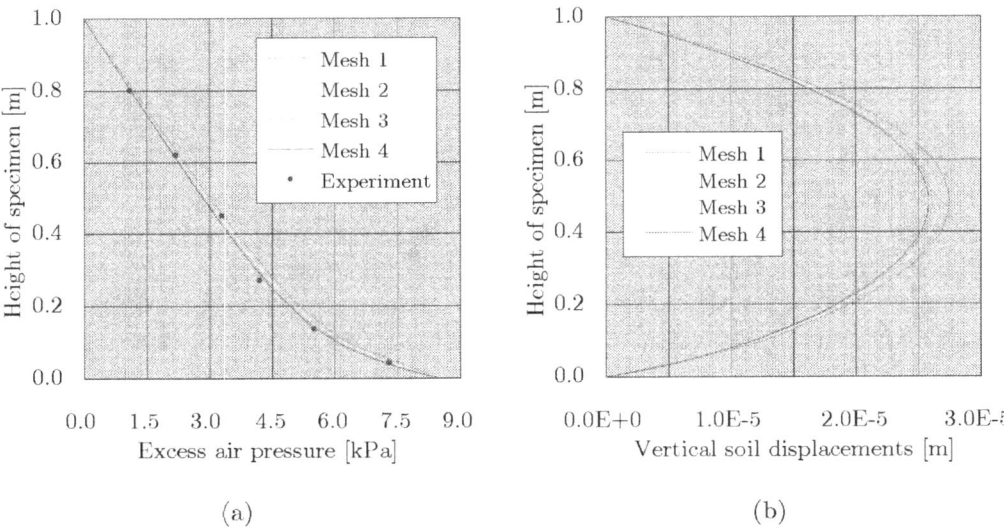

Figure 5.24: Profiles of (a) the excess air pressure and (b) the vertical soil displacements after 20.0 minutes obtained with the individual meshes shown in Figure 5.17.

excess air pressure and vertical soil displacements, is encountered for points of time throughout the whole simulation period.

5.5 One-dimensional drainage of a soil column – two-phase numerical simulation

5.5.1 Description of the problem

Liakopoulos [Liakopoulos(1965)] conducted several experiments on the drainage of water from vertical columns of Del Monte sand. One of these experiments is chosen to investigate the performance of the current numerical model.

A perspex column of 1.0 m height was packed with Del Monte sand and instrumented with a sufficient number of tensiometers to measure continuously moisture tension at various points within the specimen. Prior to the start of the experiment ($t < 0$), steady state conditions were prevailing in the sense that water was continuously added from the top of the column and allowed to drain freely at the bottom through a micropore filter. The flow was carefully regulated until all tensiometers read zero pore pressure. On obtaining this initial condition the water supply from the top was ceased. From then on ($t \geq 0$) the tensiometer readings were recorded and the flow rates at the draining boundary were measured periodically.

Thus, the dewatering process was solely driven by gravity, i.e. the unit weight of the water was the only reason for the drainage of the soil column. The physical parameters of Del Monte sand were determined by Liakopoulos by an independent set of experiments which basically yielded the hydraulic properties of the material. Unfortunately, no mechanical parameters for the soil were given.

It should be mentioned that the laboratory test under consideration has been used by other research groups to check their numerical models [Narasimhan(1978), Schrefler(1988), Schrefler(1993), Klubertanz(1999)]. Within the framework of the European Network *ALERT Geomaterials* this experiment served as basis for a benchmark study [Jommi(1997)].

5.5.2 Numerical model

The current section is concerned with a two-phase numerical simulation of the above described experiment. In contrast to the previously discussed example problems, characterised by a constant degree of saturation during the whole simulation period, the amount of one particular fluid in the voids is now allowed to change in the course of the calculation which augments the model with additional complexity. The main difference lies in the fact that the matrices in, e.g., the equations (4.48) are no longer constant but depend on the degree of saturation or on the permeability with respect to the fluid phase (which itself is a function of the degree of saturation) and thus an iterative solution procedure has to be employed (see Chapter 4).

The domain is discretised by one column of plane strain finite elements. A number of twenty equally-sized elements of 0.05 m are used along the height of the specimen. To allow a comparison of the numerical results with data presented in the literature, this discretisation is chosen according to [Lewis(1998)]. Figure 5.25 shows a plot of the finite element mesh.

Although the degree of saturation is now allowed to change during the calculation, a two-phase formulation can be employed, since the air stresses are kept at atmospheric pressure (passive air assumption). Consequently, a node of a finite element possesses a maximum number of three degrees of freedom as in a consolidation analysis, namely the displacements in the two coordinate directions and one fluid pressure degree of freedom. Displacement degrees of freedom are assigned to eight nodes per element, resulting in a quadratic interpolation of the displacement field, while fluid stresses are restricted to the four corner nodes of each finite element (bilinear interpolation of fluid stresses). As described in [Lewis(1998)], other types of interpolation give practically the same results.

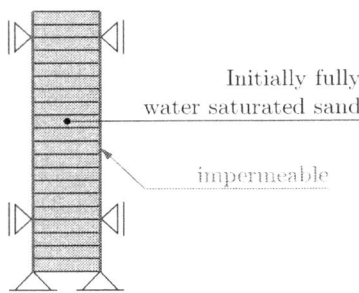

Figure 5.25: Finite element mesh employed for the numerical simulation of Liakopoulos' lab test [Liakopoulos(1965)].

The boundary conditions for both the solid and fluid phases can be recognised from Figure 5.25. With respect to the displacements of the soil skeleton the domain is fully constrained at the bottom and horizontally constrained at the vertical boundaries. For the water phase impervious vertical boundaries are assumed. At the lower boundary any water stresses are constrained to zero and thus a permeable boundary is simulated (the water is allowed to drain at the bottom of the column). Arising stresses in the air phase are not taken into account in this two-phase simulation of the experiment, the air phase is assumed to stay at atmospheric, i.e. zero, pressure (passive air assumption).

Figure 5.25 also indicates the initial conditions of the problem. At the beginning of the experiment both the pore pressure and the degree of saturation correspond to a fully saturated state of the sand, the initial pore pressure is zero and the saturation is one. There is some steady state flow under these initial conditions because the zero gradient in pore pressure does not equilibrate the specific weight of the fluid. The initial conditions for the effective stresses of the soil skeleton are calculated from the intrinsic densities of the constituents, the initial saturation, the porosity and the initial fluid pressure.

The material parameters employed for the numerical simulation of the laboratory test are summarised in Table 5.5. The values are taken from literature describing either the experiment [Liakopoulos(1965)] or its numerical simulation [Schrefler(1988), Lewis(1998)]. Since Liakopoulos did not determine any properties to define the mechanical behaviour of the soil skeleton, the two elastic constants Young's modulus and Poisson's ratio are taken from [Schrefler(1988)]. Both the porosity of the soil and the permeability with respect to the water phase are given in [Liakopoulos(1965)]. The bulk modulus of water and the densities of the constituents are again chosen according to [Schrefler(1988)]. The coefficient of earth pressure is calculated by means of Poisson's ratio, $K_0 = \nu/(1-\nu)$. The equations for the saturation-capillary stress and the relative

Parameter	Symbol	Unit	Value
Young's modulus	E	Pa	$1.3 \cdot 10^6$
Poisson's ratio	ν	-	0.40
Porosity	n	-	0.2975
Permeability water	k^{ow}	m/s	$4.40 \cdot 10^{-6}$
Bulk modulus water	K^w	Pa	$2.0 \cdot 10^9$
Density solid	ρ^s	kg/m^3	2000.0
Density water	ρ^w	kg/m^3	1000.0
Coefficient of earth pressure	K_0	-	0.67
Gravitational acceleration	g	m/s^2	9.81

Table 5.5: Material properties for Liakopoulos' lab test.

water permeability-saturation relationship (valid for degrees of saturation $S^w \geq 0.91$) derived from the experiment by Liakopoulos have the following form [Liakopoulos(1965)]:

$$S^w = 1.0 - 1.9722 \cdot 10^{-11} (p^c)^{2.4279} ,$$
$$k^{rw} = 1.0 - 2.207 (1.0 - S^w)^{1.0121} ,$$
(5.49)

with S^w, p^c and k^{rw} denoting the degree of water saturation, the capillary stress and the relative water permeability coefficient, respectively. The parameters of the implemented relations are chosen to fit these expressions.

In the experiment the water pressures were recorded for a period of 2.0 hours. Consequently, the simulated period of time also amounts to 2.0 hours. The time stepping procedure follows the one given in [Lewis(1998)], that is, for the first hour of the simulated period a time step of $\Delta t = 1.0$ sec is used which is then increased to a value of $\Delta t = 10.0$ sec for the second hour.

5.5.3 Results of the numerical simulation

In the numerical simulation the development of different quantities in the course of the experiment is investigated. Both the distribution of various variables versus the height of the column and their evolution with time are shown in the subsequent figures. Numerical results are compared with experimental data [Liakopoulos(1965)] on the one hand and calculations documented in [Lewis(1998)] on the other hand, the latter especially for the purpose of validating the finite element code. In all the subsequent figures results of the numerical simulation are represented by solid lines, while solid lines with filled circles describe experimental data. For the calculations given in [Lewis(1998)] the symbol '+' is employed.

Distribution of quantities versus column height

For the quantities plotted versus the height of the column, six instants of time in the course of the experiment are shown, each colour representing a certain point of time. Figure 5.26 shows the distribution of the tensile hydrostatic water stresses versus the height of the column. The current numerical solution (solid lines) and the calculations (symbol '+') presented in [Lewis(1998)] agree perfectly well for all points of time. However, both numerical simulations seriously overestimate the measured stresses in the water phase at the very beginning of the experiment (note the difference between the dark blue lines with and without markers). After about 20.0 minutes (green lines) the numerical results are in rather good accordance with the test data. The almost linear distribution of the tensile water stresses versus the height of the soil specimen after 2.0 hours is described by the numerical model very well. The discrepancies between the numerical results and the experimental data in the early test phase are likely to stem from non-linearities not captured by the model. Since the differences become even more pronounced if a rigid soil skeleton is used in the numerical simulation, one may conclude that some improvement of the results could be achieved if a more refined constitutive law would be employed to describe the behaviour of the soil skeleton. However, due to a lack of material parameters for the particular soil no further numerical investigations have been performed.

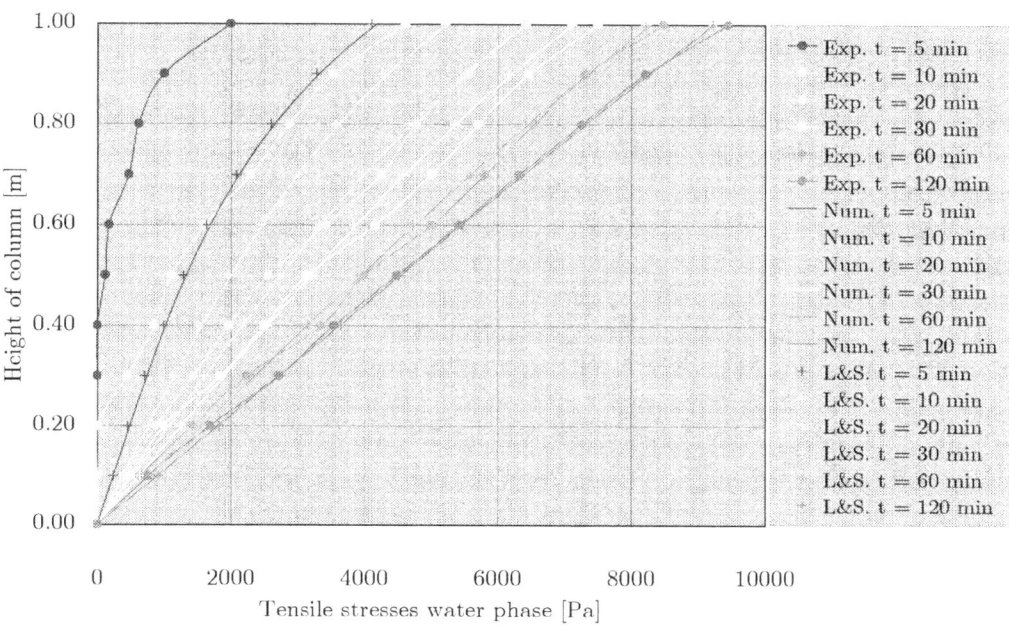

Figure 5.26: Tensile water stresses versus height of the column.

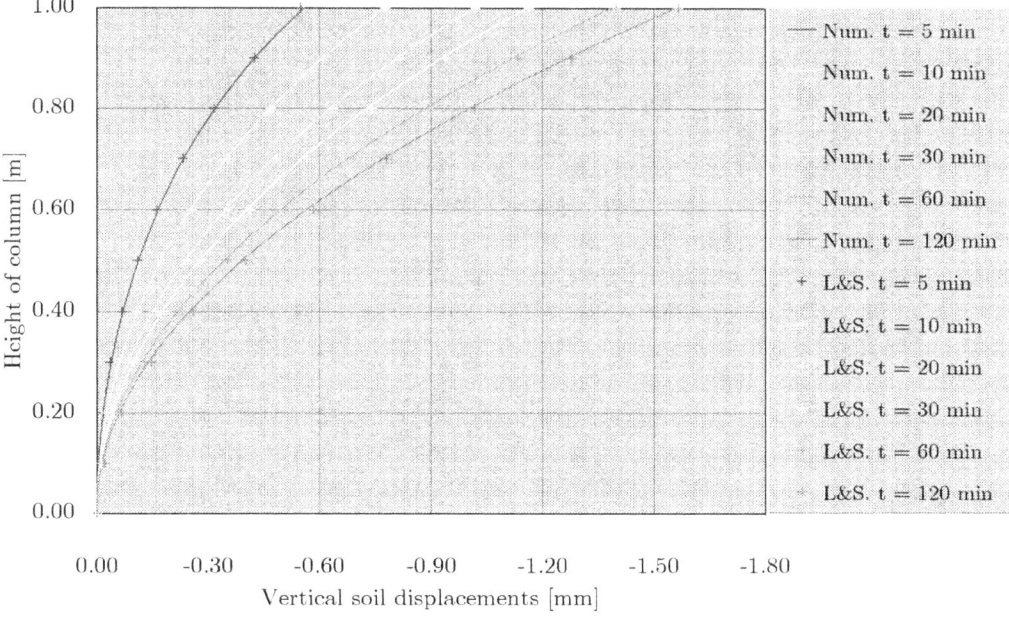

Figure 5.27: Vertical soil displacements versus height of the column.

The distribution of the vertical soil displacements versus the height of the specimen is shown in Figure 5.27 for different points of time. Negative values of displacements refer to settlements of the soil. Since no experimental data are available, the numerical results (solid lines) are compared with the solution (symbol '+') presented in [Lewis(1998)]. The agreement between both calculations is perfectly well for all instants of time. As could be expected, the dewatering process causes compaction of the soil specimen and a maximum settlement of about 1.6 mm is predicted by the numerical simulation. Concerning the distribution of the displacements versus the height of the column, parabolic curves are encountered. The time-dependent changes of the displacements are more pronounced at the beginning of the experiment, with increasing time the settlements grow slower.

Figure 5.28 depicts the evolution of the degree of water saturation versus the height of the soil column for different instants of time. Once again the numerical results (solid lines) are compared with the calculation (symbol '+') in [Lewis(1998)], yielding perfect agreement for the entire simulation period. When taking a look at the individual curves, the degree of saturation gradually decreases with increasing height of the column, no pronounced dewetting front (transition from the saturated to the unsaturated region

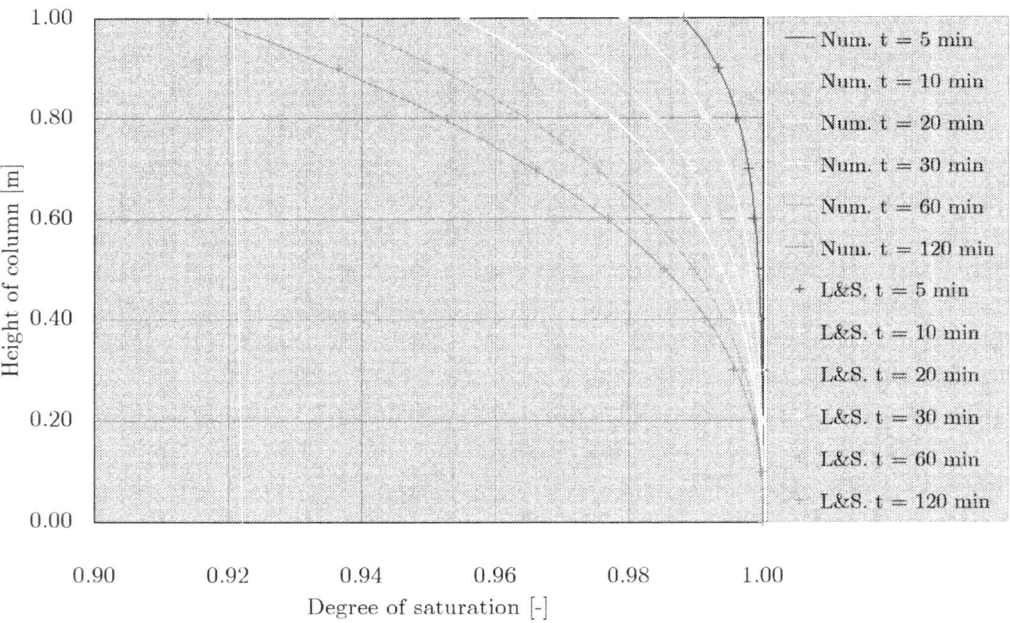

Figure 5.28: Degree of water saturation versus height of the column.

Chapter 5. Verification of the three-phase formulation

in a rather narrow layer of the soil) is encountered. Corresponding to the fluid phase boundary condition the soil stays fully saturated at the bottom of the specimen (the water stresses are constrained to zero to simulate a permeable boundary). At the top of the column the degree of saturation gradually decreases with time to a value of about 92 %. Due to the rather low hydraulic conductivity of the soil under consideration no further dewatering of the specimen is possible due to gravitational effects.

Figure 5.29 shows the distribution of the effective vertical stresses in the soil skeleton versus the height of the specimen. Positive values indicate compression. The stresses plotted are stresses in excess of the initial geostatic stress state, i.e. stresses purely induced by the dewatering process. As is easily recognised, the effective vertical compressive stresses increase with time. A rather pronounced increase occurs at the very early stage of the experiment (in the first 5.0 minutes about one half of the stress increase takes place), later on the growth becomes somewhat more moderate. These additional stresses can be explained as follows: In the course of the calculation the degree of saturation decreases due to a decrease (means an increase in absolute values) of the capillary stress, i.e. of

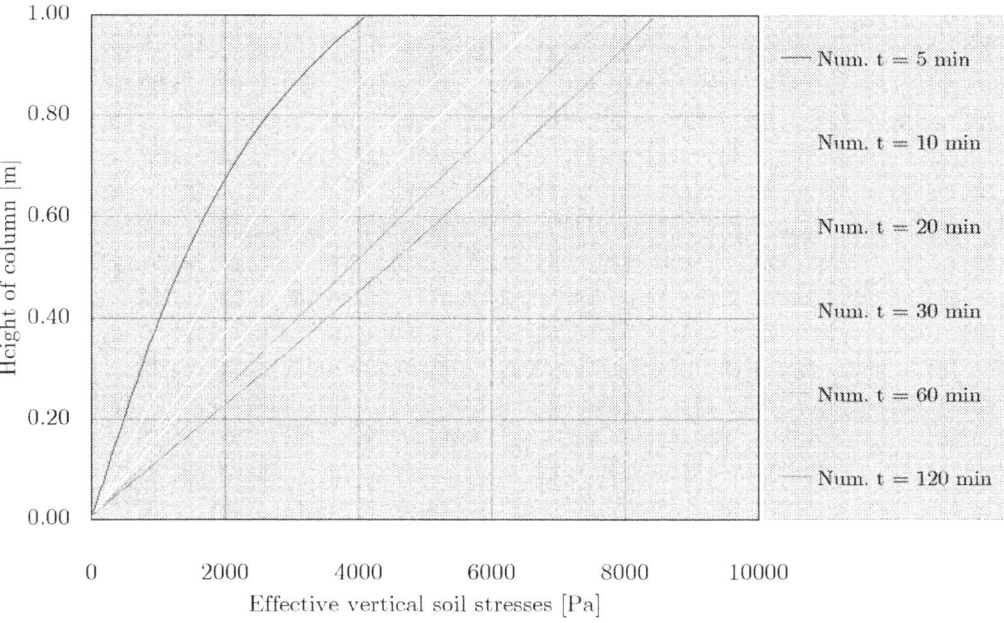

Figure 5.29: Effective vertical compressive soil stresses (stresses in excess of the initial stress state) versus height of the column.

the negative hydrostatic water stress, since the air stress is kept at atmospheric pressure. A decrease in capillary stress, however, acts like an additional hydrostatic pressure onto the soil skeleton, which is the reason for the increase of the effective vertical compressive stresses in the soil skeleton shown in Figure 5.29.

Evolution of quantities with time

If the time history of a variable is depicted, five points at various heights along the column are investigated. Distributions of the above quantities versus time are very much enlightening as will be discussed subsequently. Figure 5.30 shows the evolution of the tensile hydrostatic water stresses with time. These stresses increase both with the height of the column and with time. For all five points (one point located every 20.0 cm along the height of the column) depicted in Figure 5.30 a rather pronounced stress increase is encountered during the first stage of the experiment (about 65 % of the water stresses occur in the first 20.0 minutes of the experiment). All the curves show a very steep slope at the start of the simulation but flatten with increasing time. After 2.0 hours, however, the curves still do not run horizontal which indicates that a steady state is not yet reached. At the end of the simulated period of time five equidistant lines are obtained corresponding to the linear distribution of the hydrostatic water stresses over the height of the column.

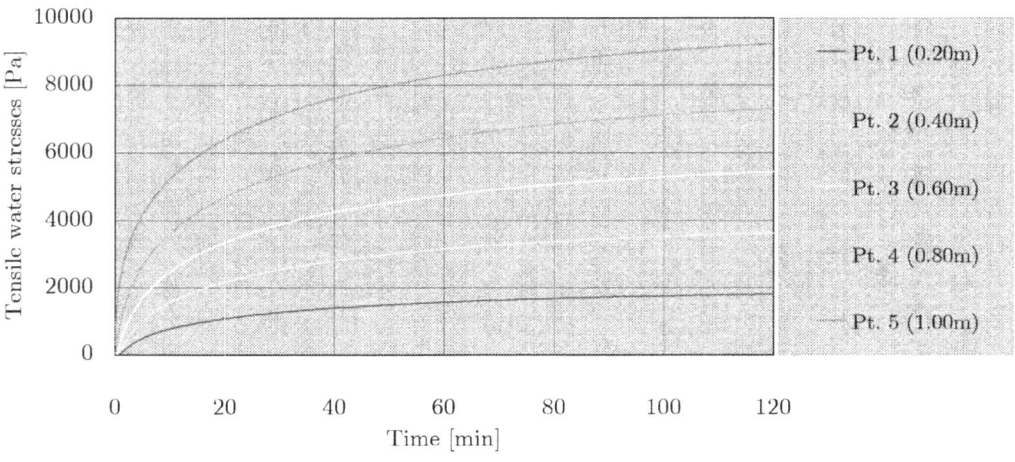

Figure 5.30: Tensile water stress history at five points along the column height.

Figure 5.31 depicts the history of the vertical displacements (settlements). Similar to Figure 5.30 there occur rather pronounced changes in the settlements at the beginning of the experiment. Furthermore, the still non-zero slope of the curves after 2.0 hours indicates that a steady state has not been reached at that time. Contrary to the hydrostatic water stresses, at the end of the experiment the curves of the individual points are not equidistant. The higher the decrease of the water saturation, the larger are of course the occurring displacements. Thus, with increasing height of the column also the distance between the individual lines increases corresponding to the parabolic shape of the vertical displacement curves shown in Figure 5.27.

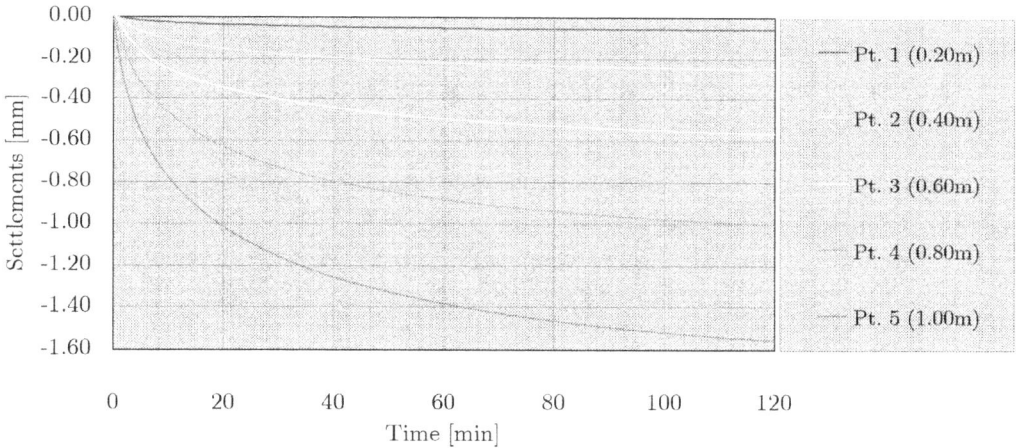

Figure 5.31: History of the vertical displacements (settlements) at five points along the column height.

The evolution of the degree of saturation with time is depicted in Figure 5.32. Compared to the hydrostatic water stresses and the displacements the behaviour of the degree of saturation exhibits a major difference at the very early stage of the experiment. The rather steep curves of the water stresses and the displacements (confer Figures 5.30 and 5.31) indicate rapid changes of these quantities at the beginning of the simulation. Concerning the degree of water saturation, its alteration in the early part of the test is less pronounced. Compared to the 60 to 65 % changes in the water stresses or the displacements, only about 35 % of the changes in the degree of saturation occur in the first 20.0 minutes of the simulated period of 2.0 hours. This indicates that the degree of saturation adjusts much slower to certain prescribed boundary conditions than the water

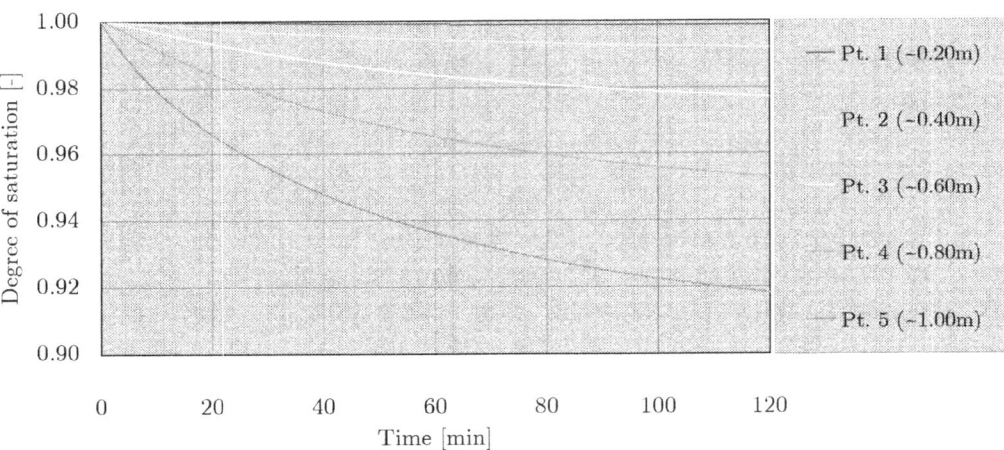

Figure 5.32: Saturation history at five points along the column height.

stresses or displacements. Or in other words, it takes considerably more time to reach a steady state when looking at the degree of saturation.

The history of the effective vertical compressive stresses in the soil skeleton is shown for five points along the height of the column in Figure 5.33. The effective soil stresses exhibit a behaviour similar to the one of the hydrostatic water stresses, that is, pronounced

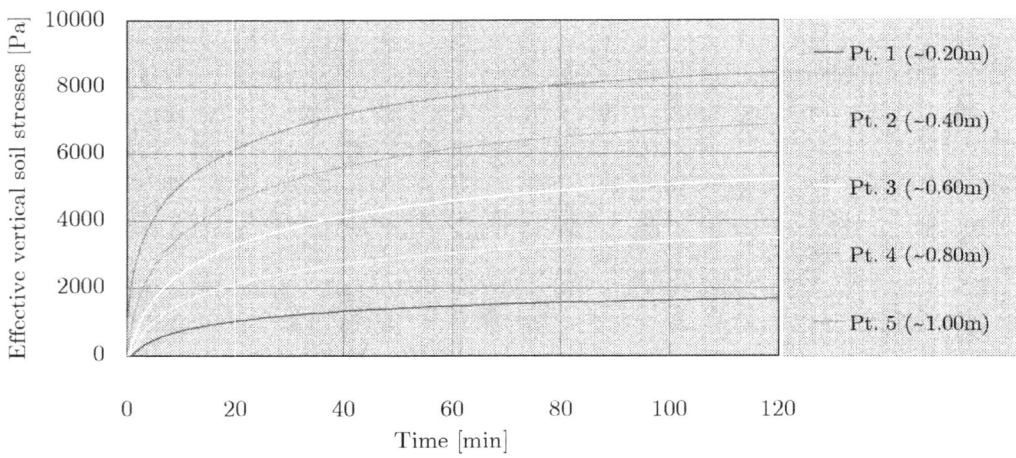

Figure 5.33: History of effective vertical soil stresses at five points along the column height (compression positive).

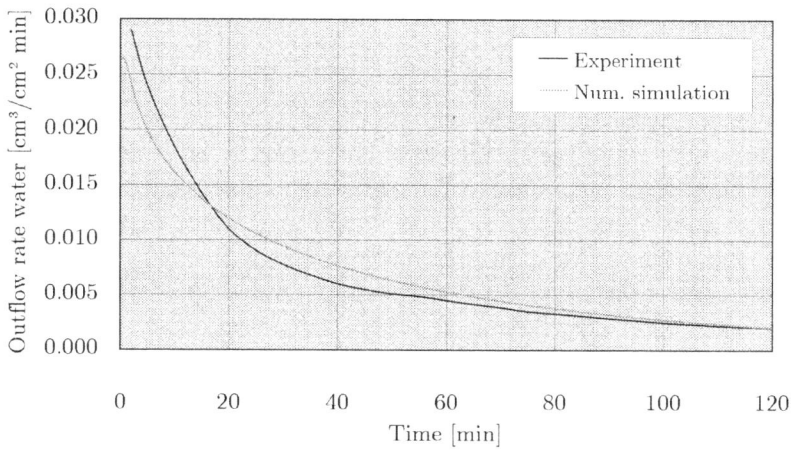

Figure 5.34: Water outflow rate at the bottom of the column.

changes at the early stage of the experiment (almost 70 % of the stress changes occur in the first 20.0 minutes) and equidistant curves at the end of the simulation period.

Finally, time histories for both water outflow rate (Figure 5.34) and water outflow (Figure 5.35) are considered. For these two quantities data from the laboratory test [Liakopoulos(1965)] are available and thus, the numerical results (red solid line) are compared to these experimental data (black solid line). Figure 5.34 shows the outflow rate

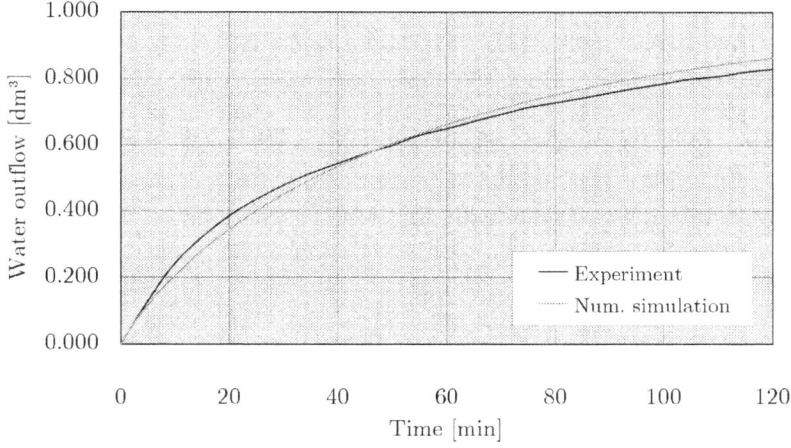

Figure 5.35: Water outflow at the bottom of the column.

of water at the bottom of the column versus time. In comparison with the experiment the computed outflow rate is somewhat too small in the early phase of the experiment (first 15.0 minutes), while it is slightly too large for the rest of the simulated period of time. However, numerical results and experimental data are in good accordance after about 100.0 minutes. The overall behaviour of the outflow rate is captured by the numerical model quite well.

Figure 5.35 shows the evolution of the water outflow (accumulated outflow rates) versus time. The numerical results agree quite well with the experimental data and a total water outflow of about 0.9 dm^3 is predicted. However, the graphs of both water outflow rate and water outflow suggest that steady state conditions are not reached after the duration of the experiment of 2.0 hours (existing non-zero slope of the curves after 2.0 hours).

Choice of different material parameters

As Figure 5.26 outlines, the agreement between numerical results and experimental data at the very early stages of the laboratory test is rather poor. Therefore, the influence of the choice of some parameters on the numerical results is investigated subsequently in brief. The value specified for Young's modulus clearly is already rather low for the sand under consideration. An increase of Young's modulus, however, would result in an even higher overestimation of the hydrostatic water stresses at the beginning of the experiment (as a calculation with a rigid soil skeleton showed) and is therefore not considered. A second parameter not determined in the experiment is Poisson's ratio. The influence of a Poisson's ratio of $\nu = 0.20$ is thus investigated and the distribution of the tensile water stresses over the height of the column is shown in Figure 5.36 for three instants of time (broken lines). The numerical results obtained with the smaller Poisson's ratio lie somewhat closer to the experimental data.

Water encountered in field situations or used in laboratory tests might include dissolved air. This inclusion of air, however, dramatically increases the compressibility of the water-air mixture (compare Figure 3.8). The bulk modulus of water of $K^w = 2.0 \cdot 10^9$ Pa, chosen according to [Lewis(1998)], is only valid for pure water. If air is dissolved in water, a bulk modulus of $K^w = 1.0 \cdot 10^6$ Pa may be considered as realistic [Klubertanz(1999)]. Thus, a further calculation is performed to investigate the influence of the fluid's bulk modulus. Figure 5.36 contains results of this simulation where both Poisson's ratio and the bulk modulus of water have been changed (solid lines). These numerical results lie

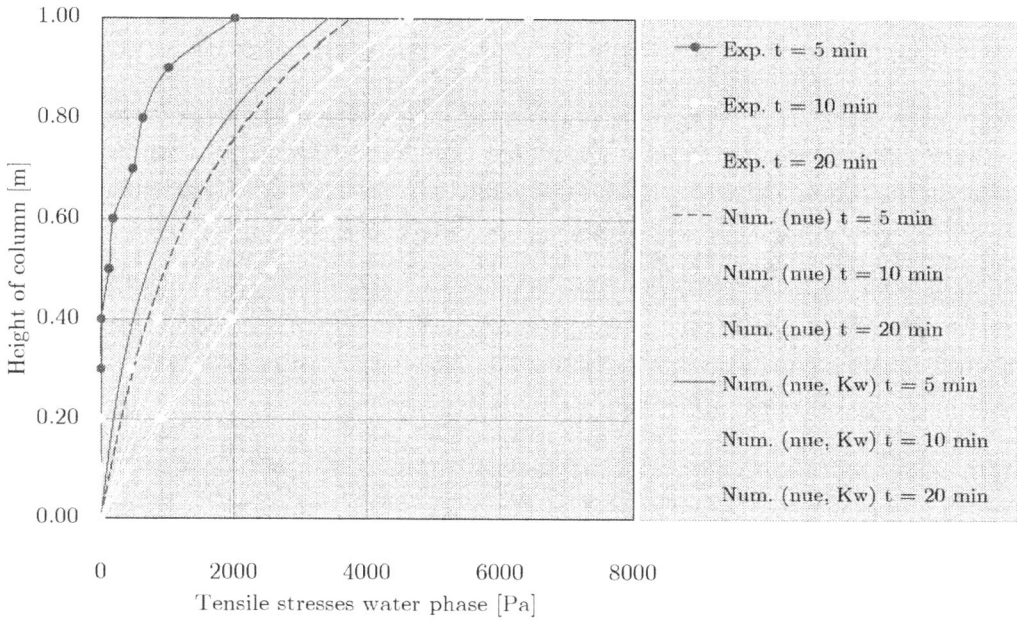

Figure 5.36: Tensile water stresses versus height of the column after changing the material parameters.

even closer to the experimental data at the very early stage of the test (5.0 minutes), however, for intermediate points of time the deviation is somewhat larger (20.0 minutes). Obviously, some improvements are possible by choosing slightly modified material parameters. Nevertheless, a more sophisticated model for the soil skeleton would probably be necessary.

5.6 One-dimensional drainage of a soil column – three-phase numerical simulation

5.6.1 Numerical model

The laboratory test described in Section 5.5.1 can also be simulated numerically employing the complete three-phase model (remember that in Section 5.5 the simpler two-phase formulation has been used implying constant atmospheric air pressure during the calculation). Hence, now the transient changes of the air stresses resulting from the dewatering process are taken into account.

The test set-up and the numerical model basically are the same as described in Sections 5.5.1 and 5.5.2. The mesh for the numerical simulation consists of twenty plane strain finite elements as shown in Figure 5.25. Since a three-phase formulation is now used, a node of a finite element possesses a maximum number of four degrees of freedom, namely the displacements in the two coordinate directions and the two fluid stresses for the constituents water and air. Displacement degrees of freedom are assigned to eight nodes per element (quadratic interpolation of displacements), fluid stress degrees of freedom to the four corner nodes (bilinear interpolation of fluid stresses). Using either eight or nine displacement nodes yields the same results.

The boundary conditions characterising the problem are shown in Figure 5.25. With respect to the displacements the domain is fully constrained at the bottom and horizontally constrained at the vertical boundaries. For both fluid phases impervious vertical boundaries are assumed. At the lower surface of the domain both fluid stresses are constrained to zero, meaning that this boundary is permeable with respect to the fluid phases. At the upper surface only the air stresses are constrained to zero. Thus, atmospheric air pressure prevails along the top and bottom boundaries during the calculation. It should be mentioned that concerning the fluid phases only stresses in excess of the atmospheric air pressure are considered. Since the atmospheric air pressure cancels out in equation (3.209) for the capillary stress, the actual value of the atmospheric pressure may be ignored.

The initial conditions of the problem, as depicted in Figure 5.25, correspond to a fully water saturated state of the soil column. Both fluid pressures are zero, the degree of water saturation is one. The initial stress state of the soil skeleton is determined according to the given material properties.

The material parameters for the example problem are summarised in Table 5.5. Additional properties necessary for the three-phase calculation, i.e. parameters describing the second fluid phase (air), are taken from the literature [Lewis(1998)] and given in Table 5.6. The permeability with respect to the air phase is calculated from the intrinsic permeability of the soil according to [Lewis(1998)]. The bulk modulus of air under atmospheric conditions and the density of the air phase are also chosen from the literature. According to [Lewis(1998)], a minimum relative air permeability coefficient of $k_{min}^{ra} = 1.0 \cdot 10^{-4}$ is specified, that is, a very small but finite value for the air permeability exists even for almost fully saturated states of the soil. This physically implies that a

Parameter	Symbol	Unit	Value
Permeability air	k^{oa}	m/s	$3.20 \cdot 10^{-7}$
Bulk modulus air	K^a	Pa	$1.0 \cdot 10^5$
Density air	ρ^a	kg/m^3	1.295

Table 5.6: Additional material properties for the three-phase numerical simulation of Liakopoulos' lab test.

minor flow of air is present even at quite high water saturations of the soil which can be imagined when thinking of the initially prevailing water inflow at the top surface of the specimen and the facts that (i) some pores might not be filled with water and (ii) a small amount of air might be dissolved in the water. Introducing this lower limit for the relative air permeability yields considerably more reasonable results for the air pressure since oscillations in the region of the dewetting front are avoided without seriously influencing any other quantities like soil or water stresses, displacements or saturations.

In the laboratory test conducted by Liakopoulos the water stresses were recorded for two hours. As can be seen from the two-phase numerical simulation, at the end of this period of time steady state conditions are approximately obtained for the hydrostatic water stresses but not for most of the other quantities of interest. Therefore, it seemed to be interesting to investigate a longer time span in the numerical simulation until all variables attain steady state conditions which turned out to be the case after about eighty hours. The time increments within this period of time are increased from $\Delta t = 1.0$ sec to $\Delta t = 3600.0$ sec in the course of the calculation.

Two main topics are discussed by means of the subsequent numerical results: (i) A comparison of the results obtained from the two-phase and the three-phase model is given and (ii) the time-dependent development of the individual variables until attaining steady state conditions is investigated.

5.6.2 Results of the numerical simulation

In the numerical simulation of the described experiment the time-dependent development of the individual variables is investigated until all quantities reach steady state conditions. Both the distribution of variables versus the height of the column and their evolution with time are plotted in the subsequent figures in analogy to Section 5.5.3. The numerical results of the current three-phase simulation (solid lines) are compared with results obtained from the two-phase model (dashed lines).

Distribution of quantities versus column height

For quantities plotted versus the height of the column the results of the current three-phase calculations are shown for nine instants of time, each colour representing a certain point of time. The first six solid lines refer to points of time within the duration of the experiment of 2.0 hours and thus can be compared with the two-phase results given by the dashed lines.

Figure 5.37 shows the distribution of the tensile hydrostatic water stresses versus the height of the column. In the early phase of the simulated period of time the differences between the two models (two-phase and three-phase formulation, respectively) are quite substantial, e.g., after 10.0 minutes the maximum hydrostatic water stress (top of the

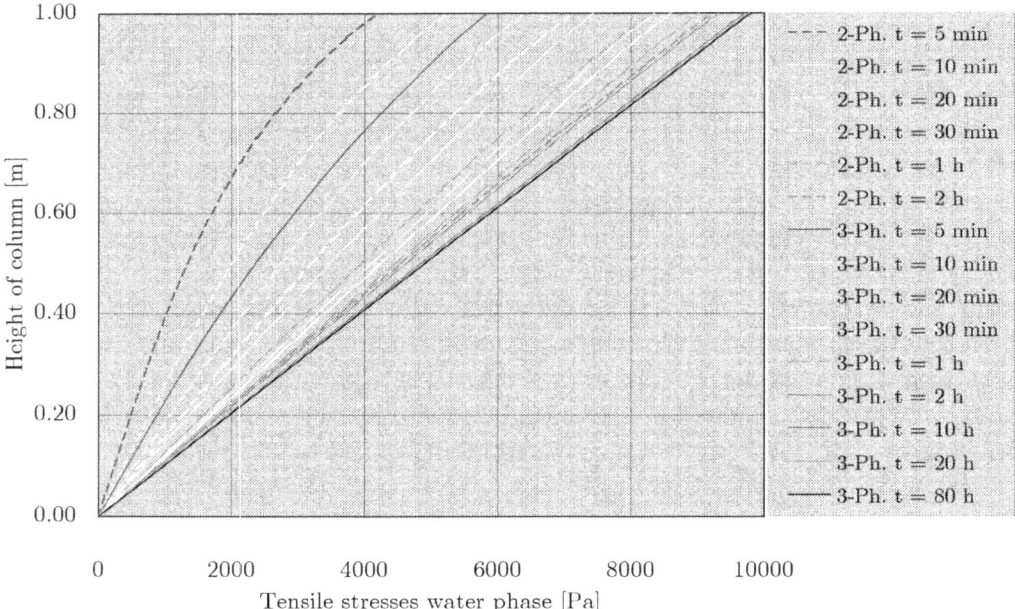

Figure 5.37: Tensile water stresses versus height of the column (three-phase model).

column) obtained with the three-phase model is about 40 % larger than the one resulting from the two-phase model (compare light blue solid and dashed lines). However, these discrepancies decay in the course of time and after 2.0 hours both formulations yield approximately the same hydrostatic water stress distribution along the height of the soil column (red lines).

When taking a look at Figure 5.26, the three-phase model overestimates the experimental data at the beginning of the experiment even more than the two-phase model. As mentioned in [Gawin(1996)], the better agreement between experimental data and the two-phase formulation may result from the fact that some material parameters were determined in [Schrefler(1988)] by a numerical 'fitting' procedure employing a two-phase model. After 2.0 hours steady state conditions are not yet attained for the hydrostatic water stresses as can be seen from Figure 5.37. A minor increase of the tensile water stresses is encountered until no further changes occur after about 80.0 hours. The final distribution of the water stresses versus the height of the column is completely linear. Along the height of the soil specimen the tensile water stresses increase from zero at the bottom boundary (prescribed boundary condition) to a value of $p^w = \rho^w g h \approx 9810.0$ Pa at the top surface.

Figure 5.38 shows the development of the hydrostatic stresses in the air phase versus the height of the column in the course of the simulated time span. It should be noted that only stresses in the air phase in excess of the atmospheric air pressure are plotted in this figure. Air stresses occurring in the calculation are tensile stresses (see Figure 5.38) and thus the hydrostatic stresses in the air phase are increased above atmospheric pressure during the simulation period (decreased below atmospheric pressure in absolute values). As can be seen, there exists a clear peak of the excess air stresses at that level of the soil specimen where the degree of water saturation starts decreasing below fully saturated conditions. Immediately below this peak oscillations of the air stresses may occur. According to [Lewis(1998)], in the current calculation these oscillations are avoided by introducing a lower limit for the relative air permeability coefficient of $k_{min}^{ra} = 1.0 \cdot 10^{-4}$, which may be admissible from a physical point of view (confer explanation in Section 5.6.1). Even in the highly water saturated regions of a soil specimen a very small but finite amount of air is assumed to be present and thus the degree of air saturation is not exactly zero (residual degree of air saturation). This existence of the air phase in the nearly saturated parts of the column not only seems to justify the assumption of a small non-zero air permeability in these regions but also explains the

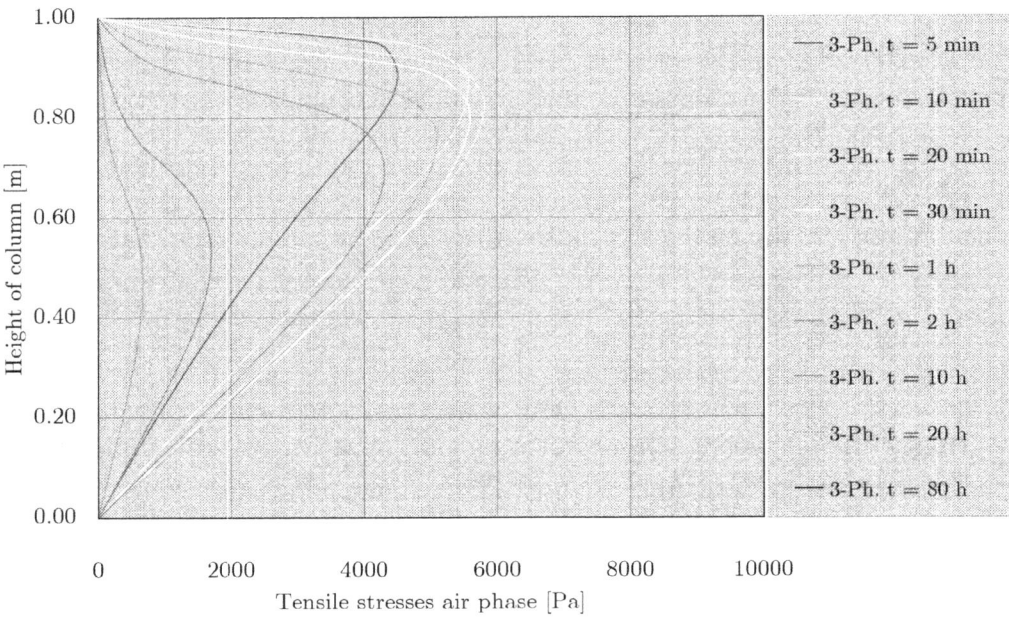

Figure 5.38: Tensile air stresses versus height of the column (three-phase model).

presence of air stresses in the water saturated area of the specimen. With increasing time the location of the peak of the excess air stresses moves down the soil column and the maximum value of the tensile air stress decreases. Hence, with proceeding time the tensile excess air stresses decay, yielding atmospheric stresses in the air phase along the whole column when a steady state is obtained. Consequently, the black solid line in Figure 5.38 (indicating the results at time $t = 80.0$ hours) coincides with the line of zero tensile air stresses.

Figure 5.39 shows the development of the capillary stress versus the height of the soil column in the course of the simulated period of time. Since, according to equation (3.209), the capillary stress is defined as the difference between the hydrostatic stresses in the air and water phases, $p^c = p^a - p^w$, this figure follows straightforward from the previous Figures 5.38 and 5.37. It should be mentioned that capillary stresses are compressive stresses which act onto the soil skeleton like an additional hydrostatic pressure. In the course of the calculation the capillary stress decreases (i.e. the capillary pressure increases), starting at the top of the specimen and spreading over the soil column with

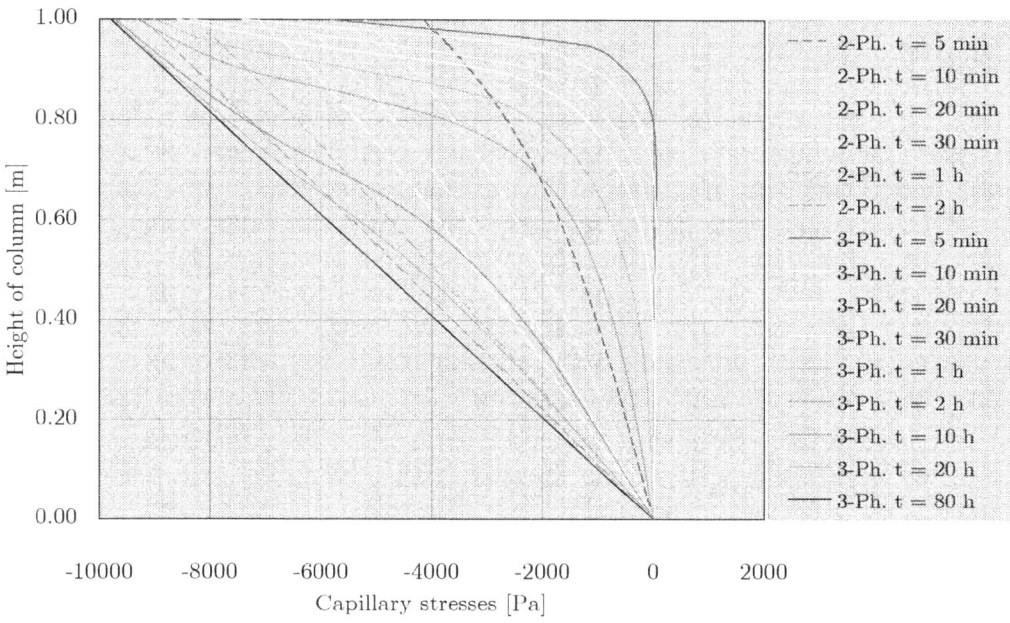

Figure 5.39: Capillary stresses versus height of the column (three-phase model).

time, which of course accompanies the spreading of the unsaturated region in the soil. After attaining steady state conditions the capillary stress is linearly distributed along the height of the specimen (black solid line in Figure 5.39) which results from the atmospheric air stresses and the linear distribution of the tensile hydrostatic water stresses at the final stage (confer Figures 5.38 and 5.37).

When considering a two-phase formulation, the air stresses are kept at atmospheric pressure (passive air assumption). Consequently, the capillary stress is equal to the negative water stress, $p^c = -p^w$. Hence, Figure 5.39 contains the capillary stress distribution obtained from the two-phase model, that is, the negative water stress distribution. Compared with the two-phase formulation the three-phase model yields a steeper gradient of the capillary pressure at the beginning of the dewatering process. The region with a capillary pressure increase is concentrated next to the upper boundary (blue solid lines), while when employing the two-phase model the whole specimen is affected by changes in the capillary stress from the start of the simulation period (dashed blue lines). This development of the capillary stress obtained from the two-phase simulation of course determines the distribution of the degree of saturation which in this case exhibits no pronounced dewetting front as can be seen from Figure 5.28.

Figure 5.40 shows the distribution of the vertical displacements (negative values denote settlements) versus the height of the column for various instants of time. Both two-phase and three-phase models yield parabolic shapes of curves when plotting the settlement versus the column height. Similar to the hydrostatic water stresses, the three-phase model predicts larger settlements than the two-phase formulation at early stages of the simulation period. When employing the three-phase model, about 55 % of the maximum settlements occur in the first 5.0 minutes, with the two-phase model only about 35 % exist after the respective time interval. Considering the displacements, the results obtained with the various models after 2.0 hours differ somewhat more than the ones of the water stresses; the former exhibit deviations of approximately 5 % while only about 1 % difference is encountered for the latter. However, concerning the actual values of the settlements, after the 2.0 hours of the experiment only very small increases occur. The maximum predicted compaction of the soil specimen amounts to about 1.7 mm.

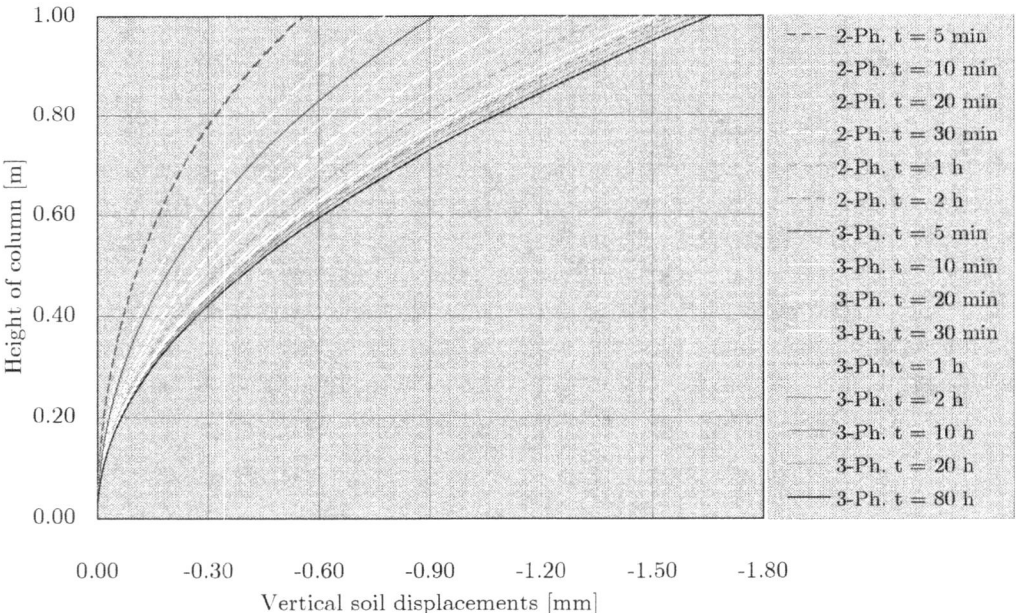

Figure 5.40: Vertical soil displacements versus height of the column (three-phase model).

Figure 5.41 depicts the development of the degree of saturation versus the height of the column. In contrast to the two-phase model, the three-phase formulation predicts a rather pronounced dewetting front, i.e. the transition from the saturated to the unsaturated region of the soil occurs in a very narrow zone (see solid lines for the first 2.0 hours). Immediately from the start of the simulation the two-phase model does not yield any dewetting front at all but the whole specimen is influenced by the dewetting process (dashed lines). This yields curves with rather smooth shapes and gradually increasing gradients obtained from the two-phase model, while the three-phase formulation predicts curves which all exhibit a very steep gradient at a certain level of the column.

As shown in Figure 5.41, the dewetting front moves down the sand column with increasing time, that is, the degree of water saturation of the soil decreases from a fully saturated state, starting at the top of the column and moving down as time passes by. The individual lines indicate the position of the dewetting front for various instants of time, e.g., 30.0 minutes after the start of the simulation (yellow solid line) the lower 80.0 cm of the column are still fully saturated. Moving down the soil column with the

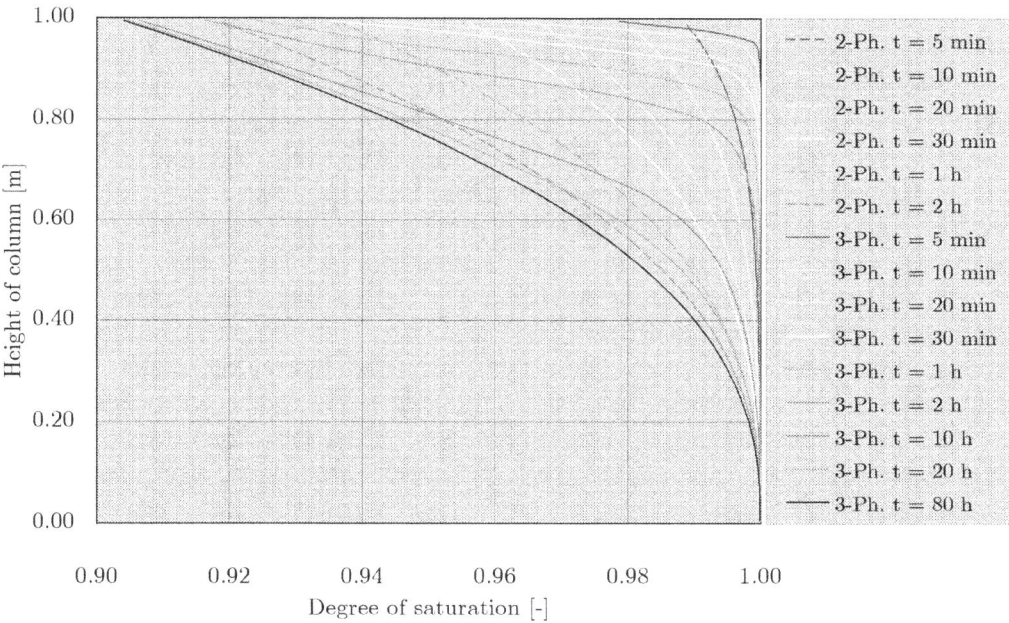

Figure 5.41: Degree of water saturation versus height of the column (three-phase model).

dewetting front is the peak in the excess hydrostatic air stresses as Figure 5.38 indicates. Since the upper boundary is permeable with respect to the air phase, air can enter the pores resulting in an increase of the hydrostatic air stresses to the atmospheric pressure, i.e. the excess hydrostatic air stresses vanish with time.

After 2.0 hours the lower 60.0 cm of the column are still fully water saturated and the degree of water saturation at the top of the specimen amounts to about 92 %. An extension of the numerical simulation to attain steady state conditions shows that for the particular soil under consideration the column remains fully saturated at the bottom (corresponding to the boundary conditions prescribing both fluid stresses to be equal to zero and thus preventing any capillary stresses) and the degree of saturation does not fall below 90 % at the top.

Figure 5.42 depicts the development of the effective vertical stresses in the soil skeleton versus the height of the specimen for various instants of time. Positive values indicate compressive stresses. The stresses plotted are stresses in excess of the initial geostatic stress state, i.e. they are purely induced by the dewatering process. The effective verti-

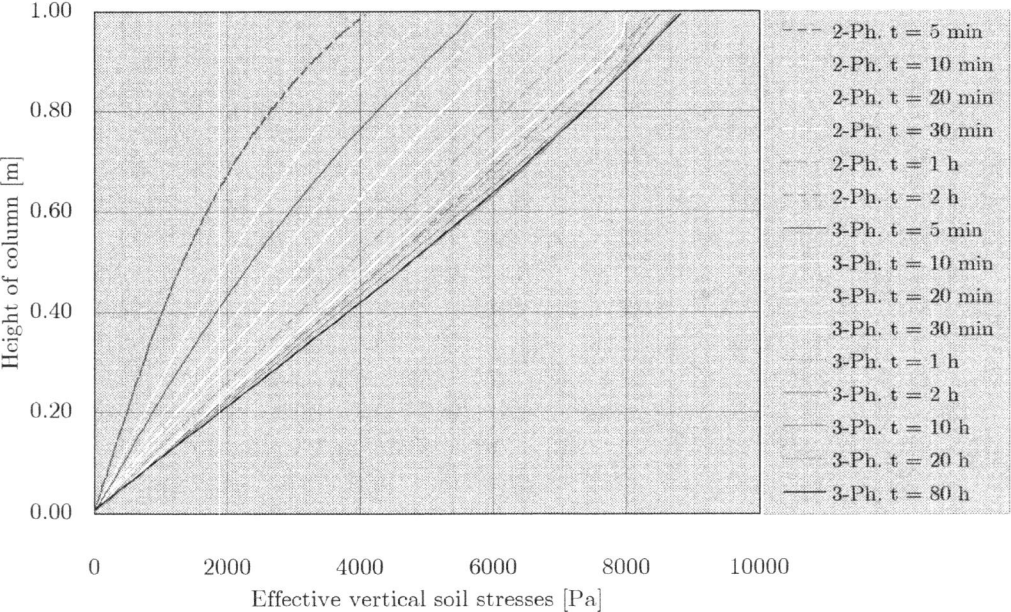

Figure 5.42: Effective vertical compressive soil stresses (stresses in excess of initial stresses) versus height of the column (three-phase model).

cal compressive stresses increase with time. The main part of this stress increase (about 65 %) occurs within the first 5.0 minutes. After 2.0 hours the changes of the effective vertical stresses are almost completed. This can be explained as follows: According to equation (3.199), the effective stresses in the soil skeleton are obtained by subtracting the hydrostatic fluid stresses (weighted by the respective degree of saturation) from the total stresses. Since the initial stress state is compressive, a subtraction of the tensile fluid stresses increases these compressive stresses. The degree of air saturation is rather low in the whole domain (even in the unsaturated region it only amounts to less than 10 %). Therefore, basically the portion of the water stresses causes the increase of the effective soil stresses. This fact can easily be recognised when comparing the solid lines for corresponding points of time in the Figures 5.42 and 5.37. These curves are practically identical, indicating that the soil stress increase is mainly due to the changes in the water stresses. Only in the topmost part of the column the corresponding curves in the Figures 5.42 and 5.37 differ somewhat, resulting from the degree of saturation which also enters the equation of the effective soil stresses. In this upper region the degree of saturation is reduced which yields a smaller increase of the effective soil stresses.

Evolution of quantities with time

If the time history of a variable is plotted, five points located at various heights along the soil column are investigated. For attaining steady state conditions for all the individual variables, a time span of approximately 80.0 hours is necessary. It should be mentioned that when considering, e.g., the hydrostatic stresses in the water phase the 2.0 hours time span of the experiment is sufficient to obtain more or less steady state conditions (confer Figure 5.37). However, quantities like the air stresses or the degree of saturation are far away from having reached a steady state at that point of time.

Figure 5.43 shows the evolution of the tensile hydrostatic water stresses with time until attaining steady state conditions. These tensile stresses increase both with the height of the column and with time. For all five points (one point located every 20.0 cm along the height of the column) the major part (almost 85 %) of the stress increase takes place in the first 20.0 minutes of the simulated time span. Thus, the three-phase model yields an even more pronounced stress increase in the first stage of the simulation period than the two-phase formulation. As can be seen, after 2.0 hours complete steady state conditions are not yet obtained, the water stresses slightly increase until a time of 80.0 hours. At the final stage the five curves exhibit zero slopes and the equidistance

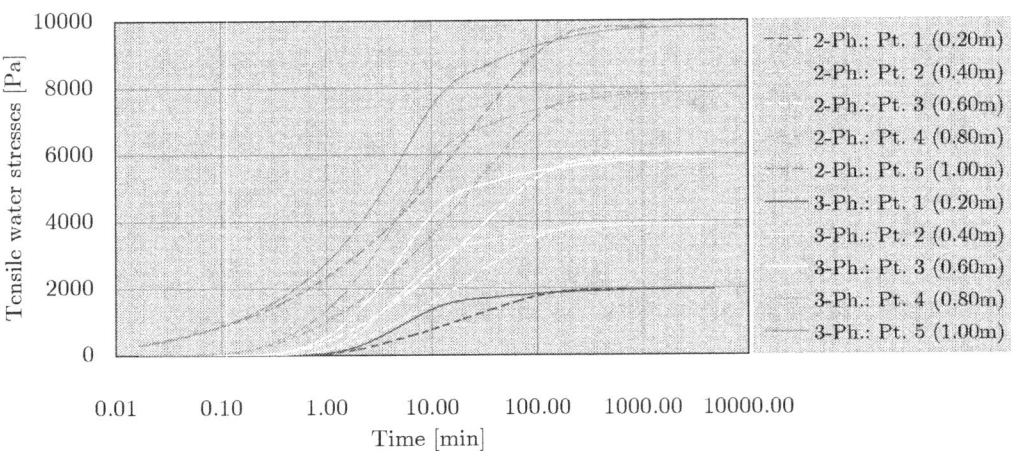

Figure 5.43: Tensile water stress history until attaining a steady state.

of the lines corresponds to the linear distribution of the hydrostatic water stresses versus the column height (see Figure 5.37). The zero slope of the curves indicates that the hydrostatic water stresses do not change with time anymore and thus steady state conditions are finally attained.

Figure 5.44 depicts the evolution of the tensile hydrostatic air stresses with time. One should remember that the plotted stresses are stresses in excess of the atmospheric

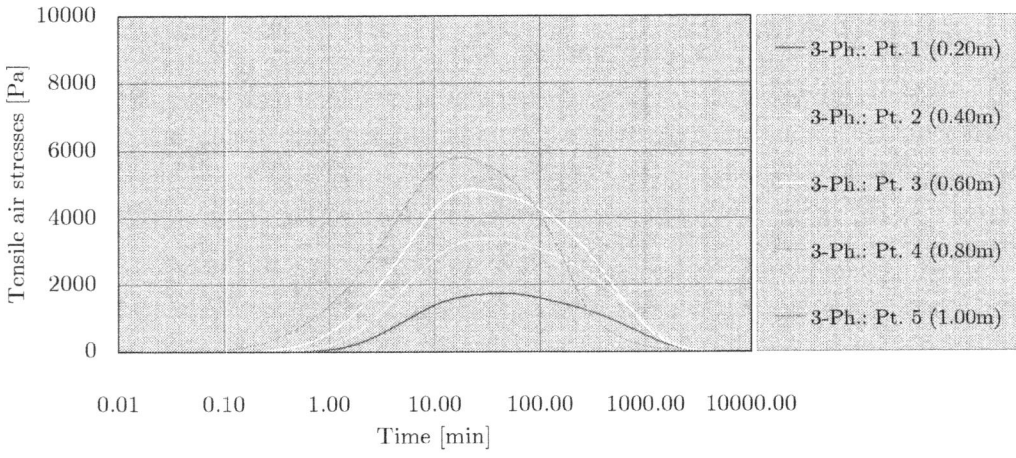

Figure 5.44: Air stress history until attaining a steady state.

air pressure. Considering the evolution with time, these excess hydrostatic air stresses are zero at the beginning of the experiment, then increase, reach a pronounced peak and finally vanish again until being zero at steady state conditions. This means that at the start of the experiment atmospheric air pressure prevails in the soil column (corresponding to the initial conditions – remember that a very small but finite amount of air is assumed to be present in the pores) which is reached again at steady state conditions. When taking a look at the peaks of the individual curves, the maximum excess tensile air stress is larger for points located in the upper part of the soil specimen (see, e.g., orange coloured curve) and becomes smaller for points positioned further down the column (e.g., blue line). The red line (point 5) indicates the zero excess air stress (or atmospheric pressure) at the top boundary of the soil column which corresponds to the specified boundary conditions. Furthermore, the closer a point is located to the upper boundary, the earlier the peak value in the excess hydrostatic air stress occurs which of course corresponds to the dewetting front moving down the soil column, starting from the top of the specimen.

Figure 5.45 shows the history of the vertical displacements (settlements) of the specimen. Similar to the water stresses, a large percentage (almost 90 %) of the final settlements occurs in the first 20.0 minutes of the simulation period. When employing the two-phase model, about 60 % of the displacements are obtained in this particular time interval. After steady state conditions are attained the individual curves are not

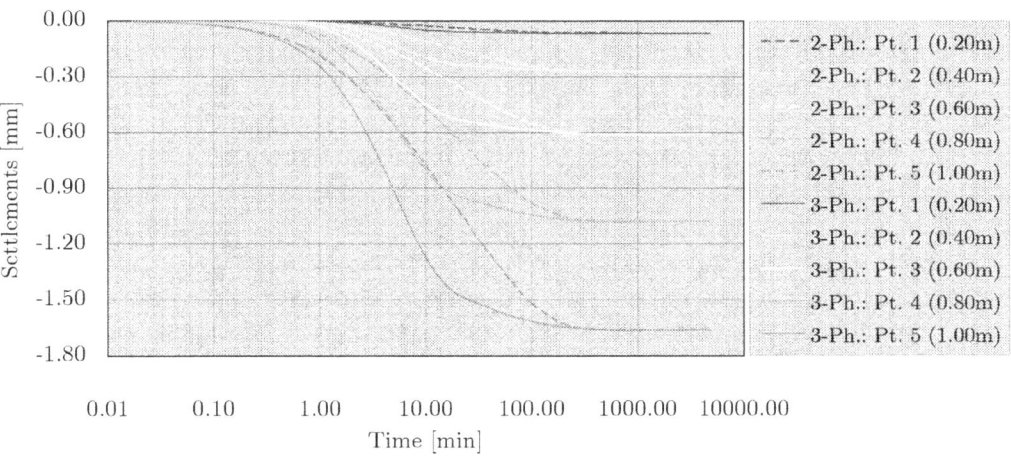

Figure 5.45: History of settlements until attaining a steady state.

equidistant but their distance increases with increasing column height which corresponds to the parabolic shape of the curves in Figure 5.40. Finally, Figure 5.45 depicts quite clearly that after 2.0 hours steady state conditions are not yet reached. A small increase in the settlements of about 2 % is encountered after 2.0 hours until a steady state is obtained.

The evolution of the degree of water saturation with time is shown in Figure 5.46. Similar to the two-phase prediction, also the three-phase model yields changes of the degree of saturation which are less pronounced in the early stage of the simulation period than, e.g., the changes of the hydrostatic water stresses. For instance, at the top of the column (red solid line) only 60 % of the final reduction of the degree of saturation occur in the first 20.0 minutes, compared to the 85 % water stress changes in the respective time interval.

However, the most interesting fact outlined by this particular figure concerns the differences between the two-phase and the three-phase model. When considering the results obtained from the two-phase formulation, e.g., after 10.0 minutes (dashed lines), the upper half of the soil column has experienced a reduction of the degree of saturation to a certain extent which gradually spreads to the overall specimen as time passes by. Thus, more or less the whole column is affected by the dewatering process. The results obtained from the three-phase model (solid lines) on the contrary are quite different. In this case, as discussed in connection with Figure 5.41, a pronounced dewetting front

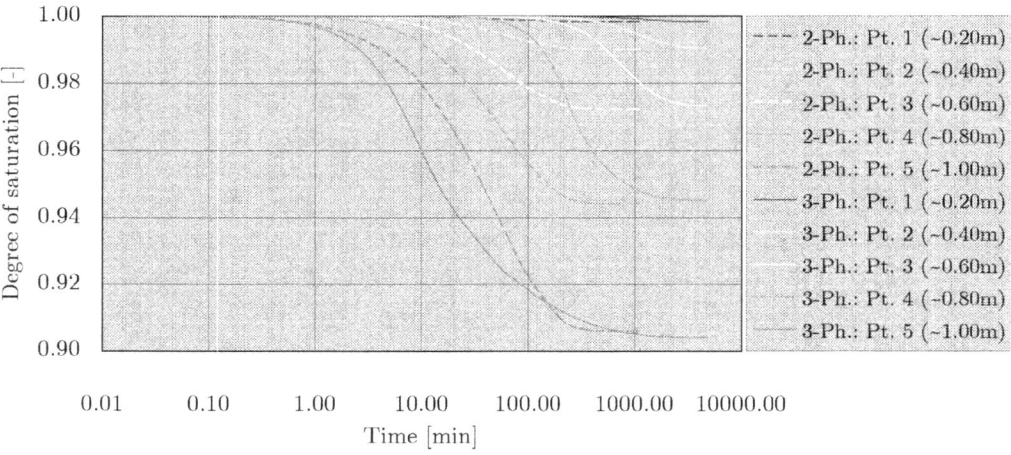

Figure 5.46: Degree of saturation history until attaining a steady state.

occurs which also can be recognised from Figure 5.46 quite well. At the topmost point 5 (red solid line) the degree of water saturation starts decreasing soon after the start of the simulation, while at the other points the soil is still fully saturated. After 20.0 minutes the degree of saturation has decreased to about 94 % at the top of the column, the dewetting front reaches point 4 (0.80 m) and the column starts dewatering at that level (orange solid line). Points 1 to 3 still exhibit a fully saturated state. In a similar manner the dewetting front proceeds and dewatering successively starts at the individual points.

Additionally, Figure 5.46 indicates quite clearly that after 2.0 hours steady state conditions are not attained at all for the degree of saturation. While for the lower 60.0 cm of the column the dewatering process did not even start, also the remaining part has not attained a steady state. At point 4 (0.80 m) after 2.0 hours about 15 % of the dewatering process are completed and at the top of the column (point 5) about 85 %. However, after approximately 80.0 hours the curves exhibit zero slopes and no further changes of the degree of saturation occur with time.

Figure 5.47 shows the history of the effective vertical compressive stresses in the soil skeleton (stresses in excess of the initial state) for five points along the height of the column. For the same reasons as given in connection with Figure 5.42, these stresses exhibit a behaviour similar to the one of the hydrostatic water stresses. Almost 90 % of the final stress changes occur in the first 20.0 minutes of the numerical simulation.

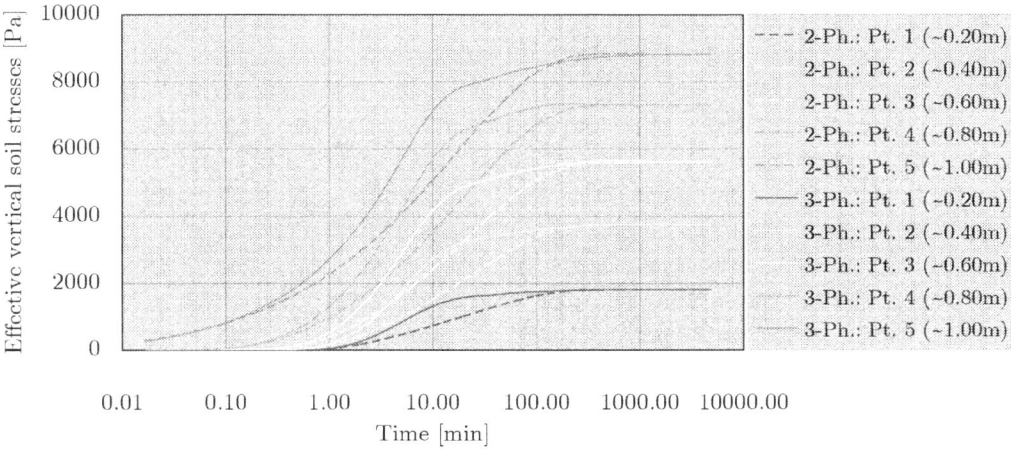

Figure 5.47: Effective soil stress history (stresses in excess of the initial stress state) until attaining a steady state.

A somewhat lower percentage is predicted by the two-phase model corresponding to the lower tensile hydrostatic water stresses in this case. After about 80.0 hours the horizontal lines indicate that also for the effective soil stresses steady state conditions are obtained.

Finally, the time history of the outflow rate of the water phase is shown in Figure 5.48. Since the outflow rate directly corresponds to the velocity of the particular fluid, an attainment of steady state conditions is most obvious from the behaviour of this variable. As soon as the dewatering process is completed, the velocity and thus the outflow rate of the water phase should be zero. As Figure 5.48 indicates, this is not the case at a point of time of 2.0 hours after the start of the simulation period where a water outflow rate of about 0.0015 cm^3/cm^2 min is encountered. As time proceeds towards 80.0 hours (red solid line), the value for the water outflow rate reaches indeed zero indicating the completion of the dewatering process.

As mentioned, the results discussed in connection with the previous figures generally show very good agreement with other numerical solutions presented in the literature [Lewis(1998), Klubertanz(1999)] for all the individual variables and for all points of time (up to 2.0 hours). However, whereas most of these calculations are restricted to a time period of 2.0 hours according to Liakopoulos' experiment, in the present calculations an extended time span until attaining steady state conditions (80.0 hours) has been considered. The main conclusions to be drawn from the numerical simulation beyond

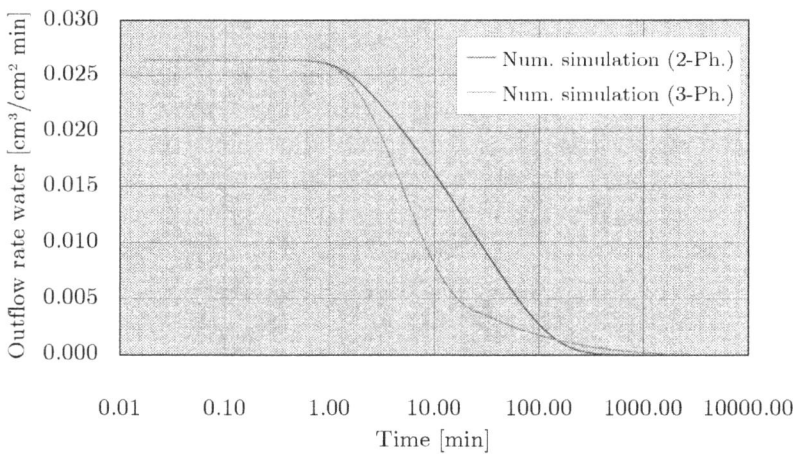

Figure 5.48: Water outflow rate at the bottom of the column until attaining a steady state.

the time span of the experiment are: (i) Finally, atmospheric air pressure prevails in the whole soil column again. (ii) For the particular type of soil under consideration (i.e. for the specified relationship between capillary stress and degree of saturation, remember that this relationship is entirely different for soils exhibiting permeabilities of different orders of magnitude as shown in Figure 3.10) it is not possible to dewater the specimen below a degree of saturation of about 90 % under the influence of gravitation. (iii) Due to the prescribed boundary conditions at the bottom surface (both fluid stresses are constrained to zero to simulate a permeable boundary, thus a zero capillary stress and fully saturated conditions prevail during the calculation) the column stays fully saturated at this boundary.

Chapter 6

Application of the three-phase formulation

6.1 Introduction

The previous Chapter 5 was dedicated to the verification of (i) the developed three-phase finite element model and (ii) two special cases (dewatering under atmospheric conditions, consolidation) contained within the proposed formulation. To this end, examples dealing with one- and two-dimensional consolidation, airflow through dry soil and the drainage of soil specimens were investigated numerically by means of the current finite element model. Results of these simulations were compared with (i) analytical solutions, (ii) experimental data and (iii) numerical results documented in the literature, which yielded reasonable agreement.

In the current chapter the applicability of the three-phase model to real civil engineering problems is demonstrated. To this end, three different examples are discussed (using either the two-phase or the complete three-phase formulation) which are of practical interest in the realm of geotechnical engineering.

The first example deals with the water flow through an earth dam. The behaviour of a homogeneous earth dam of trapezoidal cross section during reservoir filling is simulated numerically. Starting from a specified initial degree of water saturation in the dam, the flow of water through the dam, the increasing water saturation in the soil and the position of the free surface (borderline between saturated and unsaturated region, position of zero hydrostatic water stresses) due to reservoir filling are investigated until steady state conditions are obtained. Simulations are performed assuming rigid and linear elastic behaviour of the soil skeleton. In addition to that, anisotropy with respect to the permeability and an earth dam containing a clayey core are discussed.

In the second example a full-scale in-situ air permeability test conducted in connection with the subway construction in Essen, Germany, is simulated numerically. In the set of tests conducted in Essen compressed air was used to displace the groundwater in a completely saturated layer of soil. The main goal of the experiments was the investigation of the air permeability of the Essen soil. One of these tests is simulated numerically and the results are compared with experimental data.

Finally, the third example is devoted to the 2D numerical investigation of the excavation of a tunnel driven below the groundwater level where compressed air is used for displacing the groundwater in the vicinity of the working area of the tunnel during the construction period.

6.2 Water flow through an earth dam

6.2.1 Description of the problem

For the numerical simulation of the water flow through an earth dam a two-phase formulation can be employed. Thus, the air pressure is assumed to stay atmospheric in the course of the calculation and matric suction is equal to the negative value of the hydrostatic stress in the water phase. Contrary to the consolidation problems presented in Chapter 5 (which are also analysed using a two-phase model), the degree of saturation here is allowed to vary during the calculation. It should be mentioned that in classical seepage analysis only the flow problem is investigated. Hence, a rigid soil skeleton is chosen for the current simulation in a first step. The coupled model discussed in this work, however, allows to take into account the deformations of the soil skeleton. Linear elastic behaviour of the soil skeleton is assumed.

In the example problem under consideration [Fredlund(1993)] an earth dam of trapezoidal cross section, built to retain a reservoir of water, is investigated numerically. The dam is resting on an impervious foundation. According to the geometry shown in Figure 6.1, the cross section has a width of 52.0 m at the base which is reduced to 4.0 m at the top. The height of the dam is assumed to be 12.0 m, yielding slopes with a ratio of 1:2. To achieve a selective drainage of the leakage occurring through the dam and thus to prevent stability problems, an underdrain with a length of 12.0 m, consisting of gravel material, is provided in the vicinity of the downstream slope. The height of the water level at the upstream slope is assumed to be 10.0 m. Furthermore, the soil used for the construction of the earth dam is considered to be homogeneous and isotropic with respect to the coefficient of water permeability.

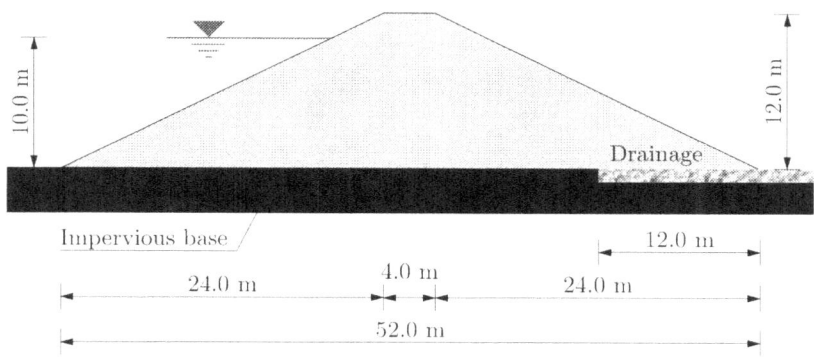

Figure 6.1: Section of the earth dam.

In a second example an earth dam of similar geometry is investigated (Figure 6.2), however, this dam possesses a rather impermeable core consisting of a clayey material. At the base of the dam the core is assumed to be 8.0 m wide which reduces to 1.85 m at a height of 10.0 m (i.e. at the top of the core). The water permeability of the core is

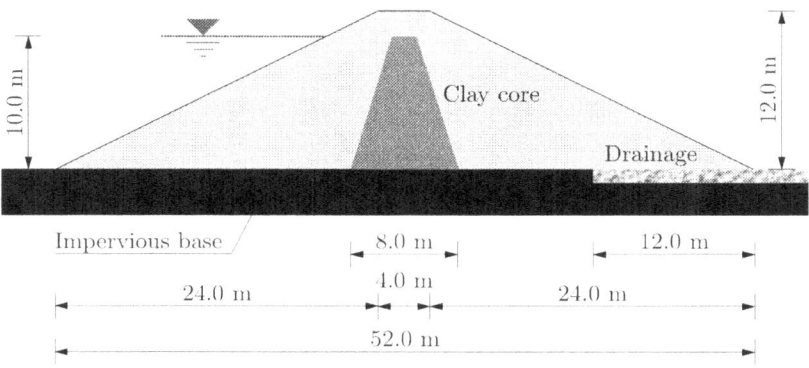

Figure 6.2: Section of the earth dam containing a clay core.

considered to be 100 times smaller than the permeability of the remainder of the earth dam. In practice, such an impervious core is built in to reduce leakage through the dam. Of course a number of other construction types for earth dams retaining water are possible [Lambe(1979), Fredlund(1993)].

6.2.2 Numerical model

In most of the numerical simulations only the earth dam is modelled, the impervious base is not taken into account. The domain is discretised by a number of 312 plane strain finite elements (26 subdivisions in horizontal and 12 subdivisions in vertical direction). Since a two-phase formulation can be used for the analysis of the problem under consideration, a node of the finite element mesh possesses a maximum number of three degrees of freedom, namely two displacement degrees of freedom in the two coordinate directions and one fluid (water) stress degree of freedom. Figure 6.3 shows a plot of the employed finite element mesh.

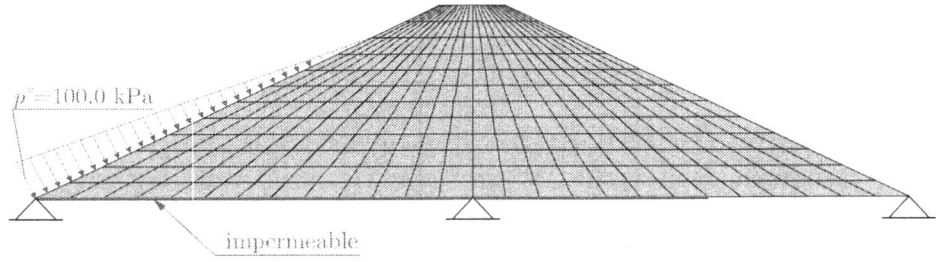

Figure 6.3: Finite element mesh employed for the numerical simulation of water flow through an earth dam.

Concerning the individual finite elements, different orders of interpolation are assigned to the individual degrees of freedom. Quadratic interpolation is used for the displacements of the soil skeleton, while bilinear interpolation characterises the hydrostatic fluid stresses. Hence, each finite element possesses eight nodes (corner and mid-side nodes) with displacement degrees of freedom and four nodes (corner nodes) with fluid stress degrees of freedom.

The boundary conditions shown in Figure 6.3 can be interpreted as follows: With respect to the displacements the bottom boundary of the domain is fully constrained. At the draining part of the bottom boundary (in the vicinity of the downstream slope) the hydrostatic fluid stresses in excess of the initial condition are constrained to zero, thus a permeable boundary is simulated. Similar boundary conditions with respect to the fluid phase are assumed for the dam at the downstream face, at the top and at the upstream slope above the water level. The fluid boundary condition below the water level at the upstream face of the dam represents the instantaneous raise of the water level in the reservoir. Therefore, the excess hydrostatic fluid pressure at this boundary is constrained to values linearly increasing from zero at a height of 10.0 m (water level of the reservoir) to an amount of 100.0 kPa at the base of the dam.

As initial condition a water saturation in the soil of 40 % is assumed in the whole domain. According to the specified relationship between degree of saturation and capillary stress, this value corresponds to an initial capillary stress (and thus tensile hydrostatic water stress because the air stress is assumed to prevail atmospheric) of -8.1 kPa. At time $t = 0$ the water level in the reservoir is raised instantaneously, thus the water pressure boundary condition at the upstream slope is applied suddenly and held constant in the course of the calculation.

The properties of the soil used as construction material for the earth dam are summarised in Table 6.1. Concerning the behaviour of the soil skeleton, two analyses are performed: a first one assuming a rigid soil skeleton and a second one where a linear elastic model is used to describe the skeleton behaviour. For the second case Table 6.1 contains the two chosen elastic constants, Young's modulus and Poisson's ratio. The saturated permeability of the soil with respect to the water phase of $k^{ow} = 8.64 \cdot 10^{-3}$ m/d (which is equal to $k^{ow} = 1.00 \cdot 10^{-7}$ m/s) is specified according to calculations presented in [Fredlund(1993)]. The remaining parameters of the upper half of Table 6.1 are taken from the literature as well.

A few words should be mentioned about the properties given in the lower half of Table 6.1. These parameters describe the hydraulic behaviour of the soil, i.e. the relationship between degree of saturation and capillary stress (hydrostatic water stress) and between degree of saturation and relative permeability coefficient, respectively. Concerning the degree of saturation expressed as a function of the capillary stress, according to equation (3.271) proposed by van Genuchten [Genuchten(1985)], the parameters given in Table 6.1 are chosen in such a way that a wetting curve similar to the one contained in Figure 3.10(b) for a fine sand is obtained. The maximum degree of water saturation is chosen to be $S_s^w = 0.95$, meaning that a minimum amount of air of 5 % is assumed to be present in the 'saturated' region of the domain due to the inclusion of air bubbles. Since

Parameter	Symbol	Unit	Value
Young's modulus	E	kPa	10000.0
Poisson's ratio	ν	-	0.25
Porosity	n	-	0.50
Permeability water	k^{ow}	m/d	$8.64 \cdot 10^{-3}$
Bulk modulus water	K^w	kPa	$2.0 \cdot 10^6$
Density solid	ρ^s	t/m^3	2.70
Density water	ρ^w	t/m^3	1.00
Gravitational acceleration	g	m/s^2	9.81
Maximum saturation	S_s^w	-	0.95
Residual saturation	S_r^w	-	0.25
Air entry value	p_b^c	kPa	-3.00
Empirical parameter	m	-	0.60

Table 6.1: Material properties for the earth dam problem.

wetting proceeding from an unsaturated state of the soil is considered, this assumption seems to be physically justified. The rather low air entry value, $p_b^c = -3.00$ kPa, is also due to the consideration of a wetting process. The parameter n in (3.271) is determined according to the relationship given in connection with this equation, $n = 1/(1-m)$. The shape of the curve expressing the relative permeability coefficient as a function of the degree of saturation (equation (3.285)) is also determined by the empirical parameter m given in Table 6.1. For the soil under consideration both relations are plotted in the subsequent Figure 6.4 (black solid lines).

In addition to the rigid and the linear elastic behaviour of the soil skeleton investigated in connection with the example problem shown in Figure 6.1, anisotropy with respect to the permeability is taken into account. As outlined in Chapter 3, the permeability for the fully saturated state, $\mathbf{k}^{ow}$, possesses a tensorial nature, i.e. different values for the saturated permeability can be specified for the individual coordinate directions. Hence, in combination with the current example a soil exhibiting different permeabilities in the horizontal and vertical directions is investigated. The saturated permeability in the horizontal direction is assumed to be five times larger than its counterpart in the vertical direction, $k_x^{ow} = 5 k_y^{ow}$.

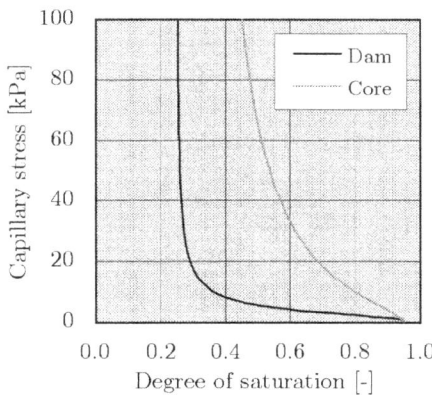 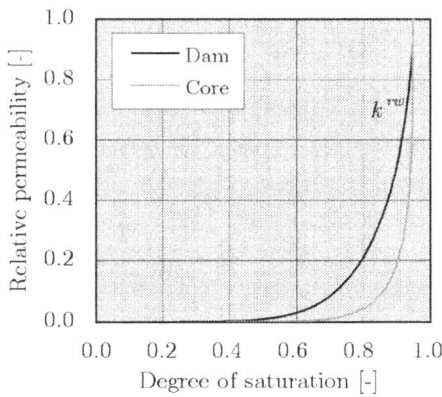

Figure 6.4: Hydraulic properties of the soil used as construction material for the earth dam and of the clayey core.

In a further example the earth dam is conceived to contain a clayey core as indicated in Figure 6.2. This particular soil is characterised by possessing a lower saturated permeability with respect to the water phase ($k^{ow} = 8.64 \cdot 10^{-5}$ m/d or $k^{ow} = 1.00 \cdot 10^{-9}$ m/s) than the surrounding material. For simplicity only some of the properties of the clay (summarised in Table 6.2) differ from the parameters assumed for the material comprising the remainder of the dam. Additionally, it should be mentioned that the initial degree of saturation of the core is chosen to be 65 %, corresponding to a capillary stress of -24.4 kPa. Similar as for the hydraulic parameters given in Table 6.1, the relations between degree of saturation and capillary stress and between degree of saturation and relative permeability coefficient for the core are plotted in Figure 6.4 (red solid lines).

Parameter	Symbol	Unit	Value
Young's modulus	E	kPa	7000.0
Poisson's ratio	ν	-	0.35
Permeability water	k^{ow}	m/d	$8.64 \cdot 10^{-5}$
Air entry value	p_b^c	kPa	-10.00
Empirical parameter	m	-	0.35

Table 6.2: Material properties for the clayey core of the earth dam.

The time span considered in each calculation is designed such that steady state conditions are obtained. For the earth dam shown in Figure 6.1 a time interval of about 3000.0 days is necessary when using a rigid soil skeleton. For the linear elastic behaviour of the soil skeleton approximately the same time span is required. Considering a five times larger value for the permeability in the horizontal direction results in a reduction of the time necessary for obtaining steady state conditions to about 1500.0 days. For the dam containing a clay core (Figure 6.2) steady state conditions are obtained only after about 30000.0 days (soil skeleton is assumed linear elastic). In each of the calculations the time increments vary between a minimum of 0.0001 day and a maximum of 500.0 days.

6.2.3 Results of the numerical simulation

In the current section results of five numerical simulations are presented. The first calculation deals with the water flow through a homogeneous earth dam considering a rigid soil skeleton. For comparison, in the second simulation the constitutive behaviour of the soil skeleton is assumed to be linear elastic. Third, a similar linear elastic analysis of the earth dam is performed taking into account both the impervious base and the underdrain in the numerical model. A homogeneous earth dam consisting of a soil exhibiting different saturated permeabilities with respect to the water phase in the two coordinate directions is analysed in a fourth step. Finally, an earth dam containing a clay core with a rather low permeability is investigated.

Homogeneous dam with rigid soil skeleton

The numerical simulation of the homogeneous earth dam shown in Figure 6.1 is discussed subsequently. Since in conventional seepage analysis only the flow problem is investigated, a rigid behaviour of the soil skeleton is assumed in a first step. Figures 6.5 and 6.6 show the evolution of the hydrostatic water stresses and the degree of saturation in the course of the analysis for four points of time. It should be mentioned that total water stresses are plotted in Figure 6.5, that is the tensile hydrostatic stresses according to the initial condition (degree of saturation of 40 %) plus the excess hydrostatic stresses due to the raise of the water level in the reservoir.

Figure 6.5(a) depicts the distribution of the water stresses in the domain at the beginning of the simulated period of time, shortly after the raise of the water level in the reservoir. In most parts of the domain tensile hydrostatic water stresses are encountered according to the specified initial stress state (the initial degree of water saturation of 40 % corresponds to a tensile water stress of 8.1 kPa). The minimum value of the legend in Figure 6.5(a) is due to some oscillations of the hydrostatic water stresses occurring in the vicinity of the proceeding water front (region of zero water stress) near the bottom boundary of the domain. At the upstream face of the dam the fluid pressure boundary condition can be recognised very clearly (linear increase of fluid pressure).

Figures 6.5(b) and 6.5(c) show the propagation of the hydrostatic fluid stresses in the domain in the course of the simulation, starting at the upstream face and moving into the direction of the downstream face of the dam with increasing time. At the upstream slope the prevailing fluid boundary condition can be identified in each of the plots.

Figure 6.5(d) depicts the distribution of the hydrostatic fluid stresses in the domain at steady state conditions which are obtained in the numerical simulation after about 3000.0 days. The conditions at the dam's upstream slope outline the linear increase of the water pressure below the water level in the reservoir. At the draining part of the base of the dam any fluid stresses in excess of the initial stress state are constrained to zero (simulation of the permeable boundary) and thus the initial tensile hydrostatic stress state in the water phase remains unchanged. The typical S-shaped curve indicating the position of zero hydrostatic water stresses, well-known from literature on seepage flow as the so-called 'phreatic surface', stretches from the position of the water level (since total fluid stresses are plotted, the curve starts somewhat below the height of 10.0 m, the difference being attributed to the initial tensile stresses of 8.1 kPa) to the upper end of the underdrain. In the region above this line the tensile water stresses according to the initial condition prevail.

Since a two-phase numerical simulation is performed in connection with the problem under consideration, the air stresses are kept atmospheric in the course of the calculation. Therefore, the capillary stress is equal to the negative hydrostatic water stress (see equation (3.209)) which in this case also governs the development of the degree of water saturation (confer equation (3.337)). The evolution of the degree of saturation, shown in Figure 6.6, closely corresponds to the hydrostatic water stresses plotted in Figure 6.5. The position of the wetting front in Figures 6.6(a) to 6.6(d) coincides with the position of the zero hydrostatic water stresses in Figures 6.5(a) to 6.5(d). In the course of the numerical simulation the area exhibiting the initial degree of saturation is reduced due to the moving wetting front which propagates from the upstream face of the dam to the upper end of the underdrain. The transition from the initial saturation of 40 % to the

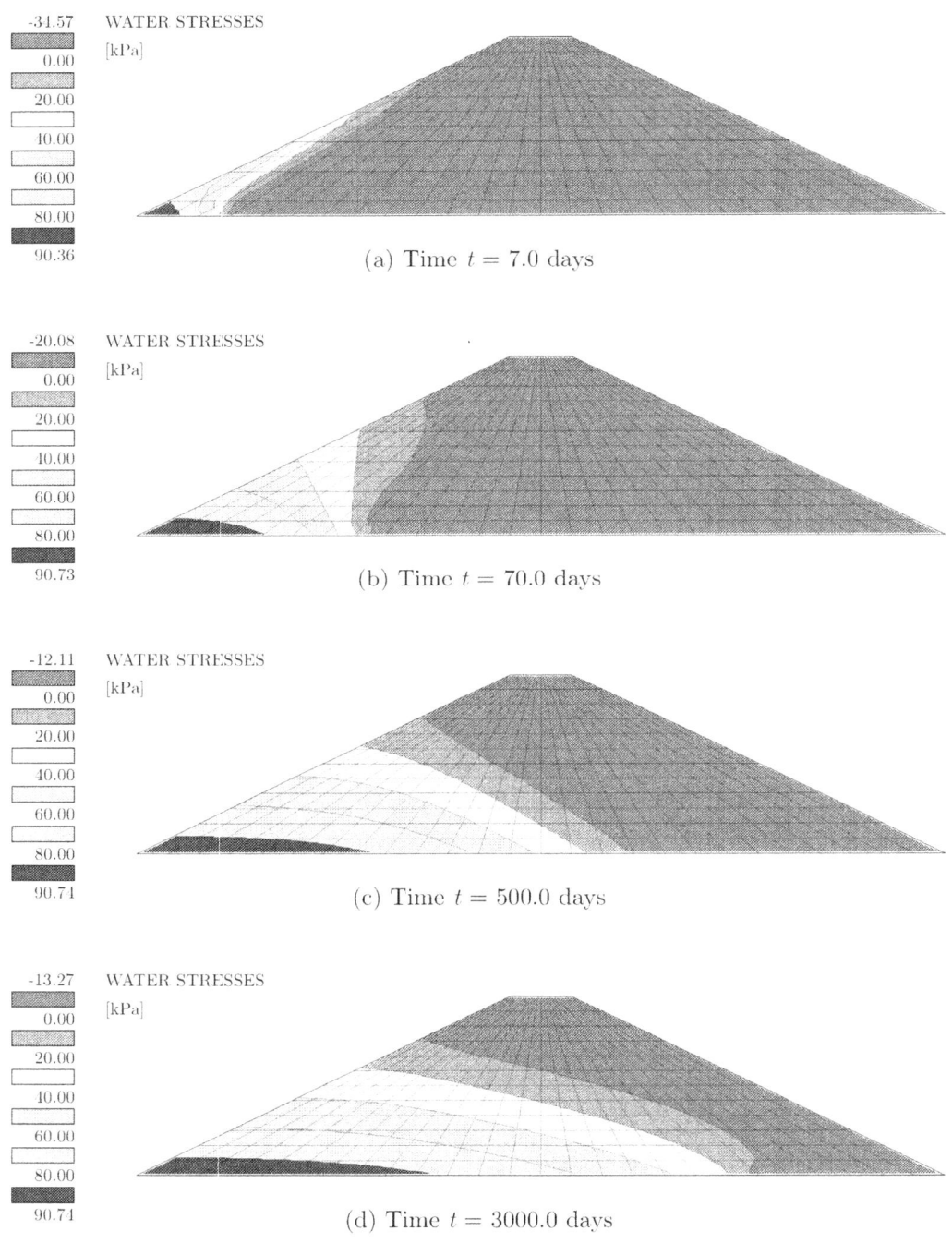

Figure 6.5: Evolution of the hydrostatic water stresses until a steady state is obtained (analysis with rigid soil skeleton).

Chapter 6. Application of the three-phase formulation

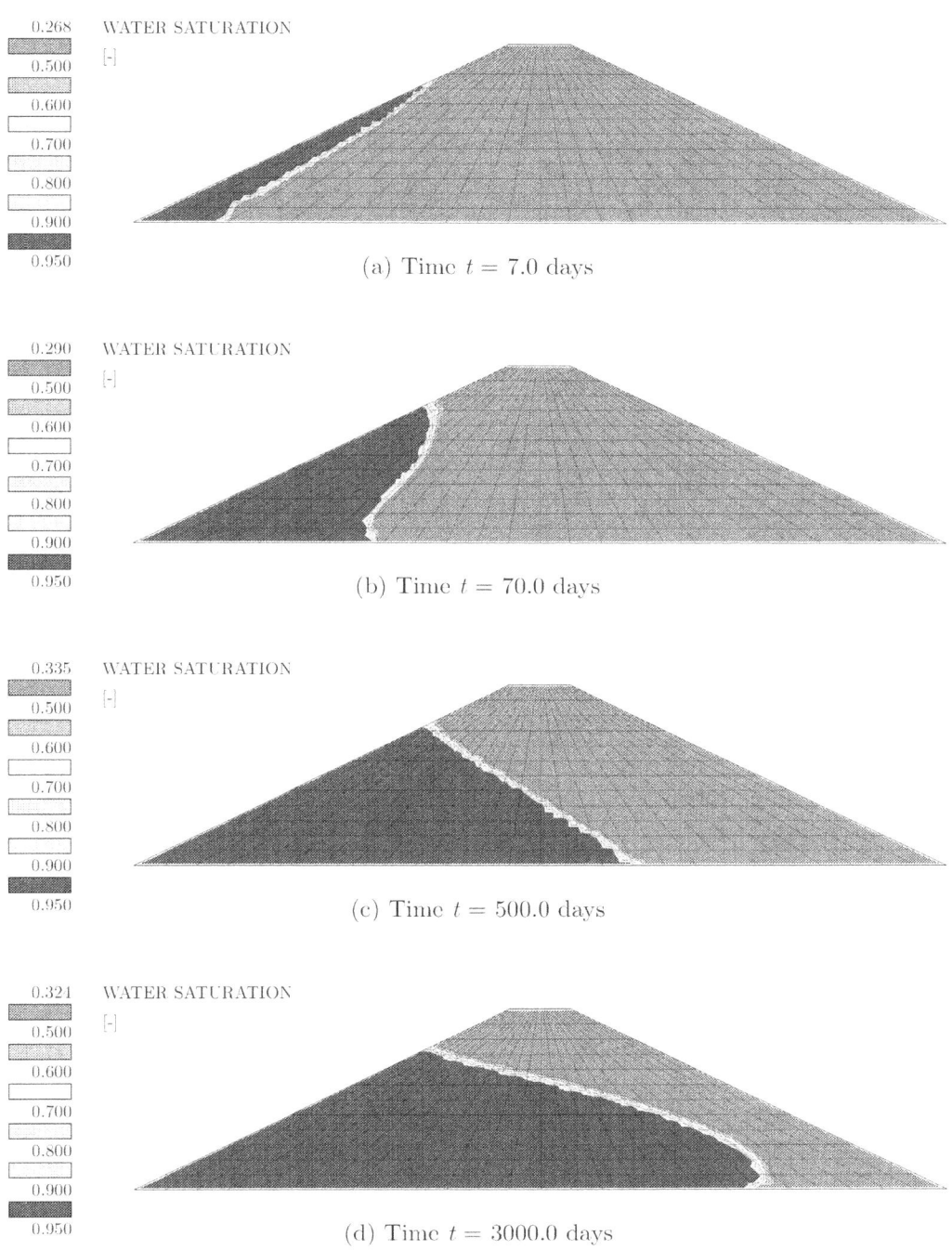

Figure 6.6: Evolution of the degree of saturation until a steady state is obtained (analysis with rigid soil skeleton).

fully saturated conditions of 95 % (in the calculation a minimum amount of air of 5 % is assumed to be present in the pores, e.g., in the form of air bubbles) is obtained by means of a narrow capillary fringe. After steady state conditions are attained, the domain is fully saturated from the upstream water level in the reservoir to the upper end of the underdrain, as Figure 6.6(d) indicates.

Homogeneous dam with linear elastic soil skeleton

In a second step, the homogeneous earth dam, discussed in the preceding section, is analysed employing a linear elastic constitutive model for the behaviour of the soil skeleton. Since exclusively unloading of the soil occurs in the whole domain, the application of this rather simple model may be justified. The computed distribution within the domain and the evolution with time of both the hydrostatic fluid stresses and the degree of saturation turn out to look very similar to the preceding Figures 6.5 and 6.6 and are thus not shown in this section. Likewise, the time span to reach steady state conditions is as long as for the analysis using a rigid soil skeleton.

Figure 6.7 depicts the computed distribution of the vertical displacements of the soil skeleton for two instants of time (70.0 days and 3000.0 days, respectively, the latter coinciding with reaching steady state conditions). Due to water flowing into the pores of the soil and yielding a saturation of increasing portions of the domain, larger parts of the dam are subject to uplift, resulting in a heave of the soil. Vertical displacements are exclusively occurring in the upward direction in the whole domain. The actual values of the computed displacements are rather small as Figure 6.7 indicates. The maximum vertical displacement predicted in the numerical simulation amounts to about 3.4 cm and is obtained at the upstream face of the earth dam.

Simulation taking into account the base of the dam

In order to investigate the influence of taking into account or neglecting the base of the dam in the numerical simulation, the previous example is re-analysed, considering an 8.0 m deep layer of the rather impervious base. The underdrain is modelled by a highly permeable layer of soil of 2.0 m height. With respect to the soil skeleton, the base of the dam and the underdrain are assumed to exhibit a rigid behaviour. The initial water level is conceived to be located at the top of the base, thus this soil layer is considered to be fully water saturated. Of course, the fluid boundary condition previously used

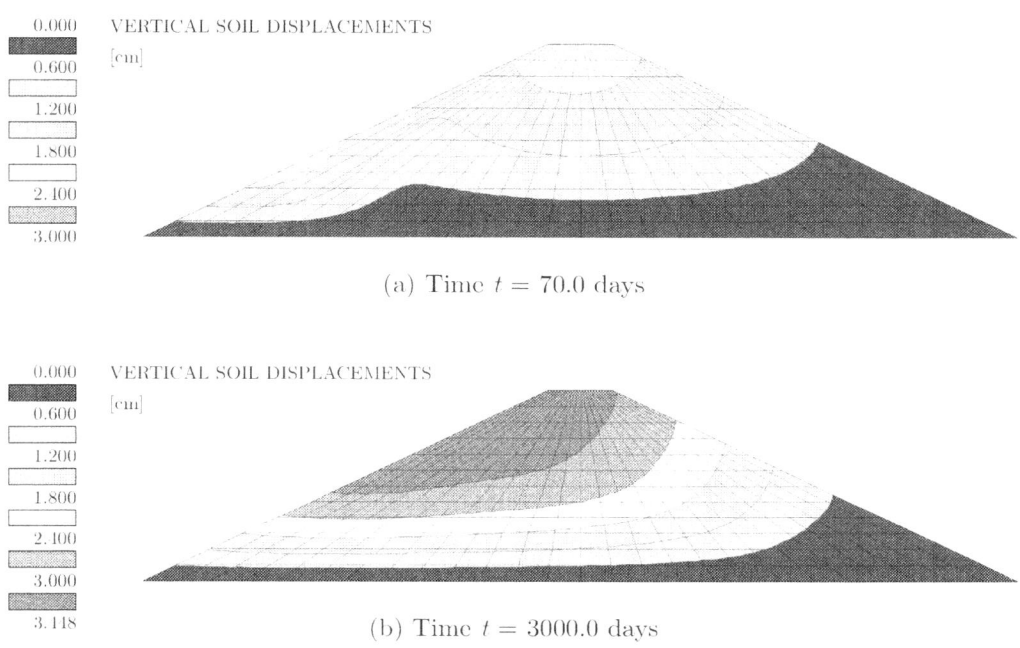

Figure 6.7: Evolution of the vertical displacements of the soil skeleton until a steady state is obtained (analysis using linear elastic behaviour of the soil skeleton).

for simulating the underdrain can be dropped in the present calculation. Other than that, the boundary and initial conditions correspond to those of the preceding analysis. Concerning the material properties, for simplicity only two parameters of the individual soils are assumed to differ, namely the permeability of the soil with respect to the water phase and the air entry value. These parameters are listed for the three soils in Table 6.3.

Parameter	Symbol	Unit	Dam	Base	Underdrain
Permeability water	k^{ow}	m/d	$8.64 \cdot 10^{-3}$	$8.64 \cdot 10^{-6}$	$8.64 \cdot 10^{-1}$
Air entry value	p_b^c	kPa	-3.00	-12.00	-1.00

Table 6.3: Hydraulic properties of the three types of soil used in the analysis of the earth dam taking into account the base.

Figure 6.8 depicts the computed distributions of the hydrostatic water stresses (Figure 6.8(a)) and the degree of saturation (Figure 6.8(b)) after steady state conditions have been attained. The plots are scaled according to the Figures 6.5 and 6.6 to allow for an easy comparison of the results. The hydrostatic water stresses (Figure 6.8(a)) in the earth dam are the same as the ones obtained without modelling the base which

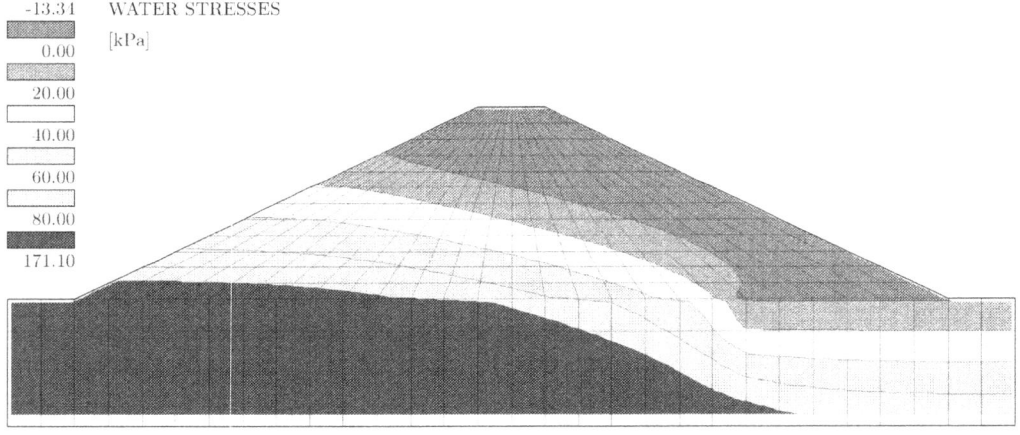

(a) Water pressure at steady state conditions

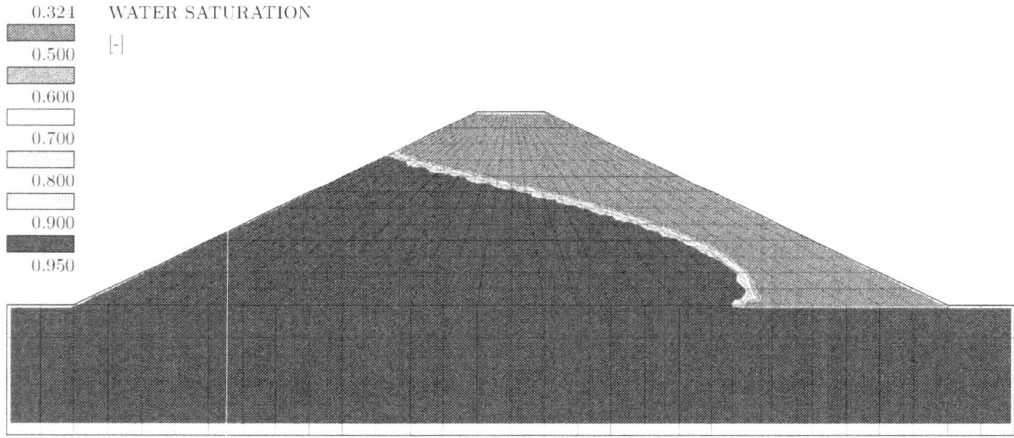

(b) Degree of saturation at steady state conditions

Figure 6.8: Distribution of (a) hydrostatic water stresses and (b) degree of saturation at steady state conditions (dam with base and underdrain).

can be recognised very clearly when comparing Figure 6.8(a) with Figure 6.5(d). In the base the linear increase of the hydrostatic water pressure below the groundwater table is encountered on the downstream side of the dam. At the upstream side of the dam the hydrostatic water stresses consist of this linearly increasing fluid pressure below the initial water table plus the additional pressure of 100.0 kPa due to the raise of the water level in the reservoir.

Figure 6.8(b) shows the corresponding degree of saturation after steady state conditions have been attained. As can be recognised, the domain is fully water saturated in the region of the base according to the assumed initial water table at the top of this layer. In the earth dam the computed distribution of the degree of saturation (and thus the position of the phreatic surface) is the same as obtained in the previous simulation.

Obviously, both the base of the dam and the underdrain not necessarily have to be taken into account in the numerical model. An appropriate specification of the boundary conditions at the bottom of the earth dam seems to provide the same results of the numerical simulation and has the additional advantage of yielding a model with a smaller number of degrees of freedom.

Anisotropy with respect to the fluid permeability

Finally, the earth dam of Figure 6.1 is analysed assuming that the soil is anisotropic with respect to the permeability of the water phase. As outlined in Chapter 3, the permeability for the fully saturated state of the soil possesses a tensorial nature and different values for the permeability can be specified for the individual coordinate directions. The impact of this kind of anisotropy is investigated by means of the current analysis. The soil constituting the earth dam thus is considered to possess a horizontal water permeability which is five times larger than the respective value in the vertical coordinate direction, that is $k_x^{ow} = 5 k_y^{ow}$. Such soil behaviour can be imagined for clayey materials consisting of platy particles which may all be bedded in a preferred direction (in this case horizontal).

Figures 6.9 and 6.10 show the evolution of the total hydrostatic water stresses and of the degree of saturation until steady state conditions are attained. These plots should be compared with the Figures 6.5 and 6.6 to realise the effect of the larger horizontal permeability with respect to the water phase. Probably most pronounced is the difference in the time span necessary for reaching a steady state. Due to the larger permeability this time interval is reduced to about 1500.0 days when considering anisotropic soil behaviour (which is about half of the time necessary for the isotropic soil).

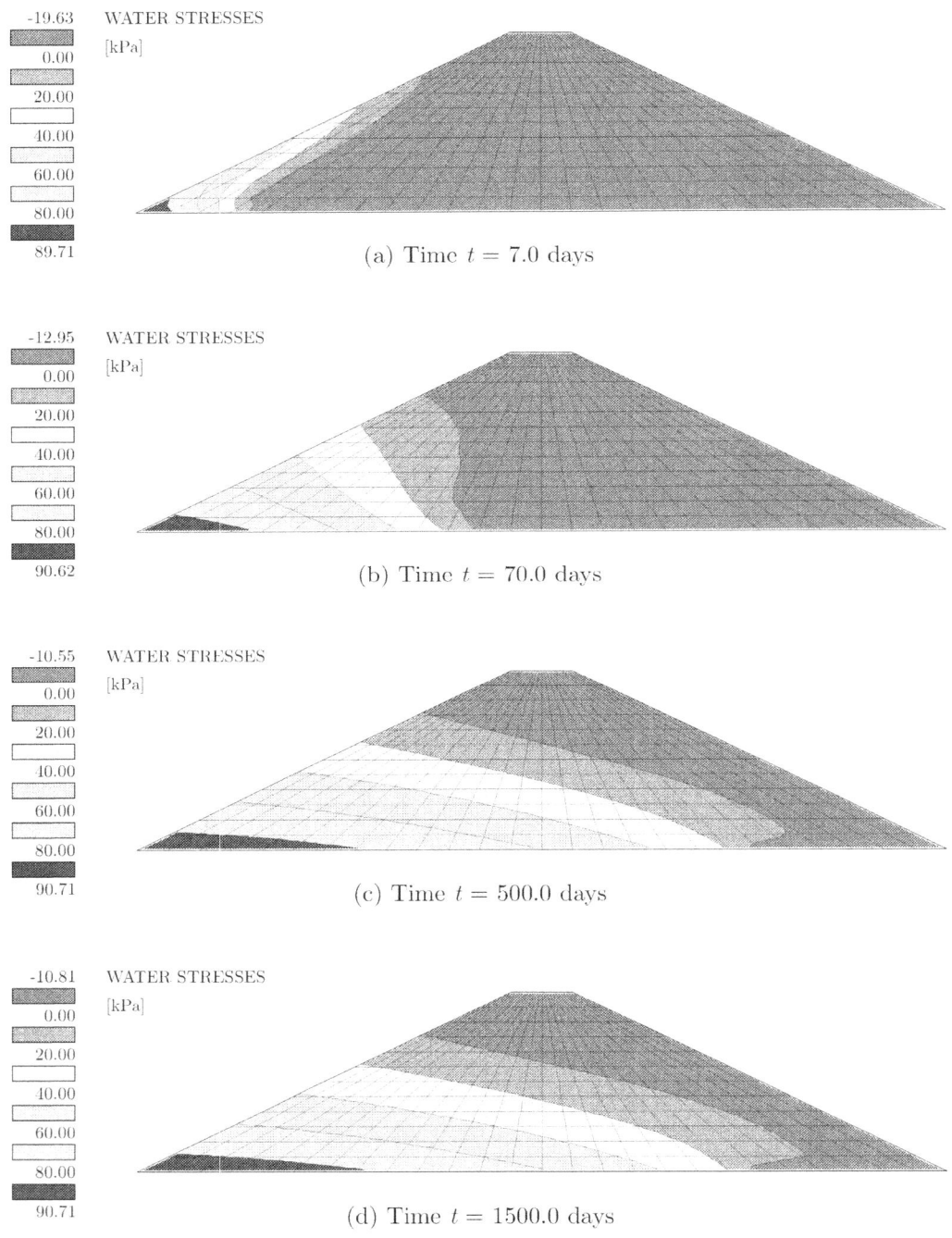

Figure 6.9: Evolution of the hydrostatic water stresses until a steady state is obtained (anisotropic permeability).

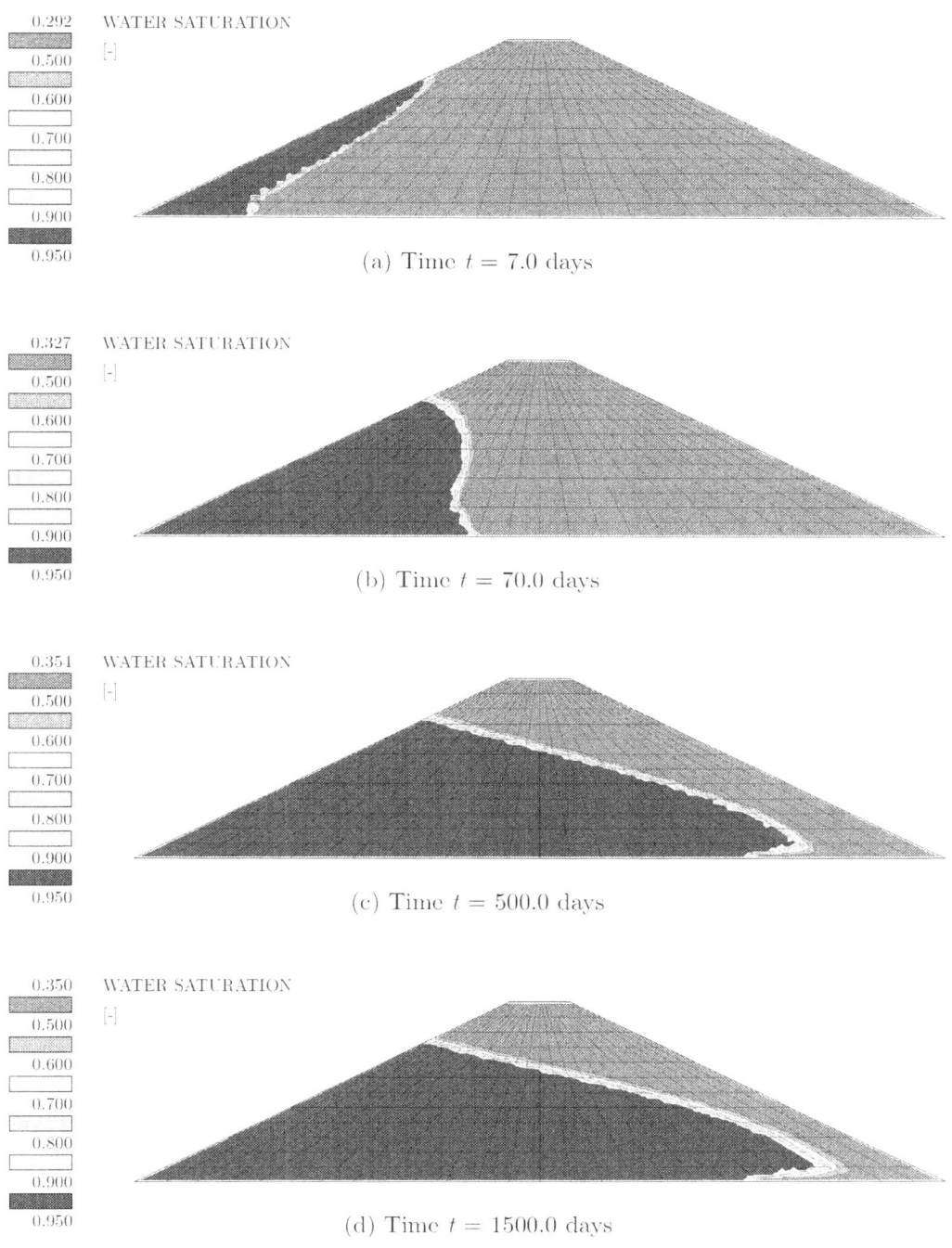

Figure 6.10: Evolution of the degree of saturation until a steady state is obtained (anisotropic permeability).

When comparing Figure 6.9 (showing the evolution of the hydrostatic water stresses with time for the anisotropic soil) with Figure 6.5 (depicting the respective quantity for the isotropic soil), it can be recognised that the distribution of the water stresses in the dam is quite similar for both cases. However, for the anisotropic soil, possessing a larger horizontal saturated permeability, the development of the water stresses takes place much faster. This fact is not very apparent at the beginning of the simulation period (see Figure 6.9(a)) but increases with time. After about 500.0 days (Figure 6.9(c)) the distribution of the hydrostatic water stresses in the anisotropic soil has progressed as far as in the isotropic soil after 3000.0 days.

Because of the larger horizontal permeability, at steady state conditions (attained after 1500.0 days) the water stresses have spread somewhat closer to the downstream face of the dam (confer Figure 6.9(d)) than in the isotropic soil. For a ratio of horizonal to vertical permeability greater than five it would even be possible for the phreatic surface to reach the downstream face of the dam (change of boundary conditions). In connection with Figure 6.9(d) it should be mentioned that the boundary condition in the region of the drainage has not been changed in the analysis assuming an anisotropic soil which is the reason for the zero hydrostatic stress isoline to run back to the upper end of the underdrain. At the upstream slope of the dam the linearly increasing water pressure below the water level in the reservoir is recognised in all plots of Figure 6.9.

Directly linked to the development of the hydrostatic water stresses is the evolution of the degree of saturation. Both quantities are related via the constitutive equation between degree of saturation and capillary stress (confer equation (3.267)). Hence, as can be seen from Figure 6.10, the wetting front in the dam proceeds together with the hydrostatic water stresses and the saturated region of the domain increases with time. At steady state conditions (see Figure 6.10(d)) the saturated area has spread somewhat more in the direction of the downstream face of the dam than in the isotropic soil.

The evolution of the vertical displacements of the dam is similar to the isotropic case depicted in Figure 6.7. However, the predicted maximum heave obtained at steady state conditions is smaller than for the isotropic soil and amounts to about 3.2 cm.

Earth dam containing a clayey core

In this section the analysis of the earth dam shown in Figure 6.2 is discussed. Now the dam is no longer homogeneous but a core, consisting of clayey material, is built in as a seal unit. In practice such a measure would be taken to reduce the loss of water through the dam. As a matter of fact, the rather low permeability of such a core not only reduces the amount of leakage but also substantially delays the time necessary for water to pass through the earth dam (which in the numerical simulation is reflected in a longer time span to reach steady state conditions).

In the present example, the permeability of the core is assumed to be 100 times smaller than the one of the surrounding material comprising the earth dam (see Table 6.2). The boundary conditions for the problem are the same as described in connection with the previous analyses. However, a difference occurs in the initial conditions. The core is assumed to possess a higher water content than the remainder of the dam. The initial degree of water saturation for the core is taken to be 65 % which, according to the hydraulic properties of this material given in Tables 6.1 and 6.2, corresponds to an initial hydrostatic water stress of 24.4 kPa.

Figure 6.11 depicts the evolution of the total hydrostatic stresses in the water phase, i.e. the initial water stresses plus water stresses due to the raise of the water level in the reservoir. As long as the wetting front does not reach the core of the dam, the development of all quantities is basically the same as for the homogeneous case. When taking a look at Figure 6.11(a), this becomes quite obvious. In the early phase of the simulation period in most parts of the domain (red coloured area) still the initial conditions prevail, that is a tensile hydrostatic stress state. In the course of time, additional hydrostatic water stresses spread into the dam, starting from the specified fluid boundary condition at the upstream slope. The water stress distribution shown in Figure 6.11(a) is practically identical to the one of Figure 6.5(b) for the homogeneous dam.

As soon as the wetting front reaches the impermeable core, the propagation of the water stresses is delayed substantially. As Figures 6.11(b) and 6.11(c) show, it takes considerable time for the water stresses to increase in the core of the dam. The isolines in the core decline rather steep.

Figure 6.11(d) depicts the distribution of the total hydrostatic water stresses after steady state conditions are attained. As can be seen, the time span is extended to 30000.0 days which is a decuple of the time span to reach a steady state necessary for the homogeneous dam. At the upstream side of the core the isolines of the hydrostatic water stresses are almost horizontal, in the core the hydrostatic water stresses are completely reduced and at the downstream side of the core almost the whole domain stays at the initial conditions (tensile stress state). Only a minor influence of the raise of the water level in the reservoir can be recognised in the downstream part of the earth dam.

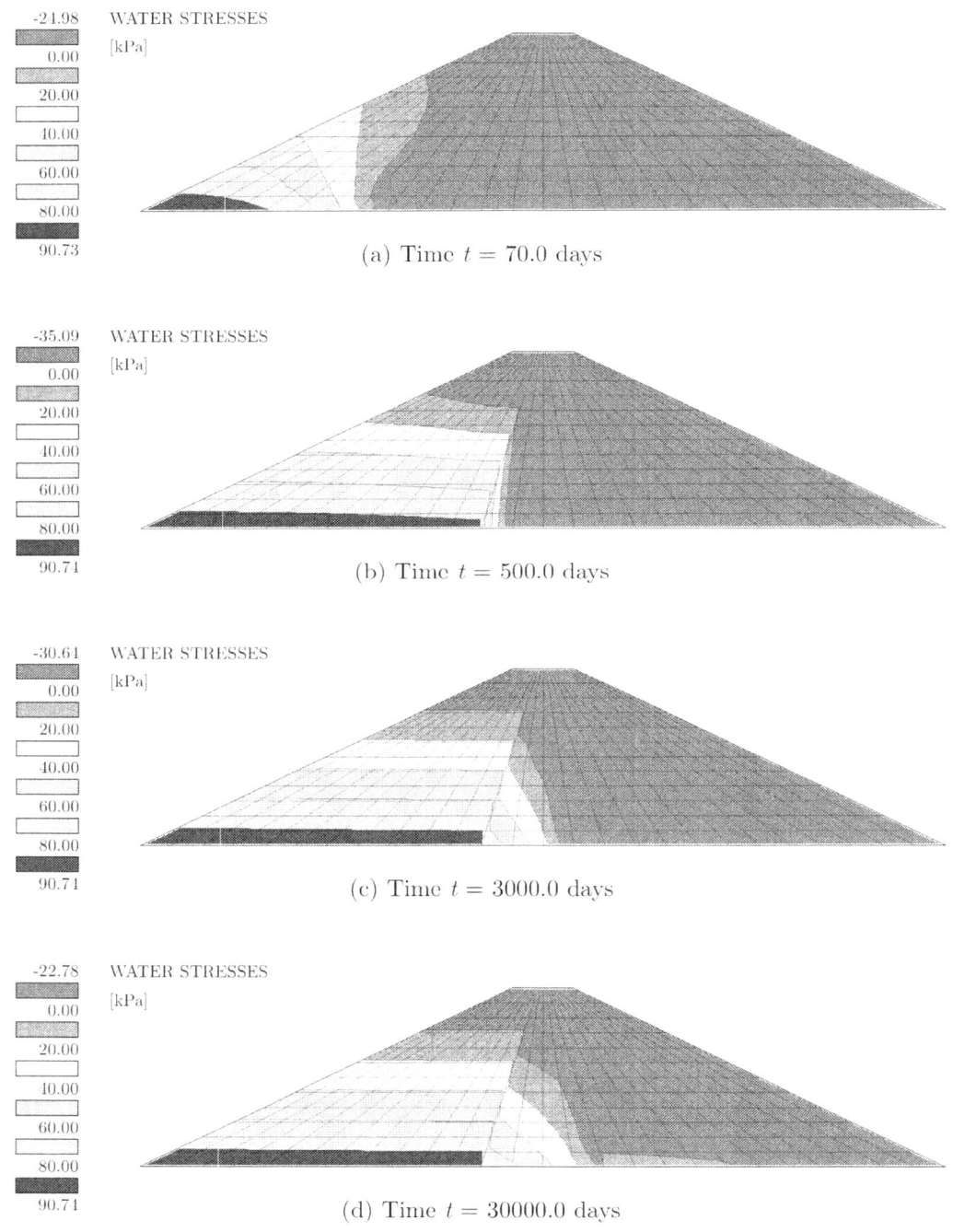

Figure 6.11: Evolution of the hydrostatic water stresses until a steady state is obtained (dam containing clayey core).

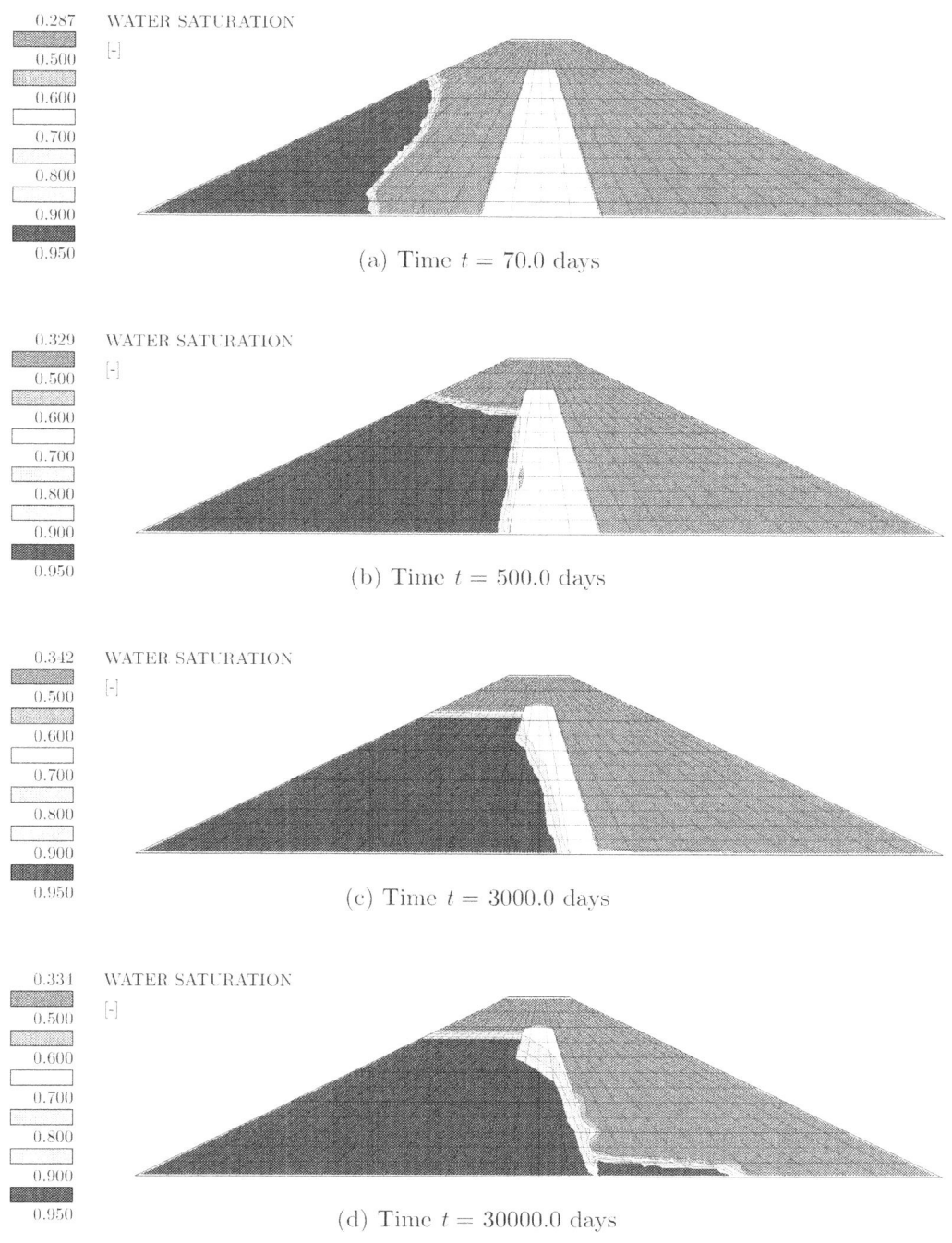

Figure 6.12: Evolution of the degree of saturation until a steady state is obtained (dam containing clayey core).

Corresponding to Figure 6.11, the evolution of the degree of water saturation in the dam with the clay core is shown in Figure 6.12. Before reaching the core, the wetting front proceeds in a way identical to the homogeneous case as a comparison of Figure 6.12(a) with Figure 6.6(b) indicates very clearly. Due to the low permeability of the clayey core, the entering of the wetting front into the core is delayed significantly. As Figures 6.12(b) and 6.12(c) depict, the propagation of the water into the core is so slow that water is built up in the upstream part of the dam such that a horizontal water level is obtained in this region (see Figure 6.12(c)). With increasing time the wetting front finally passes through the rather impermeable core, however, the isolines for the degree of saturation decline very steep in the clayey material.

Figure 6.12(d) shows the distribution of the degree of saturation in the dam after steady state conditions are attained which, according to the numerical simulation, happens after about 30000.0 days. While in the upstream part of the dam the horizontal water level in the reservoir is continued into the earth dam, the height of the water sat-

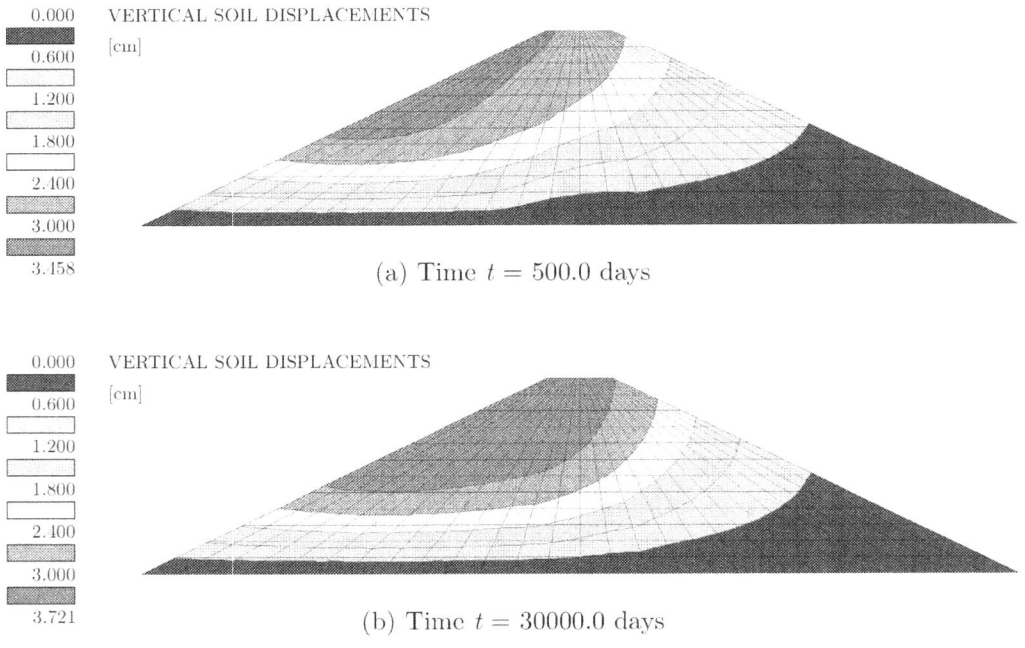

Figure 6.13: Evolution of the vertical displacements of the soil skeleton until a steady state is obtained (dam containing clayey core).

urated area is reduced considerably in the core. In most of the downstream part of the earth dam the initial conditions prevail, the degree of saturation amounts to the specified 40 %. Only a very narrow part of this side of the dam (about 1.0 m high) becomes saturated, stretching from the toe of the core to the upper end of the underdrain. This distribution of the degree of saturation clearly indicates that leakage through the dam can be reduced substantially by means of such a rather impermeable core.

Finally, Figure 6.13 shows the distribution of the vertical displacements of the soil skeleton in the dam for two instants of time (500.0 days and 30000.0 days after the raise of the water level). Compared to the homogeneous earth dam (see Figure 6.7) the deformations are slightly larger in this case which is due to the smaller Young's modulus of the core (Table 6.2). Since part of the dam now possesses a lower stiffness, the occurring displacements are larger, a fact which is more pronounced in the course of the calculation (confer size of red coloured area at point of time $t = 500.0$ days) than at the final stage (confer Figures 6.13(b) and 6.7(b)). The maximum value of the vertical displacements (heave) for the earth dam containing a clayey core predicted by the numerical simulation amounts to about 3.7 cm.

6.3 In-situ air permeability test

6.3.1 Description of the problem

In connection with the subway construction in Essen, Germany, compressed air was employed for dewatering of soils during tunnelling in aquifers. To investigate the behaviour of the outcropping types of soil with respect to the application of compressed air, a number of full-scale in-situ air permeability tests were carried out [Kramer(1989)]. The primary aims of these field experiments focussed (i) on the investigation of the air permeability of the Essen soil, i.e. on the identification of the air permeability coefficients of the various soil layers and their time- and pressure-dependent changes, (ii) on the determination of the extent of the airflow field, that is on the size of the region of the soil influenced by the compressed air and (iii) on the effect of the flow of compressed air on the deformations of the soil skeleton and in particular on the surface displacements.

According to the description of the experiment in the literature [Kramer(1989)], the soil profile at the test location consists of four distinct layers. At the ground surface a fill layer of about 3.0 m thickness is encountered, mainly composed of sand and silt but also containing parts of construction debris. Next comes a 7.0 to 8.0 m thick silt layer with portions of sand in the upper region and organic additions in the lower area. The permeability of the silt with respect to both water and air is rather low. In this layer the groundwater table is located at a depth of about 4.7 m below the ground surface. Underneath the silt, a thin highly permeable sand/gravel layer, containing partly considerable portions of silt, is encountered. At the depth where compressed air is injected into the soil, a thick layer of marl outcrops which is rather weathered in the upper region while showing pronounced joints further down. Considering the granulometric composition, the marl corresponds to a silt containing high portions of sand and clay. According to [Kramer(1989)], averaged values of the permeabilities for the marl may be used, although one should be aware of the possibility of high anisotropy between horizontal and vertical soil permeabilities. More details about the individual soil layers can be found in [Kramer(1989)].

Figure 6.14 contains a schematic diagram of the experimental set-up employed for the in-situ air permeability tests [Kramer(1989)]. It should be noted that the above-mentioned sand/gravel layer is included in the silt layer in this figure. In total, three different sets of tests were performed in which compressed air was injected into the ground via a large bore hole of 1.5 m diameter at different depths below the ground surface according to the various types of outcropping soil. In test 1A compressed air was pumped into the ground at a depth between 13.0 and 16.0 m where rather weathered marl with less pronounced joints was encountered. In experiment 1B undisturbed and jointed marl was present at the air injection level between 18.0 and 21.0 m. In experiment 2 the soil was grouted and compressed air was injected in the transition zone between marl and silt (depth of 11.0 to 14.0 m). Thus, in all sets of tests the injection region in the bore hole stretched along a height of 3.0 m. In a number of smaller bore holes, located in the vicinity of the main one, the air pressure in the soil was measured at different distances from the air injection zone and at various depths below the ground surface.

In experiment 1B under consideration here, compressed air was injected into the ground at a depth between 18.0 and 21.0 m below the ground surface with an air pressure up to 2.35 bar (air pressure in excess of the atmospheric pressure). Since the bottom end of the air injection region of the bore hole was located at a distance of about 16.0 m below the groundwater level, the minimum excess air pressure for dewatering of the soil had to be chosen as 1.60 bar. The pressure was applied in three main steps (1.60, 2.20 and 2.35 bar). Because of the time-dependent behaviour of the process, for attaining

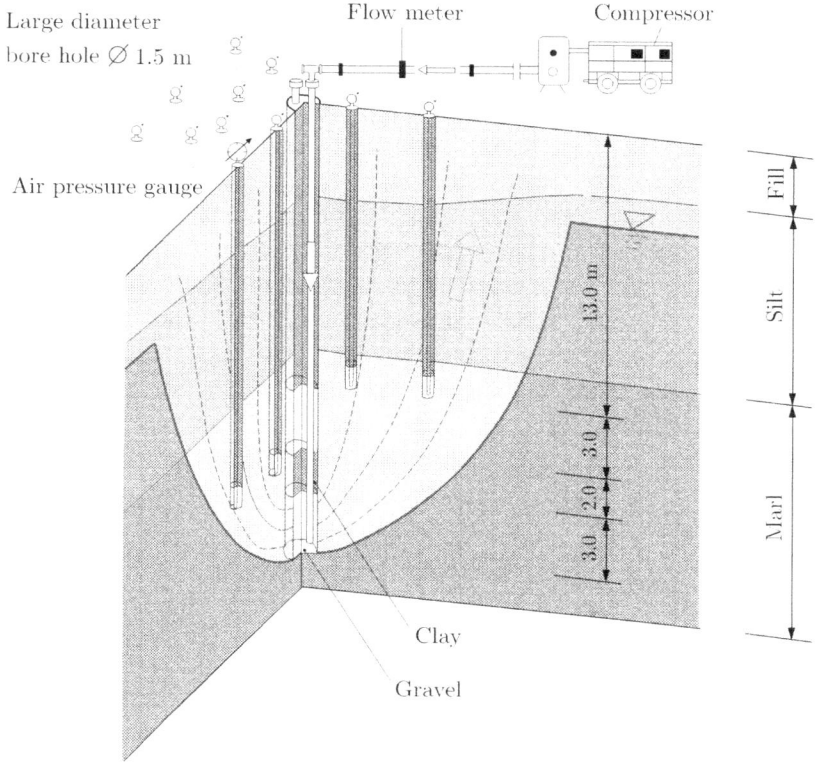

Figure 6.14: Experimental set-up for the in-situ air permeability test conducted in Essen, Germany [Kramer(1989)].

steady state conditions the air pressure was held constant for about one day at each of the three levels. Figure 6.15 shows a diagram of the three air pressure levels versus time in combination with the arising air loss. Figure 6.15 implies that at each pressure level the air loss increases with time which can be attributed to the increase of the air permeability of the soil associated with the reduction of the degree of saturation due to the displacement of the groundwater by the compressed air.

The following quantities were measured during the experiment either continuously or periodically: (i) the amount of compressed air supplied by the compressor (characterising the air loss through the soil), (ii) the excess air pressure at the level of air injection, (iii) the excess air pressure in the soil in the vicinity of the large bore hole (air pressure gauges) and (iv) the vertical displacements of the ground surface.

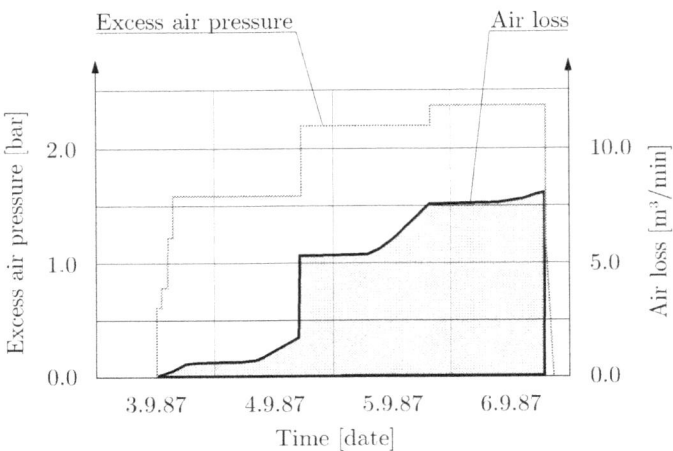

Figure 6.15: Excess air pressure and air loss versus time for test 1B.

6.3.2 Numerical model

When taking a look at Figure 6.14, the axial symmetry of the problem is easily recognised, the axis of symmetry being the axis of the large diameter bore hole. Thus, axisymmetric elements available in the finite element programme can be employed. The region to be discretised for the numerical simulation of the experiment is the soil body surrounding the bore hole, meaning that the wall of the bore hole is the left vertical boundary of the domain. The bottom and the right vertical boundaries have to be chosen such that the error induced by these artificial demarcations of the domain remains small.

The finite element mesh used for the numerical simulation of the problem is shown in Figure 6.16. The discretisation covers a region of soil with a width of 40.0 m and a height of 25.0 m. A number of 204 axisymmetric finite elements are used. While the element size is rather small in the vicinity of the bore hole and in particular in the air injection region, it significantly increases with increasing distance from the bore hole. The discretisation in the vertical direction is chosen such that the individual soil layers, the groundwater table and the capillary fringe can be taken into account.

Since the process of dewatering of soil by means of compressed air is simulated, both fluid phases water and air have to be considered in the model. Thus, a node of a finite element possesses a maximum number of four degrees of freedom, namely two displacements in the two coordinate directions and two fluid stresses for the constituents

Chapter 6. Application of the three-phase formulation 311

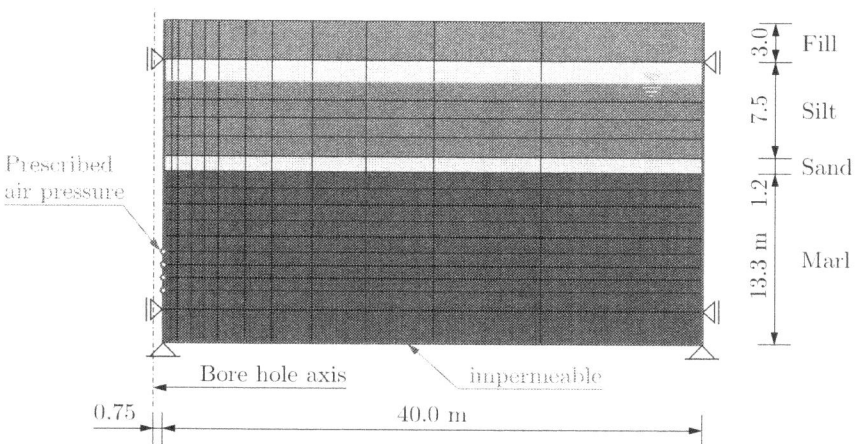

Figure 6.16: Finite element mesh employed for the numerical simulation of the in-situ air permeability test performed in Essen.

water and air. Displacement degrees of freedom are assigned to eight nodes per element (quadratic interpolation of the displacements), fluid stress degrees of freedom are assumed for the four corner nodes of each finite element (bilinear interpolation of the hydrostatic fluid stresses).

The specified initial conditions can be described as follows. Since a small amount of air (say $S^a = 0.01$) is assumed to be present in the pores of the soil even in the saturated region below the groundwater table, also hydrostatic stresses in the air phase are considered for this part of the domain. This assumption allows to avoid the transition from using a two-phase formulation for the water saturated soil to a three-phase formulation for the partially saturated soil. A linear increase of the air pressure with depth below the groundwater level is specified as initial condition, according to the assumed relationship for the water pressure below the groundwater level, $p^a = -\gamma^w z$, with γ^w being the unit weight of the water and z denoting the depth below the groundwater table. Above the groundwater level the hydrostatic stress in the air phase is assumed to be zero, i.e. atmospheric. Concerning the degree of saturation of the water phase, the initial conditions are specified as follows. Below the groundwater level an initial degree of saturation of $S^w = 0.99$ is assumed, above the groundwater table the saturation is gradually reduced. At the top of the silt layer a decrease of the degree of saturation is taken into account in a capillary fringe of 1.70 m height. Finally, for the fill layer the residual degree of saturation of about $S_r^w = 0.20$ is specified. According to this initial saturation a corresponding capillary stress is calculated from the relationship between degree of saturation and capillary stress which is then subtracted from the initial hydrostatic air stresses (which linearly increase below and are zero above the groundwater level) to obtain the initial water stresses, $p^w = p^a - p^c(S^w)$. Consequently, the initial hydrostatic water pressure linearly increases below the groundwater table, above the groundwater level tensile stresses are specified for the water phase. The initial effective stresses of the soil skeleton vary linearly with depth below the ground surface according to the unit weight of the soil layers, taking into account the hydrostatic stresses in both fluid phases and the respective degrees of saturation. At time $t = 0$ the excess air pressure along the filter part of the bore hole is applied suddenly and then kept constant in the course of the simulation.

The boundary conditions characterising the problem are depicted in Figure 6.16 and can be interpreted as follows. With respect to the displacements the domain is fully constrained at the bottom and horizontally constrained at the vertical boundaries. Concerning the two fluid phases, the subsequently described boundary conditions prevail. The wall of the bore hole, except the filter part where compressed air is injected into the ground, as well as the bottom and the right vertical boundaries are conceived to be impermeable. Along the filter part of the bore hole the air pressure applied in the experiment is prescribed as a boundary condition. It should be mentioned that due to the specified initial condition for the hydrostatic air stresses (linear increase of the air pressure with depth below the groundwater table), in the numerical simulation actually the difference between the value of 160.0 kPa (first pressure level) and the initial air pressure has to be specified as boundary condition, that is an air pressure linearly increasing from zero at the bottom end to about 30.0 kPa at the top of the 3.0 m high injection region. Concerning the water phase, any stresses in excess of the initial state are constrained to zero along the injection part of the bore hole, thus water flow normal to the boundary is possible. Along the top boundary of the domain, i.e. at the ground surface, atmospheric air pressure prevails, meaning that any excess stresses in both fluid phases water and air are constrained to zero.

The material parameters for the four soil layers as well as for the fluid phases water and air employed in the numerical simulation are summarised in Tables 6.4 to 6.6. Most parameters are determined according to the documentation of the experiment in the literature [Kramer(1987)] and data obtained from Prof. Semprich, Head of the Institute for

Soil Mechanics and Foundation Engineering at Graz University of Technology. Table 6.4 contains the properties of the four soil layers – fill, silt, sand and marl – encountered at the test site in Essen. While the mechanical parameters are given in the upper half of the table, the lower half comprises the properties necessary for the description of the hydraulic behaviour of the soils.

Parameter	Symb.	Unit	Fill	Silt	Sand	Marl
Young's modulus	E	kPa	20000.0	12466.0	21218.0	14333.0
Poisson's ratio	ν	-	0.333	0.350	0.316	0.397
Porosity	n	-	0.36	0.42	0.36	0.33
Permeability water	k^{ow}	m/s	$5.0 \cdot 10^{-5}$	$5.0 \cdot 10^{-6}$	$1.0 \cdot 10^{-4}$	$2.5 \cdot 10^{-5}$
Permeability air	k^{oa}	m/s	$3.0 \cdot 10^{-4}$	$4.5 \cdot 10^{-5}$	$4.5 \cdot 10^{-4}$	$2.5 \cdot 10^{-4}$
Density solid	ρ^s	t/m^3	2.72	2.90	2.72	2.79
Coefficient K_0	K_0	-	0.500	0.538	0.463	0.658
Max. saturation	S_s^w	-	1.00	1.00	1.00	1.00
Resid. saturation	S_r^w	-	0.20	0.20	0.05	0.15
Air entry value	p_b^c	kPa	-4.00	-30.00	-4.00	-12.00
Empir. parameter	m	-	0.80	0.50	0.65	0.60

Table 6.4: Material properties for the individual layers of the Essen soil.

Due to the rather small deformations occurring in the soil (in the experiment a maximum heave of about 0.8 cm was encountered), linear elastic behaviour of the soil skeleton is assumed for all layers and the elastic constants Young's modulus and Poisson's ratio are given in Table 6.4. Probably the most important parameters in the upper half of Table 6.4 are the permeabilities with respect to the fluids water and air. These parameters are taken from the literature describing the experiment [Kramer(1987)]. Porosities and densities as well as angles of friction (which are used to determine the coefficient of earth pressure at rest, K_0) of the individual types of soil have been provided by Prof. Semprich. It should be mentioned that the lower half of the silt layer possesses a somewhat weaker stiffness and the properties differing from the ones given in Table 6.4 are specified in Table 6.5. These parameters are used for a 3.0 m thick part of the silt layer directly overlying the sand/gravel layer.

Parameter	Symbol	Unit	Value
Young's modulus	E	kPa	9237.0
Poisson's ratio	ν	-	0.366
Density solid	ρ^s	t/m^3	2.72
Coefficient K_0	K_0	-	0.577
Air entry value	p_b^c	kPa	−28.00

Table 6.5: Material properties for the weaker part of the silt layer.

The lower half of Table 6.4 contains parameters determining the hydraulic behaviour of each soil layer. As the curves in Figure 6.17 indicate, these properties specify the relationships between degree of saturation and capillary stress and between degree of saturation and relative permeability. Especially the capillary stress versus saturation curves reflect the differences in the hydraulic behaviour of the individual types of soil.

Finally, the parameters employed for the two fluid phases water and air, that is densities and bulk moduli, are summarised in Table 6.6. Additionally, the value of the gravitational acceleration is specified.

In the numerical simulation of the experiment only the first pressure level of 1.60 bar is considered. This pressure is kept constant for a time span of 1730.0 minutes (about 29.0 hours) which is also used in the simulation. In the course of the calculation the time

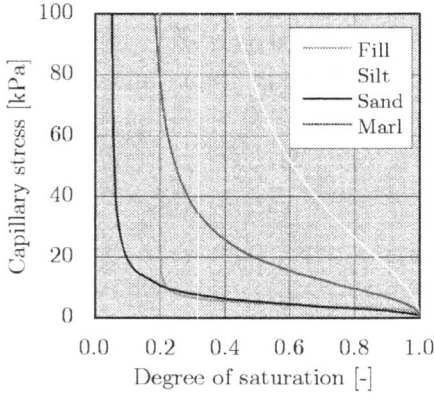
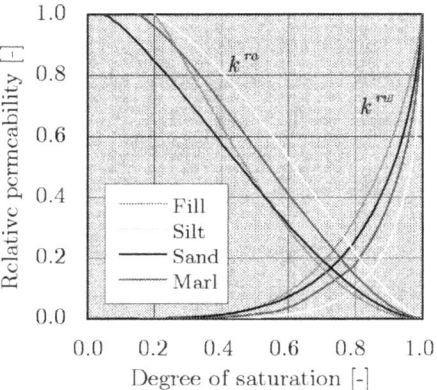

Figure 6.17: Hydraulic properties of the individual Essen soil layers.

Chapter 6. Application of the three-phase formulation   315

Parameter	Symbol	Unit	Value
Density water	ρ^w	t/m^3	1.00
Density air	ρ^a	t/m^3	$1.295 \cdot 10^{-3}$
Bulk modulus water	K^w	kPa	$2.0 \cdot 10^6$
Bulk modulus air	K^a	kPa	$1.0 \cdot 10^2$
Gravitational acceleration	g	m/s^2	9.81

Table 6.6: Material properties for the fluid phases water and air.

increments are allowed to vary between a minimum specified increment of 0.01 seconds and a maximum increment of 500.0 seconds.

6.3.3 Results of the numerical simulation

As described in Section 6.3.1, the first pressure level of 1.60 bar of experiment 1B is considered in the numerical simulation of this field test. Additionally, it should be mentioned that in all subsequent contour plots only a sector (width of 20.0 m) of the discretised domain, located in the vicinity of the bore hole, is shown.

Simulation of the first pressure level

Figure 6.18 depicts the computed distributions of the hydrostatic air stresses in the soil at four instants of time after the air pressure of 1.60 bar has been applied in the bore hole. The air stresses plotted are total values, i.e. the initially specified air pressures plus the hydrostatic air stresses caused by the application of the excess air pressure in the bore hole. Shortly after the start of the experiment only a small region close to the filter part of the bore hole is affected by the applied air pressure as Figure 6.18(a) indicates. Note that in the remaining part of the plotted domain the initial state of the hydrostatic air stresses can still be recognised, consisting in a linearly increasing pressure below and a zero (atmospheric) pressure above the groundwater table. With increasing time the influenced region becomes larger and larger, the additional air stresses in the soil, induced by the excess air pressure in the bore hole, spread to the upper and, to some extent, also to the right part of the domain (Figures 6.18(b) and 6.18(c)). Figure 6.18(d) refers to the distribution of the hydrostatic air stresses attained in the numerical simulation at

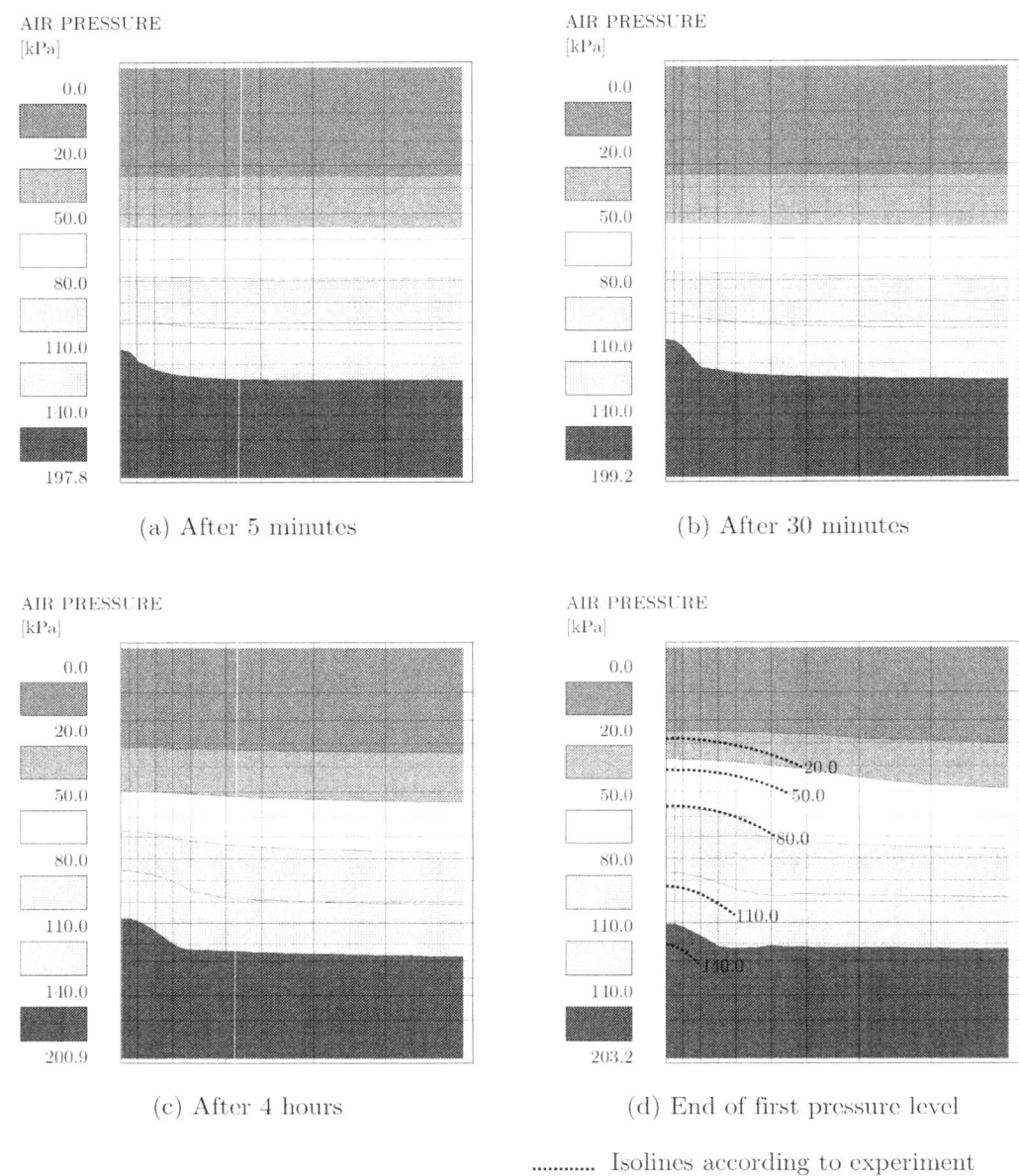

Figure 6.18: Development of air pressure during first pressure level.

the end of the first pressure level after approximately 29.0 hours. In the region close to the injection part of the bore hole the air stresses are somewhat reduced towards the end of the pressure level which is due to the substantial dewatering of the soil in this area (the void space available for the compressed air increases because of the dewatering and thus the pressure in the air phase is reduced). In order to allow for a comparison of the numerical solution with the experimental data, isolines composed from the recorded air pressure gauge values (small bore holes in Figure 6.14) are included in Figure 6.18(d). As can be recognised from this figure, the computed air pressure distribution agrees quite well with the measurements of the experiment.

Since the dewetting process and in particular the changes of the degree of saturation are governed by the capillary stress, that is by the difference between the hydrostatic air stress and the hydrostatic water stress, the computed development of the hydrostatic stresses in the water phase is depicted in Figure 6.19 for the same four points of time after the application of the excess air pressure in the bore hole. As Figure 6.19(a) indicates, shortly after the start of the experiment the initial distribution of the hydrostatic water stresses still prevails in almost the whole domain. The initial water pressure near the bottom end of the air injection region amounts to about 160.0 kPa which is the reason for designing the first pressure level to be 1.60 bar. The equidistant contour lines in the right part of this figure correspond to the linear increase of the initial hydrostatic water pressure with depth below the groundwater level, tensile stresses occur above the groundwater table due to the partially saturated state. In the course of time a water pressure increase is encountered in the domain as Figures 6.19(b) to 6.19(d) show. However, throughout the affected region the air pressure is larger than the water pressure. Consequently, the obtained compressive capillary stress, $p^c = p^a - p^w$, results in a dewatering of the soil in areas where this capillary pressure exceeds the air entry value. Along the injection part of the bore hole the boundary condition for the water stresses can be recognised in all plots (excess water stresses are constrained to zero to allow water flow normal to the wall). As Figure 6.19(d) indicates, after about 29.0 hours complete steady state conditions are not yet attained, otherwise the water pressure isolines would run horizontal according to the initial state.

The degree of water saturation, shown in Figure 6.20, develops according to the difference between the hydrostatic air stresses and the hydrostatic water stresses. Since shortly after the start of the experiment fluid pressure changes only occur in the vicinity of the air injection part of the bore hole (see Figures 6.18(a) and 6.19(a)), only a very small area of the domain is dewatered during, e.g., the first half hour (confer Figures 6.20(a) and 6.20(b)). A reduction of the degree of saturation is restricted to the immediate neighbourhood of the air injection part of the large diameter bore hole. In the remainder

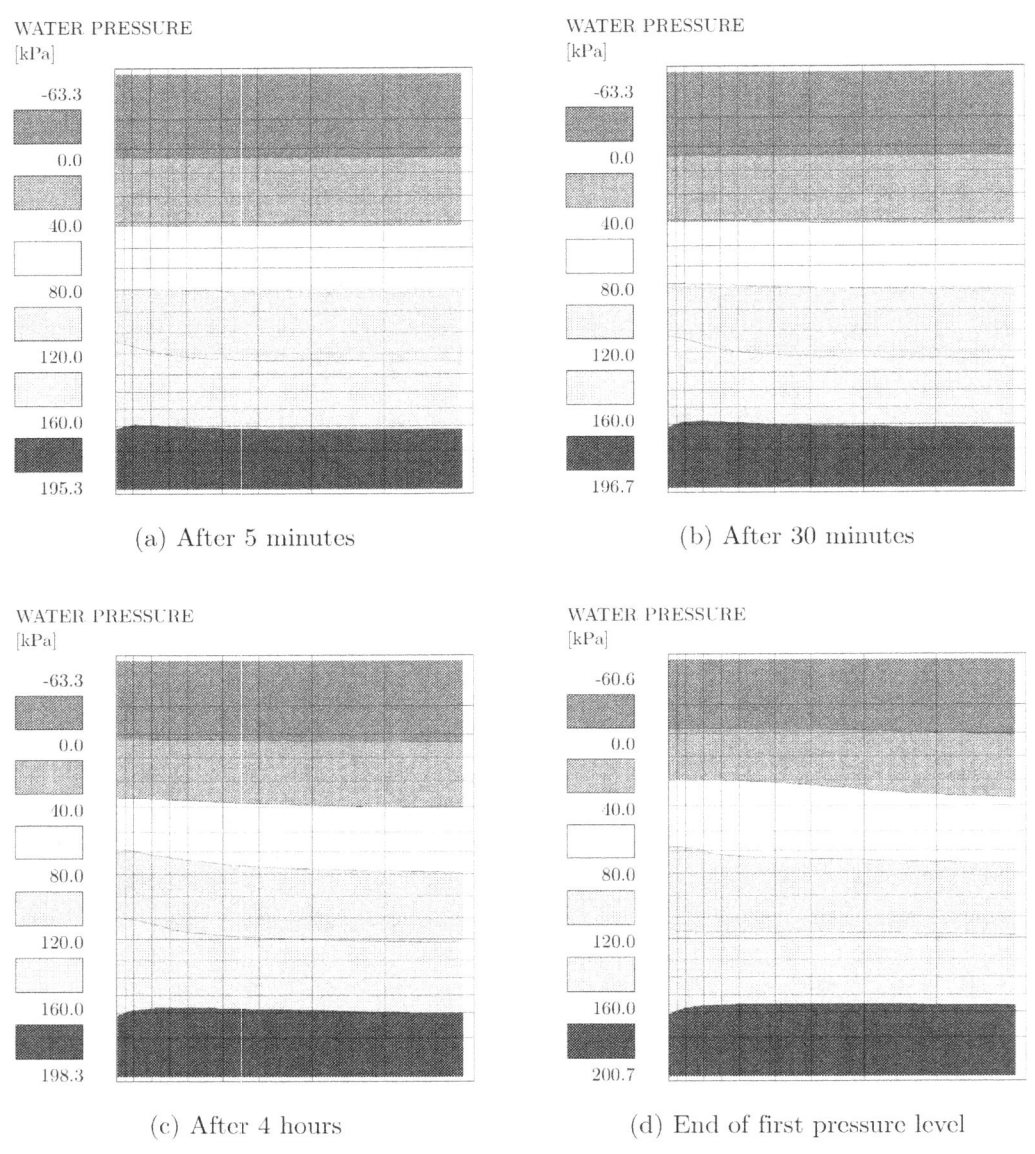

Figure 6.19: Development of water pressure during first pressure level.

of the discretised domain the initial conditions prevail, that is a degree of saturation of $S^w = 0.99$ below the groundwater table, a decreasing saturation in the capillary fringe of 1.70 m height and a residual saturation of about 20 % in the fill layer. The most pronounced dewatering of the soil is achieved near the top end of the injection

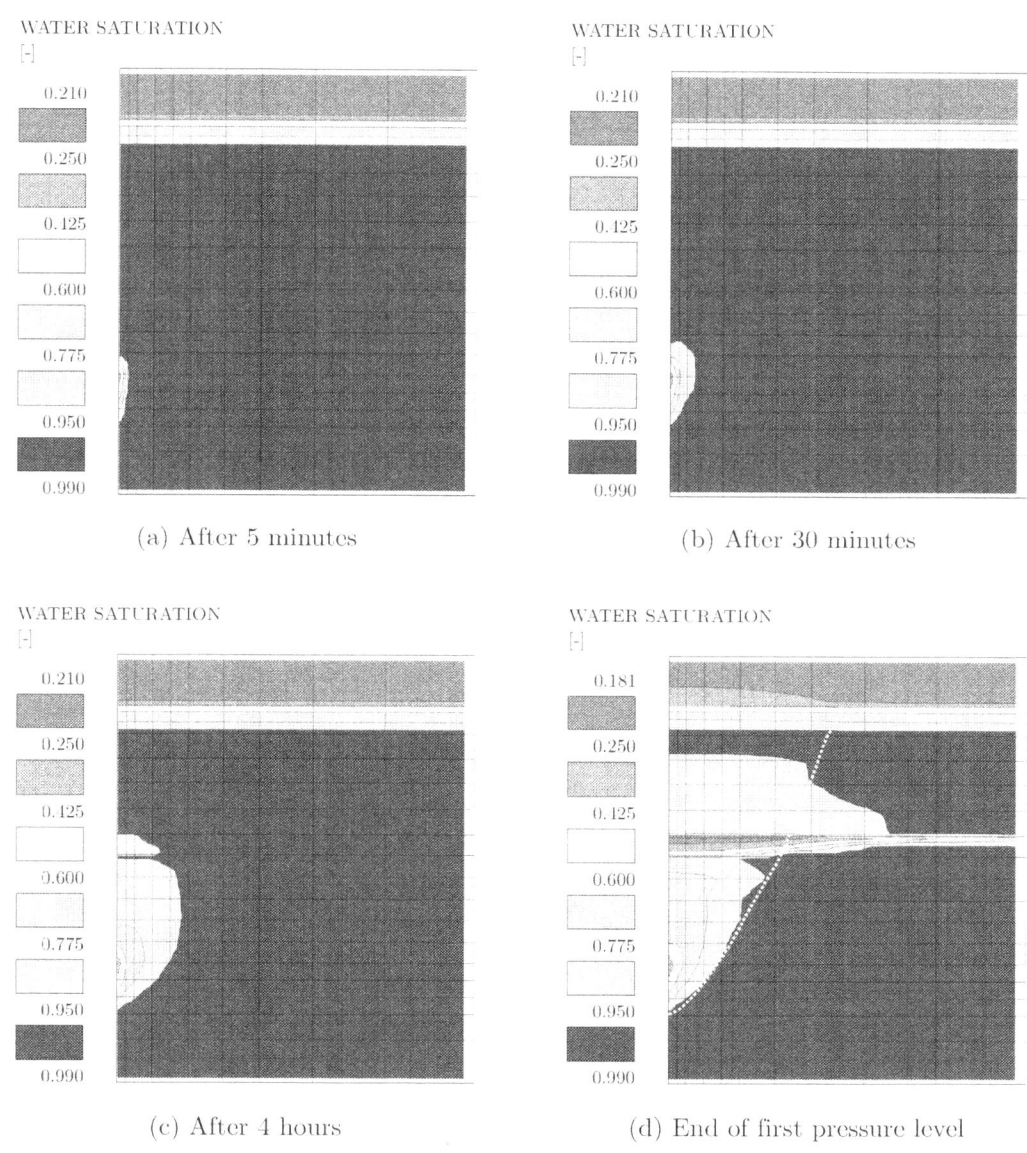

Figure 6.20: Development of water saturation during first pressure level.

region because there the difference between the excess air pressure applied in the bore hole (160.0 kPa) and the initial water pressure in the soil reaches a maximum, which is reduced to about zero at the bottom end of the injection part of the bore hole (here the air pressure equals the water pressure, being a standard assumption for dewatering of soils by means of compressed air). After about four hours the area exhibiting some dewatering has spread to the highly permeable sand/gravel layer (Figure 6.20(c)).

Figure 6.20(d) finally shows the computed distribution of the degree of saturation at the end of the first pressure level after about 29.0 hours. The maximum decrease of the water saturation is encountered in the thin highly permeable sand/gravel layer. Emanating from the air injection part of the bore hole a cone-shaped region of soil possesses a degree of saturation which is considerably lower than the initial saturation of 99 %. This cone indicates the extent of the airflow field. According to the description of test 1B under consideration [Kramer(1987)], the airflow field horizontally expanded to a distance of roughly 10.0 m from the bore hole. In Figure 6.20(d) the yellow dotted line denotes this estimated size of the airflow field obtained in the experiment. If a residual air saturation of 5 % is assumed, which might be justified according to Figure 3.14(a), for a degree of water saturation lower than 95 % a continuous air phase (and thus airflow) exists. Subdividing the saturated and the unsaturated parts of the domain by this manner, the isoline of 95 % water saturation corresponds to the experimentally determined extent of the airflow field quite well. Also obvious from Figure 6.20(d) is that due to the low permeability and the high air entry value of the silt the applied air pressure of 1.60 bar is not large enough to dewater the silt layer substantially. In the region above the groundwater table a slight increase of the degree of saturation can be recognised which is most obvious in the fill layer.

Figure 6.21 shows the displacements of the soil skeleton, occurring due to the fluid flow in the pores of the soil, for four points of time after the application of the air pressure in the bore hole. Note that there is a difference in the scale of the displacements between Figures 6.21(a), 6.21(b) and Figures 6.21(c), 6.21(d), respectively. While the displacements are restricted to the vicinity of the air injection part of the bore hole at the beginning of the experiment (only the soil in this area is affected by the flow of the fluids), they increasingly spread over the whole domain with time. In the early phase of the test (see Figures 6.21(a) and 6.21(b)), compared to the vertical components, larger horizontal components of the displacements occur than at later stages where the displacements are primarily vertical (confer Figures 6.21(c) and 6.21(d)). All plots show the effect of the flow of the fluids water and air in the pores on the deformations of the soil skeleton quite clearly. Due to the predominantly upward direction of the flow, a heave of the whole domain is encountered.

The computed deformations of the ground surface (heave) during the experiment are

Chapter 6. Application of the three-phase formulation 321

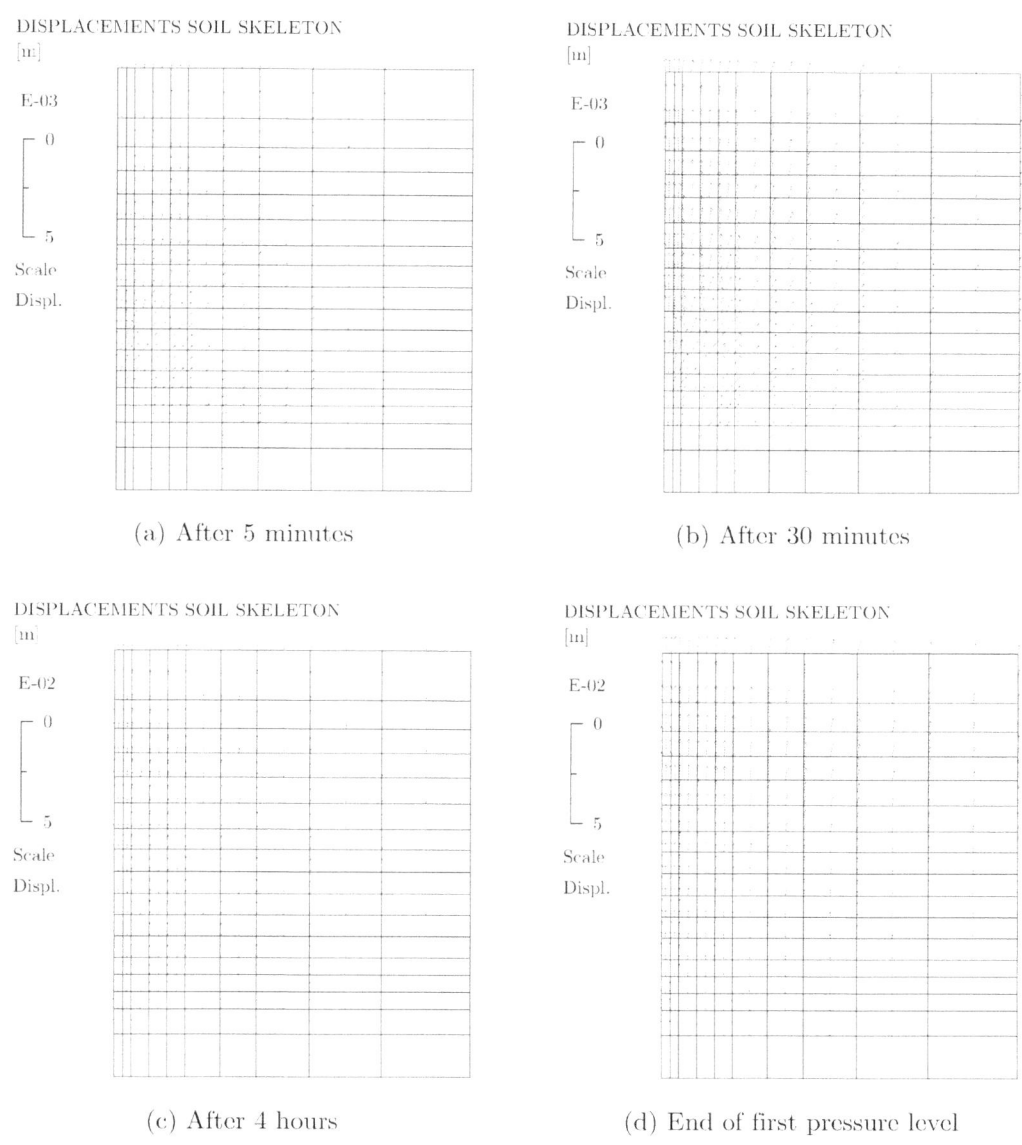

Figure 6.21: Development of soil displacements during first pressure level.

plotted in Figure 6.22 versus the distance from the large diameter bore hole. Four points of time corresponding to the preceding plots are considered. As would be expected, the surface heaves increase in the course of time. Maximum values of the vertical displacements occur directly next to the bore hole. Consequently, the displacements decrease

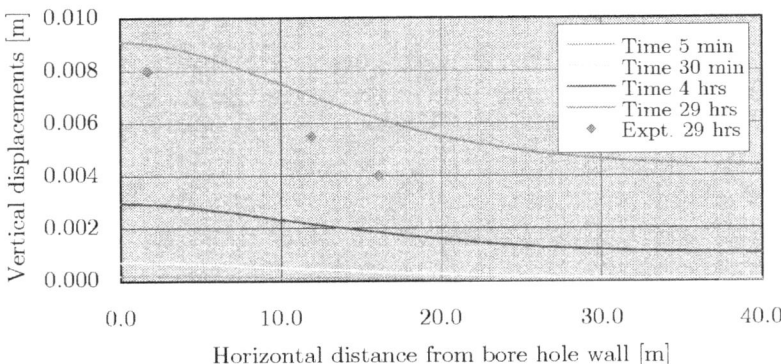

Figure 6.22: Surface heave during first pressure level.

with increasing distance from the bore hole. The surface heaves measured in the field test at the end of the first pressure level by means of geometric levelling are reported in [Kammerer(1998)]. These experimental data are included in Figure 6.22 (red diamonds) and may be compared with the topmost solid line. As can be recognised, the predictions of the numerical simulation are somewhat too large, the maximum measured value of 0.8 cm is overestimated by about 13 %, which seems to be not too bad keeping in mind the assumption of homogeneous soil layers with a simple linear elastic constitutive behaviour for the soil skeleton.

Figure 6.23 finally shows the evolution of the vertical displacements with time for three characteristic nodes of the finite element mesh distributed along the bore hole wall.

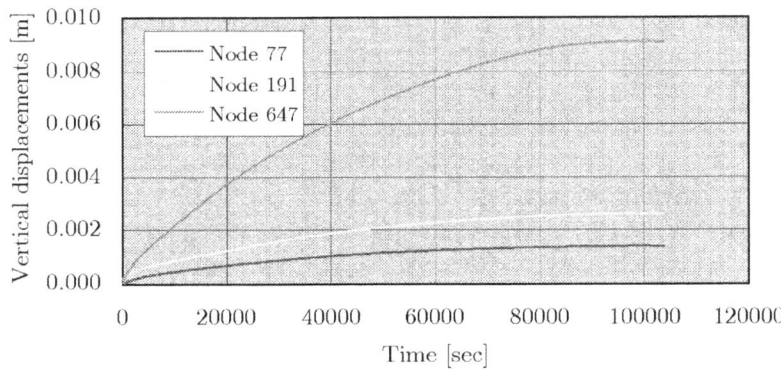

Figure 6.23: Vertical displacements for three nodes versus time.

The nodes number 77 and 191 represent the bottom and the top end of the air injection region, respectively. Node 647 is located at the ground surface. Figure 6.23 depicts a smooth and continuous increase of the deformations with time for all three nodes. The slope of the curves gradually decreases and thus the increase of the displacements (heave) gets smaller and smaller, at the end of the pressure level the curves are almost horizontal. At the ground surface (node 647) even a minor decrease of the heave is encountered near the end of the first pressure level.

Reduction of the bulk modulus of the water phase

As described in connection with Figure 3.8 or the example problem of Section 5.3, the bulk moduli of pure water ($K^w = 2.0 \cdot 10^6$ kPa) and water containing a small amount of dissolved air ($K^w = 1000.0$ kPa) differ substantially. Dealing with dewatering by means of compressed air, the presence of air dissolved in water seems to be very likely. To investigate the influence of a reduced bulk modulus for the water phase on the numerical results, the example discussed in the previous section is re-analysed using a bulk modulus of $K^w = 1000.0$ kPa instead of the one given in Table 6.6. Some results of the calculation are presented in this section and can be compared with the corresponding figures of the preceding section.

Figure 6.24 depicts the distributions of (a) the hydrostatic stress in the air phase and (b) the degree of saturation at the end of the first pressure level for the calculation using the smaller value for the bulk modulus of the water phase. A comparison of Figure 6.24(a) with Figure 6.18(d) shows only minor differences between both results. The smaller bulk modulus yields a qualitatively similar distribution of the air pressure in the whole domain, however, slightly smaller values. The results concerning the hydrostatic stress in the water phase (not shown) are characterised by deviations of the same order of magnitude.

The degree of saturation at the end of the first pressure level in Figure 6.24(b) can be compared with Figure 6.20(d). Once again only minor differences are obvious. The smaller bulk modulus results in a somewhat further expansion of the unsaturated region in the horizontal direction within the domain which is recognised in particular in the layers marl, sand and silt. Thus, the extent of the airflow field obtained from the experiment (see yellow dotted line in Figure 6.20(d)) would be approximated even better in the current calculation. The increase of the degree of saturation encountered above the groundwater table in Figure 6.20(d) is less pronounced in the present results (size of

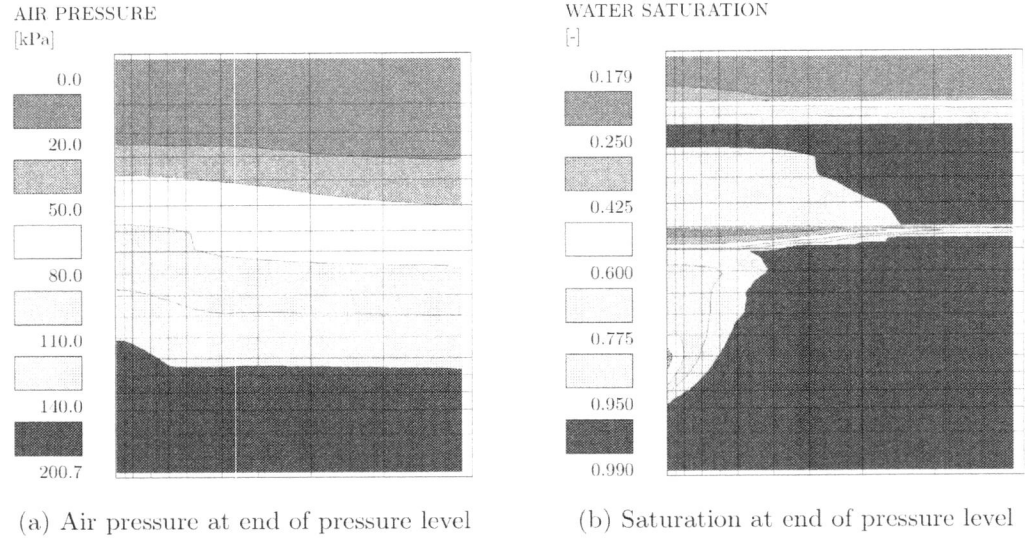

(a) Air pressure at end of pressure level (b) Saturation at end of pressure level

Figure 6.24: Final distributions of (a) air pressure and (b) degree of saturation using a reduced water bulk modulus.

the orange coloured area in the fill layer).

The displacements of the soil skeleton in the present analysis are distributed qualitatively similar to the plots shown in Figure 6.21. However, the actual values are somewhat smaller when using the smaller bulk modulus for the water phase. This seems to be plausible, since now (assuming a larger compressibility of the water phase) air entering the pores of the soil can compress the second fluid (water) present in the pores much easier. When thinking of a representative elementary volume (see Figure 3.1), the difference in the amount of air entering and leaving the volume element has not to be stored by merely increasing the volumetric strain of the soil skeleton (which is the case for the saturated area if the water phase is incompressible) but the water phase can be compressed yielding also void space available for the air. Consequently, smaller deformations of the domain (because of a smaller volumetric strain) accompanied by a reduced surface heave are obtained due to the more compressible water phase. Figure 6.25 shows the development of the surface displacements versus the distance from the bore hole in the course of the experiment. As can be recognised from this figure, the computed results (red solid line) agree much better with the measurements (red diamonds) of the experiment than for the case using the large water bulk modulus (depicted in Figure 6.22).

Chapter 6. Application of the three-phase formulation 325

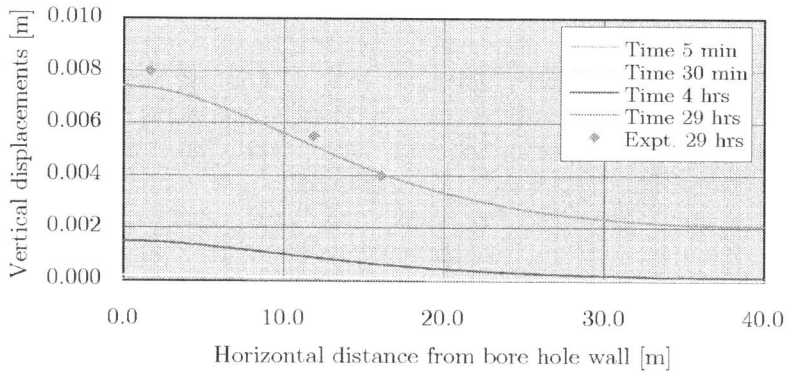

Figure 6.25: Surface heave during first pressure level using a reduced water bulk modulus.

Simulation employing a refined mesh

Finally, the first pressure level of the experiment under consideration is simulated numerically using a consistently refined mesh. Each element of the mesh shown in Figure 6.16 thus is subdivided into four elements. Consequently, the mesh for the current calculation consists of a number of 816 axisymmetric finite elements. Other than that, the degrees of freedom per node, the boundary and initial conditions and the material parameters are the same as employed for the first simulation. The results depicted in the subsequent figures can be compared with the plots shown in the Figures 6.18 to 6.20 and 6.22.

Figure 6.26 depicts the computed distribution of the hydrostatic air stresses in the domain for two instants of time after the application of the air pressure in the large diameter bore hole. For both the intermediate point of time (Figure 6.26(a)) and the end of the pressure level (Figure 6.26(b)) the hydrostatic air stresses are practically identical to the ones obtained with the coarse mesh. Only very small differences in the maximum value of the air pressure (order of magnitude of 1 %) are encountered between both calculations.

Figure 6.27 shows the computed distribution of the hydrostatic water stresses in the domain for two points of time. The accordance between calculations using both meshes is also quite perfect (maximum difference of about 1 %).

Somewhat larger deviations when employing the refined mesh occur in the computed distribution of the degree of saturation, shown in Figure 6.28 for two instants of time

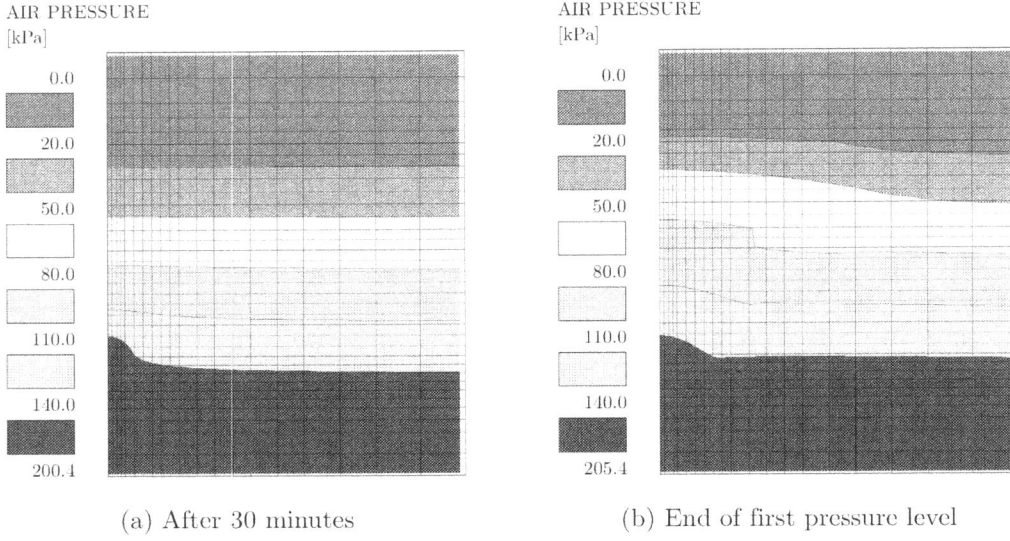

Figure 6.26: Development of air pressure employing the refined mesh.

after the application of the air pressure in the large bore hole. For the intermediate point of time (see Figure 6.28(a)) both calculations agree quite well. At the end of the first

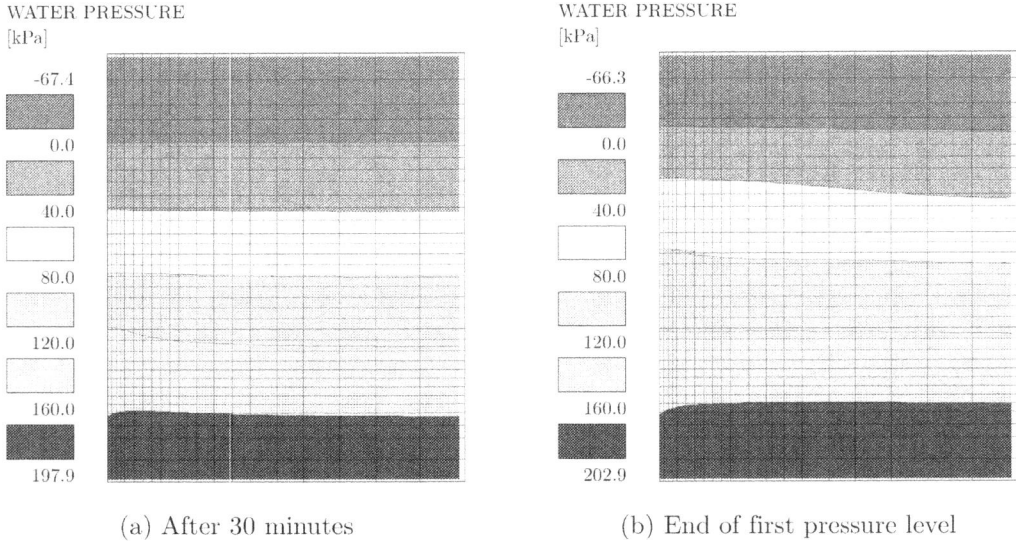

Figure 6.27: Development of water pressure employing the refined mesh.

Chapter 6. Application of the three-phase formulation 327

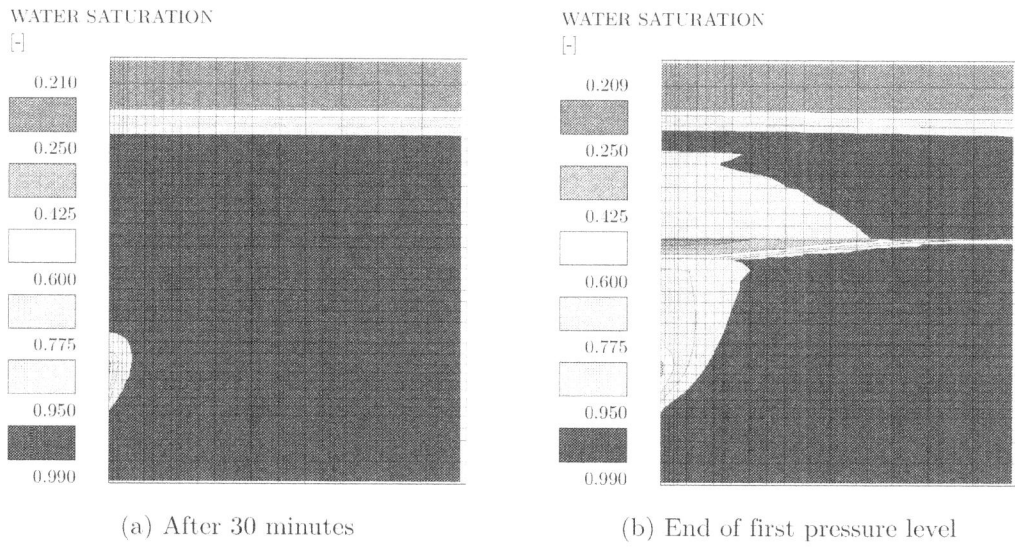

(a) After 30 minutes (b) End of first pressure level

Figure 6.28: Development of water saturation employing the refined mesh.

pressure level (Figure 6.28(b)), however, some differences are encountered in the upper soil layers, i.e. in the silt and fill layers. In the soils located below the silt (marl and sand) the computed distributions of the degree of saturation are very much the same for both meshes. In the upper half of the silt the 95 % saturation isoline stretches somewhat less in the horizontal direction in the present simulation. In the fill layer the increase of the degree of saturation is less obvious than the one shown in Figure 6.20(d). Using the refined mesh, the rise of the fully saturated part of the domain (dark blue coloured region) into the capillary fringe is more pronounced. Thus, the behaviour obtained with the refined mesh reminds of a rising groundwater table due to water displaced in underneath soil by the application of compressed air. Probably the finite element mesh shown in Figure 6.16 is too coarse in the region of the groundwater level.

Finally, Figure 6.29 depicts a plot of the vertical displacements (heave) of the ground surface versus the distance from the bore hole for four points of time. Qualitatively, the present results are similar to the ones shown in Figure 6.22. However, the actual values of the heave are somewhat larger when using the refined mesh (0.97 cm with the refined mesh compared to 0.91 cm with the coarse mesh). As a concluding remark it should be mentioned that the agreement obtained between the calculations using both finite element meshes is satisfactorily well.

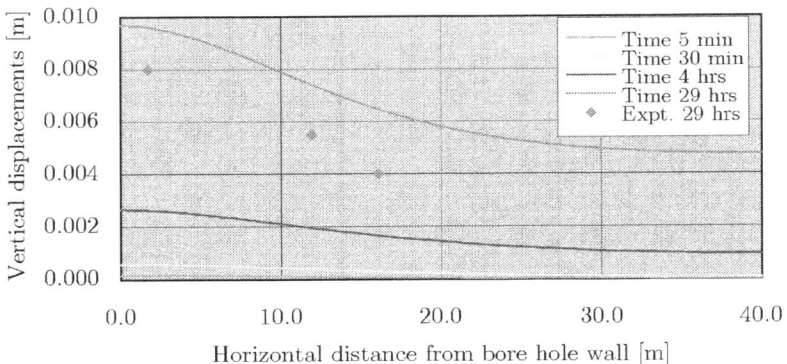

Figure 6.29: Surface heave employing the refined mesh.

6.4 Tunnelling below the groundwater level

6.4.1 Description of the problem

In the present example the proposed three-phase formulation is applied to the coupled numerical simulation of the excavation of a tunnel located below the groundwater table. During the construction of the tunnel the groundwater has to be displaced from the working area at the tunnel face. This dewatering of the soil is achieved by means of compressed air which is supplied to the tunnel. It should be emphasised that the aim of this section is not the presentation of a case study, but rather the demonstration of the principal steps of such a coupled numerical analysis of tunnelling below the groundwater level by means of compressed air.

The tunnel under consideration is assumed to be located in the Essen soil which has already been described in detail in connection with the in-situ air permeability tests discussed in the previous numerical example. The tunnel is located in soft cohesive soils, mainly consisting of silt, as the soil profile of Figure 6.30 depicts. The succession of soil layers considered in the numerical simulation consists of three types of soil (confer Figure 6.31(a)). Below the ground surface a 2.50 m thick fill layer is encountered. Following is a silt area, reaching down to a depth of 17.35 m below the ground surface. Underneath a layer of marl outcrops. The groundwater level is assumed to be located in the silt in about 4.75 m distance from the surface. As Figure 6.31(a) shows, the tunnel is driven in the transition region between silt and marl and possesses an overburden of 12.00 m.

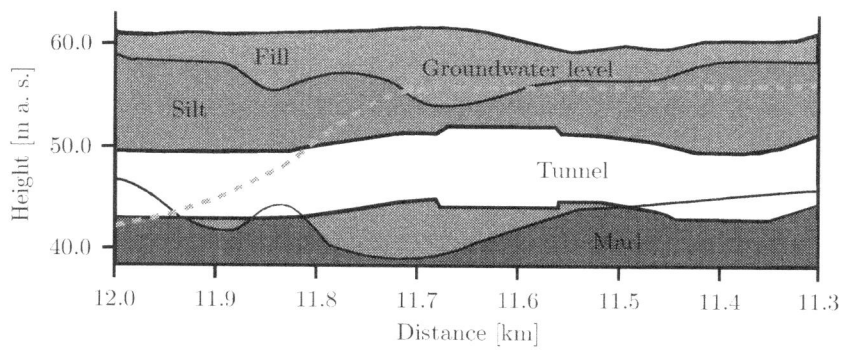

Figure 6.30: Soil profile at contract section 30 of the Essen subway.

In the analysis the standard cross section of a single track subway tunnel is investigated (geometry and excavation process are chosen according to the literature [Kropik(1994)]). Thus, the cross sectional area amounts to approximately 36.0 m² and its geometry is depicted in Figure 6.31(b). The height of the tunnel to be excavated is given as 6.85 m. Usually, such tunnels located in settlement-sensitive urban areas are driven employing a staged excavation procedure, described in more detail in the next section.

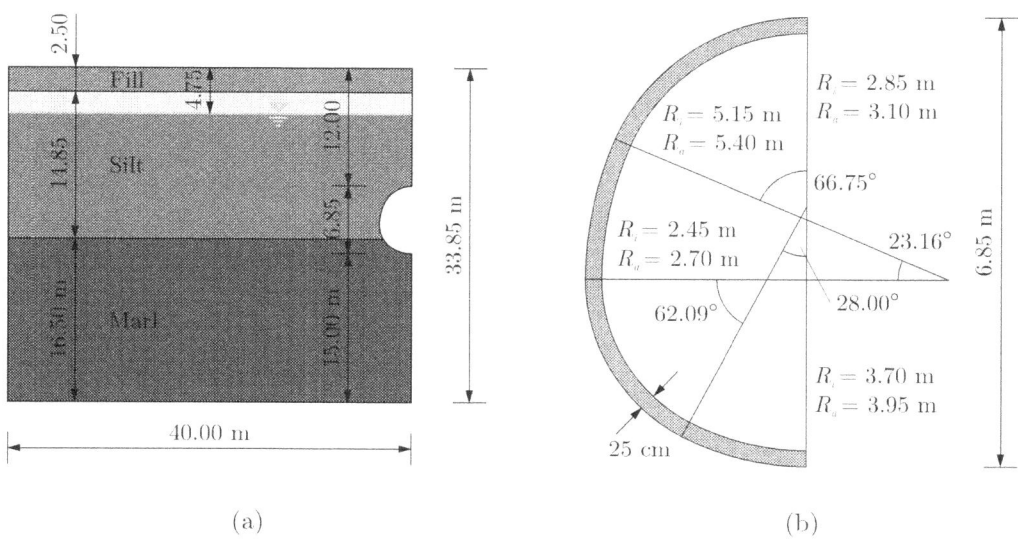

Figure 6.31: Geometry of (a) the discretised domain and (b) the tunnel cross section.

The securing of the tunnel is achieved by means of a shotcrete shell which in the current example is assumed to possess an average thickness of 25.0 cm (see Figure 6.31(b)).

In the numerical simulation of the problem both the dewatering process of the soil and the subsequent excavation of the tunnel are considered. Since a coupled model is used, deformations of the soil not only result from the excavation procedure but also the soil displacements due to the dewatering process can be taken into account in a physically consistent manner.

6.4.2 Numerical model

The numerical model for the tunnelling example relies on the assumption of plane strain conditions, allowing for a two-dimensional discretisation of the problem. Of course, the latter represents a simplification of the actual three-dimensional nature of the fluid flow in the soil as well as of the deformations and stress changes due to the advance of the tunnel face. However, the principal steps of the simulation of such a problem, which are intended to be outlined by the present example, remain the same no matter if a two-dimensional or a three-dimensional analysis is performed.

Taking advantage of the symmetry of the problem, the finite element mesh, shown in Figure 6.32, is employed for the numerical analysis. The mesh consists of a number of 496 plane strain finite elements. From the vertical symmetry plane a domain with a width of 40.00 m in each direction is considered in the simulation (however, the symmetry allows for modelling of just one half of the domain). This distance is assumed to be wide enough not to introduce an additional error into the calculation due to the truncation of the mesh. The height of the modelled region is conceived to be 33.85 m. The discretisation has to be chosen such that the individual soil layers, the groundwater table and the capillary fringe as well as the geometry of the tunnel, the different excavation stages and the shotcrete securing can be taken into account in the calculation. Rather small elements thus not only have to be used in the area of the tunnel to be excavated (and in particular for the tunnel lining to be installed after excavation) but also in the vicinity of the groundwater level (Figure 6.32).

Each finite element possesses eight nodes with displacement degrees of freedom (quadratic interpolation of the soil skeleton displacements), hydrostatic fluid stress degrees of freedom are assigned to the four corner nodes of each element (bilinear interpolation of the fluid stresses). Thus, a node of a finite element is considered to have a maximum number of four degrees of freedom, namely two displacements of the soil

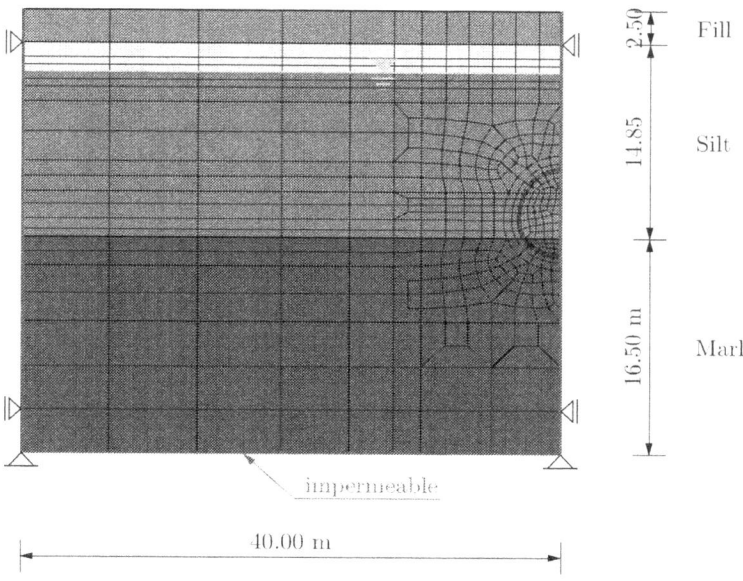

Figure 6.32: Finite element mesh employed for the numerical simulation of the tunnel excavation problem.

skeleton in the two coordinate directions and two fluid stresses for the phases water and compressed air.

Figure 6.32 also shows the boundary conditions employed for the numerical simulation of the problem. With respect to the displacements the domain is fully constrained at the bottom and horizontally constrained at the vertical boundaries. Concerning the two fluid phases water and compressed air, both vertical and the lower horizontal boundaries are assumed to be impervious. Additionally, at the ground surface atmospheric conditions prevail for both fluid phases.

At all nodes located in the interior of the tunnel, i.e. inside of the lining, both fluid stresses are constrained in the course of the analysis. The excess hydrostatic stresses in the air phase are constrained to the difference between the air pressure applied in the tunnel and the initial water stresses in the soil. Thus, the actual values of these excess air stresses increase from the invert in the direction of the crown of the tunnel. Any hydrostatic water stresses in excess of the initial stress state are constrained to zero. This assumption for the water phase is strongly connected to a three-dimensional numerical simulation of the problem, where at the tunnel face a water velocity component normal to the boundary (tunnel face) has to be permitted and thus any hydrostatic water stresses in excess of the initial stress state have to be constrained to zero.

The numerical simulation of the problem consists of three main parts, (i) the determination of the initial state, (ii) the dewatering process of the soil and (iii) the excavation and construction of the tunnel. In the first step the primary stress state for the soil skeleton and for the two fluid phases water and compressed air is computed, taking into account the given position of the groundwater level. As a consequence of the fully water saturated state of the soil below the groundwater table, the mass balance equation for the air phase could be omitted for this part of the domain. However, in order to avoid the transition between the use of a two-phase formulation in the fully saturated area and a three-phase formulation in the partially saturated region of the domain, the mass balance equation for the air phase is maintained and the degrees of saturation below the groundwater level are specified as $S^w = 0.99$ and $S^a = 0.01$. From a physical point of view the assumption of 1 % air being present even below the groundwater table may be justified due to the inclusion of air in small pores of the soil or in bubbles. In order to prevent any airflow because of this residual air saturation, the relative permeability coefficient k^{ra} for the air phase is set to zero for $S^a \leq 0.02$ according to an approach proposed in [Hochguertel(1998)].

Consequently, the initial (fictitious) hydrostatic air stresses below the groundwater table are assumed to be equal to the hydrostatic water pressure, i.e. the air stresses in excess of the atmospheric air pressure are computed from the unit weight of water and the depth below the groundwater level. This assumption may be reasonable when thinking of air bubbles trapped in the water filling the pores of the soil. Above the groundwater atmospheric air pressure, that is zero excess air pressure, is assumed. As mentioned above, the initial water saturation in the domain below the groundwater table is assumed to be $S^w = 0.99$. Above the groundwater level a capillary fringe of 2.25 m height is taken into account where the degree of saturation gradually decreases. Finally, in the fill layer the residual degree of water saturation of about 20 % is attained. The initial water stresses in the domain are then determined according to the specified initial degree of saturation, meaning that a capillary stress corresponding to the initial water saturation is subtracted from the initial hydrostatic air stress, $p^w = p^a - p^c(S^w)$. Consequently, as the hydrostatic air stresses, the initial hydrostatic water stresses linearly increase below the groundwater table, while tensile stresses prevail in the water phase above the groundwater level. The initial effective stresses in the soil skeleton are computed from the unit weight of the soil skeleton material, taking into account the presence of both fluid phases in the pores of the soil.

After determining the initial conditions for the individual constituents, the second step of the analysis consists of the numerical simulation of the dewatering of the soil. The displacement of the groundwater is achieved by applying an air pressure in excess of the atmospheric pressure of approximately 1.40 bar in the interior of the tunnel. If the tunnel is completely located below the groundwater table, the minimum excess air pressure employed in the tunnel for dewatering the soil at the tunnel face has to be chosen such that it balances the water pressure at the invert of the tunnel cross section. In the present simulation it is assumed that the excess air pressure would be equal to the water pressure at about 0.50 m below the invert of the tunnel. Thus, it is guaranteed that also the soil in the vicinity of the bottom of the tunnel is dewatered sufficiently. Since the excess air pressure applied in the tunnel is constant along the tunnel height whereas the water pressure linearly increases with depth below the groundwater level, the air pressure exceeds the water pressure above the invert of the tunnel. This excess of the air pressure over the water pressure at the invert corresponds to a pressure difference related to a height of 0.50 m (equal to 5.0 kPa), at the crown of the tunnel it amounts to about 70.0 kPa (equal to 0.70 bar). Between these two levels the excess of the air pressure over the water pressure linearly increases from the invert in the direction of the crown of the tunnel according to the linearly decreasing hydrostatic water pressure. This pressure difference initiates the flow of compressed air through the tunnel face and through the adjacent soil, resulting in a time-dependent development of an excess air pressure field in the soil and consequently in the dewatering of the soil in the vicinity of the tunnel face.

After a sufficient reduction of the degree of water saturation in the soil in the vicinity of the tunnel face has been achieved, the third step of the simulation, consisting of modelling the excavation and construction process of the tunnel, is performed. In the present example the tunnel is assumed to be constructed in three working cycles [Kropik(1994)] as depicted in Figure 6.33. The individual stages of the construction sequentially deal with the crown (Figure 6.33(a)), the bench (Figure 6.33(b)) and the invert (Figure 6.33(c)) of the tunnel. Each stage is subdivided into two further steps: first, the soil is excavated and second, a shotcrete lining of 25.0 cm thickness is installed.

Since the tunnel is assumed to be located in the Essen soil, the material parameters of the three soil layers – fill, silt and marl – are chosen according to the documentation

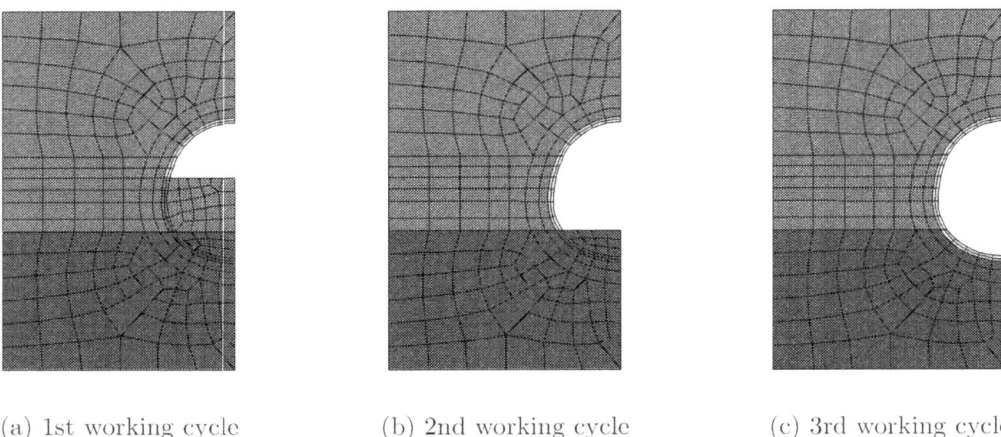

(a) 1st working cycle (b) 2nd working cycle (c) 3rd working cycle

Figure 6.33: Working cycles of the tunnel construction process.

of the in-situ air permeability tests (confer Section 6.2) in the literature [Kramer(1989)]. Both the mechanical and the hydraulic properties of the individual types of soil are summarised in Table 6.7. The upper half of Table 6.7 contains the two elastic constants (Young's modulus and Poisson's ratio), the porosity, the permeabilities with respect to

Parameter	Symbol	Unit	Fill	Silt	Marl
Young's modulus	E	kPa	20000.0	12500.0	14300.0
Poisson's ratio	ν	-	0.33	0.35	0.40
Porosity	n	-	0.36	0.42	0.33
Permeability water	k^{ow}	m/s	$5.0 \cdot 10^{-5}$	$5.0 \cdot 10^{-6}$	$2.5 \cdot 10^{-5}$
Permeability air	k^{oa}	m/s	$3.0 \cdot 10^{-4}$	$4.5 \cdot 10^{-5}$	$2.5 \cdot 10^{-4}$
Density solid	ρ^s	t/m^3	2.72	2.90	2.79
Earth pressure coefficient	K_0	-	0.500	0.538	0.658
Maximum saturation	S_s^w	-	1.00	1.00	1.00
Residual saturation	S_r^w	-	0.20	0.20	0.15
Air entry value	p_b^c	kPa	−4.00	−28.00	−12.00
Empirical parameter	m	-	0.80	0.50	0.60

Table 6.7: Material properties for the soil layers of the tunnel problem.

the two fluid phases water and air, the solid density and the coefficient of earth pressure at rest, while the lower half gives the parameters for the relationships between degree of water saturation and capillary stress, $S^w(p^c)$, as well as between degree of water saturation and relative fluid permeability, $k^{rf}(S^w)$, $f = w, a$. The curves $S^w(p^c)$ and $k^{rf}(S^w)$, $f = w, a$, are depicted in Figure 6.34 for the three layers to outline the different hydraulic behaviour of the soils.

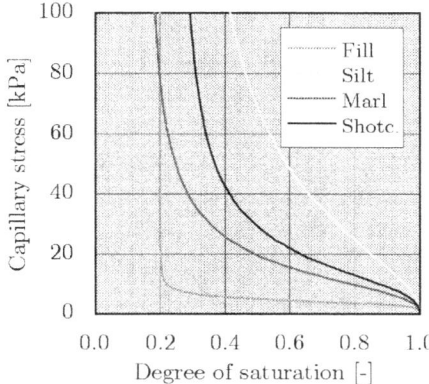 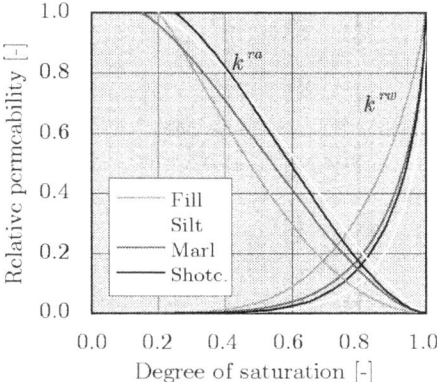

Figure 6.34: Hydraulic properties of the individual materials.

The parameters characterising the two fluid phases water and compressed air, i.e. the densities and the bulk moduli, are given in Table 6.6 of Section 6.3.2. Table 6.8 finally contains the material properties used for the shotcrete lining. The mechanical parameters (Young's modulus, Poisson's ratio, porosity and solid density) are chosen according to information obtained from Dr. Paulini from the Institute for Building Materials and Building Physics at the University of Innsbruck. The permeability of the lining with respect to the two fluids, which primarily results from cracks in the shotcrete shell [Kammerer(2000)] (otherwise the material shotcrete is rather impermeable), is considered in a smeared manner by the permeability coefficients given in Table 6.8. These parameters are determined from laboratory tests investigating the air permeability of shotcrete slabs containing cracks of defined widths [Kammerer(2000)]. From continuously recorded measurements of the air loss conducted in these experiments the value of the air permeability is calculated. The water permeability finally is obtained by means of the air permeability and the ratio of the dynamic viscosities of both fluids. The relationships between degree of water saturation and capillary stress as well as between

Parameter	Symbol	Unit	Value
Young's modulus	E	GPa	30.0
Poisson's ratio	ν	-	0.20
Porosity	n	-	0.20
Permeability water	k^{ow}	m/s	$2.7 \cdot 10^{-5}$
Permeability air	k^{oa}	m/s	$2.0 \cdot 10^{-3}$
Density solid	ρ^s	t/m^3	2.75
Earth pressure coefficient	K_0	-	0.250
Maximum saturation	S_s^w	-	1.00
Residual saturation	S_r^w	-	0.25
Air entry value	p_b^c	kPa	-15.00
Empirical parameter	m	-	0.60

Table 6.8: Material properties used for the shotcrete lining.

degree of water saturation and relative fluid permeability for the shotcrete are shown in Figure 6.34.

6.4.3 Results of the numerical simulation

Numerical results of the problem under consideration are presented according to the three main steps which the analysis consists of. The first step, concerning the specified initial state, is described in the previous section. The results shown in the current section deal with the dewatering of the soil and the excavation and construction of the tunnel. As Figure 6.30 indicates, for most parts of the contract section the tunnel is located in the transition region between silt and marl. However, due to the variation of both the soil layers and the tunnel axis in height, it is possible that the tunnel has to be driven in either silt or marl alone. Since the hydraulic behaviour of these soils is quite different, as Figure 6.34 outlines, the dewatering process is investigated numerically for these three cases of the possible tunnel location. First, the tunnel is located in the transition region between silt and marl as shown in Figure 6.32, second, the cross section only covers silt and third, the tunnel is completely located in the marl layer. For all cases a dewatering period of 24.0 hours is taken into account in the numerical simulation.

Tunnel located in the transition region between silt and marl

The computed development of the hydrostatic air stresses during the chosen dewatering period of 24.0 hours is shown in Figure 6.35 for four instants of time. It should be mentioned that the air stresses, depicted in Figure 6.35, are the specified initial air pressure plus the excess air stresses in the soil due to the application of compressed air in the tunnel. Shortly after the application of the excess air pressure of 1.40 bar in the tunnel (Figure 6.35(a)) only in the vicinity of the tunnel an increase of the air pressure

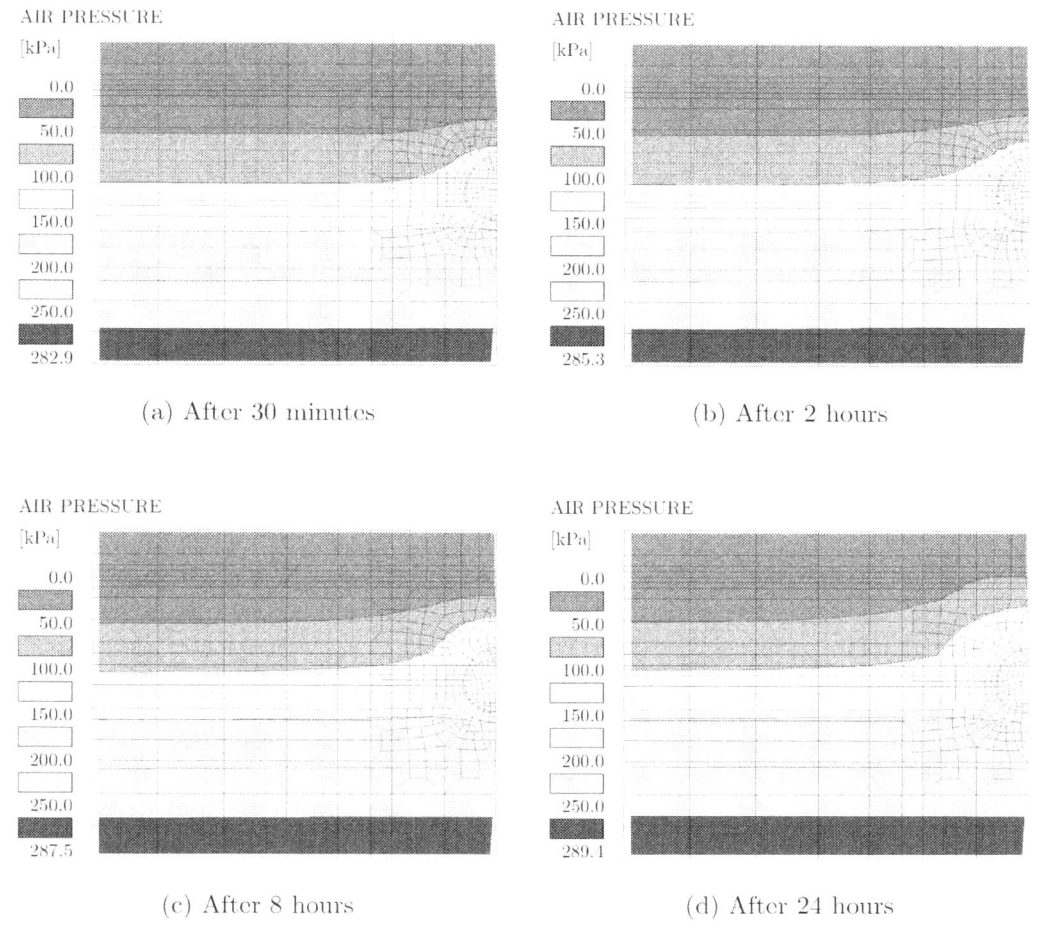

(a) After 30 minutes

(b) After 2 hours

(c) After 8 hours

(d) After 24 hours

Figure 6.35: Development of hydrostatic air pressure if the tunnel is located in the transition region between silt and marl.

is encountered. In the left part of the domain the initial conditions can be recognised, i.e. zero air pressure above and a linear air pressure increase below the groundwater level. With increasing time the air pressure increases in the domain, most pronounced in the vicinity of and in the area above the tunnel. Along the left vertical boundary of the domain basically the initial distribution of the air pressure prevails during the whole dewatering period. Additionally, the horizontal isolines indicate that the domain has been chosen sufficiently large.

The computed development of the hydrostatic water stresses is shown for the corresponding four instants of time in Figure 6.36. Again total stresses are plotted, i.e. the initial stress state plus the stresses due to the application of the air pressure in the tunnel. In the left part of all plots the initial water pressure can be recognised. Hence, there the water pressure linearly increases in the fully saturated region of the domain, the position of zero water stresses indicates the groundwater level and tensile water stresses are encountered in the partially saturated region above the groundwater table. As can be seen, in the silt layer above the groundwater level rather pronounced tensile stresses occur (minimum value of the legend). In the course of the dewatering period of the soil, in the vicinity and above the tunnel face an increase of the compressive hydrostatic water stresses is recognised. However, in the interior of the tunnel the prescribed boundary conditions for the water phase are obvious in all plots (Figures 6.36(a) to 6.36(d)) – any water stresses in excess of the initial hydrostatic water pressure are constrained to zero. Figure 6.36(d) depicts the computed distribution of the water stresses in the domain at the end of the dewatering period after 24.0 hours.

The difference between the two fluid stresses in the air and the water phase denotes the capillary stress which governs the distribution of the degree of water saturation in the domain. The computed development of the water saturation for four instants of time is shown in Figure 6.37 for a part of the discretised domain in the region of the tunnel face. Shortly after the application of the air pressure in the tunnel (Figure 6.37(a)) a change in the degree of saturation is encountered only in the close vicinity of the tunnel face (tunnel cross section). Since the applied air pressure in the tunnel is constant whereas, on the contrary, the water pressure linearly increases with depth below the groundwater table, the excess of the air pressure over the water pressure, resulting in a dewatering of the soil, linearly increases with increasing height above the tunnel invert. The increasing excess air pressure causes an increasing dewatering of the soil, or in other words, the degree of saturation is reduced more and more with increasing height above the tunnel invert.

Since the lower part of the tunnel cross section stretches into the marl whereas the upper part is located in the silt layer, a pronounced borderline in the degree of saturation is recognised at the common surface of both materials. In the region outside of the tunnel the initial conditions for the degree of saturation prevail, i.e. a degree of saturation of $S^w = 0.99$ below the groundwater table, a capillary fringe of 2.25 m in which the saturation is gradually reduced and a residual degree of saturation of $S_r^w = 0.21$ in the fill layer outcropping below the ground surface (Figure 6.37(a)).

With increasing time the dewatering of the soil outside of the tunnel cross section

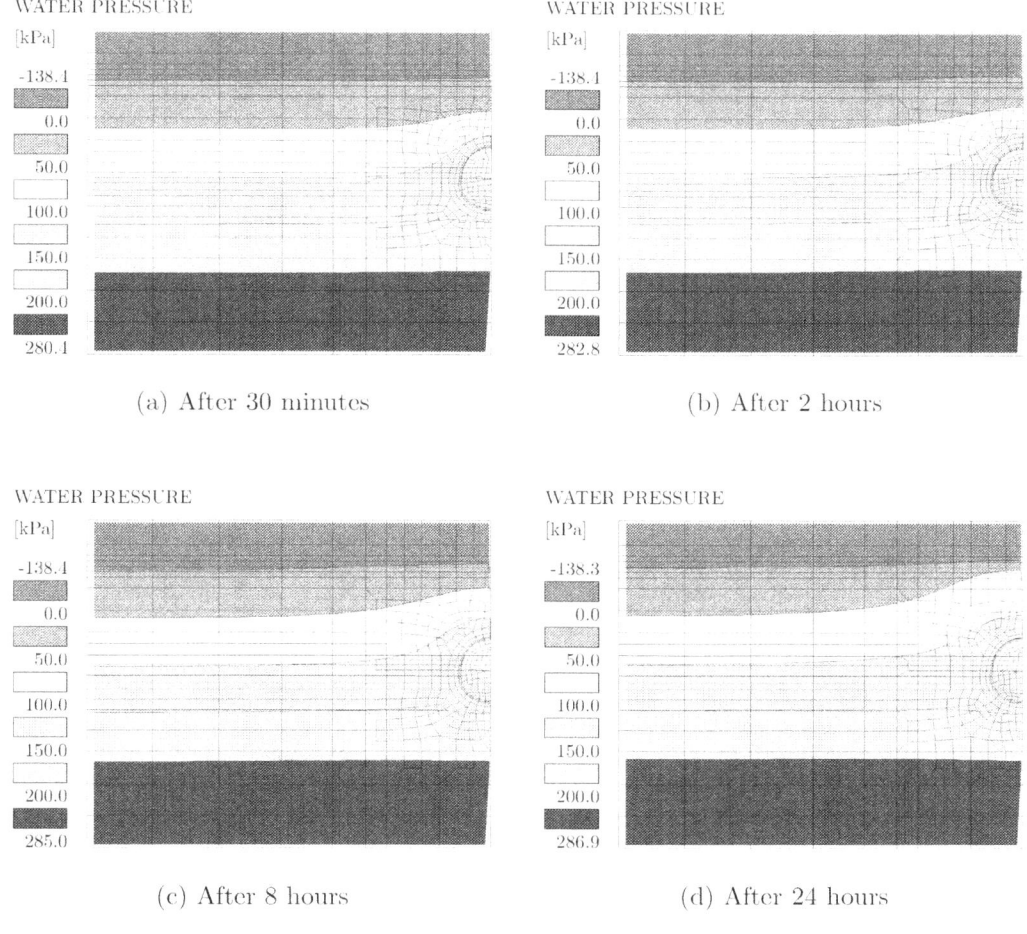

Figure 6.36: Development of hydrostatic water pressure if the tunnel is located in the transition region between silt and marl.

proceeds, the degree of saturation is reduced and the groundwater is displaced in the pores of the soil (Figures 6.37(b) and 6.37(c)). The computed distribution of the water saturation at the end of the dewatering period after 24.0 hours is shown in Figure 6.37(d). The region exhibiting some dewatering almost stretches up to the initial groundwater

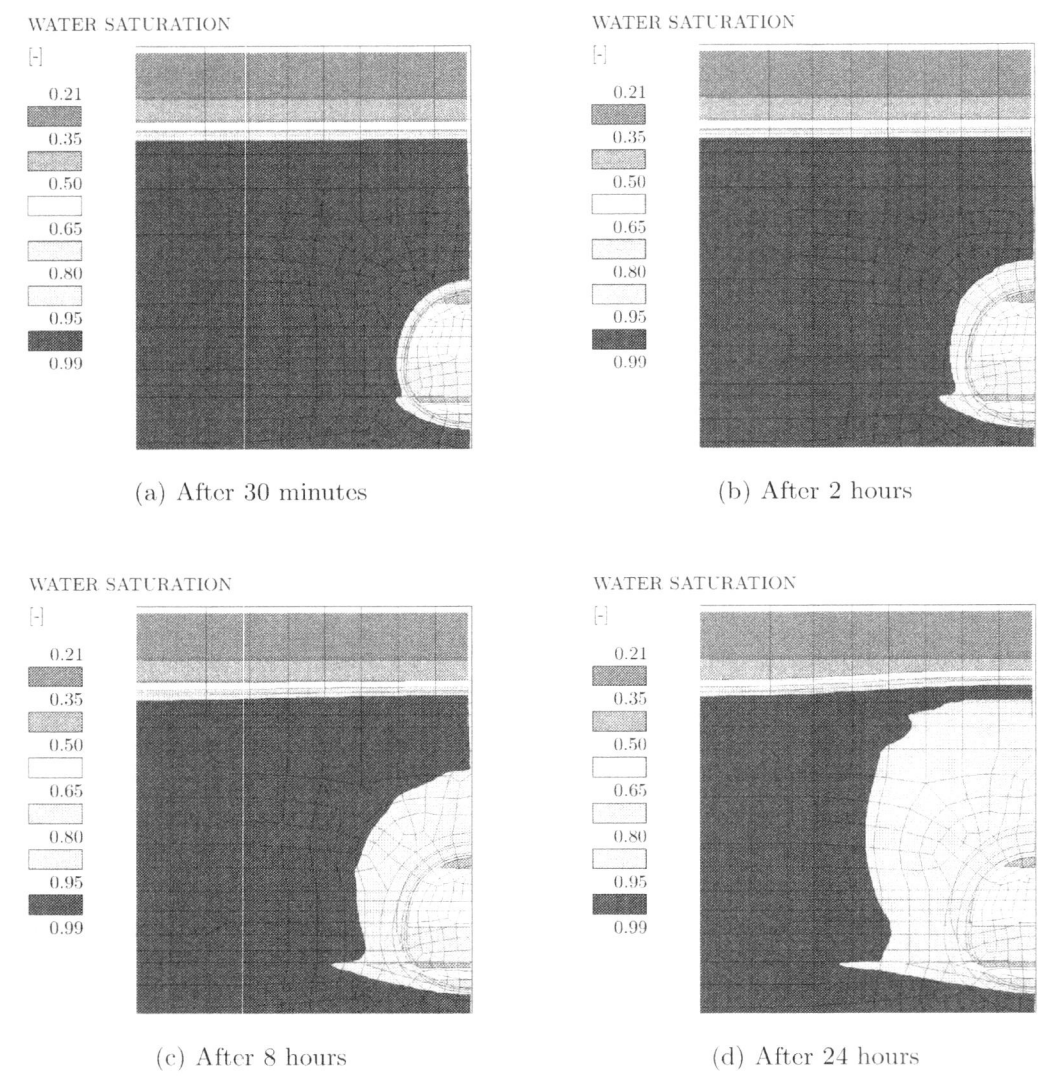

Figure 6.37: Development of degree of saturation if the tunnel is located in the transition region between silt and marl.

table. Since the permeability of the marl is larger than the one of silt, resulting in a lower (in absolute values) air entry value of the marl, a dewatering of the marl is achieved much easier than of the silt. Thus, the lower degree of saturation in the marl immediately underneath the silt layer can be explained, yielding a pronounced peak in the isolines of saturation. In the soil above the tunnel the fully saturated region moves in the direction of the ground surface, meaning that the saturation in the capillary fringe increases during the calculation. The groundwater displaced from the vicinity of the tunnel accumulates, leading to a local rise of the groundwater table above the tunnel.

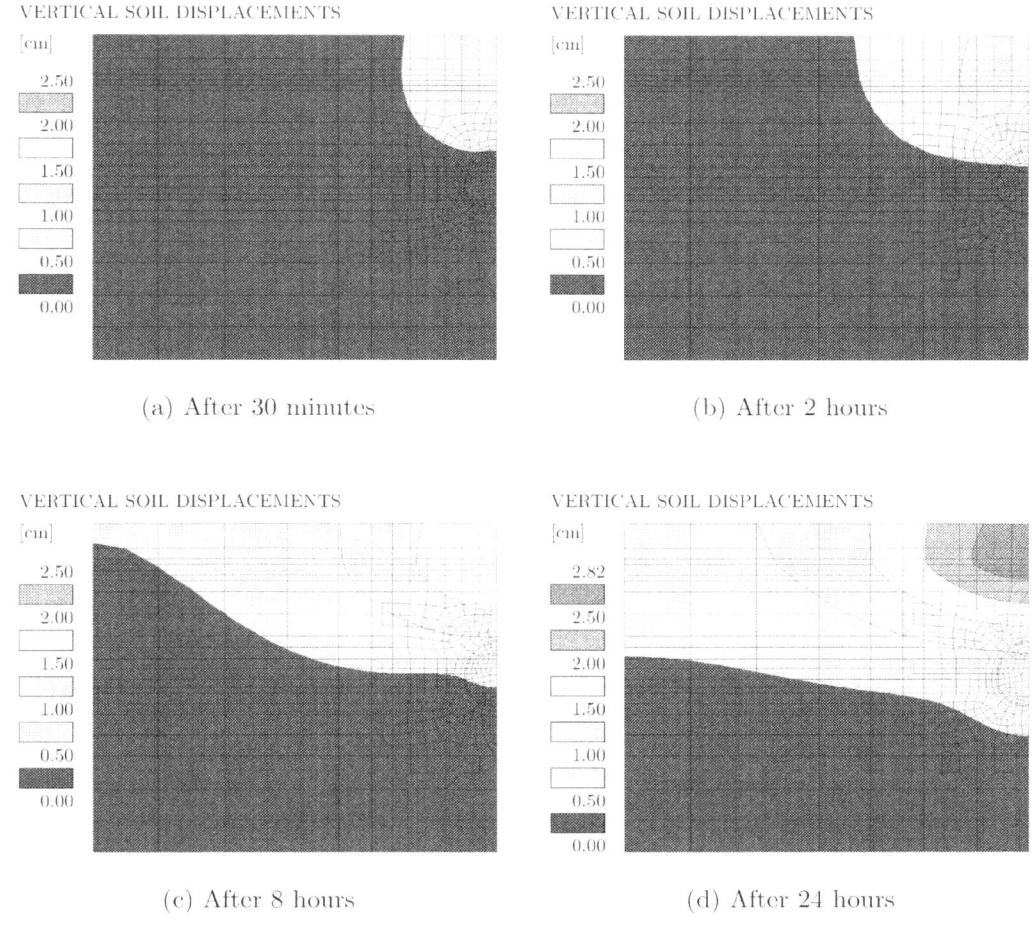

(a) After 30 minutes

(b) After 2 hours

(c) After 8 hours

(d) After 24 hours

Figure 6.38: Development of vertical soil displacements if the tunnel is located in the transition region between silt and marl.

Since a coupled numerical model is employed, the dewatering process immediately yields deformations of the soil skeleton, which are depicted for four points of time during the drainage period of 24.0 hours in Figure 6.38. The flow of the fluids water and compressed air in the pores of the soil proceeds from the tunnel face in the direction of the groundwater level and the ground surface, respectively. Consequently, primarily vertical displacements of the soil skeleton are obtained, resulting in a heave of the soil in the whole domain. Expectedly, with increasing time and increasing dewatering of the soil, also the vertical displacements increase. For all points of time the maximum values of the vertical displacements occur at the ground surface in the vertical symmetry plane above the tunnel.

The heave of the ground surface along the discretised domain is shown for the particular four points of time in Figure 6.39. This figure clearly indicates that the vertical displacements of the surface increase with time. Furthermore, the maximum values of the displacements (heaves) occur in the vertical symmetry plane through the tunnel axis. These displacements decay with increasing distance from the symmetry plane. After a time span of 24.0 hours the maximum predicted surface heave resulting from the dewatering process amounts to approximately 2.8 cm.

The effective stresses in the soil skeleton, depicted in Figure 6.40 for four instants of time during the dewatering period, are determined initially using the unit weight and the porosity of the soil as well as the hydrostatic stresses in the two fluid phases water and air and the respective degrees of saturation. Figure 6.40(a) shows the initial distribution

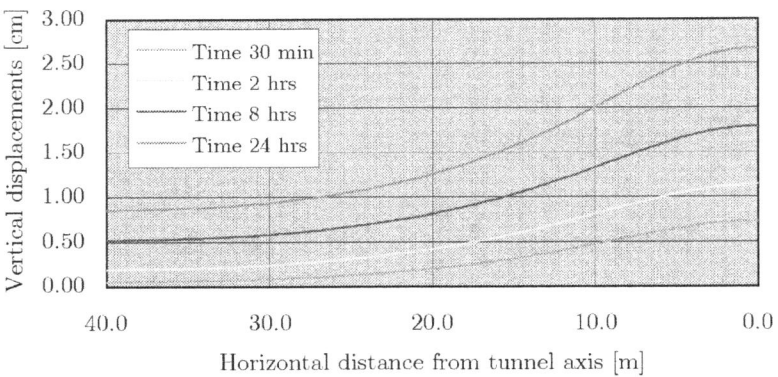

Figure 6.39: Development of surface heave if the tunnel is located in the transition region between silt and marl.

of the effective vertical soil stresses in almost the entire domain, only in the vicinity of the tunnel face some stress changes can be recognised. Since the effective stresses in the soil skeleton are obtained from the total stresses by subtracting the hydrostatic fluid stresses (weighted by the respective degrees of saturation), the increase of the fluid pressures during dewatering yields a decrease of the compressive soil stresses in the region exhibiting fluid pressures in excess of the initial state. Consequently, during the time span of dewatering, where an increase of the fluid pressures takes place, the effective compressive soil stresses decrease in the domain and in particular in the region of the

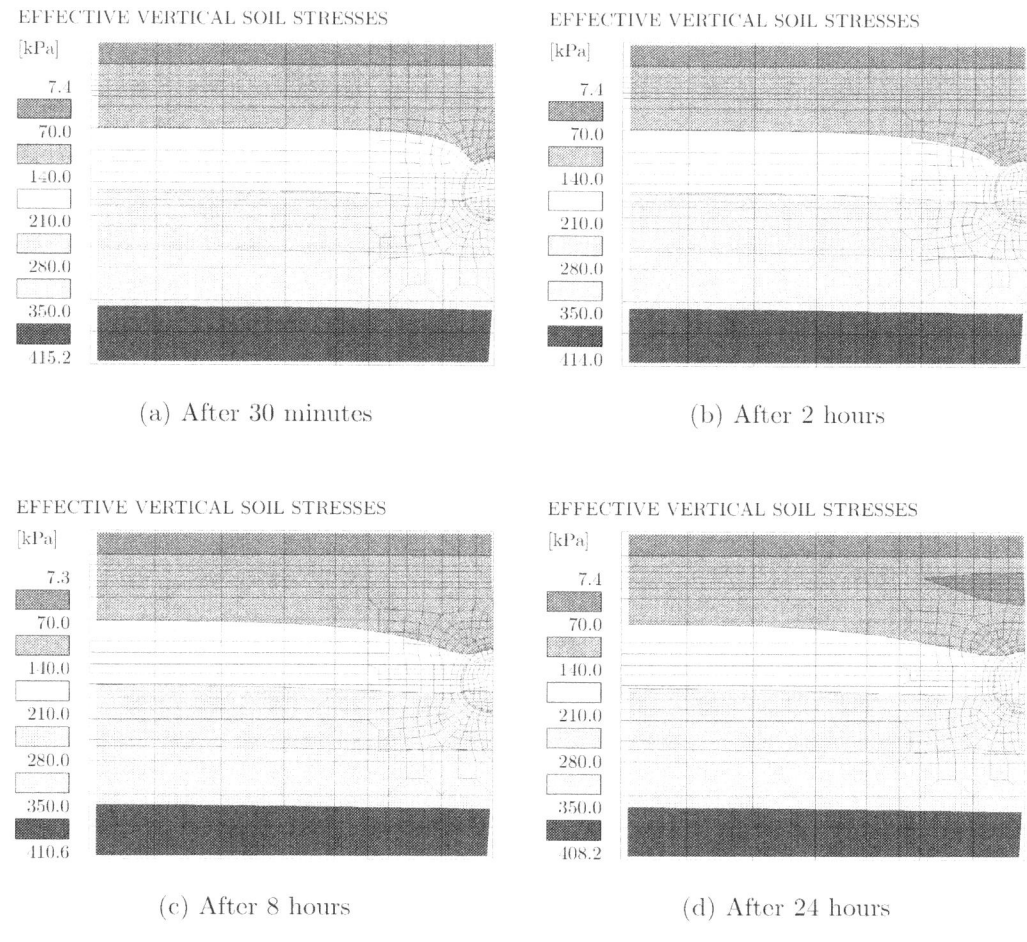

Figure 6.40: Development of effective vertical soil stresses if the tunnel is located in the transition region between silt and marl.

excess air pressure field (Figure 6.40).

Finally, Figure 6.41 shows the computed distribution of the excess air pressure field in the soil for four points of time. All plots indicate the linear increase (absolute values) of the prescribed air pressure from the invert of the tunnel in the direction of the crown quite clearly (equidistant isolines in the region of the tunnel face). Starting from the excess air pressure prescribed as boundary condition at the tunnel face, the air pressure field gradually proceeds into the soil in the vicinity of the tunnel face. For instance, after 30.0 minutes still in almost the entire domain zero excess air stresses prevail (Fig-

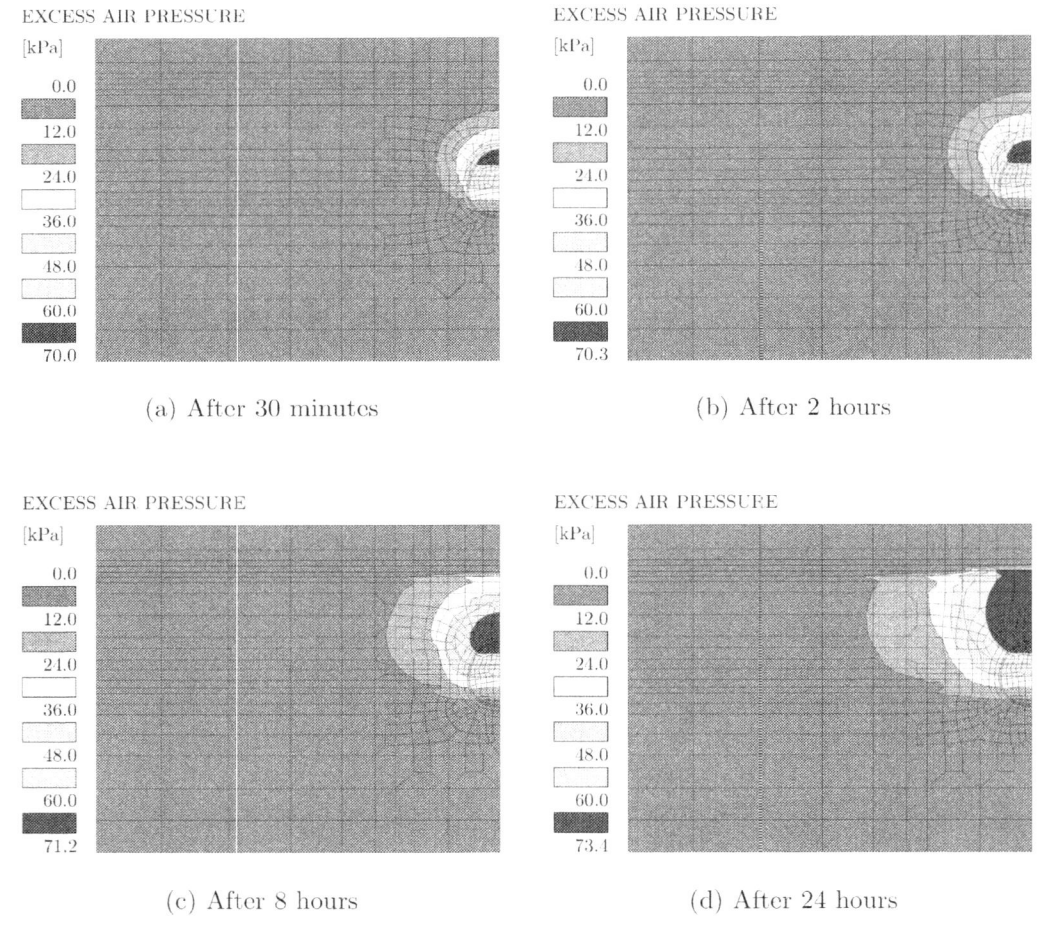

Figure 6.41: Development of excess hydrostatic air pressure if the tunnel is located in the transition region between silt and marl.

ure 6.41(a)). With increasing time of the dewatering process the excess air stress field spreads into the direction of the groundwater table and also somewhat in the horizontal direction. As the air pressure field more and more moves into the soil, the groundwater is displaced from the pores of the soil (Figure 6.37). Figure 6.41(d) shows the computed distribution of the excess air pressure in the soil after a time span of 24.0 hours. Since a narrow area of soil underneath the groundwater table is still fully water saturated (see Figure 6.37(d)), the excess air pressure is of a rather constant value from the crown of the tunnel to this saturated part of the soil. In the region of the groundwater table a very pronounced gradient of the excess air pressure is encountered. Hence, it seems that the excess air pressure decreases immediately as soon as a certain permeability of the pores with respect to the air phase is achieved, i.e. the degree of saturation is reduced below the fully saturated state. However, a horizontal distribution of the air pressure field probably takes place if the permeability with respect to the air phase is too small to allow for a dissipation of the occurring amount of compressed air.

Tunnel completely located in the silt layer

In the second analysis, it is assumed that the tunnel is completely located in the silt layer, i.e. the marl layer only covers the bottommost three rows of elements in Figure 6.32 (height of 10.0 m). The change in soil near the invert of the tunnel thus no longer exists. However, since this marl layer only reached into the tunnel for about 1.5 m in the previous analysis, no pronounced changes in the behaviour of most variables are encountered compared to this previously discussed simulation where the tunnel was located in the transition region between silt and marl.

Figure 6.42 shows plots of the computed hydrostatic stresses in the two fluid phases compressed air (Figure 6.42(a)) and water (Figure 6.42(b)) at the end of the dewatering period after 24.0 hours. Compared to Figures 6.35(d) and 6.36(d) only minor deviations are encountered due to the different material at the tunnel invert. Slightly higher fluid stresses are obtained in the vicinity of the tunnel invert, especially in the area left of the tunnel cross section. Differences of the same order of magnitude are also calculated for other variables, such as, e.g., the displacements or the effective stresses in the soil skeleton.

The most pronounced deviations can be recognised in the computed distribution of the degree of saturation, depicted in Figure 6.43 for two instants of time during the dewatering period. While the shape of the saturation isolines is of course the same as

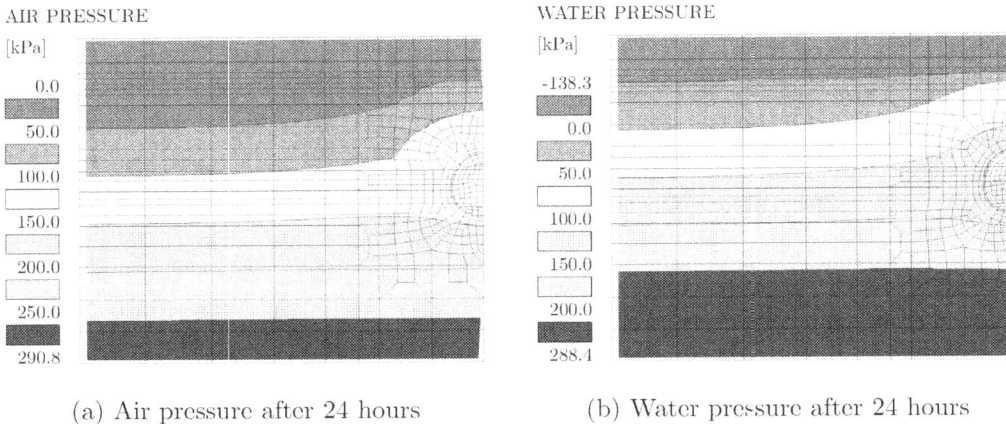

(a) Air pressure after 24 hours

(b) Water pressure after 24 hours

Figure 6.42: Distribution of hydrostatic fluid pressures if the tunnel is completely located in the silt layer.

shown in Figure 6.37 in the area of bench and crown of the tunnel as well as above the tunnel, differences occur at the tunnel invert which is now also located in the silt layer. Since the compressed air cannot enter the pores of the soil as easy as in case of the marl,

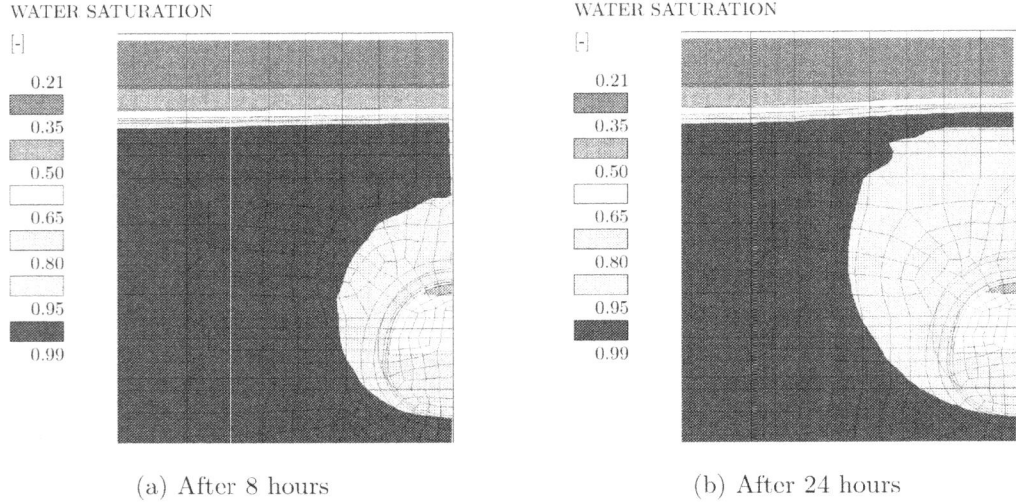

(a) After 8 hours

(b) After 24 hours

Figure 6.43: Distribution of degree of saturation if the tunnel is completely located in the silt layer.

the region of the tunnel invert is not dewatered to such an extent as if located in the marl. Consequently, the pronounced peak of the saturation isolines depicted in Figure 6.37(d) is not obtained in this case. Since the soil parameters do not change abruptly, smooth isolines are predicted in the region between invert and bench of the tunnel.

Tunnel completely located in the marl layer

The third case which is investigated concerns the tunnel being located completely in the marl layer (the marl stretches up to seven rows of elements below the ground surface). Since now a significant part of the tunnel subject to compressed air is located in the considerably more permeable marl layer, substantial differences in the behaviour during the dewatering process are expected. Plots obtained from the numerical simulation of this situation confirm this expectations as shown in the subsequent figures.

Figure 6.44 depicts the computed distribution of the hydrostatic fluid stresses in the air (Figure 6.44(a)) and the water phase (Figure 6.44(b)) at the end of the dewatering period after 24.0 hours. Now the pressure increase in the whole domain is much more pronounced for both the air and the water phase, in particular in the vicinity of and above the tunnel face. A larger region of the domain in this case is influenced by the excess air stress field emanating from the tunnel face.

The computed development of the degree of saturation in the course of the dewa-

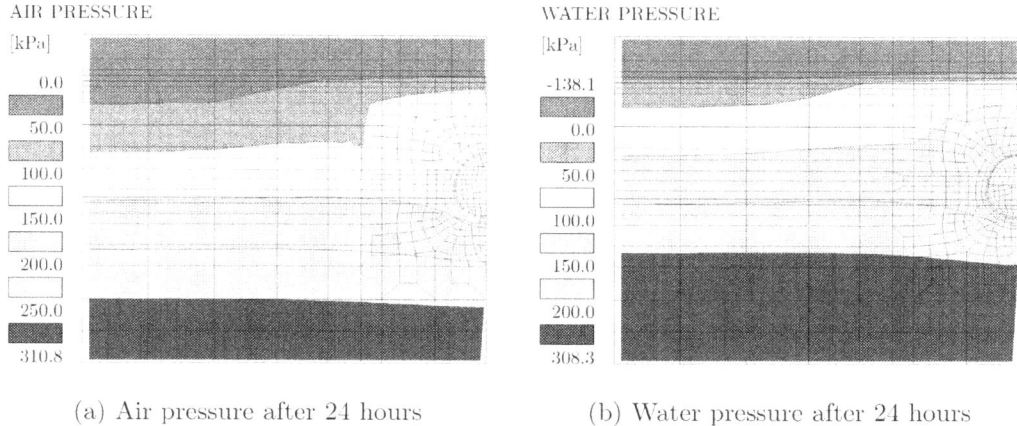

(a) Air pressure after 24 hours   (b) Water pressure after 24 hours

Figure 6.44: Distribution of hydrostatic fluid pressures if the tunnel is completely located in the marl layer.

tering period is shown in Figure 6.45 for two instants of time. Compared to the Figures 6.37(c) and 6.37(d), the dewatered region of the soil in Figure 6.45 is considerably larger. Since the air entry value of the marl is much smaller than the one of silt, the compressed air applied at the tunnel face can enter the pores of the soil much easier and the degree of saturation in the marl is reduced to a greater extent. Figure 6.45 indicates that in the present case the crown and bench of the tunnel are dewatered below a degree of saturation of 35 %. Also outside of the tunnel the degree of saturation is reduced by a considerable amount. Below the silt layer, reaching down to about 5.0 m above the crown of the tunnel, the soil exhibits a degree of saturation between 50 % and 65 %. Emanating from the invert of the tunnel a wide cone containing soil mass with a significantly lower degree of saturation is obtained. At the transition from the marl to the silt, the water saturation changes rapidly. While at the top of the marl layer (seven rows of elements below the ground surface) degrees of saturation as low as 35 % are encountered, in the silt the water saturation does not fall below a value of 90 %. In the capillary fringe, on the contrary, the degree of saturation increases as both Figures 6.45(a) and 6.45(b) indicate quite clearly. In Figure 6.45(a) the higher groundwater table can be recognised. A narrow region in the topmost part of the silt is still fully saturated (initial value of 99 %). After 24.0 hours (see Figure 6.45(b)) also the silt area above the tunnel is no

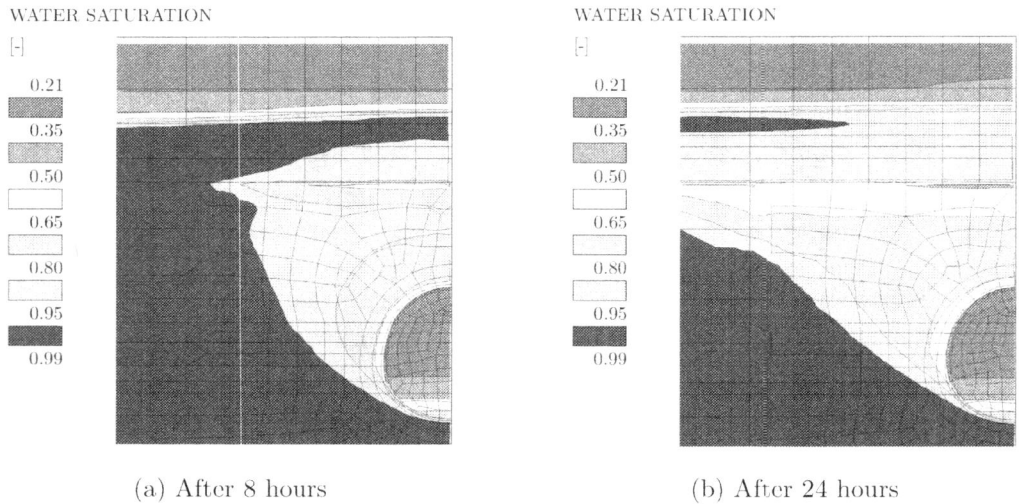

Figure 6.45: Distribution of degree of saturation if the tunnel is completely located in the marl layer.

longer fully water saturated, the degree of saturation is reduced below the initial value of 99 %. Furthermore, an increase of the water saturation is encountered in parts of the fill layer.

Since the fluid pressures in the domain are increased by a larger amount than for the case of the tunnel being located in the transition region between silt and marl, and consequently the effective stresses in the soil skeleton are reduced to a greater extent, the soil displacements, closely related to the effective soil stresses, are larger in the present calculation. Thus, the stresses of the soil are taken over by the fluid phases air and water to a larger extent, i.e. the unloading of the soil is more pronounced and the occurring heaves of the domain are larger. Figure 6.46 depicts the surface heaves of the domain for four instants of time during the dewatering period. The qualitative behaviour is of course the same as for the first case, however, the actual values of the displacements are larger. In this case the maximum predicted heave amounts to about 4.9 cm, compared to 2.8 cm obtained if the tunnel is located in the transition region between silt and marl, which decays with increasing distance from the symmetry plane through the tunnel axis.

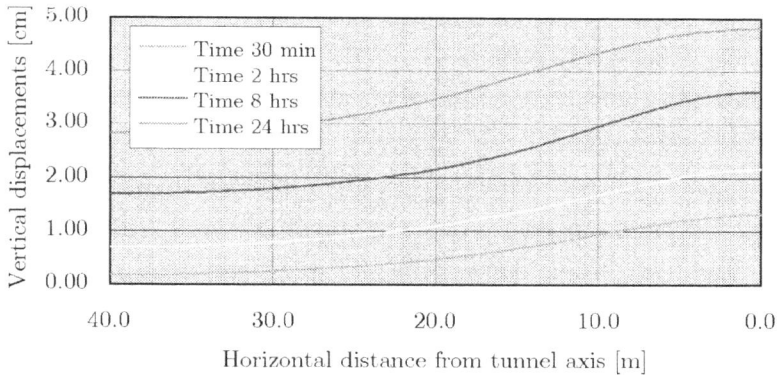

Figure 6.46: Development of surface heave if the tunnel is completely located in the marl layer.

Excavation and construction of the tunnel

After the numerical simulation of the dewatering process, the excavation and construction of the tunnel is investigated. The tunnel is located in the transition region between silt and marl in the present simulation. As mentioned in Section 6.4.2, a staged excavation and construction process is assumed for the tunnel cross section under consideration (see Figure 6.33), dealing sequentially with crown, bench and invert of the tunnel. Each of these three working cycles consists of two individual computation steps, modelling the excavation of the soil in the first step and the placement of the shotcrete lining in the second step. To take into account 3D-effects of the problem in the 2D analysis at least in an approximate manner, a partial stress relief of 50 % is assumed, yielding some deformations of the soil before the lining is installed. Thus, in the first calculation step, 50 % of the nodal forces associated with the excavation are applied. In the second calculation step, the shotcrete lining is installed before applying the full excavation load. This procedure is common practice if the three-dimensional stress redistribution at the tunnel face has to be simulated with a plane strain numerical model. The time-dependent behaviour of shotcrete is accounted for in a simplified way by using a reduced Young's modulus $E_1 = E/2$ for the 'young' shotcrete. Prior to the subsequent calculation step (excavation of the next working cycle), the material is replaced by 'hardened' shotcrete with a Young's modulus according to Table 6.8.

Figure 6.47 shows the computed distributions of both the vertical displacements (on the left) and the effective vertical stresses (on the right) of the soil in the course of the tunnel construction process. Figures 6.47(a), 6.47(b) and 6.47(c) depict the respective quantities at the end of each of the three working cycles, i.e. after excavation of the soil and the subsequent placement of the lining. It should be kept in mind that, because of the application of a coupled numerical model, the deformations and stress changes of the soil caused by the displacement of the groundwater by means of compressed air (see Figures 6.38 and 6.40), are inherent in the displacements and stresses obtained after the completion of each working cycle. The vertical displacements of the soil skeleton (confer left half of Figure 6.47) exhibit a pattern familiar from drained plane strain tunnel analyses, i.e. a settlement of the soil near the crown and a heave in the area of the tunnel invert is obtained. The maximum and minimum values of the displacements occur at the invert and the crown of the tunnel, respectively. After completion of the tunnel construction process the predicted maximum settlement at the crown of the tunnel amounts to about 6.7 cm, a maximum heave at the invert of approximately 9.4 cm is obtained in the numerical simulation.

The right half of Figure 6.47 shows the computed distributions of the effective vertical stresses in the soil at the end of each working cycle. It should be mentioned that the effective stresses in the soil skeleton are of primary importance and thus the plots are

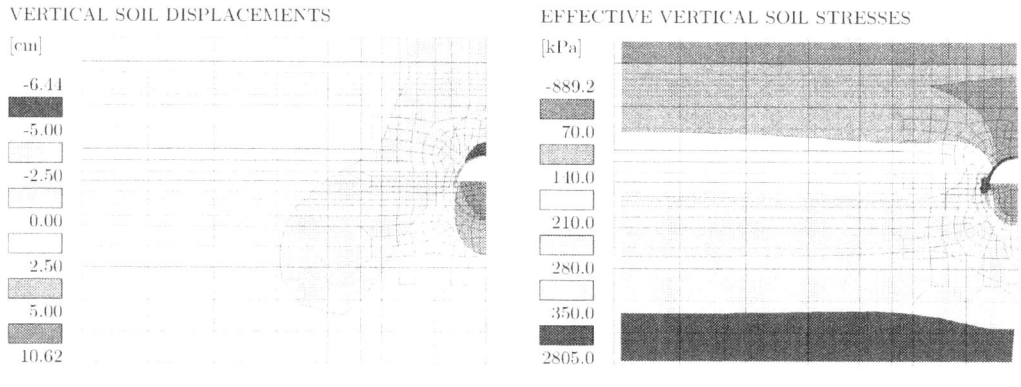

(a) 1st working cycle completed

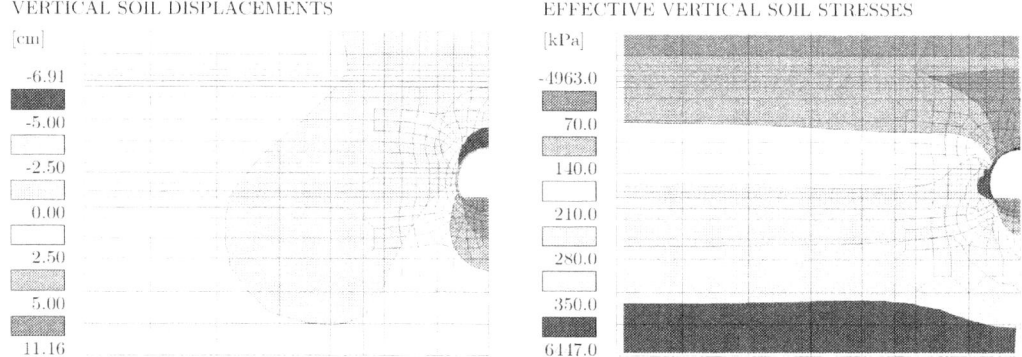

(b) 2nd working cycle completed

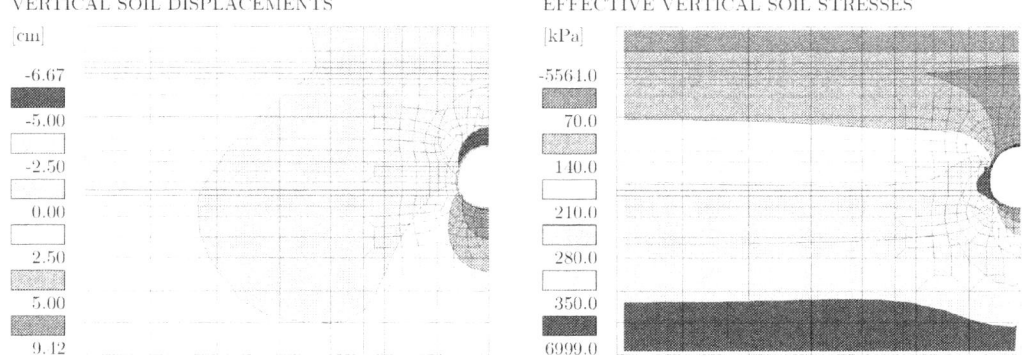

(c) 3rd working cycle completed

Figure 6.47: Distributions of vertical displacements (left) and effective vertical stresses (right) of the soil after each working cycle.

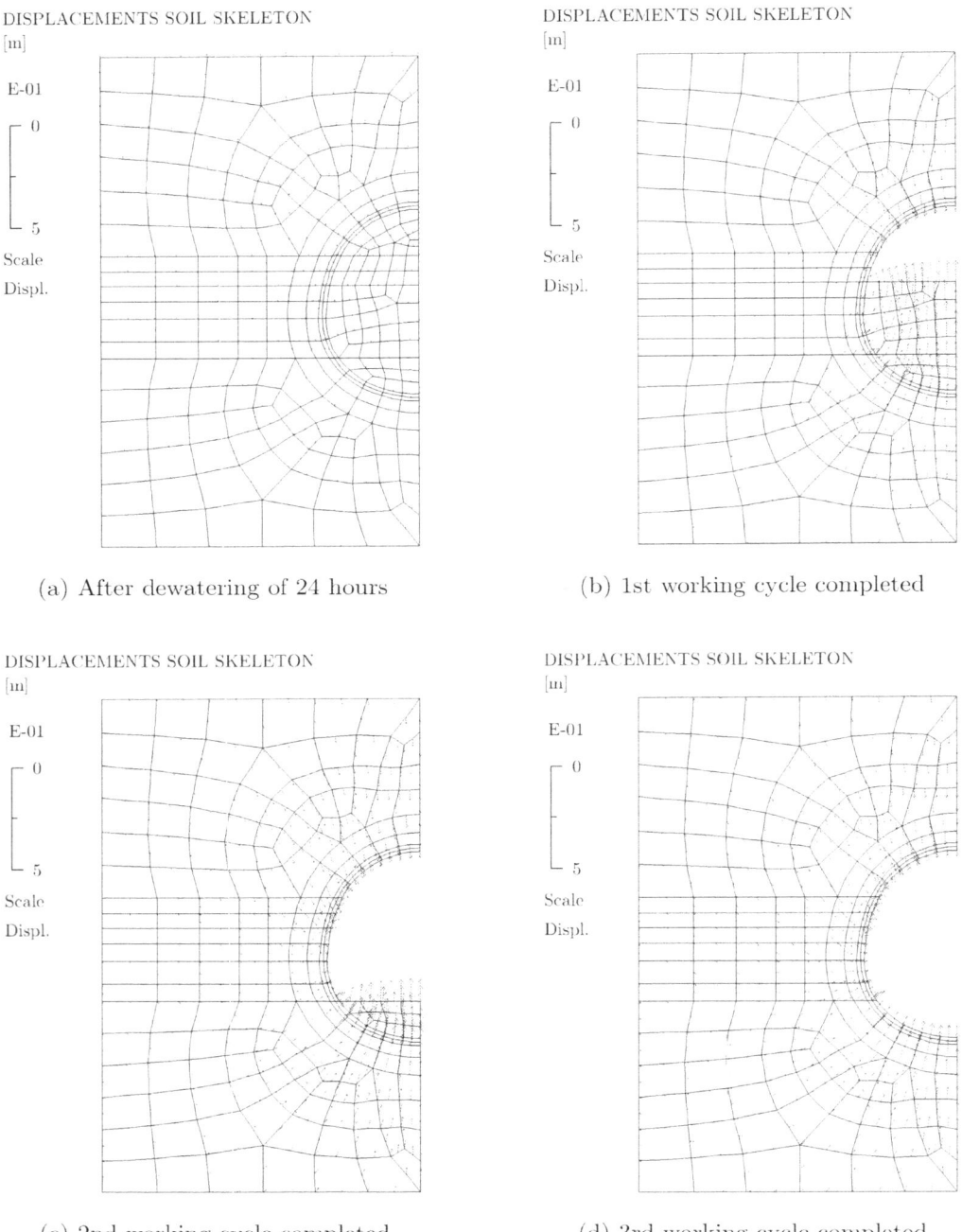

Figure 6.48: Displacements of the soil skeleton near the tunnel face (a) after dewatering and (b) to (d) after each working cycle.

scaled such that the range of occurring soil stresses is covered. The maximum and minimum values of stresses correspond to the shotcrete lining. Compared to the initial (geostatic) stress state in the soil, a decrease of the effective stresses has already occurred above the tunnel face due to the flow of the fluids in the pores of the soil (related to an increase of the fluid pressures in the water and the air phase) in this part of the domain resulting from the dewatering (compare Figure 6.40). Because of the tunnel excavation a further stress reduction is encountered in the soil above the crown and below the invert of the tunnel which can be recognised from Figure 6.47 very clearly. Next to the bench of the tunnel a pronounced increase of the effective stresses in the soil occurs.

Figure 6.48 depicts the displacement vectors of the soil in the vicinity of the tunnel cross section at the end of the dewatering period after 24.0 hours (Figure 6.48(a)) and after each of the three working cycles (Figures 6.48(b) to 6.48(d)). As Figure 6.48(a) shows, due to the dewatering process displacements in the direction of the ground surface occur. On the contrary, as can be recognised from Figures 6.48(b) to 6.48(d), due to the excavation of the tunnel this heave of the ground is reduced and settlements of the soil are obtained in the region above the tunnel face. In the area of the tunnel invert a heave of the soil is encountered because of the removal of soil during the excavation.

Figure 6.49 shows this particular behaviour for the ground surface. The vertical displacements of the ground surface are plotted versus the distance from the tunnel axis. The red solid line in this figure depicts the heave of the ground surface at the end of the dewatering period (compare Figure 6.39). A maximum heave of about 2.8 cm is

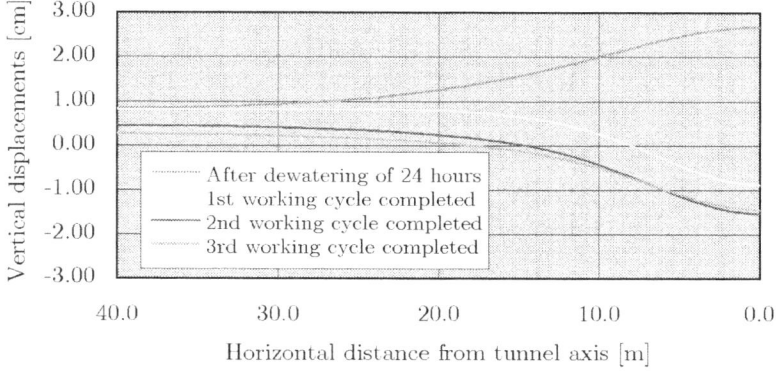

Figure 6.49: Reduction of the surface heave, caused by dewatering of the soil, during the tunnel construction process.

encountered at the symmetry plane which decays to about 1.0 cm at the left vertical boundary of the domain. Due to the excavation process the surface heave is reduced and settlements of the ground occur above the tunnel. After the excavation of the crown these settlements are restricted to a distance of about 8.0 m from the tunnel axis, the remainder of the surface of the domain still exhibits a heave. With increasing soil removal the settlements increase and spread in the horizontal direction. At the end of the construction process settlements of the ground surface are obtained for approximately half of the domain (distance of about 20.0 m from the tunnel axis). The heave of the ground surface of 2.8 cm, resulting from the dewatering of the soil, is reduced to a final settlement of about 1.5 cm at the end of the tunnel construction.

Simulation employing a Drucker-Prager model for the soil

In a final numerical simulation the influence of a linear elastic-perfectly plastic material model for the soil is investigated. For simplicity, the Drucker-Prager model, described in Chapter 3, is chosen although a number of more sophisticated non-linear constitutive models for soils are available. The tunnel once again is assumed to be located in the transition region between silt and marl. As additional material parameters, cohesion c and friction angle φ of the soil are necessary. These parameters are given in Table 6.9 for the three types of soil. By means of these common soil parameters the

Parameter	Symbol	Unit	Fill	Silt	Marl
Cohesion	c	kPa	5.0	15.0	40.0
Friction angle	φ	°	30.0	28.0	20.0

Table 6.9: Cohesion and friction angle for the three types of soil.

material constants of the Drucker-Prager model can be determined. To this end, the Drucker-Prager cone is matched to the tensile meridians of a Mohr-Coulomb model, yielding the so-called Drucker-Prager extension cone. For this case the Drucker-Prager properties μ and τ_F (confer Section 3.5.9) are calculated according to $\mu = \tan\varphi$ and $\tau_F = 6c\cos\varphi / \left[\sqrt{3}(3 + \sin\varphi)\right]$.

During the dewatering process a heave of the soil is obtained for the whole domain, i.e. the soil skeleton is unloaded. Consequently, a linear elastic-perfectly plastic material model for the soil skeleton should not yield results which substantially deviate from the ones of the linear elastic soil skeleton. As shown in the subsequent figures, these expectations are confirmed by the numerical simulation.

Figure 6.50 shows the computed hydrostatic fluid pressures (a) for the air phase and (b) for the water phase at the end of the dewatering period after 24.0 hours. The distributions of the fluid pressures resulting from the current simulation (using the Drucker-Prager model for the soil) are almost identical to the ones obtained with a linear elastic behaviour of the soil skeleton (compare Figure 6.35(d) and Figure 6.36(d)).

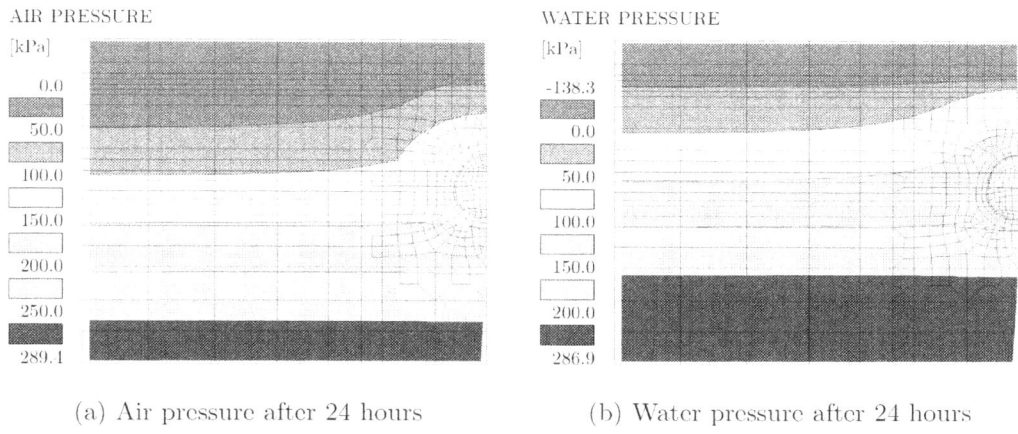

(a) Air pressure after 24 hours

(b) Water pressure after 24 hours

Figure 6.50: Distribution of hydrostatic fluid pressures if an elastic-plastic soil model is used.

Concerning the obtained vertical displacements and the effective vertical soil stresses at the end of the dewatering period (Figure 6.51) minor differences are encountered between the linear elastic and the elastic-plastic material behaviour for the soil skeleton (compare Figures 6.38(d) and 6.40(d)). In the latter case, the unloading of the soil above the tunnel is somewhat more pronounced (Figure 6.51(b)) and, closely related to that, the vertical displacements are slightly larger for the elastic-plastic model as Figure 6.51(a) indicates. The elastic-plastic model yields a maximum heave of 2.9 cm compared to 2.8 cm of the linear elastic material.

After the dewatering process the excavation and construction of the tunnel is simulated numerically. In a first attempt, a staged excavation was analysed (see Figure 6.33). However, serious convergence problems occurred after the first excavation step due to

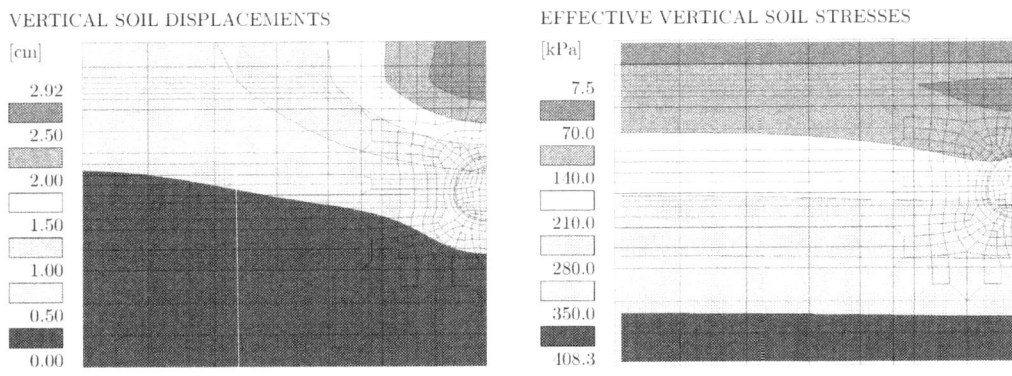

Figure 6.51: Distribution of (a) vertical displacements and (b) effective vertical stresses of the soil if an elastic-plastic material model is used.

large displacements obtained at the top of the bench of the tunnel. Similar convergence problems are also reported in the literature [Hochguertel(1998)]. In the authors opinion these problems are due to three facts: (i) Because of the employed plane strain model, an infinitely long tunnel is simulated and thus also the excavation stage is assumed to extend to infinity. On the contrary, in a three-dimensional simulation only a few meters (if at all) would possess this free boundary at the top of the bench. (ii) Due to the two fluid phases present in the pores of the soil, a major part of the hydrostatic soil stresses is carried by the fluid phases, i.e. the hydrostatic part of the stresses is reduced and consequently, the soil skeleton predominantly exhibits a deviatoric stress state. When imagining the Drucker-Prager yield surface, a smaller hydrostatic component of the stress state results in earlier yielding of the soil. (iii) In addition to that, the Drucker-Prager model does not account for the capillary stress which would affect the soil in the partially saturated region by yielding a larger cohesion (capillary stress or matric suction are sometimes termed 'apparent cohesion').

To overcome two of the above mentioned difficulties in the 2D simulation, the excavation process is modified such that the tunnel is excavated in one step (no free boundary at the bench occurs) and the value for the cohesion is increased according to the capillary stress in the partially saturated region, multiplied by the tangens of the friction angle. Such an approach is similar to the determination of the cohesion in partially saturated soils if two independent stress state variables are employed [Fredlund(1993)]. Using these two assumptions and once again taking into account a stress relief of 50 %, a solution is obtained, however, the deformations of the soil are still rather large in the vicinity of the tunnel cross section as Figure 6.52 shows.

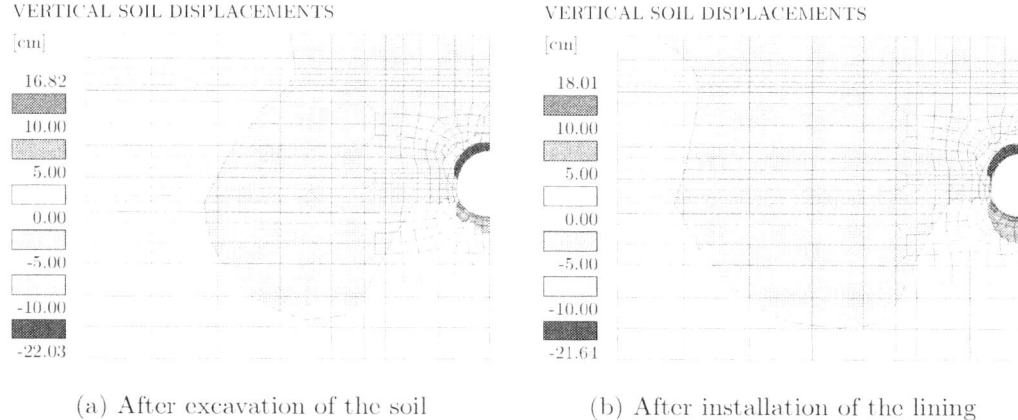

(a) After excavation of the soil   (b) After installation of the lining

Figure 6.52: Distribution of vertical displacements (a) after excavation of the soil and (b) after installing the lining.

Chapter 7

Summary, conclusions and outlook

A broad range of coupled flow and deformation problems are encountered in the field of engineering. The flow of one or several fluids through the pores of a soil causes deformations of the soil skeleton which exhibit a more or less pronounced influence on surrounding buildings or infrastructure. Thus, the risk of damage of these buildings has to be taken into account in connection with diverse engineering tasks, reaching from excavation and construction processes to the pumping of fluids from underground reservoirs.

From a mechanical point of view, uncoupled models can be distinguished from coupled models. The former are characterised by a sequential treatment of the flow and the deformation problem and thus interactions between the two phenomena are not taken into account. The latter are somewhat more sophisticated concerning the mechanical framework, however, they allow for a physically consistent consideration of the intrinsic interactions between the flow of the fluids and the deformations of the soil.

The objective of the present work consisted in the development of a coupled numerical model for the simulation of the process of dewatering of soils by means of compressed air. In particular, the dewatering of soils using compressed air was investigated in the light of tunnelling below the groundwater table, where the groundwater has to be displaced from the working area at the tunnel face during the construction process.

Chapters 1 and 2 are devoted to an introduction of the topic of the present work and an overview over the diverse subjects involved. The widely different fields of applications for multi-phase models are introduced first, reaching from 'classical' geotechnical problems such as consolidation or seepage flow, over the exploitation of underground fluid reservoirs, to modern and rather complex topics dealing with the storage of industrial waste or with contaminant transport. In the present work a three-phase formulation is employed for the numerical simulation of the dewatering of soils by means of compressed air. Hence, the soil is considered to consist of a deformable soil skeleton and pores filled with the two fluid phases water and (compressed) air.

The dewatering of soils by means of compressed air is of particular importance in connection with tunnelling below the groundwater table because in this case the groundwater has to be displaced from the working area at the tunnel face during the construction period. Therefore, a general survey over different approaches for the numerical simulation of tunnelling is presented in Chapter 2 as well. Tunnels driven in dry soil are distinguished from tunnels located in aquifers. While a broad range of models exist for tunnelling under drained conditions, only a few formulations deal with tunnelling below the groundwater table, and even less are coupled numerical approaches. This serves as a motivation for the development of a coupled numerical model in the present work.

From a mathematical/mechanical point of view, different theories are available for coupled formulations. Starting with a historical review of the development of multi-phase theories, an overview over existing modern formulations, i.e. mixture theory extended by the concept of volume fractions and averaging theory, is presented. The main aspects characterising these theories are outlined in Chapter 2 rather than giving the complete mathematical framework.

The averaging theory employed in the present work is described in detail in Chapter 3. After introducing the basic concepts of the averaging procedure and defining so-called averaging operators, a general microscopic balance equation is derived, which is then used to deduce a general macroscopic balance law. These general balance equations are specified to obtain the balance laws of mass, linear momentum and angular momentum, both on the microscale and on the macroscale. After the presentation of the necessary constitutive relations, the final balance equations, i.e. the governing equations of the three-phase model, are derived. The three-phase formulation contains a number of special cases (dewatering under atmospheric conditions, consolidation, uncoupled approach, drained case) which are discussed in the final section of Chapter 3.

The numerical formulation for the three-phase model is presented in Chapter 4. Employing a finite element approach, one usually emanates from weak formulations of the governing equations which are introduced first. The finite element framework requires a subdivision of the domain into finite elements and the choice of a set of primary variables (primary unknowns of the model). Hence, the primary variables and their approximations are discussed briefly before deriving the spatially discretised set of equations. Since the problem under consideration is time-dependent, a survey of different time integration schemes is presented. Finally, the non-linear coupled system of equations is obtained which is then solved iteratively for the incremental nodal values of the unknowns, i.e. the displacements of the soil skeleton and the hydrostatic stresses in the two fluid phases water and air.

Chapters 5 and 6 are dedicated to the verification and the application of the three-phase formulation. For verification purposes simpler two-phase problems (the two-phase formulation is a special case of the three-phase model) are discussed in Chapter 5 first. Numerical results of one- and two-dimensional (plane strain and axisymmetric) consolidation examples are presented and compared with the corresponding analytical solutions available in the literature. A laboratory test performed at Graz University of Technology, dealing with the flow of compressed air through dry soil, is investigated numerically and results are checked against measurements obtained from the experiment. Finally, the dewatering of a sand column (laboratory test conducted by Liakopoulos) is simulated numerically, using the two-phase formulation in a first and the three-phase formulation in a second step. The numerical results are verified by means of data stemming from the experiment as well as other numerical results available in the literature. Valuable insight into the difference between applying two-phase and three-phase models are deduced from a comparison of the results obtained with the respective models.

Applications of the three-phase formulation to problems of practical interest in geotechnical engineering are presented in Chapter 6. For the coupled analysis of seepage flow through an earth dam (homogeneous dam and dam containing a rather impervious core), the two-phase formulation is employed, i.e. atmospheric air pressure prevails during the whole simulation period. In a second example problem, the complete three-phase formulation is used for the numerical simulation of a full-scale in-situ air permeability test, conducted in connection with the subway construction in Essen, Germany. In this field test compressed air is employed for displacing the groundwater in the pores of the soil. Finally, the three-phase formulation is applied to the two-dimensional numerical simulation of tunnelling below the groundwater table by means of compressed air. Compressed air in this case is used for the displacement of the groundwater from the working area at the tunnel face. Since a coupled numerical model is employed, deformations resulting from the dewatering process are taken into account in a physically consistent manner in addition to displacements due to the excavation of the tunnel.

The proposed coupled numerical model allows for a simulation of the dewatering of soils by means of compressed air. To this end, the soil is treated as a three-phase medium, consisting of a deformable soil skeleton and two immiscible, compressible fluid phases water and air present in the pores of the soil. Dewatering by means of compressed air is employed in connection with tunnelling in aquifers where the groundwater has to be displaced from the vicinity of the working area at the tunnel face during the construction process. Since a coupled model is used, deformations of the soil skeleton not only result from the excavation process but also from the dewatering of the soil. The most appealing features of the proposed coupled approach can be summarised as follows:

In contrast to an uncoupled formulation, on the basis of a coupled three-phase model, the pressures in the fluid phases water and air (characterising the flow process) as well as the deformations and stresses of the soil skeleton (deformation problem) are computed simultaneously without any need for switching between flow analysis and deformation analysis.

Intrinsic interactions between the fluid flow in the pores of the soil and the deformations and stresses of the soil skeleton are taken into account in a physically consistent manner.

Apart from the numerical simulation of tunnelling below the groundwater table, employing compressed air for the displacement of the groundwater from the working area at the tunnel face, the proposed three-phase model can be used for the numerical simulation of a broad range of problems encountered in geotechnical engineering. This is possible because models for the dewatering of soils under atmospheric conditions and for consolidation of soils are contained in the three-phase formulation as special cases.

Since the theoretical framework is derived in a very general sense, adaptations of the three-phase model concerning the consideration of additional phases or improved material behaviour of individual constituents are possible in a fairly easy way. Thus, the theoretical formulation may provide a valuable basis for extensions of the model necessary to tackle entirely different subjects.

Concerning the most important material parameters, a profound knowledge of the dewetting and wetting behaviour of the soil, in particular of the relationships between degree of saturation and capillary stress (hysteresis) and between degree of saturation and permeability of the soil, is of paramount importance for the simulation of drainage and imbibition processes, respectively. The experimental determination of properties such as the air entry value or the residual degree of saturation is inevitable for a successful numerical analysis. However, usually such parameters are not included in standard soil testing programmes and thus should be required in an early stage of the investigations. Probably only the hydraulic conductivity of the soil is available fairly easy from standard experimental data.

It can be concluded that the coupled three-phase model developed in the present work can be used with a reasonable amount of confidence to analyse a range of problems arising in geotechnical engineering, as shown by means of the examples presented in the preceding chapters. The use of experimental data for a calibration of the constitutive relations thereby is of primary importance to obtain realistic results.

The proposed coupled three-phase model constitutes a valuable basis for the numerical simulation of a broad range of problems encountered in geotechnical engineering, as demonstrated by the examples presented in the current work. Improvements and further developments of the model could focus on several aspects mentioned subsequently.

In order to allow for a better and more realistic representation of the actual three-dimensional nature of the problem of tunnelling, an extension of the two-dimensional model to the three-dimensional case seems to be desirable. However, two crucial points should be emphasised in connection with the development of a coupled three-dimensional numerical model: (i) Due to the large number of degrees of freedom (a maximum number of five degrees of freedom per node, the choice of at least twenty-noded finite elements and the consideration of rather large domains because of staged excavation and construction processes contribute to that), very fast and efficient solution algorithms have to be employed. Probably a parallelisation of the finite element programme is inevitable. (ii) Further complexities occur in connection with the specification of the boundary conditions which change in each excavation or construction step. Thus, the boundary conditions have to be adapted in each step of the numerical analysis. In contrast to two-dimensional analyses, in the three-dimensional case these modifications can only be handled in an automated manner.

Another improvement of the model could probably be achieved by the application of adaptive solution strategies. Although adaptive finite element analyses may enhance the efficiency of the numerical solution, they have rarely been used in combination with tunnelling so far, even in the case of single-phase problems (characterised by drained conditions) and two-dimensional discretisations. However, 3D adaptivity which would be necessary for the simulation of tunnelling, is a complex task, in particular for tunnelling below the groundwater table where the two fluid phases water and air have to be taken into account additionally.

Clearly, the problem of dewatering is time-dependent. In the present work the rather simple but unconditionally stable Euler backward method is employed for the numerical integration in the time domain. However, as mentioned in Chapter 4, a broad range of other schemes are available which could be investigated in connection with the current three-phase formulation. With respect to the integration in the time domain automatic time stepping procedures should probably be considered in some more detail.

Another point of interest concerns the constitutive behaviour of the soil skeleton. Usually, constitutive models for unsaturated soils nowadays are based on two independent stress state variables (net normal stress and matric suction) rather than on the modified effective stress concept. However, as shown in the literature, for soils which are of interest in connection with tunnelling in aquifers by means of compressed air, this aspect does not seem to be of utmost importance. Due to the rather low capillary stresses arising in these types of soil, deviations in the results obtained with these fundamentally different models are small. Only for clays with a very low permeability, which usually exhibit a pronounced swelling behaviour, a soil model based on two stress state variables is inevitable. However, tunnelling by means of compressed air is not possible in such soils.

List of Tables

3.1	Definition of the volume fractions in a three-phase medium.	64
3.2	Specification of the general balance equation variables to obtain particular balance laws.	71
3.3	Permeability ratio for $S^w = 0.99$ and different soils.	140
5.1	Material properties for Terzaghi's problem.	219
5.2	Material properties for the footing example.	230
5.3	Results of calculations using elements with eight (8/4) or nine (9/4) displacement and four pressure degrees of freedom.	237
5.4	Material properties for the airflow test on a dry sand.	247
5.5	Material properties for Liakopoulos' lab test.	257
5.6	Additional material properties for the three-phase numerical simulation of Liakopoulos' lab test.	269
6.1	Material properties for the earth dam problem.	290
6.2	Material properties for the clayey core of the earth dam.	291
6.3	Hydraulic properties of the three types of soil used in the analysis of the earth dam taking into account the base.	297
6.4	Material properties for the individual layers of the Essen soil.	313
6.5	Material properties for the weaker part of the silt layer.	314
6.6	Material properties for the fluid phases water and air.	315
6.7	Material properties for the soil layers of the tunnel problem.	334
6.8	Material properties used for the shotcrete lining.	336
6.9	Cohesion and friction angle for the three types of soil.	354

List of Figures

2.1	Tunnelling below the groundwater table by means of compressed air.	25
2.2	Assumed distribution of hydrostatic fluid stresses.	26
2.3	Shield driving method in combination with compressed air; air lock (a) behind and (b) in front of the shield driving machine.	27
2.4	Cut-and-cover method in combination with dewatering by means of compressed air.	28
2.5	Applicability of dewatering by means of compressed air depending on the type of soil [Arz(1994)].	29
3.1	Averaging procedure for a three-phase medium: volume element (a) before and (b) after the averaging process.	53
3.2	Averaged quantity ζ versus characteristic length of the averaging volume dv [Hassanizadeh(1979a)].	61
3.3	Experimental data of the relationship between the Bishop's parameter χ and the degree of water saturation S^w for a cohesionless silt and for different types of compacted soils.	104
3.4	Total, osmotic and matric suction measurements on a compacted clay [Fredlund(1993)].	106
3.5	Physical model of the capillary phenomenon and corresponding stress distribution in the fluids.	108
3.6	(a) Various capillary heights of a soil and (b) relationship among pore radius, capillary height and matric suction.	110
3.7	(a) Effect of dissolved air on the density of water and (b) coefficient of solubility for a certain temperature range.	123

3.8   Components of the compressibility of a water-air mixture. 124
3.9   Effect of the solubility of air in water on the compressibility of a water-air mixture for different initial absolute air pressures. 125
3.10 Matric suction versus degree of saturation curves: (a) General shape of the curves usually obtained from drying-wetting cycles and (b) experimental data for three widely different types of soil [Alonso(1987)]. 127
3.11 Radius effects on the capillary height. 129
3.12 (a) Capillary stress versus degree of water saturation curve proposed by Brooks and Corey [Brooks(1964)] and (b) matric suction versus effective degree of saturation for different materials [Fredlund(1993)]. 131
3.13 (a) Capillary stress versus degree of saturation curve proposed by van Genuchten [Genuchten(1985)] and (b) matric suction versus degree of saturation data (symbols) and their approximation by van Genuchten's relationship (solid lines). 132
3.14 (a) Experimental data of the relative permeabilities for a sand [Wyckoff(1936)] and (b) approximation of experimental results by the empirical equations proposed by Mualem [Mualem(1976)]. 139
3.15 (a) Averaged stress vectors acting on the surfaces of an infinitesimal representative elementary volume (REV) and (b) decomposition of the stress vectors into the components of the averaged stress tensor. 150

4.1   Elements employed in multi-phase flow analyses: (a) triangular and (b) quadrilateral elements with the appropriate number of degrees of freedom. 196

5.1   Physical model of Terzaghi's problem. 208
5.2   Different meshes employed to simulate Terzaghi's problem. 217
5.3   Evolution of the excess pore water pressure using the equidistant mesh (Figure 5.2(a)): analytical solution (solid lines) and numerical results (dots).221
5.4   Evolution of the excess pore water pressure using the graded mesh (Figure 5.2(b)): analytical solution (solid lines) and numerical results (dots). . 222
5.5   Evolution of the excess pore water pressure using the fine graded mesh (Figure 5.2(c)): analytical solution (solid lines) and numerical results (dots).223
5.6   Evolution of the degree of consolidation for different meshes: analytical solution (solid lines) and numerical results (dots). 224
5.7   Footing on a consolidating soil layer. 226

5.8	Finite element mesh employed for the footing example.	229
5.9	Evolution of (a) excess pore water pressure and (b) soil skeleton displacements during the consolidation period.	233
5.10	Evolution of the effective vertical stresses in the soil skeleton during the consolidation period.	234
5.11	Time history of vertical deformation of the central point under the load: (a) settlement and (b) degree of consolidation.	235
5.12	Time history of vertical deformation of the central point under the load (fluid phase compressible): (a) settlement and (b) degree of consolidation.	238
5.13	Evolution of (a) excess water pressure and (b) displacements during consolidation (impermeable boundary below loading).	240
5.14	Time history of vertical deformation of the central point under the loading (impermeable boundary below the loading): (a) settlement and (b) degree of consolidation.	241
5.15	Time history of vertical deformation of the central point under the loading (circular footing): (a) settlement and (b) degree of consolidation.	242
5.16	Set-up of the laboratory test conducted at Graz University of Technology [Kammerer(2000)].	244
5.17	Three finite element meshes employed for the numerical simulation of the laboratory test conducted in Graz.	245
5.18	Evolution of the excess air pressure in dry sand specimen.	248
5.19	Evolution of the excess air pressure considering a section in the symmetry plane.	249
5.20	Evolution of the displacements of the soil skeleton.	250
5.21	Evolution of the effective vertical stresses in the soil skeleton.	251
5.22	Evolution of the velocity of the air phase.	252
5.23	Evolution of the excess air pressure within one load increment.	253
5.24	Profiles of (a) the excess air pressure and (b) the vertical soil displacements after 20.0 minutes obtained with the individual meshes shown in Figure 5.17.	254
5.25	Finite element mesh employed for the numerical simulation of Liakopoulos' lab test [Liakopoulos(1965)].	256
5.26	Tensile water stresses versus height of the column.	259
5.27	Vertical soil displacements versus height of the column.	259
5.28	Degree of water saturation versus height of the column.	260

5.29 Effective vertical compressive soil stresses (stresses in excess of the initial stress state) versus height of the column. 261
5.30 Tensile water stress history at five points along the column height. 262
5.31 History of the vertical displacements (settlements) at five points along the column height. 263
5.32 Saturation history at five points along the column height. 264
5.33 History of effective vertical soil stresses at five points along the column height (compression positive). 264
5.34 Water outflow rate at the bottom of the column. 265
5.35 Water outflow at the bottom of the column. 265
5.36 Tensile water stresses versus height of the column after changing the material parameters. 267
5.37 Tensile water stresses versus height of the column (three-phase model). . . 270
5.38 Tensile air stresses versus height of the column (three-phase model). . . . 272
5.39 Capillary stresses versus height of the column (three-phase model). 273
5.40 Vertical soil displacements versus height of the column (three-phase model). 274
5.41 Degree of water saturation versus height of the column (three-phase model). 275
5.42 Effective vertical compressive soil stresses (stresses in excess of initial stresses) versus height of the column (three-phase model). 276
5.43 Tensile water stress history until attaining a steady state. 278
5.44 Air stress history until attaining a steady state. 278
5.45 History of settlements until attaining a steady state. 279
5.46 Degree of saturation history until attaining a steady state. 280
5.47 Effective soil stress history (stresses in excess of the initial stress state) until attaining a steady state. 281
5.48 Water outflow rate at the bottom of the column until attaining a steady state. 282

6.1 Section of the earth dam. 287
6.2 Section of the earth dam containing a clay core. 287
6.3 Finite element mesh employed for the numerical simulation of water flow through an earth dam. 288
6.4 Hydraulic properties of the soil used as construction material for the earth dam and of the clayey core. 291

6.5	Evolution of the hydrostatic water stresses until a steady state is obtained (analysis with rigid soil skeleton).	294
6.6	Evolution of the degree of saturation until a steady state is obtained (analysis with rigid soil skeleton).	295
6.7	Evolution of the vertical displacements of the soil skeleton until a steady state is obtained (analysis using linear elastic behaviour of the soil skeleton).	297
6.8	Distribution of (a) hydrostatic water stresses and (b) degree of saturation at steady state conditions (dam with base and underdrain).	298
6.9	Evolution of the hydrostatic water stresses until a steady state is obtained (anisotropic permeability).	300
6.10	Evolution of the degree of saturation until a steady state is obtained (anisotropic permeability).	301
6.11	Evolution of the hydrostatic water stresses until a steady state is obtained (dam containing clayey core).	304
6.12	Evolution of the degree of saturation until a steady state is obtained (dam containing clayey core).	305
6.13	Evolution of the vertical displacements of the soil skeleton until a steady state is obtained (dam containing clayey core).	306
6.14	Experimental set-up for the in-situ air permeability test conducted in Essen, Germany [Kramer(1989)].	309
6.15	Excess air pressure and air loss versus time for test 1B.	310
6.16	Finite element mesh employed for the numerical simulation of the in-situ air permeability test performed in Essen.	311
6.17	Hydraulic properties of the individual Essen soil layers.	314
6.18	Development of air pressure during first pressure level.	316
6.19	Development of water pressure during first pressure level.	318
6.20	Development of water saturation during first pressure level.	319
6.21	Development of soil displacements during first pressure level.	321
6.22	Surface heave during first pressure level.	322
6.23	Vertical displacements for three nodes versus time.	322
6.24	Final distributions of (a) air pressure and (b) degree of saturation using a reduced water bulk modulus.	324
6.25	Surface heave during first pressure level using a reduced water bulk modulus.	325
6.26	Development of air pressure employing the refined mesh.	326

6.27 Development of water pressure employing the refined mesh. 326
6.28 Development of water saturation employing the refined mesh. 327
6.29 Surface heave employing the refined mesh. 328
6.30 Soil profile at contract section 30 of the Essen subway. 329
6.31 Geometry of (a) the discretised domain and (b) the tunnel cross section. . 329
6.32 Finite element mesh employed for the numerical simulation of the tunnel excavation problem. 331
6.33 Working cycles of the tunnel construction process. 334
6.34 Hydraulic properties of the individual materials. 335
6.35 Development of hydrostatic air pressure if the tunnel is located in the transition region between silt and marl. 337
6.36 Development of hydrostatic water pressure if the tunnel is located in the transition region between silt and marl. 339
6.37 Development of degree of saturation if the tunnel is located in the transition region between silt and marl. 340
6.38 Development of vertical soil displacements if the tunnel is located in the transition region between silt and marl. 341
6.39 Development of surface heave if the tunnel is located in the transition region between silt and marl. 342
6.40 Development of effective vertical soil stresses if the tunnel is located in the transition region between silt and marl. 343
6.41 Development of excess hydrostatic air pressure if the tunnel is located in the transition region between silt and marl. 344
6.42 Distribution of hydrostatic fluid pressures if the tunnel is completely located in the silt layer. 346
6.43 Distribution of degree of saturation if the tunnel is completely located in the silt layer. 346
6.44 Distribution of hydrostatic fluid pressures if the tunnel is completely located in the marl layer. 347
6.45 Distribution of degree of saturation if the tunnel is completely located in the marl layer. 348
6.46 Development of surface heave if the tunnel is completely located in the marl layer. 349

6.47 Distributions of vertical displacements (left) and effective vertical stresses (right) of the soil after each working cycle. 351

6.48 Displacements of the soil skeleton near the tunnel face (a) after dewatering and (b) to (d) after each working cycle. 352

6.49 Reduction of the surface heave, caused by dewatering of the soil, during the tunnel construction process. 353

6.50 Distribution of hydrostatic fluid pressures if an elastic-plastic soil model is used. 355

6.51 Distribution of (a) vertical displacements and (b) effective vertical stresses of the soil if an elastic-plastic material model is used. 356

6.52 Distribution of vertical displacements (a) after excavation of the soil and (b) after installing the lining. 357

Bibliography

[Alonso(1987)] E. E. Alonso, A. Gens and D. W. Hight. Special problem soils: General report. In: *Proceedings of the 9th European Conference on Soil Mechanics and Foundation Engineering*, vol. 3, pp. 1087–1146. 1987.

[Alonso(1990)] E. E. Alonso, A. Gens and A. Josa. A constitutive model for partially saturated soils. *Géotechnique*, **40**(3), 405–430. 1990.

[Alonso(1995)] E. Alonso and F. Batlle. Construction and impoundement of an earthdam, application of the coupled flow-deformation analysis of unsaturated soils. In: A. Gens, P. Jouanna and B. A. Schrefler (eds.), *Modern Issues in Non-Saturated Soils*, vol. 357 of *CISM Courses & Lectures*, pp. 357–395. Springer-Verlag, Wien. 1995.

[Alonso(1998)] E. E. Alonso, A. Lloret, C. H. Delahaye, J. Vaunat, A. Gens and G. Volckaert. Coupled analysis of a backfill hydration test. *International Journal for Numerical and Analytical Methods in Geomechanics*, **22**, 1–27. 1998.

[Arz(1994)] P. Arz, H. G. Schmidt, J. Seitz and S. Semprich. *Beton-Kalender*, vol. 2, chap. Arbeiten im Grundwasser – Druckluft im Tunnelbau, pp. 628–646. Verlag Ernst & Sohn. 1994.

[Bally(2002)] J. R. Bally, D. Ianus, R. Mitrache and C. Radulescu. A large waste water sewer built in Braila – Romania with an open shield under compressed air dewatering. In: *Proceedings of the 3rd International Symposium on Geotechnical Aspects of Underground Construction in Soft Ground*, pp. 87–94. 2002.

[Banerjee(1981)] P. K. Banerjee and R. Butterfield. *Boundary Element Methods in Engineering Science*. McGraw-Hill Book Co., Ltd., New York. 1981.

[Bathe(1979)] K. J. Bathe and M. R. Khoshgoftaar. Finite element free surface seepage analysis without mesh iteration. *International Journal for Numerical and Analytical Methods in Geomechanics*, **3**, 13–22. 1979.

[Bathe(1982)] K. J. Bathe. *Finite Element Procedures in Engineering Analysis*. Prentice-Hall Inc., New Jersey. 1982.

[Bear(1979)] J. Bear. *Hydraulics of Groundwater*. McGraw-Hill, New York. 1979.

[Bear(1990)] J. Bear and Y. Bachmat. *Introduction to Modeling of Transport Phenomena in Porous Media*. Kluwer Academic Publishers, The Netherlands. 1990.

[Becker(1986)] H. Becker and T. Baumann. Wasserhaltung durch Druckluft bei Spritzbetonbauweisen. *Der Bauingenieur*, **61**, 389–397. 1986.

[Bedford(1983)] A. Bedford and D. S. Drumheller. Recent advances: theories of immiscible and structured mixtures. *International Journal of Engineering Science*, **21**(8), 863–960. 1983.

[Beer(1992)] G. Beer and J. O. Watson. *Introduction to Finite and Boundary Element Methods for Engineers*. John Wiley & Sons, Chichester. 1992.

[Beer(1999)] G. Beer and J. Plank. Joint research initiative on numerical simulation in tunnelling. *Felsbau*, **17**(1), 7–9. 1999.

[Beer(2000)] G. Beer and C. Dünser. Boundary element analysis of problems in tunneling. In: D. W. Smith and J. P. Carter (eds.), *Developments in Theoretical Geomechanics, The John Booker Memorial Symposium*, pp. 103–122. Balkema, Rotterdam. 2000.

[Biot(1941)] M. A. Biot. General theory of three-dimensional consolidation. *Journal of Applied Physics*, **12**(2), 155–164. 1941.

[Biot(1954)] M. A. Biot. Theory of stress-strain relations in anisotropic viscoelasticity and relaxation phenomena. *Journal of Applied Physics*, **25**(11), 1385–1391. 1954.

[Biot(1955)] M. A. Biot. Theory of elasticity and consolidation for a porous anisotropic solid. *Journal of Applied Physics*, **26**(2), 182–185. 1955.

[Biot(1956a)] M. A. Biot. Theory of deformation of a porous viscoelastic anisotropic solid. *Journal of Applied Physics*, **27**(5), 459–467. 1956.

[Biot(1956b)] M. A. Biot. Theory of propagation of elastic waves in a fluid-saturated porous solid; i. low-frequency range, ii. higher-frequency range. *Journal of the Acoustical Society of America*, **28**, 168–191. 1956.

[Bishop(1959)] A. W. Bishop. The principle of effective stress. *Teknisk Ukeblad*, **106**(39), 859–863. 1959.

[Bishop(1963)] A. W. Bishop and G. E. Blight. Some aspects of effective stress in saturated and partly saturated soils. *Géotechnique*, **13**(3), 177–197. 1963.

[Bluhm(1996)] J. Bluhm, R. de Boer and J. Skolnik. *Allgemeine Plastizitätstheorie für poröse Medien*, vol. 73 of *Forschungsberichte aus dem Fachbereich Bauwesen*. Universität – Gesamthochschule Essen, Essen. 1996.

[Bluhm(1997a)] J. Bluhm. *A Consistent Model for Saturated and Empty Porous Media*, vol. 74 of *Forschungsberichte aus dem Fachbereich Bauwesen*. Universität – Gesamthochschule Essen, Essen. 1997.

[Bluhm(1997b)] J. Bluhm and R. de Boer. The volume fraction concept in the porous media theory. *Zeitschrift für angewandte Mathematik und Mechanik*, **77**, 563–577. 1997.

[Boer(1986a)] R. de Boer and W. Ehlers. *Theorie der Mehrkomponentenkontinua mit Anwendung auf bodenmechanische Probleme, Teil I*, vol. 40 of *Forschungsberichte aus dem Fachbereich Bauwesen*. Universität – Gesamthochschule Essen, Essen. 1986.

[Boer(1986b)] R. de Boer and W. Ehlers. On the problem of fluid- and gas-filled elasto-plastic solids. *International Journal of Solids and Structures*, **22**, 1231–1242. 1986.

[Boer(1990a)] R. de Boer and W. Ehlers. Uplift, friction and capillarity: three fundamental effects for liquid-saturated porous solids. *International Journal of Solids and Structures*, **26**, 43–51. 1990.

[Boer(1990b)] R. de Boer and W. Ehlers. The development of the concept of effective stresses. *Acta Mechanica*, **83**, 77–92. 1990.

[Boer(1991a)] R. de Boer. *Theorie poröser Medien – Historische Entwicklung und gegenwärtiger Stand*, vol. 53 of *Forschungsberichte aus dem Fachbereich Bauwesen*. Universität – Gesamthochschule Essen, Essen. 1991.

[Boer(1991b)] R. de Boer, W. Ehlers, S. Kowalski and J. Plischka. *Porous Media, a Survey of Different Approaches*, vol. 54 of *Forschungsberichte aus dem Fachbereich Bauwesen*. Universität – Gesamthochschule Essen, Essen. 1991.

[Boer(1996)] R. de Boer. Highlights in the historical development of the porous media theory: Toward a consistent macroscopic theory. *Applied Mechanics Reviews*, **49**(4), 201–261. 1996.

[Booker(1974)] J. R. Booker. The consolidation of a finite layer subject to surface loading. *International Journal of Solids and Structures*, **10**, 1053–1065. 1974.

[Booker(1975)] J. R. Booker and J. C. Small. An investigation of the stability of numerical solutions of Biot's equations of consolidation. *International Journal of Solids and Structures*, **11**, 907–917. 1975.

[Booker(1982a)] J. R. Booker and J. C. Small. Finite layer analysis of consolidation I. *International Journal for Numerical and Analytical Methods in Geomechanics*, **6**, 151–171. 1982.

[Booker(1982b)] J. R. Booker and J. C. Small. Finite layer analysis of consolidation II. *International Journal for Numerical and Analytical Methods in Geomechanics*, **6**, 173–194. 1982.

[Borja(1989)] R. I. Borja. Linearisation of elasto-plastic consolidation equations. *Engineering Computations*, **6**, 163–168. 1989.

[Borja(1991a)] R. I. Borja. One-step and linear multistep methods for nonlinear consolidation. *Computer Methods in Applied Mechanics and Engineering*, **85**, 239–272. 1991.

[Borja(1991b)] R. Borja and S. Kishnani. On the solution of elliptic free-boundary problems via Newton's method. *Computer Methods in Applied Mechanics and Engineering*, **88**, 341–361. 1991.

[Borja(1995)] R. I. Borja and E. Alarcón. A mathematical framework for finite strain elastoplastic consolidation – Part 1: Balance laws, variational formulation, and linearization. *Computer Methods in Applied Mechanics and Engineering*, **122**, 145–171. 1995.

[Borja(1998)] R. I. Borja, C. Tamagnini and E. Alarcón. Elastoplastic consolidation at finite strain – Part 2: Finite element implementation and numerical examples. *Computer Methods in Applied Mechanics and Engineering*, **159**, 103–122. 1998.

[Bowen(1976)] R. M. Bowen. Theory of mixtures. In: A. C. Eringen (ed.), *Continuum Physics*, vol. III. Academic Press, New York. 1976.

[Bowen(1980)] R. M. Bowen. Incompressible porous media models by use of the theory of mixtures. *International Journal of Engineering Science*, **18**, 1129–1148. 1980.

[Bowen(1982)] R. M. Bowen. Compressible porous media models by use of the theory of mixtures. *International Journal of Engineering Science*, **20**, 697–735. 1982.

[Brebbia(1978)] C. A. Brebbia. *The Boundary Element Method for Engineers*. Pentech Press Limited, Plymouth Devon, England. 1978.

[Britto(1987)] A. M. Britto and M. J. Gunn. *Critical State Soil Mechanics via Finite Elements*. John Wiley & Sons, Chichester. 1987.

[Brooks(1964)] R. H. Brooks and A. T. Corey. Hydraulic properties of porous media. *Hydrology Papers*, Colorado State University, Fort Collins, **3**, 1–27. 1964.

[Brutsaert(2000)] W. Brutsaert. A concise parameterization of the hydraulic conductivity of unsaturated soils. *Advances in Water Resources*, **23**, 811–815. 2000.

[Buchmaier(1985)] R. F. Buchmaier. *Zur Berechnung von Konsolidationsproblemen bei nichtlinearem Stoffverhalten*. Ph.D. thesis, Baugrundinstitut Stuttgart, Stuttgart. 1985.

[Burdine(1953)] N. T. Burdine. Relative permeability calculations from pore-size distribution data. *Petroleum Transactions*, American Institute of Mining, Metallurgical, and Petroleum Engineers, **198**, 71–77. 1953.

[Callen(1985)] H. B. Callen. *Thermodynamics and an Introduction to Thermostatistics*. John Wiley & Sons, New York, 2nd edn. 1985.

[Carter(1979)] J. P. Carter, J. C. Small and J. R. Booker. The analysis of finite elasto-plastic consolidation. *International Journal for Numerical and Analytical Methods in Geomechanics*, **3**, 107–129. 1979.

[Carter(1995)] J. P. Carter and N. P. Balaam. *AFENA User's Manual, Version 5.0*. Centre for Geotechnical Research, University of Sydney, Australia. 1995.

[Celia(1992)] M. A. Celia and W. G. Gray. *Numerical Methods for Differential Equations*. Prentice-Hall, Englewood Cliffs, New Jersey. 1992.

[Chen(1990)] Z. S. Chen and H. A. Mang. Über ein Randelementeverfahren zur dreidimensionalen Berechnung des abgesenkten Grundwasserspiegels und der stationären Luftströmung in heterogenem, anisotropem Boden beim Tunnelvortrieb unter Druckluft. *Der Bauingenieur*, **65**, 523–532. 1990.

[Chen(1991)] Z. S. Chen, G. Hofstetter and H. A. Mang. 3D-boundary element analysis of the lowered groundwater level for tunnels driven under compressed air. *International Journal for Numerical and Analytical Methods in Geomechanics*, **15**, 735–752. 1991.

[Cividini(1984)] A. Cividini and G. Gioda. An approximate F.E. analysis of seepage with a free surface. *International Journal for Numerical and Analytical Methods in Geomechanics*, **8**, 549–566. 1984.

[Cividini(1989)] A. Cividini and G. Gioda. On the variable mesh finite element analysis of unconfined seepage problems. *Géotechnique*, **39**(2), 251–267. 1989.

[Coleman(1962)] J. D. Coleman. Stress strain relations for partly saturated soil. *Correspondence to Géotechnique*, **12**(4), 348–350. 1962.

[Coleman(1963)] B. D. Coleman and W. Noll. The thermodynamics of elastic materials with heat conduction and viscosity. *Archive for Rational Mechanics and Analysis*, **13**, 168–178. 1963.

[Desai(1972)] C. S. Desai. Seepage analysis of earth banks under drawdown. *Journal of the Soil Mechanics and Foundations Division ASCE*, **98**(SM11), 1143–1162. 1972.

[Desai(1976)] C. S. Desai. Finite element residual schemes for unconfined flow. *International Journal for Numerical Methods in Engineering*, **10**, 1415–1418. 1976.

[Dluzewski(2001)] J. M. Dluzewski. Nonlinear problems during consolidation process. In: D. V. Griffiths and G. Gioda (eds.), *Advanced Numerical Applications and Plasticity in Geomechanics*, vol. 426 of *CISM Courses & Lectures*. Springer-Verlag, Wien. 2001.

[Ehlers(1989)] W. Ehlers. *Poröse Medien, ein kontinuumsmechanisches Modell auf der Basis der Mischungstheorie*, vol. 47 of *Forschungsberichte aus dem Fachbereich Bauwesen*. Universität – Gesamthochschule Essen, Essen. 1989.

[Ehlers(1994)] W. Ehlers and J. Kubik. On finite dynamic equations for fluid-saturated soils. *Acta Mechanica*, **105**, 101–117. 1994.

[Eriksson(1996)] K. Eriksson, D. Estep, P. Hansbo and C. Johnson. *Computational Differential Equations*. Cambridge University Press, Cambridge. 1996.

[Eringen(1964)] A. C. Eringen and E. S. Suhubi. Nonlinear theory of simple microelastic solids. *International Journal of Engineering Science*, **2**, 189–203. 1964.

[Eringen(1989)] A. C. Eringen. *Mechanics of Continua*. Robert E. Krieger Publishing Company, Inc., Florida, 2nd edn. 1989.

[Fernandez(1972)] R. T. Fernandez. *Natural Convection from Cylinders Buried in Porous Media*. Ph.D. thesis, University of California. 1972.

[Fillunger(1914)] P. Fillunger. Neuere Grundlagen für die statische Berechnung von Talsperren. *Zeitschrift des Österreichischen Ingenieur- und Architekten Vereines*, **23**, 441–447. 1914.

[Fillunger(1936)] P. Fillunger. *Erdbaumechanik?* Selbstverlag des Verfassers, Wien. 1936.

[Fredlund(1977)] D. G. Fredlund and N. R. Morgenstern. Stress state variables for unsaturated soils. *Journal of the Geotechnical Engineering Division, ASCE*, **103**(GT5), 447–466. 1977.

[Fredlund(1979)] D. G. Fredlund. Appropriate concepts and technology for unsaturated soils. *Canadian Geotechnical Journal*, **16**, 121–139. 1979.

[Fredlund(1993)] D. G. Fredlund and H. Rahardjo. *Soil Mechanics for Unsaturated Soils*. John Wiley & Sons, New York. 1993.

[Fredlund(1996)] D. G. Fredlund, A. Xing, M. D. Fredlund and S. L. Barbour. The relationship of the unsaturated soil shear strength to the soil-water characteristic curve. *Canadian Geotechnical Journal*, **33**(3), 440–448. 1996.

[Fuchsberger(1988)] M. Fuchsberger, B. Strobl, N. Ayaydin and R. Heinrich. Tunnelling under compressed air in granular soils – research, design and site experience. In: J. M. Serrano (ed.), *Proceedings of the International Congress on Tunnels and Water*, vol. 1, pp. 137–143. Balkema, Rotterdam. 1988.

[Fung(1998)] T. C. Fung. Complex-time-step Newmark methods with controllable numerical dissipation. *International Journal for Numerical Methods in Engineering*, **41**, 65–93. 1998.

[Fung(1999)] T. C. Fung. Complex-time-step methods for transient analysis. *International Journal for Numerical Methods in Engineering*, **46**, 1253–1271. 1999.

[Gawin(1995)] D. Gawin, P. Baggio and B. A. Schrefler. Coupled heat, water and gas flow in deformable porous media. *International Journal for Numerical Methods in Fluids*, **20**, 969–987. 1995.

[Gawin(1996)] D. Gawin and B. A. Schrefler. Thermo-hydro-mechanical analysis of partially saturated porous materials. *Engineering Computations*, **13**(7), 113–143. 1996.

[Gens(1998)] A. Gens, A. J. Garcia-Molina, S. Olivella, E. E. Alonso and F. Huertas. Analysis of a full scale in situ test simulating repository conditions. *International Journal for Numerical and Analytical Methods in Geomechanics*, **22**, 515–548. 1998.

[Genuchten(1980)] M. T. van Genuchten. A closed-form equation for predicting the hydraulic conductivity of unsaturated soils. *Soil Science Society of America Journal*, **44**, 892–898. 1980.

[Genuchten(1985)] M. T. van Genuchten and D. R. Nielsen. On describing and predicting the hydraulic properties of unsaturated soils. *Annales Geophysicae*, **3**(5), 615–627. 1985.

[Ghaboussi(1973)] J. Ghaboussi and E. L. Wilson. Flow of compressible fluid in porous elastic media. *International Journal for Numerical Methods in Engineering*, **5**, 419–442. 1973.

[Gioda(1987)] G. Gioda and C. Gentile. A nonlinear programming analysis of unconfined steady-state seepage. *International Journal for Numerical and Analytical Methods in Geomechanics*, **11**, 283–305. 1987.

[Gioda(1988)] G. Gioda and A. Desideri. Some numerical techniques for free-surface seepage analysis. In: G. Swoboda (ed.), *Numerical Methods in Geomechanics*, pp. 71–84. Balkema, Rotterdam. 1988.

[Golser(1999)] H. Golser. Application of numerical simulation methods on site. *Felsbau*, **17**(1), 21–25. 1999.

[Goodman(1972)] M. A. Goodman and S. C. Cowin. A continuum theory for granular materials. *Archive for Rational Mechanics and Analysis*, **44**, 249–266. 1972.

[Grasberger(2000)] S. Grasberger and G. Meschke. A hygro-thermal-poroplastic damage model for durability analyses of concrete structures. In: *Proceedings of the European Congress on Computational Methods in Applied Sciences and Engineering, ECCOMAS*, pp.: 18. Barcelona. 2000.

[Grasberger(2002)] S. Grasberger. *Gekoppelte hygro-mechanische Materialmodellierung und numerische Simulation langzeitiger Degradation von Betonstrukturen*. Ph.D. thesis, Institute for Structural Mechanics, Ruhr-University Bochum. 2002.

[Gray(1977)] W. G. Gray and P. C. Y. Lee. On the theorems for local volume averaging of multiphase systems. *International Journal of Multiphase Flow*, **3**, 333–340. 1977.

[Gray(1991a)] W. G. Gray and S. M. Hassanizadeh. Paradoxes and realities in unsaturated flow theory. *Water Resources Research*, **27**(8), 1847–1854. 1991.

[Gray(1991b)] W. G. Gray and S. M. Hassanizadeh. Unsaturated flow theory including interfacial phenomena. *Water Resources Research*, **27**(8), 1855–1863. 1991.

[Gray(1998)] W. G. Gray and S. M. Hassanizadeh. Macroscale continuum mechanics for multiphase porous-media flow including phases, interfaces, common lines and common points. *Advances in Water Resources*, **21**, 261–281. 1998.

[Gray(1999)] W. G. Gray. Thermodynamics and constitutive theory for multiphase porous-media flow considering internal geometric constraints. *Advances in Water Resources*, **22**(5), 521–547. 1999.

[Green(1966)] A. E. Green and T. R. Steel. Constitutive equations for interacting continua. *International Journal of Engineering Science*, **4**, 483–500. 1966.

[Green(1970)] A. E. Green and P. M. Naghdi. The flow of fluid through an elastic solid. *Acta Mechanica*, **9**, 329–340. 1970.

[Guelzow(1994)] H. G. Gülzow. *Dreidimensionale Berechnung des Zweiphasenströmungsfeldes beim Tunnelvortrieb unter Druckluft in wassergesättigten Böden*. Ph.D. thesis, Institut für Grundbau, Bodenmechanik, Felsmechanik und Verkehrswasserbau, RWTH-Aachen. 1994.

[Harsch(2002)] W. Harsch and P. Menetry. Hydrogeologische und geotechnische Probleme und Lösungen beim Bau des Eisenbahntunnels Emmequerung. In: H. Schad (ed.), *3. Kolloquium Bauen in Boden und Fels*, pp. 129–135. 2002.

[Hassanizadeh(1979a)] S. M. Hassanizadeh and W. G. Gray. General conservation equations for multi-phase systems: 1. Averaging procedure. *Advances in Water Resources*, **2**, 131–144. 1979.

[Hassanizadeh(1979b)] S. M. Hassanizadeh and W. G. Gray. General conservation equations for multi-phase systems: 2. Mass, momenta, energy and entropy equations. *Advances in Water Resources*, **2**, 191–203. 1979.

[Hassanizadeh(1980)] S. M. Hassanizadeh and W. G. Gray. General conservation equations for multi-phase systems: 3. Constitutive theory for porous media flow. *Advances in Water Resources*, **3**, 25–40. 1980.

[Hassanizadeh(1990)] S. M. Hassanizadeh and W. G. Gray. Mechanics and thermodynamics of multiphase flow in porous media including interphase boundaries. *Advances in Water Resources*, **13**(4), 169–186. 1990.

[Hassanizadeh(1993)] S. M. Hassanizadeh and W. G. Gray. Toward an improved description on the physics of two-phase flow. *Advances in Water Resources*, **16**, 53–67. 1993.

[Heinrich(1961)] G. Heinrich and K. Desoyer. Theorie dreidimensionaler Setzungsvorgänge in Tonschichten. *Ingenieur-Archiv*, **30**(4), 225–253. 1961.

[Helmig(1997)] R. Helmig. *Multiphase Flow and Transport Processes in the Subsurface, A Contribution to the Modeling of Hydrosystems*. Springer-Verlag, Berlin, Heidelberg. 1997.

[Herbert(1972)] R. Herbert and M. Zytynski. A new technique for time-variant ground water flow analysis. *Journal of Hydrology*, **16**, 77–92. 1972.

[Hochguertel(1998)] T. Hochgürtel. *Numerische Untersuchungen zur Beurteilung der Standsicherheit der Ortsbrust beim Einsatz von Druckluft zur Wasserhaltung im schildvorgetriebenen Tunnelbau*. Ph.D. thesis, Institut für Grundbau, Bodenmechanik, Felsmechanik und Verkehrswasserbau, RWTH-Aachen. 1998.

[Holzapfel(2000)] G. A. Holzapfel. *Nonlinear Solid Mechanics, A Continuum Approach for Engineering*. John Wiley & Sons, Chichester. 2000.

[Hsi(1992)] J. P. Hsi. *Analysis of Excavation Involving Drawdown of the Water Table*. Ph.D. thesis, School of Civil and Mining Engineering, University of Sydney, Australia. 1992.

[Hughes(1983)] T. J. R. Hughes. Analysis of transient algorithms with particular reference to stability. In: T. J. R. Hughes and T. Belytschko (eds.), *Computational Methods for Transient Analysis*, pp. 67–155. North-Holland, Amsterdam. 1983.

[Hughes(1987)] T. J. R. Hughes. *The Finite Element Method, Linear Static and Dynamic Finite Element Analysis*. Prentice-Hall, Englewood Cliffs, New Jersey. 1987.

[Hutter(1999)] K. Hutter, L. Laloui and L. Vulliet. Thermodynamically based mixture models of saturated and unsaturated soils. *Mechanics of Cohesive-Frictional Materials*, **4**, 295–338. 1999.

[Javadi(2001)] A. A. Javadi and C. P. M. Snee. The effect of air pressure on the shear strength of soil as a consequence of compressed air tunnelling. *Canadian Geotechnical Journal*, **38**(6), 1187–1200. 2001.

[Jodl(1995)] H. G. Jodl and B. Strobl. Baupraktische Luftmengenberechnungen für den Druckluftvortrieb. *Felsbau*, **13**(2), 100–110. 1995.

[Jommi(1997)] C. Jommi, J. Vaunat, A. Gens, B. A. Schrefler and D. Gawin. Multiphase flow in porous media: A numerical benchmark. In: *Nafems World Congress 97 on*

Design, Simulation & Optimisation, vol. 2, pp. 1338–1349. Nafems Ltd., Glasgow. 1997.

[Jouanna(1995)] P. Jouanna and M. A. Abellan. Generalized approach to heterogeneous media. In: A. Gens, P. Jouanna and B. A. Schrefler (eds.), *Modern Issues in Non-Saturated Soils*, vol. 357 of *CISM Courses & Lectures*, pp. 1–127. Springer-Verlag, Wien. 1995.

[Kammerer(1998)] G. Kammerer and S. Semprich. Settlements due to tunnelling under compressed air. In: *Proceedings of the International Conference on Soil-Structure Interaction in Urban Civil Engineering*, vol. 4 of *Darmstadt Geotechnics*, pp. 85–96. Darmstadt University of Technology. 1998.

[Kammerer(2000)] G. Kammerer. *Experimentelle Untersuchungen von Strömungsvorgängen in teilgesättigten Böden und in Spritzbetonrissen im Hinblick auf den Einsatz von Druckluft zur Wasserhaltung im Tunnelbau*. Ph.D. thesis, Institut für Bodenmechanik und Grundbau, Technische Universität Graz, Austria. 2000.

[Kenyon(1976a)] D. E. Kenyon. Thermostatics of solid-fluid mixtures. *Archive for Rational Mechanics and Analysis*, **62**, 117–129. 1976.

[Kenyon(1976b)] D. E. Kenyon. The theory of an incompressible solid-fluid mixture. *Archive for Rational Mechanics and Analysis*, **62**, 131–147. 1976.

[Kezdi(1976)] A. Kezdi. *Fragen der Bodenphysik*. VDI-Verlag. 1976.

[Klausner(1991)] Y. Klausner. *Fundamentals of Continuum Mechanics of Soils*. Springer-Verlag, London. 1991.

[Klubertanz(1999)] G. Klubertanz. *Zur hydromechanischen Kopplung in dreiphasigen porösen Medien*. Ph.D. thesis, École Polytechnique Fédérale de Lausanne, Switzerland. 1999.

[Kramer(1987)] J. Kramer. U-Bahn Bau in Essen, Baulos 30 – Spritzbetonbauweise unter Druckluft: Luftverbrauch und Spannungsumlagerungen im Boden, Senkungen. In: *Proceedings der STUVA Tagung, Essen*, vol. 32 of *Forschung + Praxis*, pp. 193–199. 1987.

[Kramer(1989)] J. Kramer and S. Semprich. Erfahrungen über Druckluftverbrauch bei der Spritzbetonbauweise. *Taschenbuch für den Tunnelbau*, **13**, 91–153. 1989.

[Kreyszig(1993)] E. Kreyszig. *Advanced Engineering Mathematics*. John Wiley & Sons, New York, 7th edn. 1993.

[Kropik(1994)] C. Kropik. *Three-Dimensional Elasto-Viscoplastic Finite Element Analysis of Deformations and Stresses Resulting from the Excavation of Shallow Tunnels*. Ph.D. thesis, Institute for Strength of Materials, Vienna University of Technology, Austria. 1994.

[Lacy(1987)] S. J. Lacy and J. H. Prevost. Flow through porous media: a procedure for locating the free surface. *International Journal for Numerical and Analytical Methods in Geomechanics*, **11**, 585–601. 1987.

[Lambe(1979)] T. W. Lambe and R. V. Whitman. *Soil Mechanics, SI Version*. John Wiley & Sons, New York. 1979.

[Lewis(1976)] R. W. Lewis, G. K. Roberts and O. C. Zienkiewicz. A non-linear flow and deformation analysis of consolidation problems. In: C. S. Desai (ed.), *Numerical Methods in Geomechanics*, vol. 2, pp. 1106–1118. ASCE, New York. 1976.

[Lewis(1989)] R. W. Lewis and D. V. Tran. Numerical simulation of secondary consolidation of soil: finite element application. *International Journal for Numerical and Analytical Methods in Geomechanics*, **13**, 1–18. 1989.

[Lewis(1993)] R. W. Lewis and Y. Sukirman. Finite element modelling of three-phase flow in deforming saturated oil reservoirs. *International Journal for Numerical and Analytical Methods in Geomechanics*, **17**, 577–598. 1993.

[Lewis(1998)] R. W. Lewis and B. A. Schrefler. *The Finite Element Method in the Static and Dynamic Deformation and Consolidation of Porous Media*. John Wiley & Sons, Chichester, 2nd edn. 1998.

[Li(1990)] X. Li, O. C. Zienkiewicz and Y. M. Xie. A numerical model for immiscible two-phase fluid flow in a porous medium and its time domain solution. *International Journal for Numerical Methods in Engineering*, **30**, 1195–1212. 1990.

[Li(1992)] X. Li and O. C. Zienkiewicz. Multiphase flow in deforming porous media and finite element solutions. *Computers & Structures*, **45**(2), 211–227. 1992.

[Li(2000)] X. Li, W. Wu and S. Cescotto. Contaminant transport with non-equilibrium processes in unsaturated soils and implicit characteristic Galerkin scheme. *International Journal for Numerical and Analytical Methods in Geomechanics*, **24**, 219–243. 2000.

[Liakopoulos(1965)] A. C. Liakopoulos. *Transient Flow Through Unsaturated Porous Media*. Ph.D. thesis, University of California, Berkeley. 1965.

[Maidl(1995)] B. Maidl, M. Herrenknecht and L. Anheuser. *Maschineller Tunnelbau im Schildvortrieb*. Verlag Ernst & Sohn, Berlin. 1995.

[Malvern(1969)] L. E. Malvern. *Introduction to the Mechanics of a Continuous Medium*. Prentice-Hall, Englewood Cliffs, New Jersey. 1969.

[Mang(2000)] H. A. Mang and G. Hofstetter. *Festigkeitslehre*. Springer, Wien. 2000.

[Marle(1982)] C. M. Marle. On macroscopic equations governing multiphase flow with diffusion and chemical reactions in porous media. *International Journal of Engineering Science*, **20**, 643–662. 1982.

[Marsden(1983)] J. E. Marsden and T. J. R. Hughes. *Mathematical Theory of Elasticity*. Prentice-Hall, Englewood Cliffs, New Jersey. 1983.

[Meschke(2001)] G. Meschke, F. Bangert, S. Grasberger and D. Kuhl. Computational durability analysis of concrete structures considering damage, moisture transport and chemical dissolution processes. Report RUB/SM-01/10, Institute for Structural Mechanics, Ruhr-University Bochum. pp.: 20, 2001.

[Mills(1966)] H. Mills. Incompressible mixture of Newtonian fluids. *International Journal of Engineering Science*, **4**, 97–112. 1966.

[Morland(1972)] L. W. Morland. A simple constitutive theory for a fluid-saturated porous solid. *Journal of Geophysical Research*, **77**, 890–900. 1972.

[Mualem(1974)] Y. Mualem. A conceptual model of hysteresis. *Water Resources Research*, **10**(3), 514–520. 1974.

[Mualem(1976)] Y. Mualem. A new model for predicting the hydraulic conductivity of unsaturated porous media. *Water Resources Research*, **12**(3), 513–522. 1976.

[Narasimhan(1978)] T. N. Narasimhan and P. A. Witherspoon. Numerical model for saturated-unsaturated flow in deformable porous media: 3. Applications. *Water Resources Research*, **14**(6), 1017–1034. 1978.

[Navarro(2000)] V. Navarro and E. E. Alonso. Modeling swelling soils for disposal barriers. *Computers and Geotechnics*, **27**, 19–43. 2000.

[Neuman(1971)] S. P. Neuman and P. A. Witherspoon. Analysis of nonsteady flow with a free surface using the finite element method. *Water Resources Research*, **7**, 611–623. 1971.

[Nunziato(1980)] J. W. Nunziato and E. K. Walsh. On ideal multiphase mixtures with chemical reactions and diffusion. *Archive for Rational Mechanics and Analysis*, **73**, 285–311. 1980.

[Nunziato(1981)] J. W. Nunziato and S. L. Passman. A multiphase mixture theory for fluid-saturated granular materials. In: A. P. S. Selvadurai (ed.), *Mechanics of Structured Media*, pp. 243–254. Elsevier, Amsterdam. 1981.

[Oden(1980)] J. T. Oden and N. Kikuchi. Recent advances: theory of variational inequalities with applications to problems of flow through porous media. *International Journal of Engineering Science*, **18**, 1173–1284. 1980.

[Oettl(1998)] G. Oettl, R. F. Stark and G. Hofstetter. A comparison of elastic-plastic soil models for 2D FE analyses of tunnelling. *Computers and Geotechnics*, **23**, 19–38. 1998.

[Pariseau(1999)] W. G. Pariseau. Poroelastic-plastic consolidation – analytical solution. *International Journal for Numerical and Analytical Methods in Geomechanics*, **23**, 577–594. 1999.

[Passman(1977)] S. L. Passman. Mixtures of granular materials. *International Journal of Engineering Science*, **15**, 117–129. 1977.

[Poulos(1991)] H. G. Poulos and E. H. Davis. *Elastic Solutions for Soil and Rock Mechanics*. Centre for Geotechnical Research, University of Sydney, Australia, 2nd edn. 1991.

[Prevost(1980)] J. H. Prevost. Mechanics of continuous porous media. *International Journal of Engineering Science*, **18**, 787–800. 1980.

[Prevost(1982)] J. H. Prevost. Nonlinear transient phenomena in saturated porous media. *Computer Methods in Applied Mechanics and Engineering*, **20**, 3–18. 1982.

[Rodriguez(1993)] E. Ruiz Rodriguez. *Bodenluftströmung in teilgesättigten Böden*. Wasserbau – Mitteilungen, TH Darmstadt. 1993.

[Rose(1989)] W. Rose. Data interpretation problems to be expected in the study of coupled fluid flow in porous media. *Transport in Porous Media*, **4**, 185–198. 1989.

[Sandhu(1969)] R. S. Sandhu and E. L. Wilson. Finite element analysis of flow in saturated porous media. *Journal of the Engineering Mechanics Division ASCE*, **95**(EM3), 641–652. 1969.

[Schmettow(2002)] T. Graf von Schmettow and J. Gattermann. Tunnel Siegaue – Auffahren eines Großquerschnitts unter Druckluft. *Compress, Verlagssupplement des Verlages Glückauf für die Fachzeitschriften Geotechnik und Felsbau*, pp. 23–32. 2002.

[Schrefler(1988)] B. A. Schrefler and L. Simoni. A unified approach to the analysis of saturated-unsaturated elastoplastic porous media. In: G. Swoboda (ed.), *Numerical Methods in Geomechanics*, pp. 205–212. Balkema, Rotterdam. 1988.

[Schrefler(1990)] B. A. Schrefler, L. Simoni, X. Li and O. C. Zienkiewicz. Mechanics of partially saturated porous media. In: C. S. Desai and G. Gioda (eds.), *Numerical Methods and Constitutive Modelling in Geomechanics*, vol. 311 of *CISM Courses & Lectures*, pp. 169–209. Springer-Verlag, Wien. 1990.

[Schrefler(1991)] B. A. Schrefler and L. Simoni. Comparison between different finite element solutions for immiscible two-phase flow in deforming porous media. In: G. Beer, J. R. Booker and J. P. Carter (eds.), *Computer Methods and Advances in Geomechanics*, pp. 1215–1220. Balkema, Rotterdam. 1988.

[Schrefler(1993)] B. A. Schrefler and X. Y. Zhan. A fully coupled model for water flow and airflow in deformable porous media. *Water Resources Research*, **29**(1), 155–167. 1993.

[Schrefler(1994)] B. A. Schrefler, L. D'Alpaos, X. Y. Zhan and L. Simoni. Pollutant transport in deforming porous media. *European Journal of Mechanics, A/Solids*, **13**(4-Suppl.), 175–194. 1994.

[Schrefler(1995)] B. A. Schrefler. F.E. in environmental engineering: Coupled thermo-hydro-mechanical processes in porous media including pollutant transport. *Archives of Computational Methods in Engineering*, **2**(3), 1–54. 1995.

[Schuller(1999)] H. Schuller and H. F. Schweiger. Application of multilaminate model for shallow tunnelling. *Felsbau*, **17**(1), 44–47. 1999.

[Schweiger(2000)] H. F. Schweiger and H. Schuller. New developments and practical applications of the multilaminate model for soils. In: D. W. Smith and J. P. Carter (eds.), *Developments in Theoretical Geomechanics, The John Booker Memorial Symposium*, pp. 329–350. Balkema, Rotterdam. 2000.

[Simoni(1989)] L. Simoni and B. A. Schrefler. FE solution of a vertically averaged model for regional land subsidence. *International Journal for Numerical Methods in Engineering*, **27**, 215–230. 1989.

[Simoni(1991)] L. Simoni and B. A. Schrefler. A staggered finite-element solution for water and gas flow in deforming porous media. *Communications in Applied Numerical Methods*, **7**, 213–223. 1991.

[Siriwardane(1981)] H. J. Siriwardane and C. S. Desai. Two numerical schemes for non-linear consolidation. *International Journal for Numerical Methods in Engineering*, **17**, 405–426. 1981.

[Sloan(1999a)] S. W. Sloan and A. J. Abbo. Biot consolidation analysis with automatic time stepping and error control – Part 1: Theory and implementation. *International Journal for Numerical and Analytical Methods in Geomechanics*, **23**, 467–492. 1999.

[Sloan(1999b)] S. W. Sloan and A. J. Abbo. Biot consolidation analysis with automatic time stepping and error control – Part 2: Applications. *International Journal for Numerical and Analytical Methods in Geomechanics*, **23**, 493–529. 1999.

[Small(1976)] J. C. Small, J. R. Booker and E. H. Davis. Elasto-plastic consolidation of soil. *International Journal of Solids and Structures*, **12**, 431–448. 1976.

[Smith(1998)] I. M. Smith and D. V. Griffiths. *Programming the Finite Element Method*. John Wiley & Sons, Chichester, 3rd edn. 1998.

[Smith(2000)] D. W. Smith. Thermodynamics in civil engineering. In: D. W. Smith and J. P. Carter (eds.), *Developments in Theoretical Geomechanics, The John Booker Memorial Symposium*, pp. 267–293. Balkema, Rotterdam. 2000.

[Snee(1994)] C. P. M. Snee. Modelling compressed air losses from a tunnel in sand. In: *Proceedings of the International Conference on Engineering and Health in Compressed Air Work*, pp. 250–260. 1994.

[Snee(1996)] C. P. M. Snee and A. A. Javadi. Estimation of air flow in compressed air tunnelling. *Journal of Tunnelling and Underground Space Technology*, **11**(2), 189–195. 1996.

[Snee(1997)] C. P. M. Snee and A. A. Javadi. A new procedure for compressed air tunnelling. *Ground Engineer*, **30**(7), 34. 1997.

[Strobl(1986)] B. Strobl. NATM under compressed air – geomechanical aspects and numerical modelling. In: *Proceedings of the Conference on Large Underground Openings*, vol. 2. 1986.

[Strobl(1991)] B. Strobl. *Die NATM in Böden in Kombination mit Druckluft*. Ph.D. thesis, Institut für Bodenmechanik und Grundbau, Technische Universität Graz, Austria. 1991.

[Sukirman(1993)] Y. Sukirman and R. W. Lewis. A finite element solution of a fully coupled implicit formulation for reservoir simulation. *International Journal for Numerical and Analytical Methods in Geomechanics*, **17**, 677–698. 1993.

[Terzaghi(1923)] K. von Terzaghi. Die Berechnung der Durchlässigkeitsziffer des Tones aus dem Verlauf der hydrodynamischen Spannungserscheinungen. *Akademie der Wissenschaften in Wien, Sitzungsberichte der mathematisch-naturwissenschaftlichen Klasse, Abt. IIa*, **132**(3/4), 125–138. 1923.

[Terzaghi(1925)] K. von Terzaghi. *Erdbaumechanik auf bodenphysikalischer Grundlage*. Franz Deuticke, Leipzig. 1925.

[Thomas(1995)] H. R. Thomas and Y. He. Analysis of coupled heat, moisture and air transfer in a deformable unsaturated soil. *Géotechnique*, **45**, 677–689. 1995.

[Thomas(1997)] H. R. Thomas and Y. He. A coupled heat-moisture transfer theory for deformable unsaturated soil and its algorithmic implementation. *International Journal for Numerical Methods in Engineering*, **40**, 3421–3441. 1997.

[Thomas(1999)] H. R. Thomas and H. Missoum. Three-dimensional coupled heat, moisture and air transfer in a deformable unsaturated soil. *International Journal for Numerical Methods in Engineering*, **44**, 919–943. 1999.

[Trapp(1976)] J. A. Trapp. On the relationship between continuum mixture theory and integral averaged equations for immiscible fluids. *International Journal of Engineering Science*, **14**, 991–998. 1976.

[Truesdell(1984)] C. Truesdell. *Rational Thermodynamics*. Springer Verlag, New York, 2nd edn. 1984.

[Tuli(2001)] A. Tuli, K. Kosugi and J. W. Hopmans. Simultaneous scaling of soil water retention and unsaturated hydraulic conductivity functions assuming lognormal pore-size distribution. *Advances in Water Resources*, **24**, 677–688. 2001.

[Turska(1993)] E. Turska and B. A. Schrefler. On convergence conditions of partitioned solution procedures for consolidation problems. *Computer Methods in Applied Mechanics and Engineering*, **106**, 51–63. 1993.

[Turska(1994)] E. Turska, K. Wisniewski and B. A. Schrefler. Error propagation of staggered solution procedures for transient problems. *Computer Methods in Applied Mechanics and Engineering*, **114**, 177–188. 1994.

[Vermeer(1981)] P. A. Vermeer and A. Verruijt. An accuracy condition for consolidation by finite elements. *International Journal for Numerical and Analytical Methods in Geomechanics*, **5**, 1–14. 1981.

[Verruijt(1995)] A. Verruijt. *Computational Geomechanics*, vol. 7 of *Theory and Applications of Transport in Porous Media*. Kluwer, Dordrecht. 1995.

[Vogel(2001)] T. Vogel, M. T. van Genuchten and M. Cislerova. Effect of the shape of the soil hydraulic functions near saturation on variably-saturated flow predictions. *Advances in Water Resources*, **24**, 133–144. 2001.

[Wheeler(1995)] S. J. Wheeler and D. Karube. Constitutive modelling. In: E. E. Alonso and P. Delage (eds.), *Unsaturated Soils, offprint: Proceedings of the First International Conference on Unsaturated Soils*, pp. 1323–1356. Balkema, Rotterdam. 1995.

[Wyckoff(1936)] R. D. Wyckoff and H. G. Botset. The flow of gas-liquid mixtures through unconsolidated sands. *Physics*, **7**, 325–345. 1936.

[Zienkiewicz(1977)] O. C. Zienkiewicz, C. Humpheson and R. W. Lewis. A unified approach to soil mechanics problems including plasticity and visco-plasticity. In: G. Gudehus (ed.), *Finite Elements in Geomechanics*, pp. 151–177. Wiley & Sons, London. 1977.

[Zienkiewicz(1984)] O. C. Zienkiewicz and T. Shiomi. Dynamic behaviour of saturated porous media; the generalized Biot formulation and its numerical solution. *International Journal for Numerical and Analytical Methods in Geomechanics*, **8**, 71–96. 1984.

[Zienkiewicz(1989)] O. C. Zienkiewicz and R. L. Taylor. *The Finite Element Method*, vol. 1. McGraw-Hill Book Company, England, 4th edn. 1989.

[Zienkiewicz(1990a)] O. C. Zienkiewicz, A. H. C. Chan, M. Pastor, D. K. Paul and T. Shiomi. Static and dynamic behaviour of soils: a rational approach to quantitative solutions. I. Fully saturated problems. *Proceedings of the Royal Society of London*, **A**(429), 285–309. 1990.

[Zienkiewicz(1990b)] O. C. Zienkiewicz, Y. M. Xie, B. A. Schrefler, A. Ledesma and N. Bićanić. Static and dynamic behaviour of soils: a rational approach to quantitative solutions. II. Semi-saturated problems. *Proceedings of the Royal Society of London*, **A**(429), 311–321. 1990.

[Zienkiewicz(1991)] O. C. Zienkiewicz and R. L. Taylor. *The Finite Element Method*, vol. 2. McGraw-Hill Book Company, England, 4th edn. 1991.

www.ingramcontent.com/pod-product-compliance
Lightning Source LLC
Chambersburg PA
CBHW082320220526
45470CB00008B/2362